Molecular systematics is revealing that viruses are the master explorers of evolutionary space; they exploit both RNA and DNA worlds using all the tricks of the trade, and sometimes with devastating speed. This book is of invited essays on various aspects of virus evolution, especially clues to their phylogenetic histories as revealed by gene sequence analysis and other new techniques. The early chapters give examples of the impact of viruses and their control. Further chapters discuss the genetic variation of viruses and their interactions with other organisms at various levels, from molecule to population. The largest section, of 17 chapters, focusses on views of the evolution of particular groups of viruses, and describes their genetic origins, molecular variation, population genetics and interactions with hosts. Finally, there are chapters on phylogenetic analysis to aid newcomers. This work is not only for virologists but for all who are interested in practical studies of biological evolution.

T0171802

MOLECULAR BASIS OF VIRUS EVOLUTION

MOLECULAR BASIS OF VIRUS EVOLUTION

Edited by

ADRIAN J. GIBBS
Australian National University, Canberra, Australia

CHARLES H. CALISHER
Colorado State University, Fort Collins, USA

FERNANDO GARCÍA-ARENAL
Universidad Politécnica, Madrid, Spain

CAMBRIDGE UNIVERSITY PRESS
Cambridge, New York, Melbourne, Madrid, Cape Town, Singapore, São Paulo

Cambridge University Press
The Edinburgh Building, Cambridge CB2 2RU, UK

Published in the United States of America by Cambridge University Press, New York

www.cambridge.org
Information on this title: www.cambridge.org/9780521455336

First published 1995
Reprinted 1997
This digitally printed first paperback version 2005

A catalogue record for this publication is available from the British Library

Library of Congress Cataloguing in Publication data
Molecular basis of virus evolution / edited by Adrian J. Gibbs. Charles H. Calisher,
Fernando García-Arenal.
p. cm.
Includes index.
ISBN 0 521 45533 2 (hardback)
1. Molecular virology. 2. Viruses–Evolution. 3. Molecular evolution. I. Gibbs, A.J.
(Adrian J.) II. Calisher, Charles H. III. García-Arenal, Fernando.
QR389.M64 1995 94-30349
576'.64–dc20 CIP

ISBN-13 978-0-521-45533-6 hardback
ISBN-10 0-521-45533-2 hardback

ISBN-13 978-0-521-02289-7 paperback
ISBN-10 0-521-02289-4 paperback

Contents

Contents

Contributors

Miguel A. Aranda
ETSI Agrónomos Ciudad, Departmento de Patalogia Vegetal, Universidad Politecnica, 28040 Madrid, Spain

Charles R.M. Bangham
Molecular Immunology Group Institute of Molecular Medicine, John Radcliffe Hospital, Oxford OX3 9DU, UK

Pascale Barbier
INRA, Centre de Recherches de Colmar, 28 Rue de Herrlisheim, BP 507, F-68021 Colmar, France

W. J. Bean
Scientific Review Branch, NIAID-NJH, Solar Building, Room 4C32, 6003 Executive Boulevard, Bethesda MD 20892, USA and Dept. of Virology and Molecular Biology, St Jude Children's Research Hospital, PO Box 318, Memphis TN 38101–0318, USA

Françoise Bernardi
Institut Jacques Monod, 2 Place Jussieu, Tour 43 F-75251 Paris Cedex 05, France

Christof Biebricher
Max-Planck-Institut für Biophysikalische Chemie, D-3400 Göttingen, Germany

Rafael Blasco
Centro de Investigación – Sanidad Animal – INIA, 28130 Valdeolmos, Madrid, Spain

Jan Blok
Sir Albert Sakzewski Virus Research Laboratory, Royal Children's Hospital, Herston Road, Herston QLD 4029, Australia

Conal J. Burgess
Department of Genetics, Trinity College, Dublin 2, Ireland now: King Faisal Military Hospital, Box 101, Khamis Mushayt, Saudi Arabia

Robert D. Burk
Department of Paediatrics, Microbiology and Immunology, Albert Einstein College of Medicine, 1300 Morris Park Avenue, Bronx NY 10461, USA

C.H. Calisher
ArthropodBorne Infectious Diseases Laboratory, Foothills Campus, Colorado State University, Fort Collins CO 80523, USA

Mark W. Chase
Department of Biology, Wake Forest University, PO Box 7325, Reynolda Station, Winston-Salem NC 27109, USA

Andrew J. Davison
MRC Virology Unit, Institute of Virology, Church Street, Glasgow G11 5JR, UK

Estaban Domingo
Centro de Biologia Molecular, UAM CSIC, Madrid, Spain

Joaquín Dopazo
Departmento de Sanidad Animal CIT – INIA, Minesterio de Agricultura Pesca y Alimentacion, Madrid, Spain now Centro Nacional de Bio-technologia, CSIC, Universidad Autónoma, 28049 Cantoblanco, Madrid, Spain

Manfred Eigen
Max-Planck-Institut für Biophysikalische Chemie, D-3400 Göttingen, Germany

Richard M. Elliott
Department of Virology, Institute of Virology, University of Glasgow, Church Street, Glasgow G11 5JR, UK

Frank Fenner
John Curtin School of Medical Research, Australian National University, GPO Box 4, Canberra ACT 2601, Australia

Aurora Fraile
ETSI Agrónomos Ciudad, Departmento de Patalogia Vegetal, Universidad Politécnica, 28040 Madrid, Spain

Fernando García-Arenal
ETSI Agrónomos Ciudad, Departmento de Patalogia Vegetal, Universidad Politécnica, 28040 Madrid, Spain

Adrian J. Gibbs
Research School of Biological Sciences, Australian National University, GPO Box 4, Canberra ACT 2601, Australia

Mark Gibbs
Institute of Virology and Environmental Microbiology, Mansfield Road, Oxford OX1 3SR, UK

Alexander E. Gorbalenya
Institute for Poliomyelitis and Viral Encephalitides, Russian Academy of Medical Sciences, Moscow, Moscow Region 142782, Russia

O.T. Gorman
Glen Canyon Environmental Studies, 835 East David Drive, Flagstaff, AZ 86001 USA and Department of Virology and Molecular Biology, St Jude Children's Research Hospital, PO Box 318, Memphis, TN 38101–0318, USA

Jaap Goudsmit
Human Retrovirus Laboratory, Department of Medical Microbiology, University of Amsterdam, Meibergdreef 15 – 1105 AZ, Amsterdam, The Netherlands

Anne-Lise Haenni
Institut Jacques Monod, 2 Place Jussieu, Tour 43 F-75251 Paris Cedex 05, France

Beatrice H. Hahn
Departments of Medicine and Microbiology, University of Alabama at Birmingham, Birmingham, Alabama 35294, USA

John Holland
Department of Biology, University of California, San Diego, LaJolla, CA 92037 USA

Wolfgang Joklik
Department of Microbiology and Immunology, Duke University Medical Centre, PO Box 3020, Durham, NC 27710, USA

Linda D. Jones
NERC Institute of Virology and Environmental Microbiology, Mansfield Road, Oxford OX1 3SR, UK

Jeffrey B. Kaplan
Lederle Laboratories American Cyanamid Co. 401 N. Middletown Rd. Pearl River, NY 10965, USA

Paul K. Keese
Division of Plant Industry, CSIRO Acton, Canberra, ACT 2601, Australia

Kathleen A. Kron
Department of Biology, Wake Forest University, PO Box 7325, Reynolda Station, Winston-Salem, NC 27109, USA

Gael Kurath
Department of Plant Pathology, College of Natural and Agricultural Sciences, University of California, Riverside, California 92521–0122, USA

Michael M.C. Lai
Department of Microbiology, University of Southern California, School of Medicine, 2011 Zonal Avenue, Los Angeles, CA 90033, USA

Thomas L. Lentz
Department of Cell Biology, Yale University Sterling Hall of Medicine, PO Box 3333, New Haven, Connecticut 06510–8002, USA

C. López-Galíndez
Instituto de Salud Carlos III, Centro National de Biologia Cellular y Retrovirus, Majadahouda, Madrid 28220, Spain

Marcella A. McClure
Department of Ecology and Evolutionary Biology, University of California, CA 92717, USA

Kathleen A. McGann
Department of Microbiology, University of Pennsylvania Medical Centre, Philadelphia, PA 19104–6076, USA

Duncan McGeoch
MRC Virology Unit, Institute of Virology, Church Street, Glasgow G11 5JR, UK

Robert M. May
Department of Zoology, University of Oxford, South Parks Road, Oxford OX1 3PS, UK

Gregor Meyers
Federal Research for Virus Diseases of Animals, PO Box 1149, D-7400 Tübingen, Germany

Mary A. Morse
Virology Group, Glaxo Research Group Limited, Greenford Road, Greenford, Middlesex UB6 0HE, UK

Andrés Moya
Department de Genetica, Facultad de Biologia, C/-Moliner 50, 46100 Burjassot, Valencia, Spain

Neal Nathanson
Department of Microbiology, University of Pennsylvania Medical Centre, Philadelphia, PA 19104–6076, USA

Patricia Nuttall
NERC Institute of Virology and Environmental Microbiology, Mansfield Road, Oxford OX1 3SR, UK

Agustin Portela
Instituto de Salud Carlos III, Centro Nacional de Microbiología, Majadahonda 28220, Madrid, Spain

Craig Pringle
Department of Biological Sciences, University of Warwick, Coventry CV4 7AL, UK

Bertha-Cecilia Ramírez
Institut Jacques Monod, 2 Place Jussieu, Tour 43 F-75251 Paris Cedex 05, France

Christophe Robaglia
Laboratoire de Biologie Cellulaire, INRA F78026 Versailles Cédex, France

María J. Rodrigo
Unidad de Biología Molecular y Cellular de Plantas, Instituto de Agroquímica y Technología de los Alimentos, CSIC c/Jaume Roig 11, 46071 Valencia, Spain

Ana Rodríguez
Yale University School of Medicine, Department of Cell Biology, New Haven, Connecticut 06515, USA

J.M. Rojas
Instituto de Salud Carlos III, Centro National de Biologia Cellular y Retruvio, Majadahouda, Madrid 28220, Spain

Michael R. Roner
Department of Microbiology and Immunology, Duke University Medical Centre, PO Box 3020, Durham NC 27710, USA

Juan C. Sáiz
Centro de Biología Molecular 'Severero Ochoa', CSIC, Universidad Autónoma, 28049 Cantoblanco, Madrid, Spain

Karin Séron
Institut Jacques Monod, 2 Place Jussieu, Tour 43 F-75251, Paris Cedex 05, France

Paul M. Sharp
Department of Genetics, Trinity College, University of Dublin, Lincoln Place Gate, Dublin 2 Eire now Department of Genetics, University of Nottingham, Queens Medical Centre, Nottingham NG7 2UH, UK

Dharma D. Shukla
Division of Biomolecular Engineering, CSIRO Parleville Laboratory, 343 Royal Parade, Parkville VIC 3052, Australia

F. Sobrino
Centro de Investigación en Sanidad Animal, INIA., 28130 Valdeolmos, Madrid, Spain

Erko Stackebrandt
Department of Microbiology, University of Queensland, St Lucia QLD 4067, Australia

John P. Sundberg
The Jackson Laboratory, 600 Main Street, Bar Harbor, ME, USA

Norbert Tautz
Federal Research for Virus Diseases of Animals, PO Box 1149 D-7400 Tübingen, Germany

Heinz-Jürgen Theil
Federal Research for Virus Diseases of Animals, PO Box 1149, D-7400 Tübingen, Germany

Marc A. Van Ranst
Dept. of Microbiology & Immunology, Rega Institute for Medical Research University of Leuven, Minderbroedersstraat 10, B-3000 Leuven, Belgium

Colin W. Ward
Division of Biomolecular Engineering, CSIRO Parkville Laboratory, 343 Royal Parade, Parkville, VIC 3052, Australia

Scott C. Weaver
Department of Biology, University of California San Diego, La Jolla, California 92093, USA

Robert G. Webster
Dept. of Virology and Molecular Biology, St Jude Children's Research Hospital, PO Box 318 Memphis, TN 38101–0318, USA

Georg F. Weiller
Molecular Evolution and Systematics, Australian National University, PO Box 475, Canberra ACT 2601, Australia

John Wilesmith
Department of Microbiology, University of Pennsylvania Medical Centre, Philadelphia, PA 19104–6076, USA

Editors' preface

The idea to collate and edit a book on virus evolution occurred to each of us quite independently, but when we discovered that we had this common desire, it seemed reasonable to pool our resources, experiences, and energies. We assumed that, whereas none of us had a firm grasp of either the questions or the answers, collectively we might be able to attract as participants many people who have given virus evolution more than passing thoughts. Further, we did not want to edit a book containing 'great thoughts by great people'. From the outset, we wanted to bring together current data-based views of the evolution of various viruses. Considering that thousands of viruses have been recognized, it was an obvious impossibility to try to cover the entire field. Indeed, we have made a concerted effort not to do so. Keeping in mind Duncan McGeoch's dictum that 'Viruses may be looked on as mistletoe on the Tree of Life', we fully expected that evolutionary practice would vary, and vary considerably from one family to another.

Therefore, we thought it appropriate to convene a meeting of people we expected would provide a basis for organizing a book on the subject and who might provide at least a glimmer of a consensus. We organized a symposium 'Co-evolution of viruses, their hosts and vectors' in Madrid from December 9–11, 1991, and invited some of the colleagues we knew to be working directly or indirectly on virus evolution, and advertisements attracted others who wished to join us.

The meeting proved that we were naive, in so far as recognizing how varied are the evolutionary patterns, mechanisms, and possibilities of viruses and how little is known about virus evolution. Each of us is at least comfortable that he has 'a feel' for the evolution of a particular virus or group of viruses, but the larger picture is not so clear. Part of the problem is that reliable techniques to examine viral genomes at the molecular level have only recently become available, and the body of data required for meaningful evolutionary conclusions is still being amassed. There is, however, no lack of opinions on these matters. For many years, discussions of virus evolution were relegated to informal discussion among virologists with fanciful imaginations. The problem, of course, was that there were no hard data to distinguish between the myriad of possibilities. In those days, resources and energies were only put to examining phenotypic characteristics (viral biology, ecology, epidemiology); however, dreams were realized when techniques permitted the focus to turn to an examination of genotypic characters (gene sequences, relationships between structure and function). A by-product of the accumulation of such data was the opportunity to compare sequences of many viruses using computer programs. The plethora of molecular data began cascading down on to us, and the problem began to show promise of providing solutions.

Our intent for this book is that it be used as a textbook on the state-of-the-art, such as it is. There seems to be both more known than one would expect and less known than one would hope. That being the case with essentially everything, we are encouraged by the product. We trust readers of this book will learn something, realize how complex the issue is, and have some fun in the bargain.

We also hope that this book will interest those studying the evolution of cellular organisms. Viruses are just another way that genes get about and, in their entrepreneurial lifestyle, viral genes have explored a considerably larger proportion of 'evolutionary space' than most cellular genes. Furthermore, viruses can provide experimentally accessible exemplars for evolutionary processes that are so much more difficult to study in cellular genes.

The editors thank Ms Tanya Joce for skilled and careful secretarial help. The editors also thank the *Fundación Juan March*, Madrid, Spain for its generosity in providing a splendid milieu in which to hold the symposium and for its financial support. In particular, we thank Mr Andrés González, Director of Administrative Services, *Fundación Juan March*, for his attention to details and for repairing the problems we caused.

Established in 1955 by the Spanish financier Don Juan March Ordinas, the *Fundación Juan March* is a cultural, scientific and charitable institution, with numerous special programmes devoted to the study of culture and science. Its International Meetings on Biology aim to promote close co-operation among Spanish and foreign scientists. It therefore supports all fields of biological study, giving particular attention to the most current advances. Thus it was a very appropriate sponsor and host for our conference, which was attended by 30 biologists from Spain and abroad. We are pleased to dedicate this book to the *Fundación*.

Adrian J. Gibbs
Charles H. Calisher
Fernando García-Arenal

Conference participants

A.A. Agranovsky, M.A. Aranda, C.R.M. Bangham, R. Blasco, R.D. Burk, P. Caciagli, C.H. Calisher, W.F. Carman, J.M. Coffin, A.J. Davison, E.L. Delwart, E. Domingo, J. Dopazo Blazquez, F. Fenner, A. Fraile, F. García-Arenal, I. García Barreno, I. García Luque, A.J. Gibbs, M.J. Gibbs, A.E. Gorbalenya, J. Goudsmit, C.M. Kearney, K.A. Kron, G. Kurath, T.L. Lentz, C. López Galíndez, M.A. Martínez, M.A. McClure, D.J. McGeoch, G. Meyers, A. Moya, N. Nathanson, P.A. Nuttall, J. Ortín, C.R. Pringle, C. Robaglia, S. Rochat, M.N. Rozanov, R.R. Rueckert, R.E. Shope, E. Stackebrandt, V. Stollar, D. Stuart, C.W. Ward, R.G. Webster, G.F. Weiller, R. Yáñez.

1

Introduction and guide

ADRIAN GIBBS, CHARLES CALISHER
AND FERNANDO GARCÍA-ARENAL

This chapter briefly outlines the lines of evidence of virus evolution that are discussed in this book; hopefully it augments the information given in the Table of Contents by indicating where those lines anastomose.

Evolution is the process whereby the population of an organism changes genetically over a period of time. It occurs when genetic variation within a population, combined with selection from among the mixture of genotypes, results in change. Viruses, especially those with RNA genomes, sometimes evolve very rapidly and so their evolution can be studied as it occurs. By contrast, all cellular organisms evolve more slowly and so their evolution is deduced by comparing extant forms and, for some, the fossils of their ancestors. No fossils of viruses are known.

Pre-molecular evidence of virus evolution

It has been realized for a long time that pathogens evolve. The fact that new epidemics of disease appear (Chapter 3), and that some viral diseases, such as measles and smallpox, induce life-long immunity in those individuals that survive, whereas others, such as the common cold or influenza, do not, all indicate that some viruses change. Of course, such differences influence choice of control strategies because stable viruses are amenable to control by vaccination, whereas rapidly changing viruses are not. Another type of evidence of virus evolution is the fact that avirulent strains of some viruses can be selected for use as vaccines, usually by growing them in unusual hosts, or under unusual conditions.

One of the first deliberate attempts to study natural virus evolution was the classic study of myxoma leporipoxvirus when it was liberated in Australia to attempt to control European rabbits (Chapter 2).

1

Molecular evidence of virus evolution

Since the 1970s the development of methods for sequencing genes and determining the structures of proteins has transformed the study of evolution, particularly that of viruses. Although viruses had previously been classified into groups using various characters, such as their host range, virion characteristics and the serological specificity of their proteins, there was no evidence that the resulting groupings reflected evolutionary relationships, nor whether any of the groups were phylogenetically related. However, sequences of viral genes revealed for the first time which genes are shared and homologous, and how closely related they are. This is because there is redundant information at all levels of biological systems; 61 nucleotide triplets encode 20 amino acids, and innumerable different amino acid combinations can give the same fold in a protein. As a consequence, genes encode information essential for life mixed with clues of their evolutionary history. Thus, by comparing the nucleotide sequences of genes, or the sequences or structure of proteins, from different lineages, one can attempt to separate the different components of change (Chapter 36). In this way, the mode and tempo of evolution of viral populations, species, genera and even higher taxa can be deduced. It has been found, most surprisingly, that many viral genes are widely distributed among viruses that seemed quite unrelated (Chapters 4, 5 and 6), some are related to those of their hosts and may have very ancient origins, and there is clear evidence of genetic recombination in all viral genomes (Chapter 9).

Evolution of viral species

The need to define and name viral species caused much controversy in the past. Viruses reproduce asexually and, in early studies, only a few gave evidence of genetic recombination. Thus it was unclear how viruses could maintain apparently stable species because it was thought that sexual reproduction, involving gene shuffling among individuals constituting a 'gene pool', was the cohesive force that maintained the genetic integrity of species. None the less some viruses are known to be genetically stable, notably some of the viral strains used as vaccines. This paradox was resolved when the genomic sequence of Qβ levivirus, a bacteriophage, was first determined. During this work, Qβ was passaged 55 times, and only a single variant nucleotide was found in the genomic sequence over a two-year period. However, when individual clones were prepared from inocula saved during the work, it was found that their genomic nucleotide sequences differed by an average of 1.6 sites from that of the unselected stock. The variant nucleotides were apparently distributed at random in the genome, and thus the stock gave the appearance of being a single genomic sequence, the 'master copy'. When variant clones were mixed with the stock population and passaged, they rapidly disappeared, and the 'master copy' became dominant. Thus stabilizing ('purifying') selection operating on the variants of a genome that arise from a common ancestor usually favours a single master copy and close variants of it, resulting in a stable 'quasispecies' or 'mutant spectrum' (Chapter 13).

It seems that most variant genomes produced by point mutation are less fit than their parents, as in the Qβ study, but not all are, especially when selection pressures change and these allow the evolution of populations (Chapters 14 and 15). Some variants may be favoured by extant conditions; for example, it was a single point mutation in the haemagglutinin gene of a wild fowl plague orthomyxovirus in 1982 that resulted in a changed glycosylation site and caused an epidemic of virulent disease in American poultry farms resulting in the death or slaughter of 18 million birds (Chapter 34). Similar variants allow antigenic drift, and this enhances the survival of many viruses of vertebrates.

Patterns of relatedness in viral populations

It is now becoming evident that all viral populations, except those that have just been cloned, vary to a greater or lesser extent (Chapters 13, 35 and 36). Individual isolates in a population can be compared, their relationships inferred and represented, usually as a network or 'tree'. Such a tree can give useful information on the origin and the mode and extent of selection affecting that population. In many viral populations the between-isolate relationships fit a quasi-species normal distribution, and there is no obvious correlation between those relationships and the time or place that the isolates were obtained. Such populations are probably stable quasi-species that have established a balance between mutation and selection. This pattern has been found in populations of, for example, influenza A orthomyxovirus in wild bird populations (Chapter 34), foot-and-mouth disease aphthovirus (Chapter 21) and tobacco mild green mosaic tobamovirus (Chapter 23) and its satellite virus (Chapter 26). Very few of these populations have been proven to be quasi-species by competition experiments, and it is possible that some of the variation in them is maintained by host selection.

Another population pattern is that of a 'shrub-like' tree with the most recent isolates furthest from the centre of the shrub. In such populations, many independent lineages have evolved from a common ancestor, and there has, as yet, been little or no selection against any lineage. One of the first studies reporting this pattern was of an epidemic of enterovirus 70 (Miyamura *et al.*, 1986); the current epidemic of HIV in the human population seems to be of this type (Chapter 30). By contrast, influenza A orthomyxovirus epidemics in human populations give trees with a single narrow dominant apex and few short branches, which indicates that, during the epidemic, there has been selection against all but one lineage (Chapter 34), probably the result of herd immunity selecting for the line that is most antigenically novel. A similar pattern is also found in successive isolates obtained from individual animals persistently infected with lentiviruses (Chapters 11, 12).

Sequence variation

Mutations (Chapter 8) apparently occur in random positions in viral genomes, though the different types of mutation, or their survival, may not be random. This is shown, for example, in biased nucleotide useage in some viral lineages, but not others; evidence that the bias is genetically determined by the virus. In ORFs, third codon positions are usually redundant and vary most frequently, transitions (changes from purine to purine or from pyrimidine to pyrimidine) occur more frequently than transversions (changes from purine to pyrimidine or vice versa), and insertions/deletions least frequently. There are large differences in the rate of change of different parts of each gene caused by differences in the ease with which their function is conserved despite changes. Mutations appear at a rate of about 10^{-3}/ nucleotide position/replication cycle in RNA genomes, and this allows rapid nucleotide sequence changes, up to 1% /year, in some viral populations, such as HIV (Chapter 30) or influenza A virus (Chapter 34) in the human population. It also permits speculation on the timing of recent changes (Chapter 33). By contrast, DNA genomes change at least a million times more slowly, and there is evidence that the observed mutation rate per genome per replication is virtually constant (.0033), not the mutation rate per nucleotide, which is fastest in the smallest genomes (Drake, 1991, 1993).

Few studies, however, have determined which changes are adaptive, and which are mere evolutionary noise. Clearest evidence comes from viruses with virion proteins of known structure, like the influenzas and FMDV, where the changes can be shown to have occurred in antigenically important surface regions, or to have been avoided in receptor sites. However, little is known of the molecular basis of host range and of the changes that enable viruses to acquire and adapt to new hosts; a feature of great evolutionary significance.

Evolution of viral genera or groups

As mentioned above, most of the taxonomic groupings of viral species, that were defined long before the genomic sequences of the viruses were known, were confirmed by genomic sequence analysis. The relatedness of species of each viral genus usually vary greatly, perhaps because the criteria for defining genera are artificial and because new species can probably arise at any time.

Species of the same viral genus often have different natural hosts, though their experimental host range may be very much wider (Selling, Allison & Kaesberg, 1990; Ball, Amann & Garrett, 1992). This implies that adapting to a new natural host and speciation are linked; however, the genomic sequences of the viruses give few, if any, clues to the molecular basis of host adaptions. In some viral genera, for example, the papillomaviruses (Chapter 31), the poxviruses (Chapter 18) and the tobamoviruses (Chapter 23), the molecular taxonomy of the viral species is mostly congruent with the taxonomy of the natural hosts. This suggests that the origins of such viral groups and their hosts pre-date their present divergences, also that the viruses and hosts have mostly co-evolved and co-speciated. However, incongruous host affiliations in such groupings indicate that taxonomically unrelated hosts can be acquired occasionally. For example, all the tobamoviruses found naturally in solanaceous plants form a single close-knit group, but that group also includes a virus of orchids and another of cacti.

Past studies of viral host ranges (Bald & Tinsley, 1967) have been hampered by poor knowledge of the real evolutionary relationships of viral hosts, especially at the higher taxonomic levels. However, molecular studies are now producing a flood of information, especially for 'prokaryotes' (Chapter 16) and higher plants (Chapter 17), and, more slowly, for animals (Graur, 1993; Marshall & Schultze, 1992). Soon it should be possible to search sensibly for correlations between the taxonomies of hosts, viruses and their vectors and genes (Chapters 19, 20, 22, 25 and 32).

The species in each viral genus have genomes with unequivocally related sequences, but these differ in ways, and on a scale, that is greater than that found between different isolates of each viral species. First, there are sequence alterations that often produce amino acid differences, usually conservative, in the encoded proteins. Secondly, there are also length differences. These result from insertions, additions or deletions in individual genes, and also from the acquisition or loss of genes; for each gene, comparative taxonomic analysis may show which of these options is correct. Many are acquired by recombination from other genomes, both viral and cellular (Chapter 7). For example, recombination has clearly been common among the retroviruses (Chapter 27) and among the luteo-like viruses (Chapter 24), but it has also occurred, though less frequently, in other taxa. For example, it is clear that western equine encephalitis alphavirus (WEEV) originated by recombination (Chapter 33) indeed, recombination may be very common but the recombinants usually less fit than their parents.

Pseudo-recombination, the reassortment of the segments of multi-partite genomes, is especially common among some viruses. It is important in the success of orthomyxoviruses in the human population (Chapter 34). The major epidemics of influenza A in the human population in 1957 (Asian 'flu) and 1968 (Hong Kong 'flu) were caused by strains of the virus that had acquired novel virion surface antigens. This probably resulted when genome segments re-assorted during mixed infection of a 'bridging' host, perhaps a pig, by bird and human isolates.

Although most recombinant viruses acquire their genes from viral parents, sometimes they acquire them from their hosts (Chapters 7 and 27). Best known are the oncogenes acquired by some retroviruses to yield defective variants that require help from the parental virus but which rapidly transform susceptible cells into a cancerous state.

Comparative taxonomic analysis also shows that some genes have arisen *de novo* (Chapter 6). For example, the methyl transferase gene of tymoviruses and the virion protein gene of luteoviruses are shared with a large number of closely or distantly related viruses; however, the genes that overlap them are only found in tymo- and luteoviruses respectively, indicating that they arose *de novo* in the progenitors of these groups. Similarly, the overlapping spliced regulatory genes of lentiviruses can be shown to have arisen *de novo*, because whereas the associated *env* gene is related to the *env* genes of all other retroviruses, only the lentiviruses have *tat* and *rev* genes.

Origins of viral genera or groups

Sequence comparisons show that the virion protein genes and many of the basic metabolic enzymes of viruses, such as those involved in nucleic acid and protein metabolism, have evolved from a limited number of parental genes and are shared by many genera (Chapter 4). For example, many of the viruses with RNA genomes have replicases, nucleotide binding or helicases and methyl transferases that probably evolved from single ancestral genes, and virion protein genes that evolved from, perhaps, three genes; Qβ levivirus has a virion protein with a structure that is a β-sheet with an α-helical clip, whereas all the others that have isometric virions about 30 nm in diameter have virion proteins that are eight-stranded anti-parallel β-barrels, and, by contrast, many, if not all, with rod-shaped or filamentous virions have four-stranded α-helical bundle virion proteins. Sequence similarities also link many of the genes of viruses with DNA genomes, and some of these genes are also unequivocally related to genes of their hosts (Chapters 5 and 18).

The viral phylogeny indicated by one gene may, however, be quite different from that indicated by another. This indicates that major viral groups arise by 'modular evolution' (Botstein, 1980), the assembly by recombination of gene modules from several ancestral sources. Viral genomes also contain genes unique to each group, such as the *rev* and *tat* genes of lentiviruses. Some of these genes arise *de novo*, like the overlapping genes of tymo- and luteoviruses, but the sources of many such genes are unknown.

The fact that some genes are shared between viral genera/families has been used to propose higher viral taxonomic groupings. The taxonomic value of such groups is clear when its members share most of their genes (Chapter 29), but is of less obvious utility when few genes are shared as the taxonomist then has to decide on the 'importance' of different genes. However, such groupings are interesting as indicators of past biological linkages that have permitted genetic recombination, and they are not comparable to higher groupings in cellular organisms, whose genes are mostly linked by descent and not acquired by genetic recombination from disparate sources; the genomes of the different cellular organelles probably came from different sources, and there has been limited movement of genes between them.

However, an overview indicates that multicellular organisms are hosts of several major clusters of viruses that have one or more related genes. These are the viruses with single-stranded positive-sense (messenger) RNA genomes, those that are single-stranded negative-sense RNA, those that are double-stranded RNA, those that are DNA and those whose genomes alternate between RNA and DNA. However, even these major 'viral gene pools' are not genetically isolated from one another as is shown most dramatically by the clear homology (c. 20%) of the envelope glycoproteins of acariviruses and the gp64 membrane proteins of baculoviruses (Chapter 28); acariviruses have a negative-sense RNA genome and three of their genes are clearly related to those of orthomyxoviruses, whereas the baculoviruses have a double-stranded DNA genome.

Origins of viruses and rates of evolution

Viruses are clearly polyphyletic in origin, but there is no unequivocal evidence about when they originated and how quickly they evolve, although comparative taxonomic analyses consistently indicate that their origins are ancient, and their rates of evolution very variable.

Some viral genes, especially those of viruses with DNA genomes, are related to those of their hosts. For example, there are two families of viral thymidine kinases (TKs) (Chapters 18 and 20) and one of these, that found in poxviruses, African swine fever virus and T4 'phage, also includes the TKs of their hosts. The sequences of the poxvirus TKs and the central region of the other TKs are clearly homologous, although the non-poxvirus TKs have unique extra portions at their N- and C-termini. It could be that the TK genes have moved from viruses to animals or vice versa, or that both have inherited them from a common ancestor, or that there has been some mix of 'horizontal' and 'vertical' transmission. Taxonomic analysis of the TK sequences to distinguish between these possibilities is complicated by their differences in length, by the likelihood that the viral sequences are evolving much more quickly than the animal sequences, and by nucleotide biases.

Comparative taxonomy also indicates the great antiquity of the retroviruses and related 'retroid elements' (Chapter 27), which include the hepadna- and caulimoviruses, transposons and other mobile genetic elements that have genomes which alternate between DNA and RNA phases. Hybridization analysis, for example, has shown that there are virogenes derived from baboon endogenous retrovirus (BaEV) in all Old World monkey species, and that they have co-evolved with their hosts (Benveniste, 1985). Thus BaEV is at least 35 million years old, yet it is just one of many distinct retroviruses that form a small twig of the tree of retroid elements, that can be calculated from comparisons of their reverse transcriptase genes. Retroid elements are found in all types of cellular organisms, giving credence to the suggestion that they are very ancient, and that their reverse transcriptase may have been involved in the earliest phases of life on earth when genomes based on DNA evolved from those based on RNA.

The fast mutation of some viruses with RNA genomes has led some virologists to conclude that 'most (RNA genome) viruses we know today probably arose since the last Ice Age'! However, there is no evidence for this and although some viruses with RNA genomes do evolve at extraordinary rates, most, like orthomyxoviruses in birds or retroviruses in their wild hosts, give little evidence of current change, their large mutation rate matched by equivalently stringent selection.

The most convincing evidence that viruses with RNA genomes have origins at least as ancient as those with DNA genomes is that, like the latter, most of their genera or groups, and even higher taxa, are 'phylogenetically contained'; they infect a restricted group of phylogenetically related hosts, and, when their molecular taxonomy is known, it often largely mirrors that of their hosts or vectors. Some are probably very ancient, for example, the rhabdoviruses (Chapter 29), whose two types, the lyssa-like viruses and vesiculo-like viruses, are found in both animals and plants. When more of the viruses of lower organisms are known, these links may be better understood.

The ancient origins of viruses with RNA genomes are also indicated by sequence or structural similarities of some of their genes and those of cellular organisms. For example, viral helicases share sequence motifs with some cellular enzymes, and the β-barrel fold found in some virion proteins is also found in the catabolic activator and other cellular proteins. However, these similarities are very tenuous and may have arisen by convergence.

In summary, viruses are fascinating and are starting to attract more attention than they have in the past, from those interested in the study of evolution (Szathmary, 1993). Viruses are an alternative genetic lifestyle to that of cellular organisms, not so much a family, more a way of life, and are probably as ancient as life itself.

References

Bald, J.G. & Tinsley, T.W. (1967). *Virology*, **32**, 328–336.
Ball, L.A., Amann, J.M. & Garrett, B.K. (1992). *J. Virol.*, **66**, 2326–34.
Benveniste, R.E. (1985). In *Molecular Evolutionary Genetics*, R.J. MacIntyre (ed.). Plenum Press, New York.
Botstein, W. (1980). *Ann. NY Acad. Sci.*, **354**, 484–91.
Drake, J.W. (1991). *Proc. Natl. Acad. Sci. USA*, **88**: 7160–4.
Drake, J.W. (1993). *Proc. Natl. Acad. Sci. USA*, **90**, 4171–5.
Graur, D. (1993). *Trends in Ecology and Evolution*, **8**: 141–7.
Marshall, C. & Schultze, H.-P. (1992). *J. Mol. Evol.*, **35**, 93–101.
Miyamura, K., Tanimura, M., Takeda, N., Kono, R. & Yamazaki, S. (1986). *Arch. Virol.*, **89**, 1–14.
Selling, B.H., Allison, R.F. & Kaesberg, P. (1990). *Proc. Natl. Acad. Sci. USA*, **87**, 434–8.
Szathmary, E. (1993). *Trends in Ecology and Evolution*, **8**, 8–9.

Part I

The impact of viral diseases

2

Classical studies of virus evolution
FRANK FENNER

'Classical' has connotations of 'ancient', but for this book the word relates to studies on virus evolution dating from the time before the sequencing of viral genomes became a way of life for evolutionary virologists. Classical ideas on the origins of viruses were pure speculation and are not worth pursuing further; it was suggested that viruses had arisen either from host cell components in some unknown way, or that some of the more complex viruses may have evolved by progressive degeneration along the sequence bacterium, mycoplasma, rickettsia, chlamydia, poxvirus. Since the focus of this book is the role of molecular biology in understanding viral evolution, I have selected as examples the only two viruses that were discussed in an evolutionary context in books on animal virology published before the emergence of recombinant DNA technology in the early 1970s, namely influenza A virus and myxoma virus (Burnet, 1955, 1960; Andrewes, 1967; Fenner *et al.*, 1974).

Early studies on the evolution of influenza A virus

Here I describe our knowledge of the evolution of influenza viruses as it appeared in 1975, when the only methods of study were serology and peptide mapping of selected proteins. Chapter 34 takes over from that time and tells how molecular biological methods have expanded our understanding of the evolution of these very interesting viruses.

Structure of the virion

First, a little background is needed. There are three species of influenza virus, A, B, and C, each of which can produce respiratory disease in

13

humans. Influenza A and B viruses are very similar in structure and each has eight genome segments; influenza C virus has seven segments and a different envelope glycoprotein peplomer; it will not be further considered. Both influenza A and B viruses cause influenza outbreaks in most countries every winter, but only influenza A virus causes pandemic influenza, i.e. epidemics that sweep around the world and infect a large proportion of the population wherever they are introduced. Significantly, influenza A viruses occur as infections of other animals, notably birds, swine and horses, whereas influenza B viruses seem to have no host other than humans.

By 1975 it was known that the virion of influenza A virus had a lipoprotein envelope which enclosed a nucleoprotein composed of eight segments of RNA and a type-specific protein. Immediately beneath the lipoprotein envelope is a matrix protein which is also type-specific, and projecting from the envelope are two morphologically and functionally different glycoprotein peplomers, the haemagglutinin and the neuraminidase. In contrast to the nucleoprotein and matrix protein, the haemagglutinin and neuraminidase are subtype-specific. Infection with influenza viruses results in development of antibodies to various components of the virion; the important ones from the point of view of virus evolution, as determined by serology, being antibodies to the haemagglutinin and the neuraminidase. The haemagglutinin is the attachment protein and antibodies to it are protective. Because the structure of the protective antigen changes from year to year in such a way that the specific immunity established in response to a particular strain may give little or no protection against viruses circulating three or four years later, influenza due to both A and B viruses continues to be a major epidemic disease of man.

Antigenic drift in influenza virus

The pioneer in studies of the evolution of influenza viruses was Macfarlane Burnet. When he published *Principles of Animal Virology*, Burnet (1955) introduced the term 'immunological or antigenic drift' to describe the progressive antigenic changes in the haemagglutinin observed when strains isolated between 1933 and 1951 were examined. He ascribed this to 'a technique for survival that depended on mutation to novel antigenic patterns which would allow the virus to pass readily through populations still partially immune and carrying antibody as a result of relatively recent infection'. When he came to publish a second edition of that book (Burnet, 1960), the Asian strain of influenza virus had spread around the world. In spite of the fact that Burnet had spent most of the period between 1950 and 1955 studying what he called 'high frequency recombination' (now called re-assortment) between strains of influenza A virus, all he said of the appearance of the Asian strain in 1957 was 'the much sharper discrepancy between the 1956–57 A' strains and Asian A, in other characters as well as serological ones, is more readily interpreted as a re-entry into the human species of a virus that had been dormant (? since the 1890s) in some animal reservoir'.

Antigenic shift in influenza virus

The modern view of the origin of pandemic strains derives primarily from the work of Andrewes (1959), Kilbourne (1968) and Webster and Laver (1971), all of whom studied the antigenic relationships of the haemagglutinins of various strains of influenza A viruses. Besides confirming the concept of antigenic drift in strains found after a pandemic, serological studies showed the Asian virus, which appeared in China in 1957 and affected more people than any other epidemic that the world had seen, had two 'new' antigens, both the haemagglutinin and the neuraminidase being antigenically quite different from those of the strains circulating in 1956. The radical nature of the differences recognized serologically was confirmed by peptide mapping of the isolated haemagglutinins and neuraminidases (Laver & Kilbourne, 1966). Then in 1968 a strain that was first isolated in Hong Kong spread around the world as another pandemic strain. It was found to differ radically from the then dominant Asian strain in its haemagglutinin, but not in its neuraminidase.

Such strains were said to arise by 'antigenic shift', a major change in the protective antigen, the haemagglutinin, which could not be explained by the sequential mutations that occurred in antigenic drift, but might be due to genetic re-assortment, a process which had been demonstrated experimentally in the developing egg (Burnet & Lind, 1951), but never in nature, or experimentally in natural hosts of influenza A virus, such as swine or chickens. The critical experiments in intact animals were done by Webster, Campbell and Granoff (1973) and can be briefly summarized as follows. Re-assortment, resulting in the production of antigenic hybrids that had exchanged haemagglutinins, could be isolated from sentinel turkeys that were exposed to natural infection with viruses from other turkeys that had been infected with avian

F. Fenner

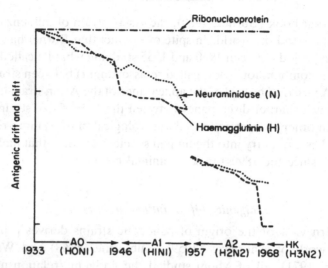

Fig. 2.1. Diagram illustrating antigenic drift, antigenic shift, and the appearance of new subtypes of influenza A virus. 'Antigenic drift and shift' (ordinate) represents the serological relatedness of the antigens of influenza A virus recovered from humans between 1933 and 1972. The internal ribonucleoprotein antigen (–·–·–) has not changed over the whole period. Both the haemagglutinin (– – – –) and the neuraminidase (·······) have shown antigenic drift, i.e. annual but independent changes in antigenicity, and antigenic shift, i.e. larger changes in antigenicity of the haemagglutinin (H0 to H1 in 1946, H1 to H2 in 1957, and H2 to H3 in 1968), and the neuraminidase (N1 to N2 in 1957). (From Fenner *et al.*, 1974.)

influenza viruses with antigenically different haemagglutinins. Similar experiments in sentinel swine exposed to swine infected with either swine influenza virus or Hong Kong influenza virus demonstrated that antigenic hybrids, containing the haemagglutinin of one parent and the neuraminidase of the other, could be isolated from some of the sentinel swine, i.e. re-assortment had occurred in a quasi-natural situation. This led Webster and Kilbourne independently to propose that two kinds of genetic processes were involved in the natural history of human influenza. Pandemic strains that swept through the totally non-immune human population were due to antigenic shift, which was postulated to occur between human strains and animal or bird strains in some host animal, probably domestic swine, that lived in close proximity to humans. Then the new strain underwent progressive mutations in antigenic sites on its haemagglutinin sufficient to ensure that it could continue to spread in a partially immune human population (Fig. 2.1).

Since influenza B virus has no host other than humans it maintains itself in nature by antigenic drift, it does not undergo antigenic shift, and it never causes pandemics.

This was the situation just before the introduction of gene and protein sequencing and monoclonal antibodies made it possible to dissect the nature of different influenza virus strains with exquisite refinement. Although such work has greatly expanded our knowledge, the essential features of the evolution of influenza A virus, both as regards the occurrence of pandemics at intervals of decades or more and the subsequent annual winter epidemics, were worked out by virologists working in the period 1950–1975, using classical methods of serology, peptide mapping and genetic recombination.

Studies on the evolution of myxoma virus and its host

The other example of evolution in action that was studied before the molecular biological revolution is myxomatosis in the European wild rabbit (Fenner & Ratcliffe, 1965; Fenner & Ross, 1994). I am extending the time-frame for discussing myxomatosis into the 1980s, because so far there has been virtually no molecular biological research on its evolution. In contrast to influenza virus, myxoma virus is a poxvirus, with a genome consisting of a single large molecule of double-stranded DNA.

Myxoma virus in the Americas

The natural host of myxoma virus is the *Sylvilagus* rabbit, several species of which occur in the Americas. Since 1896 the disease myxomatosis has been known as a lethal infection of European laboratory and hutch rabbits (Fig. 2.4A), 'spontaneous' outbreaks having been reported in Uruguay, Brazil, Colombia, Panama, and the United States (California and Oregon). In 1943, Aragão, at the Oswaldo Cruz Institute in Rio de Janeiro, showed that the source of the virus was the tropical forest rabbit *Sylvilagus brasiliensis*, in which the virus caused only a trivial localized fibroma. He also showed that the virus was transmitted mechanically by insects that probed through the skin lesions. Twenty years later Marshall, Regnery and Grodhaus (1963) showed that the natural host of the strain of myxoma virus that occurred in California was the brush rabbit, *Sylvilagus bachmani* (Fig. 2.2).

Fig. 2.2. Small fibroma produced in the Californian brush rabbit (*Sylvilagus bachmani*) by the Californian strain of myxoma virus. (Courtesy Dr D. Regnery.) The Brazilian strain of myxoma virus produces a similar lesion in its natural host, *Sylvilagus brasiliensis*.

European rabbits in new ecosystems: America and Australia

Having identified myxoma virus as a virus that had emerged from the tropical forest rabbit to present as a lethal disease of the European rabbit, we need to turn briefly to a consideration of the history of the European rabbit, especially its history as an emerging pest when it invaded new ecosystems. This happened to only a limited extent in the Americas, on a few offshore islands and in Chile and Patagonia. In part, the abundant predators on the American mainland played a role in preventing an explosion of rabbit numbers. However, even more important was the presence of myxoma virus in the leporids in otherwise favourable habitats (but not in Chile or the offshore islands), which prevented *Oryctolagus* rabbits from colonizing the American mainland in the same way that trypanosomiasis kept cattle out of large regions in Africa. In Australia, however, wild *Oryctolagus* rabbits met with no competition from native predators or viruses, and soon after their introduction in 1859 they rapidly spread over the southeastern half of the continent and became Australia's principal agricultural pest. Their effect on pastures was dramatic, but they were even more devastating in causing severe erosion problems in the semidesert which constitutes the greater part of inland Australia.

Myxoma virus for biological control of rabbits

Although attempts at biological control of the rabbit pest were attempted as early as 1888 by none less than Louis Pasteur, and although myxomatosis was suggested as a method of rabbit control by the Brazilian microbiologist Aragão as early as 1918, it was not until 1950 that serious attempts were made to infect Australian wild rabbits with the virus. Field trials begun in the Murray Valley in May 1950 continued through until November, that is, from autumn to late spring. The results were disappointing; although myxomatosis spread within rabbit warrens, the numbers of rabbits in these warrens continued to increase and there was no spread of the virus between warrens. Just as the scientists involved were about to give up, hundreds of rabbits were found affected with myxomatosis in places many miles from the trial sites. Over the summer of 1950–51 the disease spread dramatically over the whole of southeastern Australia, principally along the major rivers, where mosquitoes were most numerous. Few cases were seen during the winter months, but the following summer was wet and the disease spread both along and between the rivers. Investigations at a trial site called Lake Urana showed that the case-fatality rate was fantastically high, over 99% (Fig. 2.3). A year later it was found that a less virulent variant had replaced the original highly virulent strain.

This change in virulence set the stage for investigations of the evolutionary changes in myxomatosis carried out in collaboration by workers in the Commonwealth Scientific and Industrial Research Organization (CSIRO) and the Australian National University (ANU) over the next 15 years. The changes in the virulence of the virus, the resistance of the rabbit, and the disease which was the result of this evolving interaction are summarized below.

Changes in the virulence of myxoma virus

The first problem facing the ANU team was to devise a way by which it would be possible to determine the virulence of field strains using only

F. Fenner

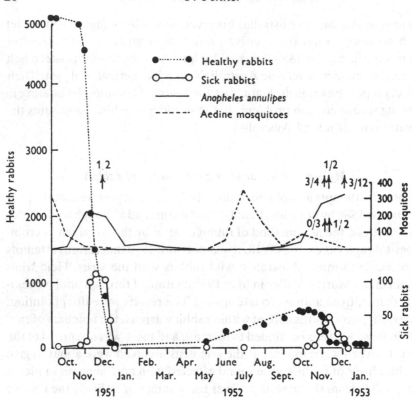

Fig. 2.3. Population counts of rabbits at a trial site at Lake Urana. The epidemic in November–December 1951 was initiated by inoculating several hundred rabbits with myxoma virus of Grade I virulence (Moses strain) in September and October and releasing them. The arrows indicate the occasions on which virus was isolated from batches of mosquitoes. *Anopheles annulipes*, but not aedine mosquitoes, were efficient vectors. The first epidemic was caused by the Grade I virus and the calculated case-fatality rate was 99.8%. The second epidemic was caused by a naturally occurring mutant virus of Grade III virulence; the case-fatality rate in genetically unselected laboratory rabbits was 90%. (From Myers, Marshall & Fenner, 1954.)

a small number of rabbits. Tests with a few strains of widely differing virulence showed that the mean survival times in groups of six rabbits were correlated with death rates tested in hundreds of rabbits. Using this method, five 'virulence grades' were established. Typical clinical pictures of three of the prototype strains and the highly virulent strain 'Lausanne', which was used to introduce the disease into Europe, are shown in Fig. 2.4. The virus first used in Australia (Fig. 2.4B) was provided by Richard Shope of the Rockefeller Institute of Medical

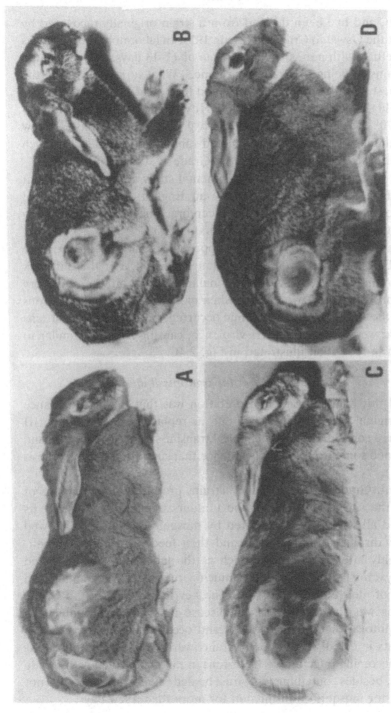

Fig. 2.4. The appearance of normal (genetically unselected) laboratory rabbits infected with various strains of myxoma virus. A. Lausanne strain, Grade I virulence, 10 days after inoculation. B. Standard laboratory (Moses) strain, Grade I virulence, 10 days after inoculation. C. Field strain of Grade III virulence, 21 days after inoculation. D. Field strain of Grade V virulence, 16 days after inoculation. (From Fenner & Ross, 1994.)

Research, and had been derived from a strain originally recovered by Moses at the Oswaldo Cruz Institute in 1911. In laboratory rabbits it was always lethal, with a mean survival time of 11–13 days. We designated such strains as being of Grade I virulence. What turned out to be the commonest kind of virus to be recovered from field cases since the first release (Fig. 2.4C) was virus of Grade III virulence, with a case-fatality rate of 70% to 95% and a mean survival time of 17–28 days. A few strains killed less than 50% of infected rabbits and produced a much milder disease; these were designated Grade V strains (Fig. 2.4D).

There was also available for experimental study a much less virulent strain which had been developed by an English virologist by serial passage of the Moses strain by the intracerebral inoculation of laboratory rabbits, the 'neuromyxoma' strain (Hurst, 1937).

Let us now look again at what happened at our trial site at Lake Urana (Fig. 2.3), noting especially what happened after the outbreak in the summer of 1951–52. An occasional infected rabbit was seen every month throughout the winter, and when mosquitoes became numerous again in the spring another epidemic occurred, but this time the strains recovered were all of Grade III virulence, causing a disease similar to that seen in the rabbit illustrated in Fig. 2.4C.

Natural selection for transmissibility

Why did this happen? Our interpretation was that the disappearance of the original very virulent strain and its replacement by a Grade III strain was a consequence of mechanical transmission by mosquitoes, and we obtained some experimental evidence that supported this hypothesis (Fig. 2.5).

Grade I, Grade III and Grade V strains produced large amounts of virus in the superficial parts of the tumours that give the disease its name and all were well transmitted by mosquitoes that were allowed to probe through these tumours and then feed on a normal rabbit. Very highly attenuated strains, such as the laboratory variant called 'neuromyxoma', did not produce much virus in the skin and had a low rate of mosquito transmission. The key to selection of the Grade III strains was that they caused a disease that was highly infectious for mosquitoes for up to two weeks and occasionally longer, whereas the rabbits infected with Grade I strains were dead within about four days of becoming infectious. The lesions in rabbits infected with Grade V strains, besides containing less virus, healed rapidly and were thus not available for mosquito transmission for more than a few days.

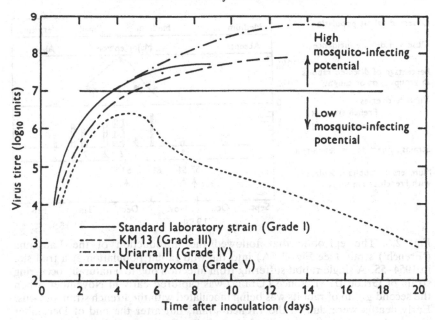

Fig. 2.5. Correlation between the titre of myxoma virus in skin lesions in *Oryctolagus cuniculus* and the suitability of the lesion as a source of virus for mechanical transmission by mosquitoes. Except for a few Grade V strains from which some infected rabbits rapidly recovered, the Grade I virus used for inoculation campaigns and all field strains produced lesions that were highly infectious for rabbits. However, the rapid death caused by the highly virulent strains removed the infectious source within a few days, whereas strains of Grade III virulence produced lesions that were infectious for two weeks and occasionally much longer. With more attenuated strains the lesions healed within 10 days. The laboratory variant 'neuromyxoma' was so attenuated and replicated so poorly in the skin that mosquito transmission was rarely achieved. (From Fenner & Ratcliffe, 1965.)

The validity of this interpretation was revealed, unwittingly, when we conducted a field experiment to study the effectiveness of the highly virulent strain of virus that had been recently recovered from a naturally infected laboratory rabbit in Brazil, and had been used to introduce myxomatosis into Europe in 1952 (the 'Lausanne' strain, Fig. 2.4.A). Unexpectedly, soon after the inoculation of wild rabbits with this virus at the field site, an endemic Grade III strain was recovered from a captured rabbit. The lesions that the Lausanne strain produced after intradermal inoculation were much more protuberant than those produced by the Moses strain that we had been using in Australia (and the less virulent strains derived from it), so it was possible to use a simple intradermal

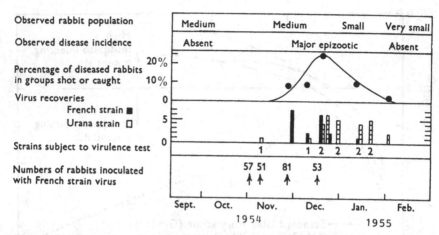

Fig. 2.6. The epizootic that followed the introduction of the Lausanne ('French') strain (see Fig. 2.4A) into the wild rabbit population at a trial site in 1954–55. A single rabbit suffering from infection due to a naturally occurring strain of Grade III virulence (Urana) was captured early in November, when the second group of rabbits was being inoculated with the French strain of virus. Early deaths were due to the French strain, but after the end of December almost all strains of virus obtained were similar to the naturally occurring strain of Grade III virulence. (From Fenner *et al.*, 1957.)

inoculation test to distinguish between the Lausanne virus and the virus that happened to be spreading naturally in the population, a Grade III strain. The results of tests carried out through the transmission season showed that although the highly virulent Lausanne virus, which had been introduced in a small area on a massive scale, caused many early deaths, by the end of the season it was completely replaced by the endemic Grade III strain (Fig. 2.6), and this Grade III strain was the only kind of virus recovered in the outbreak that occurred in the following summer.

Selection for resistance in rabbits

The existence of virus strains that allowed 10% or more of infected rabbits to survive and to breed provided an opportunity for Darwinian selection for increased resistance to occur in the rabbit population. To investigate genetic changes in field strains of virus, we had collected large numbers of strains from field cases each year and tested them in laboratory rabbits from a stock that had never been exposed to myxomatosis. To test for changes in innate resistance, the process was reversed. We caught young rabbits from endemic areas each spring, before myxomatosis had broken out, and raised them to young adulthood in

Table 2.1. *Responses to inoculation with myxoma virus of Grade III virulence of non-immune wild rabbits captured in succeeding years, after exposure of the population to various numbers of epidemics*[a]

Number of epidemics	Case-fatality rate (%)	Clinical signs		
		Severe	Moderate	Mild
0	90	93	5	2
2	88	95	5	0
3	80	93	5	2
4	50	61	26	12
5	53	75	14	11
7	30	54	16	30

Note: [a] From Fenner & Ross, 1994.

the laboratory, by which time they had lost maternal antibody. They were then tested for their innate resistance to myxomatosis by inoculating them with a particular sample of Grade III virus, many ampoules of which had been stored frozen. Over the next seven years the results were dramatic: the case-fatality rate fell from 90% to about 30% (Table 2.1). With continuing selection for genetic resistance it was necessary to change to Grade I virus for tests of increasing resistance, which was shown to increase most rapidly in a region (Mallee) where weather conditions favoured frequent epidemics (Table 2.2). Interestingly, tests in these resistant rabbits showed that the strain of virus used in Europe (Fig. 2.4A) was substantially more virulent than our Australian Grade I strain (Fig. 2.4B), although the virulence of these two strains had appeared to be the same when they were tested in laboratory rabbits.

Co-evolution of virus and host

What of the virulence of field strains of virus over the next 20 years? Although after 1959 the tests were conducted in other laboratories and on a smaller scale, assays of field strains demonstrated the continuing dominance of Grade III and the somewhat less virulent Grade IV strains (Table 2.3).

What might be expected from the dynamic interaction between the changing virulence and the developing genetic resistance? Reviewing the situation in 1965, we suggested that if selection of the virus strain was dependent on transmissibility, then as rabbits became more resistant somewhat more virulent viruses than Grade III might be better

Table 2.2. *Responses to inoculation with myxoma virus of Grade I virulence of non-immune wild rabbits captured in various parts of Victoria where the exposure to myxomatosis differed*[a]

Place and time of capture	Case-fatality rate (%)	
	Standard laboratory strain (Figure 2.4B)	Lausanne strain (Figure 2.4A)
Gippsland[b]		
1961–66	94	
1967–71	90	
1972–75	85	*c.* 100
1975–81	79	*c.* 100
Mallee[c]		
1961–66	68	
1967–71	66	
1972–75	67	*c.* 100
1975–81	60	98

Note: [a] Data from Fenner & Ross, 1994. [b] Gippsland is a part of Victoria where epidemics of myxomatosis are infrequent. [c] Mallee is a part of Victoria where there have been annual outbreaks of myxomatosis since 1951.

transmitted (Fenner & Ratcliffe, 1965). Subsequently, analysis of the virulence of strains of virus recovered from wild rabbits from two parts of Victoria where the rabbits had different levels of genetic resistance (Table 2.3) appeared to support this hypothesis (Table 2.4). While Grade III strains remained dominant through the 20-odd years over which tests were carried out, a higher proportion of more virulent strains were recovered from the Mallee region, where the rabbits had the highest innate resistance. This picture was even better illustrated in tests of changing virus virulence and rabbit resistance carried out in England, where the virus that had originally been introduced was the highly virulent 'Lausanne' strain (Fenner & Ross, 1994).

Summary

I have described the only two examples of the evolution of animal viruses on which data were available before the advent of molecular biology, namely influenza A virus and myxoma virus. In both cases, natural selection favoured transmissibility. With influenza virus, continued transmission in the human population depends upon continuing spread between susceptible humans. This exerts pressure for serial changes

Table 2.3. *The virulence of strains of myxoma virus recovered from wild rabbits in Australia between 1951 and 1981, calculated on the basis of survival times (expressed as percentages)*[a]

| Virulence grade | I | II | III | IV | V | Number of |
| Case-fatality rate (%) | >99 | 95–99 | 70–95 | 50–70 | <50 | samples |
Mean survival times (days)	<13	14–16	17–28	29–50	–	
1950–51	100[b]				–	1
1952–55	13.3	20.0	53.3	13.3	0	60
1955–58	0.7	5.3	54.6	24.1	15.5	432
1959–63	1.7	11.1	60.6	21.8	4.7	449
1964–66	0.7	0.3	63.7	34.0	1.3	306
1967–69	0	0	62.4	35.8	1.7	229
1970–74	0.6	4.6	74.1	20.7	0	174
1975–81	1.9	3.3	67.0	27.8	0	212

Note: [a] From Fenner & Ross, 1994. [b] Although only one field strain was tested, this extrapolation is justified by the very high mortality rates in the initial outbreaks.

Table 2.4. *The virulence of strains of myxoma virus recovered between 1959 and 1981 from wild rabbits in two different parts of Victoria, where the genetic resistance differed substantially*[a]

| Virulence grade | I | II | III | IV | V | Number of |
| Case-fatality rate (%) | >99 | 95–99 | 70–95 | 50–70 | <50 | samples |
Mean survival times (days)	<13	14–16	17–28	29–50	–	
Mallee region (high genetic resistance)						
1959–63	0	4.3	57.1	34.3	4.3	70
1964–66	2.0	0	64.7	31.3	2.0	51
1967–69	0	0	68.1	31.9	0	31
1970–74	1.0	6.9	77.5	14.7	0	102
1975–81	3.0	5.8	67.8	23.4	0	121
Elsewhere in Victoria (lower genetic resistance)						
1959–63	2.1	12.4	61.2	19.5	4.7	379
1964–66	0.4	0.4	63.5	34.5	1.2	255
1967–69	0	0	61.6	36.4	2.0	198
1970–74	0	1.4	69.4	29.2	0	72
1975–81	0	0	65.8	34.2	0	91

Note: [a] From Fenner & Ross, 1994.

in the protective antigens, notably those of the haemagglutinin and to a lesser extent those of the neuraminidase. This process is called antigenic drift. With influenza A virus, of which various birds are the reservoir hosts, another kind of change occurs at intervals of decades and leads to pandemics. These changes, called antigenic shift, are due to re-assortment of genes of human and bird viruses, which is possible because the genome consists of eight separate RNA molecules.

Myxoma virus occurs naturally as a benign fibroma in American rabbits, transmitted mechanically from one animal to another by biting arthropods. When the European rabbit is infected with virus from these animals, an extremely severe and almost invariably fatal disease ensues. This led to the deliberate introduction of myxoma virus into Australia to control wild European rabbits, which were a major agricultural pest. Changes in the virulence of the virus, the genetic resistance of host rabbits, and the interactions between virus and rabbit provided a unique natural experiment of evolution in action and an opportunity to observe these rapid changes. Because of advantages in transmission during periods of low mosquito numbers, there was first a selection for less virulent but not highly attenuated viruses that would produce infectious lesions lasting for many days. The original highly virulent virus killed rabbits within 3–4 days of their lesions becoming infectious, whereas highly attenuated strains produced too little virus in the skin and the rabbits recovered quickly. The change from less than 1% recovery to 10% recovery that followed the change in the virulence of the dominant viruses provided the opportunity for the selection of resistance among the rabbits. Subsequently, there was co-evolution of virulence and resistance, such that in regions where rabbit resistance was high there was selection for viruses of higher virulence, because they were more efficiently transmitted.

References

Andrewes, C.H. (1959). *Perspect. Virol.*, 184–91.

Andrewes, C.H. (1967). *The Natural History of Viruses*, Weidenfeld and Nicholson, London.

Aragão, H. de B. (1943). *Mem. Inst. Oswaldo Cruz*, **38**, 93–9.

Burnet F.M. (1955). *Principles of Animal Virology*, Academic Press, New York.

Burnet F.M. (1960). *Principles of Animal Virology*, 2nd edn., Academic Press, New York.

Burnet, F.M. & Lind, P.E. (1951). *J. Gen. Microbiol*, **5**, 67–82.

Fenner, F., McAuslan, B.R., Mims, C.A., Sambrook, J. & White, D.O.

(1974). *The Biology of Animal Viruses*, 2nd edn., Academic Press, New York, p. 628.

Fenner, F., Poole, W.E., Marshall, I.D. & Dyce, A.L. (1957). *J. Hyg. Camb.*, **55**, 192–206.

Fenner, F. & Ratcliffe, F.N. (1965). *Myxomatosis*, Cambridge University Press, Cambridge, pp. 335–347.

Fenner, F. & Ross, J. (1994). In *The Rabbit in Britain, France, and Australasia: the Ecology of a Successful Colonizer*, H.V. Thompson and C. King (eds.). Oxford University Press, Oxford.

Hurst, E.W. (1937). *Br. J. Exp. Path.*, **18**, 15–22.

Kilbourne, E.D. (1968). *Science*, **160**, 74–6.

Laver, W.G. & Kilbourne, E.D. (1966). *Virology*, **30**, 493–501

Marshall, I.D., Regnery, D.C. & Grodhaus, G. (1963). *Am. J. Hyg.*, **77**, 195–204.

Myers, K., Marshall, I.D. & Fenner, F. (1954). *J. Hyg. Camb.*, **52**: 337–60.

Webster, R.G., Campbell, C.H. & Granoff, A. (1973). *Virology*, **51**, 149–62.

Webster, R.G. & Laver, W.G. (1971). *Progr. Med. Virol.*, 271–338.

3

The evolution of virus diseases: their emergence, epidemicity, and control

NEAL NATHANSON, KATHLEEN A. McGANN AND JOHN WILESMITH

Introduction

The evolution of virus diseases is distinct from, although related to, the evolution of viruses in general. The emergence of a virus disease can reflect the 'evolution' of the causal agent, but a disease can emerge in the absence of any change in the agent. To maintain this distinction we will use a special vocabulary and refer to the 'emergence, epidemicity, and control' of virus diseases (Nathanson, 1990).

Propagated infections

In some instances, a new virus disease is the direct consequence of the appearance of an agent which is truly new to the population affected. Recent striking instances are the appearance of canine parvovirus and of human immunodeficiency virus (HIV), both of which were signalled by a global pandemic of disease. It appears that canine parvovirus represents an authentic instance of virus evolution, in which a pre-existing agent, feline panleukopenia virus, underwent a few key point mutations which made it very infectious and pathogenic for the canine population (Parish, 1990). These mutations presumably led to the emergence of canine parvovirus disease. Likewise, it seems probable that HIV evolved from a viral ancestor, such as simian immunodeficiency virus (SIV), which was circulating in subhuman primates. It is likely that this highly plastic agent has undergone some evolution since originally infecting humans, which has enhanced its ability to replicate in human CD4-positive mononuclear cells. A mass of circumstantial data strongly suggests that the virus first invaded the human population in Africa, and that it has spread worldwide only in the last 15 years (Desrosiers, 1990; Getchell *et al.*, 1987; Karpas, 1990; Levy *et al.*, 1986).

The evolution of a virus which is already present in the population

31

can also lead to the emergence of epidemic disease. A familiar example is the introduction of an influenza virus re-assortant, bearing a new haemagglutinin or a new neuraminidase gene, into a host population which has not been immunized by prior infection with the antigenic determinants encoded by the new gene segments. This provides a highly susceptible population through which the re-assortant virus can spread, often with epidemic consequences (Webster & Murphy, 1990). In some instances, evolution of an existing agent may be associated with enhancement of virulence resulting in an increase in the ratio of cases of clinical disease to infections. One dramatic example is the pandemic of avian influenza which devastated the poultry industry of the State of Pennsylvania in the early 1980s. In this example, a single point mutation in the haemagglutinin of a relatively innocuous enzootic virus enormously enhanced its virulence, thereby initiating a major epizootic (Chapter 34; Webster *et al.*, 1986).

However, disease may also emerge without any change in the causal virus. For example, epidemic disease may follow the entry of an agent into a highly susceptible population where the virus has long been absent. A classical example is the introduction of measles into a remote island population, such as produced the epidemic of 1846 in the Faroe Islands (Panum, 1940). Alternatively, emergence of epidemic disease may occur in a population where the virus is already highly prevalent, due to a host-specific change increase in the case:infection ratio. An instance is the emergence of poliomyelitis, which will be discussed in detail below.

Common source infections

Existing agents, if artificially introduced from a common source into a large population, can cause a virus disease to emerge in epidemic proportions. The disease may already occur at low endemic frequency as in the case of the outbreak of hepatitis B which followed the use of contaminated yellow fever vaccine at the outset of World War II (Sawyer *et al.*, 1944). However, the disease may be completely new to the host population, as in the instance of bovine spongiform encephalopathy discussed below. Common source outbreaks may be due to contamination of water, food, air, or parenteral injectables.

Zoonotic or vector-borne infections

The very large number of arboviruses are maintained by vector–vertebrate cycles of transmission. In addition, there are many zoonotic infections, such as rabies, hantaviruses or arenaviruses, which are maintained by animal-to-animal transmission, but which can be transmitted to humans under special circumstances. Humans usually act as dead-end hosts and are not involved in maintenance of the cycle, with a few exceptions, such as urban yellow fever and dengue.

For this category of agents, the emergence of disease in the human population can occur under several different circumstances. If humans enter a new habitat, they may be exposed to infection by the vector, as in the instance of jungle yellow fever. Likewise, haemorrhagic fever with renal syndrome may occur in persons exposed to Hantaan virus, transmitted by contaminated aerosolized urine directly from infected field mice.

Viruses may invade an existing ecosystem, with the emergence of a new or rarely seen disease. Examples are outbreaks of Rift Valley fever in Africa, Oropouche disease in Brazil, and St. Louis encephalitis in the United States. In such instances, there are usually special circumstances, such as conditions resulting in massive increases of the vector population, which are responsible for the outbreak. Another instance is the invasion of communities in the mid-Atlantic United States by raccoon rabies, leading to potential infection of cats and dogs, and possible exposure of humans, and the spread of fox rabies in Europe.

Finally, an ecosystem may be artificially altered, as in the instance of major dam projects or jungle clearance, which leads to the introduction of new vectors and arboviruses into an existing habitat.

We have selected two specific examples, poliomyelitis and bovine spongiform encephalopathy (BSE) to illustrate virus infections which have evolved dramatically and emerged as major new epidemic diseases. In these examples, evolution has come full circle and the disease has been brought under control and declined, or is predicted to decline.

Poliomyelitis

It is likely that poliomyelitis has occurred as a sporadic disease since ancient times. During the eighteenth and first half of the nineteenth century, a number of individual cases and a few clusters of cases were reported. About 1870, poliomyelitis burst on the paediatric scene as a cause of epidemics of 'infantile paralysis'. The first scattered epidemics were in rural areas of the northern United States and Sweden. But shortly they became more common, so that epidemics were reported annually in the United States, from about 1905, as shown in Fig. 3.1.

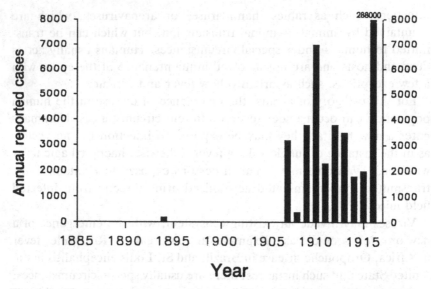

Fig. 3.1. The appearance of epidemic poliomyelitis in the United States, 1885–1916. Based on reported cases, most of which were paralytic, during an era when reporting was estimated at about 50%. (After Lavinder *et al.*, 1918.)

Epidemics of poliomyelitis disease have occurred many times since, first in developed and then in developing countries. Paul (1971) has reviewed the history of poliomyelitis in an authoritative and fascinating monograph.

What accounts for the emergence of epidemic poliomyelitis? There is no evidence that it is due to the appearance of a virus strain of increased virulence. Evidence against a strain difference comes from several sources, but one of the most convincing is a study conducted in Casablanca, Morocco, in the 1950s. At that time, there was a substantial European population in Casablanca, and this population experienced a poliomyelitis attack rate about 20-fold that of the native Moroccan population, as shown in Table 3.1. Furthermore, a clue to the explanation is provided by the striking differences in age distribution; the European cases were among children and teenagers but almost all the Moroccan cases were among very young infants. It appears that the same wild polioviruses were circulating among both populations, but that the relative incidence and age selection of disease was very different.

Direct evidence for this view is provided by a serological survey, also conducted in the early 1950s, which compared Cairo, where poliomyelitis

Table 3.1. *Poliomyelitis attack rates and age distribution for the European and Moroccan populations of Casablanca, 1947–1953 (after Paul & Horstmann, 1955)*

Population	European	Moroccan
Paralytic cases 1947–1953	125 000	530 000
Average annual attack rate per 100 000	13.4	0.7
1953 cases by age (years)		
0–1	8	9
2–9	15	2
10–39	5	0

was very rare, with Miami, where it produced regular outbreaks (Fig. 3.2). Clearly, poliovirus infection was ubiquitous in Cairo, and essentially the whole population had acquired antibody by early childhood. By contrast, in Miami infection was also widespread, but was often delayed until teenage years or later. The attack rates, as in Casablanca, were very

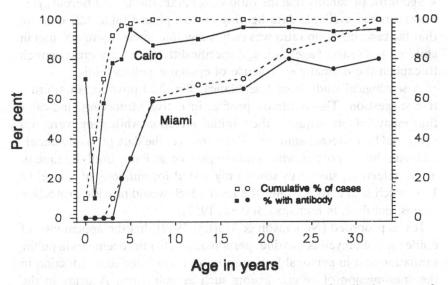

Fig. 3.2. Comparison of the age-specific prevalence of poliovirus type 2 neutralizing antibodies in Cairo and Miami, and of the cumulative numbers of poliomyelitis cases in the same populations. (After Paul, 1955.)

Table 3.2. *Age-specific case : infection ratio for poliomyelitis estimated from a prospective study in North Carolina, USA, 1948 (after Melnick & Ledinko, 1953)*

Age group	Popu- lation	Sero- converters	Estimated total converters	Paralytic cases	Cases per 100 converters
<1	1800	5/20	450	3	0.67
1–2	3900	10/39	1000	10	1.00
3–4	3600	7/34	741	12	1.62
5–9	7300	8/56	1043	25	2.40
10–14	6300	5/44	716	13	1.82

different (Miami, high, and Cairo, very low) and the age distribution of cases reflected the serological profile.

These data suggest that there was an important correlation between the age of first infection with poliovirus and the risk of a paralytic case. Since the early epidemics were almost exclusively concentrated in young children, under age 5 ('infantile paralysis'), the difference must be sought within the first 5 years of life. It is well established that the case : infection ratio for poliomyelitis is about 1 case per 100 infections, and this suggests the possibility that the ratio varies dramatically at different ages. However, the available data bearing on this point (Table 3.2) indicate that the case:infection ratio was only a few fold lower in infants than in children 1–4 years of age. This age-specific difference is not great enough to explain the dramatic emergence of epidemic poliomyelitis.

A serological study from Casablanca (Fig. 3.3) provides an alternative suggestion. This antibody profile, in native Moroccans, indicates that many infants acquired their initial infection while they were still protected by maternal antibody. Furthermore, the data probably underestimate the proportion who retained passive antibody until the time of active infection, since this study only tested for antibody at a level of 1:10 which is at least 10 times the level which would provide protection against paralysis in humans (Bodian, 1952).

It was proposed (Nathanson & Martin, 1979) that the appearance of epidemic poliomyelitis was due, paradoxically, to improvements in public sanitation and in personal hygiene, both of which led to a reduction in the transmission of enteric agents such as poliovirus. A delay in the age of initial infection with poliovirus beyond the age when infants were protected by passively acquired maternal antibody is postulated

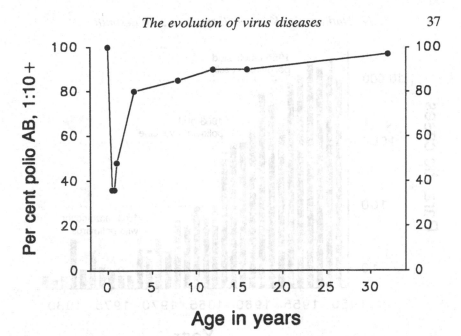

Fig. 3.3 Age-specific proportion of native Moroccans with type 1 poliovirus neutralizing antibody at a 1 : 10 titre or greater, Casablanca, 1953. (After Paul & Horstmann, 1955.)

to increase the risk of clinical disease. A secondary contribution may have been an intrinsically somewhat lower case : infection ratio in infancy compared to early childhood.

Poliomyelitis continued to be a scourge in the United States from about 1905 to 1960. Between 1960 and 1972 the disease was gradually brought under control (Fig. 3.4), first by the use of inactivated poliovirus vaccine (IPV) and then by oral poliovirus vaccine (OPV). Remarkably, and unexpectedly, control measures actually led to the eradication of wild polioviruses from the United States in 1973. How did this occur?

There was no expectation that immunization programmes would eradicate poliomyelitis because experience had indicated that, in the United States, it was impossible to immunize over 90–95% of the population. In fact, immunization surveys conducted after the introduction of OPV indicated that at least 5% of the population had received no vaccine. Serosurveys indicated that the proportion lacking antibody for any given serotype of poliovirus was probably considerably higher. As a minimum estimate, there were well over 5 000 000 unprotected persons in the United States during the 1970s. It is clear that this group

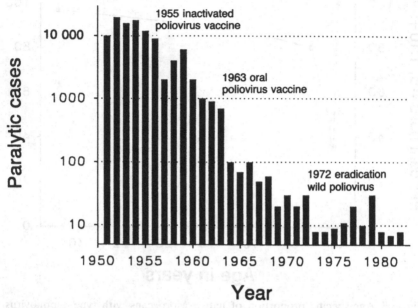

Fig. 3.4. Annual reported cases of paralytic poliomyelitis in the United States, 1951–1982. For the years 1972–1982, residual cases are either vaccine-associated or imported, with the exception of an outbreak in 1979 in the unvaccinated Amish population. (After Centers for Disease Control, 1982.)

would register the circulation of wild polioviruses, by the occurrence of paralytic poliomyelitis, even at a case : infection ratio of 1 : 100. And yet, in spite of intense surveillance, there were no cases of poliomyelitis in the United States after 1972, except for vaccine-associated cases and imported cases acquired in Mexico or elsewhere. The one exception was a small outbreak in the unvaccinated Amish population in 1979, which demonstrates that the population would indeed register the circulation of wild polioviruses.

It was suggested (Nathanson, 1984) that the disappearance of wild polioviruses was due to two factors, the seasonality of the infection and the relatively poor transmissibility of the infection. In the pre-vaccine era, the seasonality of poliomyelitis was very striking, particularly in the northern United States. Fig. 3.5 shows that, in New England, about 0.3% of the annual total incidence occurred in each of the low-incidence months of the year. It can be calculated that this was equivalent to only 20 infections per generation period in a population of 1 000 000. In the post-vaccine era, with a reduction of the susceptible

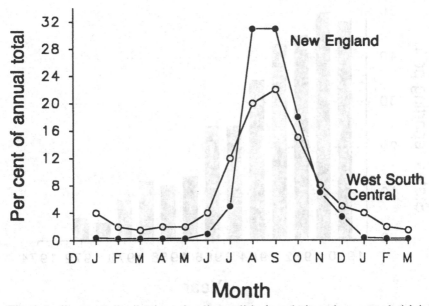

Fig. 3.5. Seasonal distribution of poliomyelitis (paralytic and non-paralytic) in New England and in the West South Central region of the United States, 1942–1951. (After Serfling & Sherman, 1953.)

population to about 1/20 that prior to vaccine, there would be only a few infections per generation period in the seasonal trough. Under these circumstances, the virus might totally disappear from the population of a city of considerable size. The seasonality of poliomyelitis continued during 1960–1972, which was the period of poliovirus eradication in the United States. Concomitantly, there was a stepwise reduction in the number of States reporting any cases of poliomyelitis, from 50 (100%) States in 1961 to 0 States in 1973 (Fig. 3.6). It is implicit in these data that, following the local fadeout of infection, poliovirus was not re-introduced into each virus-free area. This contrasts with measles, where fadeouts of virus occurred with regularity but virus was frequently reintroduced. It has been postulated (Nathanson, 1984) that this difference explains the failure to eradicate measles in spite of immunization programs which were at least as thorough as those for poliomyelitis.

In summary, it is proposed that poliovirus was eradicated in spite of a large residual unimmunized susceptible population because of fadeout of virus during the seasonal trough in individual localities, leading to a stepwise disappearance of wild poliovirus from the continental United

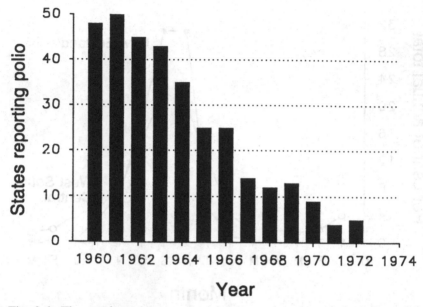

Fig. 3.6. The number of States reporting any cases of paralytic poliomyelitis each year, United States, 1960–1973. (After Centers for Disease Control, 1961–1974.)

States. A similar phenomenon has now taken place in South and Central America, so that the whole western hemisphere was recently rendered free of poliomyelitis, and global eradication of poliovirus can now be contemplated.

Bovine spongiform encephalopathy (BSE)

Bovine spongiform encephalopathy (BSE) is a scrapie-like disease of cattle (Wilesmith & Wells, 1991), which has occurred primarily in Great Britain, although cases have now been reported from Europe. BSE is a newly recognized entity, and it is questionable whether it ever occurred prior to the recent epizootic in England. In retrospect, the first cases of BSE probably began in England in April, 1985 and the disease was officially recognized in November, 1986. The epizootic curve is shown in Fig. 3.7 with cases through 1990, which indicate a dramatic increase in reported incidence from about 50 cases in 1985, before the disease became notifiable, to over 12 000 cases in 1990. A series of detailed investigations have been conducted in the UK (Wilesmith,

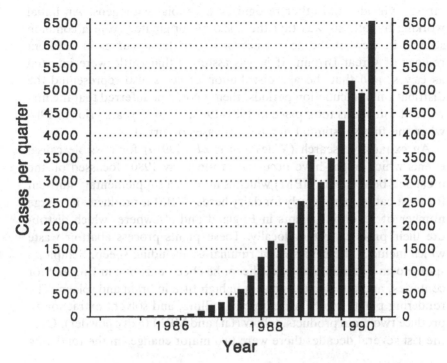

Fig. 3.7. The number of cases of BSE by quarter of onset, United Kingdom, 1985–1990. (After Wilesmith, 1991; Wilesmith & Wells, 1991.)

1991; Wilesmith & Wells, 1991; Wilesmith, Ryan & Atkinson, 1991*a*; Wilesmith *et al.*, 1991*b*), and these have illuminated a number of salient epidemiological features of the outbreak. BSE has occurred throughout Great Britain, and the initial cases were observed almost simultaneously in many different areas. Cumulative data have revealed a geographic gradient of incidence within Great Britain, with the highest rates in southern England (Table 3.3) and lowest rates in Scotland. Rates have been consistently higher in dairy than in beef herds. Cumulative attack rates within affected herds have averaged about 1–2% indicating that only a small proportion of animals develop disease. The age distribution of cases is shown in Fig. 3.8, and indicates that most onsets occurred between 3 and 8 years of age with a peak at 4–5 years of age.

What was the cause of this dramatic epizootic? The pathological lesions were confined to the brain and were apparently identical to scrapie of sheep, a subacute spongiform encephalopathy caused by an unconventional transmissable agent, which appears to differ from known viruses, viroids, and other recognized microbial pathogens. An initial working hypothesis was that the epizootic originated from a common source, in view of the near synchronous epidemic curves in different regions of Great Britain. If it was assumed that cattle were infected as calves, and that the age distribution of cases also represented the distribution of incubation periods, then it could be inferred that the first wave of infections took place in 1981–1982, i.e. 4–5 years prior to 1986 when the first substantial numbers of cases occurred.

An exhaustive search (Wilesmith *et al.*, 1991*a*) for possible causal events which might have occurred in the early 1980s focussed on the meat and bone meal (MBM) which is used as a supplementary nutrient for most cattle, particularly for dairy herds. MBM is produced in a large number of rendering plants in England and elsewhere, which distribute their products rather locally. These plants process abattoir waste which includes the carcasses of ruminants, including sheep. Scrapie, a spongiform encephalopathy of sheep, has been enzootic in England for centuries, and the agent is present in high titre in brain and spleen. The rendering plants use heat treatment, milling, and solvent extraction to produce two main products, tallow (fat) and MBM (a dry powder). Over the last several decades there were two major changes in the rendering

Table 3.3. *The frequency of bovine spongiform encephalopathy (BSE) by region, compared with methods used for processing of meat and bone meal (MBM), in the United Kingdom, 1986–1989 (after Wilesmith et al., 1991a)*

Region	Per cent of dairy herds with at least one case of BSE	Proportion of MBM produced by solvent processing	Per cent of MBM produced by reprocessing of greaves
Southern England	12.6%	0	0.2%
Midlands	3.9%	0	8.6%
Northern England	2.8%	0	25.5%
Scotland	1.8%	BULK	39.0%

Note: Greaves are a partially processed intermediate which is then further processed to yield MBM, and this two-step procedure probably enhances inactivation of the scrapie agent. Most rendering plants in the United Kingdom did not use solvent extraction, but the two plants in Scotland, which supplied the bulk of MBM in that region, continued to use solvents. It is likely that solvent extraction would markedly reduce the infectivity of the scrapie agent.

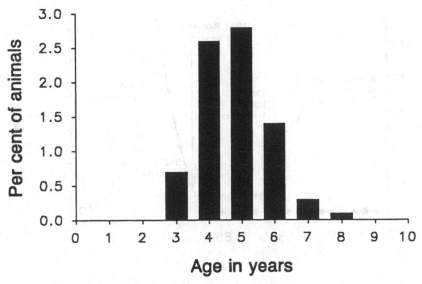

Fig. 3.8. Age-specific incidence of BSE in affected herds of cattle, Great Britain, 1989. (After Wilesmith *et al.*, 1991*b*.)

process. There had been a change from a batch to a continuous process, which had occurred gradually in England between 1970 and 1985; by 1985 about 75% of MBM was produced by the continuous process. The second change, which had occurred more abruptly, mainly between 1977 and 1982, was the abandonment of solvent extraction, so that by 1982 only 10% of MBM produced in England involved solvent extraction (Fig. 3.9). Support for the importance of solvent extraction came from the low rate of BSE in Scotland (Table 3.3), where solvent extraction has continued to be used to the present time. Also, work with experimental scrapie in mice has established the fact that this agent is quite sensitive to fat solvents.

The tentative hypothesis, that the epizootic of BSE was caused by contamination of MBM supplements with scrapie-infected sheep tissues, leaves unanswered another question. Why did the epizootic of BSE occur in Great Britain, while other cattle-raising countries, in Europe and North America, have experienced only a few scattered cases? The proposed explanation (Wilesmith & Wells, 1991) is that there were several critical requirements for the occurrence of the BSE outbreak, and all of these came together only in the United Kingdom. These postulated requirements were: first, a large sheep population relative

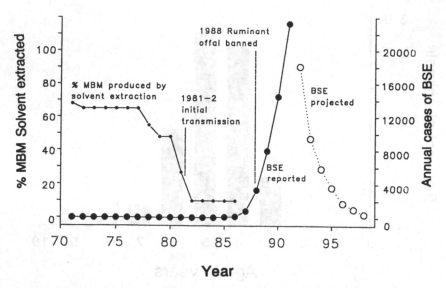

Fig. 3.9. The chronology of bovine spongiform encephalopathy and related events in Britain, 1975–2000. The composite graph shows: the percentage of meat and bone meal (MBM) which was processed by solvent extraction; the calculated time of initial transmission of BSE (1981–1982); the prohibition in July, 1988 of the use of ruminant protein in dietary supplements fed to ruminants; the annual incidence of BSE in the United Kingdom, 1985–1990; and the projected incidence of BSE from 1991–2000, assuming that transmission ceased in 1988. (After Wilesmith & Wells, 1991, Wilesmith *et al.*, 1991*a*.)

to other ruminants, so that sheep contributed a substantial proportion of the raw material entering rendering plants; second, a fairly high enzootic level of scrapie in sheep; third, intensive feeding of MBM to dairy cattle; and, finally, changes in the rendering process, including solvent extraction, which reduced the inactivation of scrapie infectivity during the production of MBM.

Experiments to test some of these proposed factors are currently under way. However, the most significant 'experiment' was begun in July, 1988, when a ban was placed on the use of any ruminant tissue-derived protein in dietary supplements destined for feeding to ruminants. At the time of writing, there are no reported cases of BSE in cattle born after introduction of the ban. If it is assumed that transmission of BSE ceased in mid-1988, and that there will be no transmission from affected cattle in the field, either vertically to their calves or laterally to their contacts, then the epizootic would be predicted to wane quite rapidly in the next 5–10 years, as projected in Fig. 3.9.

Comment

The evolution of virus diseases, both their emergence and disappearance, involves complex interactions between the agent, the host, and the environment. There are some instances of emergence of major new virus diseases which clearly involve evolution in the virus genome. Examples include canine parvovirus, pandemic influenza in humans and, probably, AIDS in humans. Evolution may also play a secondary role in some instances. For example, it has been postulated that sheep scrapie, once introduced into cattle as BSE, may have adapted so that after the first cattle passage it is now more infectious for cattle. Such a postulated change in the agent might have amplified the ongoing BSE epizootic and perhaps contributed to its extent.

However, in many instances, the emergence of a new viral disease is determined by host or environmental factors with no alteration in the viral agent. Zoonotic and arthropod-borne virus diseases of humans are caused by exposure to agents which are not maintained in human populations, and their emergence is associated with special circumstances leading to human infection. Epidemics which are transmitted from a common source usually involve a special set of conditions leading to contamination of food, water, air, or therapeutic agents. Viruses which are maintained by propagation in human populations can also cause emergence of disease, when they are introduced into susceptible populations from which they have been absent, as occurred repeatedly with measles in isolated islands. Also, propagated viruses can cause emergence of disease if there is an increase in the case : infection ratio, as illustrated above by the emergence of poliomyelitis.

The reduction or disappearance of virus diseases usually involves human intervention, as exemplified by immunization or eradication programmes for diseases such as smallpox, poliomyelitis, and measles. Naturally occurring immunity, with an exhaustion of the susceptible population and a fadeout of the virus disease, can also occur, as has been seen with measles in isolated island populations or with pandemic influenza on a global basis. In all of these instances, host and environmental factors are crucial. Waning of a virus disease because of evolution of the causal agent is probably rare, but the history (Fenner & Ratcliffe, 1965) of myxomatosis in rabbits in Australia suggests that it can occur.

Study of the evolution of virus disease, and the mechanisms which underlie it, is both an interesting scientific exercise and one which carries lessons relevant to the control of important diseases of humans, animals, and plants.

Acknowledgements

This review was inspired by discussions on emerging infectious diseases with Frank Fenner, Ashley Haase, Richard Krause, Donald Krogstad, and Thomas Monath, which took place at a workshop sponsored by the NIAID in Hamilton, MT, July, 1991.

References

Bodian, D. (1952). *Am. J. Hyg.*, **56**, 78–89.

Desrosiers, R.C. (1990). *Nature*, **345**, 288–9.

Fenner, F. & Ratcliffe, F.N. (1965). *Myxomatosis*. Cambridge University Press, London.

Getchell, J.P., Hicks, D.R., Srinivasan, A., Heath, J.L., York, D.A., Malonga, M., Forthal, D.N., Mann, J.M. & McCormick, J.B. (1987). *J. Infect. Dis.*, **156**, 833–7.

Karpas, A. (1990). *Nature*, **348**, 587.

Levy, J.A., Pan, L.-Z., Beth-Giraldo, E., Kaminsky, L.S., Henle, G., Henle, W. & Giraldo, G. (1986). *Proc. Natl. Acad. Sci. USA*, **83**, 7935–7.

Melnick, J.L. & Ledinko, N. (1953). *Am. J. Hyg.*, **58**, 207–22.

Nathanson, N. (1984). *Rev. Infec. Dis.* **6**, S308–12.

Nathanson, N. (1990). In *Epidemiology*, B.N. Fields *et al.* (Eds). *Virology*. Raven Press, pp. 267–291.

Nathanson, N. & Martin, J.R. (1979). *Am. J. Epidemiol.*, **110**, 672–92.

Panum, P. (1940). *Observations made during the Epidemic of Measles Virus on the Faroe Islands during the Year 1846*. Hatcher, A.S., translator. Reprint published by the American Public Health Association, New York.

Parish, C. (1990). *Adv. Virus Res.*, **38**, 403–50.

Paul, J. (1971). *A History of Poliomyelitis*. Yale University Press, New Haven, CN.

Paul, J. & Horstmann, D.M. (1955). *Am. J. Trop. Med Hyg.*, **4**, 512–24.

Sawyer, W.A., Meyer, K.F., Eaton, M.D., Bauer, J.H., Putnam, P. & Schwentleer, F.F. (1944). *Am. J. Hyg.*, **39**, 337–87.

Webster, R.G., Kawaoka, Y. & Bean, W.J. Jr. (1986). *Virology*, **149**, 165–73.

Webster, R.G. & Murphy, B.R. (1990). In *Orthomyxoviruses*, B.N. Fields *et al.*, (Eds.), *Virology*, Raven Press, New York, pp. 1099–1154.

Wilesmith, J.W. (1991). *Seminars in Virology*, Saunders, London, in press.

Wilesmith, J.W. & Wells, G.A.H. (1991). *Curr. Top. Microbiol. Immunol.* **172**, 21–38.

Wilesmith, J.W., Ryan, J.B.M. & Atkinson, M.J. (1991a). *The Veterinary Record*, **128**, 199–202.

Wilesmith, J.W., Ryan, J.B.M., Hueston, W.D. & Hoinville, L.J. (1991b) *The Veterinary Record*, in press.

Part II
Origins of viruses and their genes

4

Origin of RNA viral genomes; approaching the problem by comparative sequence analysis

ALEXANDER E. GORBALENYA

Introduction

One of the notable features of viruses is the exceptional diversity of their genome organization and of the modes of their genome expression. Viruses whose genome is RNA rather than DNA, as in all cellular organisms (Fig. 4.1), contribute greatly to virus diversity. The origin and evolution of RNA viruses has intrigued researchers for decades.

Three classes of RNA viruses can be distinguished, depending on the type of genomic molecule (Baltimore, 1971; Fig. 4.1). These are RNA viruses that employ positive messenger-sense molecules as genomes (+RNA), those that have genome molecules with complementary polarity (negative) to that of their mRNAs (−RNA), and those that have genomes of double-stranded (ds) RNA molecules. A unique virus-encoded enzyme RNA-dependent RNA polymerase (RdRp) directs the synthesis of all of these RNA species. Sometimes, viruses with RNA genomes which are replicated through a DNA stage, the retroviruses, are also regarded as RNA viruses (Chapter 27).

In order to sort out the diversity of viruses, a classification has been introduced which uses about a dozen different phenotypic characteristics (Francki *et al.*, 1991). Thus, more than 60 RNA viral families and separate groups have been delineated, but this classification does not attempt to reflect the evolutionary relationships between the taxa.

A lack of virus fossils, and the apparent high rate of turnover of genome sequences, seemed to preclude our ability to understand RNA virus evolution. However, through the last decade, thanks primarily to the enormous increase in our knowledge of viral genome sequences, it became evident that some progress in evolutionary inference might be possible. Comparative sequence analysis showed unexpected similarities between RNA viruses themselves and between viral

A.E. Gorbalenya

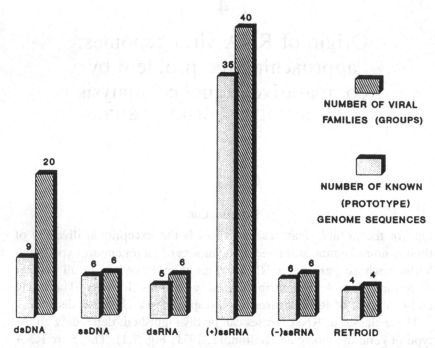

Fig. 4.1. Numbers of viral families/groups, and the numbers of them for which sequence information is available. Viral classes are those described by Baltimore (1971) with additions, and the viral families/groups described by Francki *et al.* (1991) with minor modifications. Data on the sequences of viral genomes are from published reports up to, and including, 1992.

and cellular proteins, indicating that there are relationships between the viruses and also between the RNA viruses and DNA-based organisms. These data will be briefly reviewed below with the aim of assembling evidence that RNA viruses could have originated from a common ancestor which might have inhabited the ancient RNA world.

Monophyletic versus polyphyletic origin of RNA viruses

Viruses with a positive RNA genome, the +RNA viruses, are the most abundant virus class comprising close to 80% of the total RNA virus families or groups (Fig. 4.1). Thus, it is not surprising that proteins of viruses of this class have been the subject of the most thorough analysis which have yielded the most illuminating results so far (see also Gorbalenya & Koonin, 1993; Koonin & Dolja, 1993).

Picorna-like supergroup

Studies of picornaviruses, which belong to the most 'simple-set' of well-characterized +RNA viruses, illustrate well the achievements of comparative sequence analysis. This family encompasses mammalian viruses of five established genera: entero-, rhino-, hepato-, cardio- and aphthoviruses, and also several other insect and plant viruses of potentially novel genera (Stanway, 1990; Turnbull-Ross *et al.*, 1993). Picornavirus RNA directs synthesis of a single giant protein (polyprotein) containing approximately 2200 amino acid residues. Limited proteolysis of the polyprotein mediated by virus-encoded proteases gives rise to the mature products. Capsid proteins are derived from the N-terminal one-third part of the polyprotein, and the non-structural proteins come from the rest of it. Polyproteins of a representative set of picornaviruses were the first to be shown to share a similar domain organization and clear sequence similarity (Argos *et al.*, 1984). The common origin of picornaviruses was unequivocally indicated by these observations. Moreover, the pattern of conserved amino acid residues allows one to distinguish picornaviruses from all other viruses.

More unexpected results have been obtained when the genome sequences of viruses of other groups have also been included in computer comparisons. The patterns of conserved amino acid residues of the polyprotein of picornaviruses, and of those encoded by the genomes of como-, nepo-, poty- and bymoviruses, that infect plants, and caliciviruses, that infect animals, were revealed to be similar in the genomic region encoding the non-structural proteins (King Lomonossoff & Ryan, 1991; Neill, 1990). The presence of a VPg (5'-terminal RNA-bound protein) is characteristic of this group. Its gene may occupy the same position in the viral genomes of different families (Fig. 4.2), although its sequence is not conserved among them.

It is also very likely that the capsid proteins of como-, nepo-, calici- and picornaviruses have common origin as they have clear structural and/or sequence resemblance (Rossmann & Johnson, 1989; Tohya *et al.*, 1991). However, viruses of these families have a different number of capsid genes, whose exact position in the genome, and/or the mode of their expression, are not conserved (Fig. 4.2). There is an even greater difference between the capsid proteins of picornaviruses and poty/bymoviruses as they belong to two different structural classes (Rossmann & Johnson, 1989). Furthermore, picornavirus capsid proteins are encoded in the 5'-part of the genome, while the single capsid protein gene of poty/bymoviruses is encoded in the 3'-region of the RNA molecule.

The results obtained provide solid ground to determine the evolutionary relationships among viruses of the above-mentioned families, and puts them together in an independent unit of high taxonomic rank: the picorna-like supergroup (Goldbach & Wellink, 1988; King *et al.*,

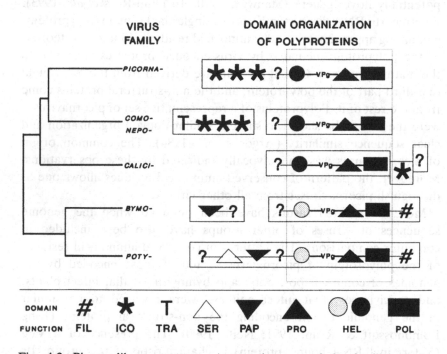

Fig. 4.2. Picorna-like supergroup. A tentative tree of the families of the supergroup is composite and not in scale. It was constructed from the trees for POL (RdRp), HEL (helicase) and PRO (chymotrypsin-like) domains of the viruses (unpublished data), and depicts the most probable order of divergence of the viral groups. Other domains drawn are FIL and ICO: filamentous and icosahedral capsid proteins, respectively; TRA: domain ensuring virus spread throughout plant; SER and PAP, unclassified and papain-like proteases, respectively; ? – domain(s) with an unassigned function. The position of the calicivirus branch is somewhat uncertain. The domain organization of the viral polyproteins is presented in simplified form, not to scale. The domains derived from the same polyprotein are enclosed in common box, and those encoded in the different RNA segments are spaced. The capsid domain and the 3'-encoded domain with an unassigned function of caliciviruses appear to be produced from distinct subgenomic RNA(s). The position of the VPg domain in calici- and bymovirus polyproteins is uncertain. The conserved domain combination of picorna-like viruses is highlighted by the dotted box.

1991). It has been suggested that the genome of the ancient RNA infectious agent, the common ancestor of the viruses of this supergroup, encoded a combination of the conserved replicative domains typical of contemporary representatives of the supergroup (Fig. 4.2). However, it remains unclear whether this hypothetical virus resembled any of the known viruses or was unique for all of the other important features of the genome coding and expression strategy; some other relevant results will be discussed below.

Obviously, intra- and intertypical recombination contributed much to the divergence within this virus supergroup. The genes for capsid and some 'accessory' proteins were probably relocated and changed most frequently in the course of virus evolution. The close relationships of viruses with monopartite (picorna-, calici- and potyviruses) and bipartite (como-, nepo- and bymoviruses) genomes (Fig. 4.2) may also be another result of the recombination process.

Other supergroups of +RNA viruses

Equally interesting results have also been obtained by comparative sequence analysis of other virus families. In the majority of instances, viruses of each family have been found to share conserved proteins and hence can be considered to be evolutionarily related. In a few cases, no close sequence similarity has been found between the proteins of all viruses belonging to one family; instead, some were found to be more similar to the proteins of viruses of other families. These observations have led to a re-evaluation of some viral families/groups. Surprisingly, a reliable sequence resemblance was noticed between the (putative) replicative proteins encoded by the genomes of viruses that undoubtedly belong to different families/groups (for review, see Dolja & Carrington, 1992), and this promoted the invention of supergroups (Goldbach *et al.*, 1991).

In total, six supergroups that differ in the set of conserved domains encoded by their genomes were established (Gorbalenya & Koonin, 1993; Fig. 4.3). The flavi-, sobemo- and corona-like supergroups are the smallest divisions, each containing not more than three virus families/groups. The alpha-like supergroup is biggest and includes no less than 13 families/groups. The carmo-like supergroup was delineated on the basis of the specific pattern of conserved amino acids of the RdRp domain, whereas the domain array of the corona-like supergroup probably has six domains. Curiously, the sobemo- and carmo-like supergroups

A.E. Gorbalenya

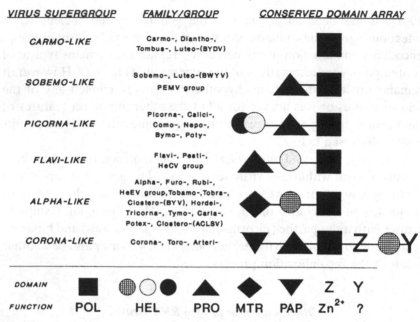

Fig. 4.3. Supergroups of +RNA viruses as presented by Gorbalenya and Koonin (1993). ACLSV: apple chlorotic leaf spot virus; BYDV: barley yellow dwarf virus; BYV: beet yellows virus; BWYV: beet western yellows virus; HeCV and HeEV: hepatitis C and E viruses, respectively; PEMV: pea enation mosaic virus. The conserved domains were designated as Z: *Cys*-rich domain presumably involved in RNA-binding through Zn^{2+}-chelating; Y: domain with an unassigned function; for other designations see legend to the Fig.4.2 and text. The helicase domain is hatched in three different ways in order to reflect the reported groupings of the helicases (Fig. 4.4). The composition of the conserved domain arrays is depicted in a simplified form and not to scale. In hordei- and tricornaviruses of the alpha-like supergroup, the conserved domain array is split between HEL and POL domains. The presence of the conserved PAP and PRO domains in the polyprotein of toroviruses is tentative as the 5'-part of its genome is not yet sequenced.

comprise exclusively plant viruses, the corona- and flavi-like supergroups animal viruses, and the picorna- and alpha-like supergroups encompass viruses infecting plants or animals.

Insect viruses of the Nodaviridae and bacterial viruses (phages) of Leviviridae are not included in the six postulated supergroups (Gorbalenya & Koonin, 1993; Fig. 4.3); these families may eventually become prototypes of two extra supergroups.

There may be differences, and often gross ones, in the phenotypic features of viruses of the different distinct families belonging to the same supergroup. Again, as discussed for the picorna-like viruses, RNA recombination may have been the mechanism that generated much of the current diversity of viruses (Morozov, Dolja & Atabekod, 1989). Importantly, RNA exchange probably occurs not only between the genomes of viruses of the same family, or even supergroup, but more widely. Revelation of an unexpectedly high level of similarity between the capsid proteins of the poty/bymoviruses and the potex/carlaviruses (Dolja *et al.*, 1991), which belong to the picorna- and alpha-like supergroups, respectively (Fig. 4.3), is one of a number of examples that favour the hypothesis of genome recombination between viruses of different RNA supergroups (Chapter 8). Moreover, in some instances, RNA virus genomes undoubtedly captured cellular genes (Gorbalenya, 1992; Chapter 7).

Origin of the most conserved domains

It is evident that the domains found in different viruses come from a limited set of domains (Fig. 4.3). Three most frequently used domains are the core of the RdRp common to all RNA viruses (Koonin, 1991), a chymotrypsin-like protease (PRO) and helicase (HEL) domain, each found in the four combinations of the conserved domain (Gorbalenya & Koonin, 1993). In addition, there are a RNA-methyltransferase domain, which is a hallmark of the alpha-like supergroup viruses (Rozanov, Koonin & Gorbalenya, 1992), and two functionally uncharacterized domains specific for the corona-like supergroup viruses (den Boon *et al.*, 1991). The conserved domain combination of the latter supergroup probably also includes one extra domain, a papain-like protease which is also found in the distinct representatives of the alpha-and picorna-like supergroups (Gorbalenya, Koonin & Lai, 1991). It should be noted that not all domains have been unequivocally proven to have particular functions, so I should use the word 'putative' when describing them; however, this will be omitted below for simplicity.

The extant diversity of the conserved domain combinations might have been generated by domain shuffling using one pool of domain genes. This suggests a common ancestral route for viruses of the different supergroups. Alternatively, the conserved domain combinations might have originated independently from distinct, albeit related, cellular protein genes. Obviously, evaluation of both hypotheses requires an analysis of the evolutionary relationships within the families of the most abundant proteins, so I will analyse three.

RNA-dependent RNA polymerases

The RdRps of each viral supergroup are characterized by a specific pattern of conserved amino acid residues. None the less, an amino acid pattern, common for the RdRps of the whole set of +RNA viruses, can be derived (Kamer & Argos, 1984; Koonin, 1991). This pattern has not been found in any cellular protein, although it, or a closely similar form, has been found in the putative RdRps of some dsRNA viruses (Koonin, 1992). An obvious interpretation of these results is that RdRps of +RNA viruses and, possibly, some dsRNA viruses, comprise a distinct protein superfamily. Importantly, each of the main branches of the RdRps' tree unites the enzymes of viruses of only one particular supergroup (unpublished data), strengthening the claim that supergroups are a valid evolutionary rank in virus classification.

RdRps of the other dsRNA viruses, especially dsRNA bacteriophages, and of all the various viruses with negative sense RNA genomes (−RNA) are related to those of +RNA viruses too. This relationship is certainly more remote than that between the RdRps of different +RNA viruses. The biological significance of this relationship is supported by the observation that the RdRps of −RNA viruses are the only conserved proteins of this virus class (Poch *et al.*, 1989). A phylogeny of RdRps splits −RNA viruses into two subclasses, those with segmented and those with non-segmented genomes, respectively (Tordo *et al.*, 1992).

Data obtained from comparative sequence analysis favours the hypothesis that a gene encoding RdRp, the pivotal enzyme for RNA virus genome replication, emerged only once in the course of evolution. The RdRp superfamily includes enzymes encoded by eukaryotic and prokaryotic viruses. Thus RNA molecules encoding the ancestral RdRp may have arisen earlier than the first simple eukaryotic cells (Kamer & Argos, 1984; Poch *et al.*, 1989).

RNA helicases

Helicases of RNA viruses are proteins sharing a pattern of conserved residues involved in the binding and hydrolysis of purine nucleotides (NTP-binding pattern), and, probably, one or two extra sequence motifs (Gorbalenya & Koonin, 1989; Walker *et al.*, 1982). Based on sequence similarities, viral helicases have been sorted into three superfamilies. In contrast to the results obtained with RdRps, the grouping of RNA viral helicases that can be derived from the phylogenetic analysis is not identical to that obtained by sorting +RNA viruses between supergroups (Fig. 4.4). Thus, helicases of the so-called superfamily 1 (SF1) are encoded by genomes of viruses of alpha- and corona-like supergroups, whereas those of the SF2 belong to the flavi-like viruses and poty/bymoviruses, a subset of the picorna-like supergroup (Gorbalenya *et al.*, 1989*b*). The third division, SF3, consists of helicases of the remaining picorna-like viruses (Gorbalenya *et al.*, 1990; Fig. 4.4). The Helicases SF1 and SF2 that are encoded by genomes of viruses of the two different RdRp supergroups occupy two distinct, albeit closely related, branches. Other members of each superfamily are represented by DNA-encoded helicases of cellular and/or viral origin. Also, helicases of the SF1 and SF2 groups show clear similarity between the entire set of their conserved seven motifs (Gorbalenya *et al.*, 1989*b*), and, hence they probably diverged more recently, compared to the SF3 helicases (Fig. 4.4).

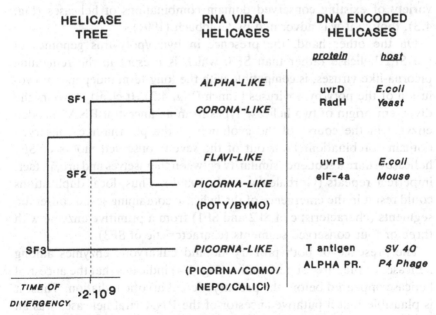

Fig. 4.4. Helicases. A tentative order of the phylogeny of three helicase superfamilies determined by comparative sequence analysis (Gorbalenya & Koonin, 1989; and unpublished data). The helicase tree is not to scale. To illustrate the antiquity of origin of the three helicase branches, pairs of DNA-encoded helicases from prokaryotic or eukaryotic organisms and/or from their viruses belonging to each of the superfamilies are included in the figure. A plausible time of appearance of the ancestral SF1 and SF2 helicases was calculated based on an estimate of the time of divergence of the prokaryotic/eukaryotic lineages (Chapter 16).

Obviously, the ancestors of the RNA viral helicases of each super-family were different enzymes, probably of RNA virus origin. More surprisingly, despite the presence of DNA-encoded helicases in the three main branches of the helicase tree, a common precursor for all known RNA viral helicases (and, thereby, for all helicases of the three SFs) might have been a RNA-encoded enzyme too.

To examine the validity of this rather bold hypothesis let us analyse briefly the possible ways that the helicases of SF2 and of SF3 could come into the same position of the conserved domain array of the picorna-like supergroup (Fig. 4.2). There appear to be two different ways. One is that two different helicase domains could have been placed into the same conserved domain array driven by convergent evolution. Alternatively, the two helicases could have arisen from a common ancestral sequence in a stepwise manner driven by divergent evolution.

The convergent option seems to be less attractive in view of the wide variety of existing conserved domain combinations of helicases (Fig. 4.3), though it was advocated by Goldbach (1991).

On the other hand, the presence in bymo/potyvirus genomes of the SF2 helicase rather than SF3, which is present in the remaining picorna-like viruses, is compatible with the long-term independent evo-lution of the poty/bymoviruses branch (Fig. 4.2). It clearly favours the divergent origin of two helicase types from an ancestral RNA-encoded enzyme in the course of the evolution of the picorna-like conserved domain combination. Four out of the seven conserved motifs of SF2 helicases share a sequence similarity between themselves and are, in fact, imperfect repeats (Gorbalenya et al., 1989b). Thus, local duplications could result in the emergence of the helicase containing seven conserved segments (characteristic of SF2 and SF1) from a primitive enzyme with three or four conserved segments (characteristic of SF3).

The presence of both prokaryotic and eukaryotic enzymes among helicases of the three superfamilies (Fig. 4.4) indicates that the ancestral helicase appeared before the prokaryotic/eukaryotic radiation. Thus, it is plausible that a putative ancestor of the RNA viral helicases was an RNA-encoded and rather ancient enzyme.

Chymotrypsin-like proteases

Because it is difficult to obtain reliable multiple sequence alignments of the viral chymotrypsin-like proteases, their phylogeny, let alone the phylogeny of the whole set of these proteases, has not yet been inferred.

Fortunately, there are other ways of obtaining plausible clues as to the time when a putative ancestral viral protease may have emerged, and as to what it was like.

The catalytic *Ser* residue of the cellular chymotrypsin-like proteases was shown to be encoded by two different sets of codons (Brenner, 1988). Diversification of the codons for the catalytic *Ser* residue seems to require explanation through some unusual mechanism, as the simultaneous replacement of two nucleotides in the catalytic *Ser* codon is an event of very low probability (Brenner, 1988; albeit see Koonin & Gorbalenya, 1989). It was postulated that the ancestor of the cellular chymotrypsin-like proteases contained a *Cys* instead of a *Ser* residue (Brenner, 1988), as changing from the *Cys* codon set to either of the *Ser* codon sets can be accomplished by replacement of a single, albeit different, nucleotide. The possible existence of a *Cys*-containing ancestor of chymotrypsin-like proteases seems to be also attractive in view of relative simplicity of cysteine protease catalysis (Polgar & Asboth, 1986).

Until now, cellular cysteine chymotrypsin-like proteases stand as purely hypothetical enzymes. However, a fraction of the +RNA viral chymotrypsin-like proteases are indeed such enzymes (Bazan & Fletterick, 1988; Gorbalenya *et al.*, 1989*a*). The majority of cysteine-containing proteases are encoded by the genomes of picorna-like viruses whose conserved domain combination includes exclusively this sort of the protease. Thus, it is tempting to speculate that RNA viral proteases, specifically those of the picorna-like viruses, resemble most a putative ancestor of the chymotrypsin-like proteases that emerged before the prokaryotic and eukaryotic lineages separated (Gorbalenya *et al.*, 1986*a*). It would be logical to assume that such an enzyme was RNA, encoded.

An RNA plasmid as plausible progenitor of primitive RNA viruses

To conclude this chapter, it must be stressed that the analysis of the existing data on the three most conserved RNA viral enzymes indicates that each of them probably had a monophyletic and ancient origin. The ancestral genome encoding RdRp and other conserved enzymes appears to have already possessed one of the most important properties of the extant RNA viruses, an ability to self-replicate; thus it resembles an RNA plasmid. Such an RNA molecule could precede RNA viruses in their evolution, and provide a 'replicative domain framework' for first true RNA viruses which could emerge when ancestral RNA plasmids have acquired genes for capsid and, possibly, some other proteins.

The possible evolutionary relations between RNA viruses and plasmids is supported by the results of a recent computer sequence analysis of the HyAV genome, a virus-like double-stranded RNA associated with reduced virulence of the chestnut blight fungus, *Cryphonectria parasitica*. Two polyproteins encoded by HyAV genome were demonstrated to contain counterparts of a papain-like protease, a helicase, and a RdRp of poty/bymoviruses, albeit in a different order in their polyproteins (Koonin *et al.*, 1991).

Did RNA virus genome ancestors inhabit the pre-cellular world?

In the framework of a hypothesis on the ancient origin of ancestor(s) of RNA viruses, it seems to be most important to figure out how and when stable combinations of all three conserved domains, especially that of the picorna-like viruses, arose.

In this respect, the discovery in the conserved regions of the poliovirus polyprotein of several types of tandem repeats, whose size is varied between 88 and 440 amino acid residues, seems to be highly suggestive (Gorbalenya *et al.*, 1986*b*; Fig. 4.5). The presence of several types of repeats in one and the same region of the polyprotein was explained by the existence of a latent principal 11-meric periodicity that could give rise to all other repeats and to the polyprotein as a whole, as the repeats are multiples of 11 in size.

There is a striking correlation between the size and the position of some of the tandem repeats, and those of certain mature proteins and their precursors that originate from this part of the polyprotein (Fig. 4.5). This result strongly indicates the non-fortuitous character of the periodicity. The smallest poliovirus protein VPg consists of only 22 (dimer of 11) amino acid residues, and its aphthovirus homologue being slightly larger in size is represented by three imperfect tandem copies (Forss & Schaller 1982; Fig. 4.5). Other clear examples of imperfect repeats are proteases of 2A and 3C of entero/rhinoviruses (Blinov, Donchenko & Gorbalenya, 1985), three capsid proteins of the picornaviruses (Rossmann & Johnson, 1989), and tandem repeats in the 5'-non-coding region of enterovirus RNAs (Pilipenko, Blinov Agol, 1990).

Overlapping open reading frames (ORFs) found in the genomes of a number of the +RNA viruses (Keese & Gibbs, 1992; Chapter 6) could potentially be a particular manifestation of the latent 11-meric or other

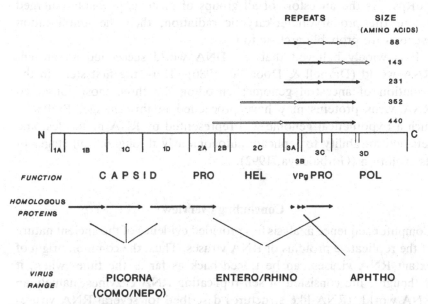

Fig. 4.5. Repeats in picornavirus polyproteins. In the middle part of the Figure the position of proteins in the poliovirus polyprotein along with their function is shown (for abbreviations see legend to Fig. 4.2 and text). The main cleavage sites are drawn as long vertical lines, whereas alternative and putative cleavage sites are shown as short lines. The protein composition, and the position of cleavage sites in the polyproteins of other picornaviruses is often slightly different. The position of the tandem repeats in the poliovirus polyprotein is as reported by Gorbalenya *et al.* (1986*b*) and is marked by arrowheads above the polyprotein. Filled arrowheads distinguish the repeats flanked by two (putative) cleavage sites. Homologous proteins found in the polyproteins of different picornaviruses are shown in the lower part of the Figure.

periodicity as well. Indeed, Ohno (1984) revealed that all three frames of periodicity-ordered polynucleotides, whose size of repeats is not divisible by 3 (which is the case for a segment 11 long), are open. Hence, such polynucleotides possess an enormously large coding capacity, that could be partially preserved in the form of overlapping ORFs even when much mutated.

These facts encourage one to take seriously the possibility that the ancestor of RNA viral genomes was a short gene sequence that amplified multiple times throughout evolution (Gorbalenya *et al.*, 1986*b*). This scenario suggests concerted evolution from the very beginning of such vital and highly conserved proteins of +RNA viruses and/or cells as chymotrypsin-like proteases, NTP-pattern-containing proteins, and

RdRps. As the ancestors of all groups of proteins probably emerged before the prokaryotic/eukaryotic radiation, then the amplification events were probably archaic too.

It is widely believed that the DNA world succeeded an ancient RNA world (Darnell & Doolittle, 1986). Thus, the first steps in the evolution of ancestral genomes encoding the three most conserved RNA virus proteins may have proceeded in this change. Probably, such a hypothetical genome was represented by RNA molecules, and retained an ability to replicate autonomously throughout all stages of its evolution (Gorbalenya, 1992).

Concluding overview

Computer sequence analysis has provided evidence of the ancient nature of the replicative proteins of RNA viruses. Thus, the common origin of extant RNA viruses can be traced back as far as the times when, it is thought, life consisted of self-replicating RNA enzymes, namely an RNA world. tRNA-like structures described for several RNA viruses could be descendants of such ancient RNA molecules (Weiner & Maizels, 1987).

All extant viral RdRps probably originated from one ancestral enzyme. It can be assumed that a genome encoding the ancestral RdRp already possessed the most vital property of viruses, namely the ability to self-replicate; hence it was an RNA plasmid.

RdRps of positive and negative RNA viruses appear to be two distinct branches of the same tree (Poch *et al.*, 1989). The evolution of the RdRps of at least some dsRNA viruses is rather closely interconnected to the evolution of some of their counterparts encoded by genomes of +RNA viruses (Koonin, 1992).

The greatest event in the evolution of −RNA viruses appears to be the emergence of ancestors of viruses with segmented and unsegmented genomes. Plausibly, similar events have occurred quite frequently in different branches of the +RNA virus lineage. The ancestral split of the +RNA virus class into its main divisions may have been achieved by generation of several stable combinations of the replicative domains.

It is logical to propose that the emergence of true RNA viruses encapsidated in their own genome-encoded protein(s) was connected to the appearance of cells. Genes for proteins ensuring virus life outside the cell could have been captured by viral ancestors from the cellular genome, and/or could have arisen in the course of divergent evolution of duplicated viral genes for non-structural proteins (Gorbalenya *et al.*, 1986*b*).

Acknowledgements

I am grateful to Drs Vadim Agol and Adrian Gibbs for critical reading of the manuscript.

References

Argos, P., Kamer, G., Nicklin, M.J.H. & Wimmer, E. (1984). *Nucl. Acids Res.*, **12**, 7251–67.

Baltimore, D. (1971). *Bacteriol. Rev.*, **35**, 235–41.

Bazan, J.F. & Fletterick, R.J. (1988). *Proc. Natl. Acad. Sci. USA*, **85**, 7872–6.

Blinov, V.M., Donchenko, A.P. & Gorbalenya, A.E. (1985). *Dokl. Akad. Nauk USSR*, **281**, 984–7.

Brenner, S. (1988). *Nature*, **334**, 528–30.

Darnell, J.E. & Doolittle, W.F. (1986). *Proc. Natl. Acad. Sci. USA*, **83**, 1271–5.

den Boon, J.A., Snijder, E.J., Chirnside, E.D., de Vries, A.A.F., Horzinek, M.C. & Spaan, W.J.M. (1991). *J. Virol.*, **65**, 2910–20.

Dolja, V.V., Boyko, V.P., Agranovsky, A.A. & Koonin, E.V. (1991). *Virology*, **184**, 79–86.

Dolja, V.V. & Carrington, J.C. (1992). *Semin. Virol.* **3**, 315–26.

Forss, S. & Schaller, H. (1982). *Nucl. Acids Res.*, **10**, 6441–50.

Francki, R.I.B., Fauquet, C.M., Knudson, D.L. & Brown, F. (eds) (1991). Classification and nomenclature of viruses. *Fifth Report of the International Committee on Taxonomy of Viruses, Arch. Virol.* (Suppl. 2). Springer-Verlag, Wien, New York.

Goldbach, R. (1991). *Arch. Virol. Suppl.*, **3**, ??–??

Goldbach, R., Le Gall, O. & Wellink, J. (1991). *Semin, Virol.*, **2**, 19–25.

Goldbach, R. & Wellink, J. (1988). *Intervirology*, **29**, 260–7.

Gorbalenya, A.E. (1992). *Semin. Virol.*, **3**, 359–71.

Gorbalenya, A.E., Blinov, V.M. & Donchenko, A.P. (1986*a*). *FEBS Lett.*, **194**, 253–7.

Gorbalenya, A.E., Donchenko, A.P. & Blinov, V.M. (1986*b*). *Molek. Genetika No 1*, 36–41.

Gorbalenya, A.E., Donchenko, A.P., Blinov, V.M. & Koonin, E.V. (1989*a*). *FEBS Lett.*, **243**, 103–14.

Gorbalenya, A.E. & Koonin, E.V. (1989). *Nucl. Acids Res.*, **17**, 8413–40.

Gorbalenya, A.E. & Koonin, E.V. (1993). In *Soviet Sci. Rev. D.*., Physico-chemical Biology Series, V.P. Skulachev (Ed.). Harwood Acad. **11**, 1–87, London, in press

Gorbalenya, A.E., Koonin, E.V., Donchenko, A.P. & Blinov, V.M. (1989*b*). *Nucl. Acids Res.*, **17**, 4713–30.

Gorbalenya, A.E., Koonin, E.V., & Lai M.M.C. (1991). *FEBS Lett.*, **288**, 201–5.

Gorbalenya, A.E., Koonin, E.V. & Wolf, Yu.I. (1990). *FEBS Lett.*, **262**, 145–8.

Kamer, G. & Argos, P. (1984). *Nucl. Acids Res.*, **12**, 7269–82.

Keese, P.K. & Gibbs, A. (1992). *Proc. Natl. Acad. Sci. USA*, **89**, 9489–93.

King, A.M.Q., Lomonossoff, G.P. & Ryan, M. (1991). *Semin. Virol.*, **2**, 11–17.
Koonin, E.V. (1991). *J. Gen. Virol.*, **72**, 2197–206.
Koonin, E.V. (1992). *Semin. Virol.*, **3**, 327–39.
Koonin, E.V. & Dola, V.V. (1993). *Crit. Rev. Biochem. Mol. Biol.*, **28**, 375–430.
Koonin, E.V. & Gorbalenya, A.E. (1989). *Nature*, **338**, 467–8.
Koonin, E.V., Choi, G.H., Nuss, D.L., Shapira, R. & Carrington, J.C. (1991). *Proc. Natl. Acad. Sci. USA*, **88**, 10647–51.
Morozov, S.Yu., Dolja, V.V. & Atabekov, J.G. (1989). *J. Mol. Evol.*, **29**, 52–62.
Neill, J.D. (1990). *Virus Res.*, **17**, 145–60.
Ohno, S. (1984). *Proc. Natl. Acad. Sci. USA*, **81**, 2421–5.
Pilipenko, E.V., Blinov, V.M. & Agol, V.I. (1990). *Nucl. Acids Res.*, **18**, 3371–5.
Poch, O., Sauvageut, I., Delarue, M. & Tordo, N. (1989). *EMBO J.*, **8**, 3867–74.
Polgar, L. & Asboth, B. (1986). *J. Theor. Biol.*, **121**, 323–6.
Rossmann, M.G. & Johnson, J.E. (1989). *Annu. Rev. Biochem.*, **58**, 533–73.
Rozanov, M.N., Koonin, E.V. & Gorbalenya, A.E. (1992). *J. Gen. Virol.*, **73**, 2129–34.
Stanway, G. (1990). *J. Gen. Virol.*, **71**, 2483–501.
Tohya, Y., Taniguchi, Y., Takahashi, E., Utagawa, E., Takeda, N., Miya-mura, K., Yamazaki, S. & Mikami, T. (1991). *Virology*, **183**, 810–14.
Tordo, N., De Haan, P., Goldbach, R. & Poch, O. (1992). *Semin. Virol.*, **3.**, 341–57.
Turnbull-Ross, A.D., Reavy, B., Mayo, M.A. & Murant, A.F. (1993). *J. Gen Virol.*, **74**, 3203–11.
Walker, J.E., Saraste, M., Runswick, M.J. & Gay, N.J. (1982). *EMBO J.*, **2**, 945–51.
Weiner, A.M. & Maizels, N. (1987). *Proc. Natl. Acad. Sci. USA*, **84**, 7383–7.

Further reading

Agol, V.I. (1976). *Origin Life*, **7**, 119–32.
Agranovsky, A.A., Boyko, V.P., Karasev, A.V., Lunina, N.A., Koonin, E.V. & Dolja, V.V. (1991). *J. Gen. Virol.*, **72**, 15–23.
Ahlquist, P., Strauss, E.G., Rice, C.M., Strauss, J.M., Haseloff, J. & Zimmern, D. (1985). *J. Virol.*, **53**, 536–42.
Argos, P. (1988). *Nucl. Acids Res.*, **16**, 9909–19.
Bazan, J.F. & Fletterick, R.J. (1989). *Virology*, **171**, 637–69.
Blinov, V.M., Gorbalenya, A.E. & Donchenko, A.P. (1984). *Doklady Akad. Nauk USSR*, **279**, 502–5.
Boccara, M., Hamilton, W.D.O. & Baulcombe, D.C. (1986). *EMBO J.*, **5**, 223–9.
Bouzoubaa, S., Quillet, L., Guilley, H., Jonard, G. & Richards, K. (1987). *J. Gen. Virol.*, **68**, 615–26.
Candresse, T. Morch, M.D. & Dunez, J. (1990). *Res. Virol. Institut Pasteur*, **141**, 315–29.
Choi, H.-K., Tong, L., Minor, W., Dumas, P., Boege, U., Rossmann, M.G. & Wengler, G. (1991). *Nature*, **354**, 37–43.

Collett, M.S., Anderson, D.K. & Retzel, E. (1988). *J. Gen. Virol*, **69**, 2637–43.
Cornelissen, B.J.C. & Bol, J.F. (1984). *Plant Mol. Biol.*, **3**, 379–384.
Delarue, M., Poch, O., Tordo, N., Moras, D. & Argos, P. (1990). *Protein Eng.*, **3**, 461–7.
Demler, S.A. & de Zoeten, G.A. (1991). *J. Gen. Virol.*, **72**, 1819–34.
Dolja, V.V. & Koonin, E.V. (1991). *J. Gen. Virol.*, **72**, 1481–6.
Domier, L.L., Shaw, J.G. & Rhoads, R.E. (1987). *Virology*, **158**, 20–7.
Dominguez, G., Wang, C.-Y. & Frey, T.K. (1990). *Virology*, **177**, 225–38.
Franssen, H., Leunissen, J., Goldbach, R., Lomonosoff, G.P. & Zimmern, D. (1984). *EMBOJ.*, **3**, 855–61.
Geigenmuller-Gnirke, U., Weiss, B., Wright, R. & Schlesinger, S. (1991). *Proc. Natl. Acad. Sci. USA*, **88**, 3253–7.
Gibbs, A. (1987). *J. Cell. Sci. (Suppl.)*, **7**, 319–37.
German, S., Candresse, T., Lanneau, M., Huet, J.C., Pernollet, J.C. & Dunez, J. (1990). *Virology*, **178**, 104–2.
Goldbach, R. (1986). *Annu. Rev. Phytopathol*, **24**, 289–310.
Goldbach, R. (1987). *Microbiol. Sci.*, **4**, 197–202.
Gorbalenya, A.E., Blinov, V.M., Donchenko, A.P. & Koonin, E.V. (1989). *J. Mol. Evol.*, **28**, 256–8.
Gorbalenya, A.E., Blinov, V.M. & Koonin, E.V. (1985). *Molek. Genetika*, **11**, 30–6.
Gorbalenya, A.E., Donchenko, A.P., Koonin, E.V. & Blinov, V.M. (1989). *Nucl. Acids Res.*, **17**, 3889–7.
Gorbalenya, A.E., Koonin, E.V., Donchenko, A.P. & Blinov, V.M. (1988). *FEBS Lett.*, **239**, 16–24.
Gorbalenya, A.E., Koonin, E.V., Donchenko, A.P. & Blinov, V.M. (1988). *FEBS Lett.*, **236**, 287–90.
Gorbalenya, A.E., Koonin, E.V., Donchenko, A.P. & Blinov, V.M. (1989). *Nucl. Acids Res.*, **17**, 4456–9.
Greif, C., Hemmer, O. & Fritsch, C. (1988). *J. Gen. Virol.* **69**, 1517–29.
Gustafson, G., Armour, S.L., Gamboa, G.C., Burgett, S.G. & Shepherd, J.W. (1989). *Virology*, **170**, 370–7.
Hardy, W.R. & Strauss, J.H. (1989). *J. Virol.*, **63**, 4653–64.
Haseloff, J., Goelet, P., Zimmern, D., Ahlquist, P., Dasgupta, R. & Kaesberg, P. (1984). *Proc. Natl. Acad. Sci. USA*, **81**, 4358–62.
Hodgman, T.C. (1988). *Nature*, **333**, 22–23, 578.
Kashiwazaki, S., Minobe, Y., Omura, T. & Hibino, H. (1990). *J. Gen. Virol.*, **71**, 2781–90.
Kato, N., Hijikata, M., Nakagawa, M., Ootsuyama, Y., Muraiso, K., Ohkoshi, S. & Shimotohno, K. (1991). *FEBS Lett.*, **280**, 325–8.
Koonin, E.V. & Gorbalenya, A.E. (1989). *J. Molec. Evol.*, **28**, 524–7.
Koonin, E.V. & Gorbalenya, A.E. (1992). *FEBS Lett.*, **297**, 81–6.
Koonin, E.V., Gorbalenya, A.E. & Chumakov, K.M. (1989). *FEBS Lett.*, **252**, 42–6.
Koonin, E.V., Gorbalenya, A.E., Purdy, M.A., Rozanov, M.N., Reyes, G.R. & Bradley, D.W. (1992). *Proc. Natl. Acad. Sci. USA*, **89**, 8259–63.
Lain, S., Riechman, J.L., Martin, M.T. & García, J.A. (1989). *Gene*, **82**, 357–62.
Mans, R.M.W., Pleij, C.W.A. & Bosch, L. (1991). *Eur. J. Biochem.*, **201**, 303–24.
Marsh, L.E. & Hall, T.C. (1987). *Cold Spring Harbor Symp. Quant. Biol.*, **52**: 331–41.

Miller, R.M. & Purcell, R.M. (1990). *Proc. Natl. Acad. Sci. USA*, **87**, 2057–61.
Miller, W.A., Waterhouse, P.M. & Gerlach, W.L. (1988). *Nucl. Acids Res.*, **16**, 6097–111.
Morozov, S. Yu., Lukasheva, L.I., Chernov, B.K., Skryabin, K.G. & Atabekov, J.G. (1987). *FEBS Lett.*, **213**, 438–42.
Oh, C.S. & Carrington, J.C. (1989). *Virology*, **173**, 692–9.
Poch, O., Blumberg, B.M., Bougueleret, L. & Tordo, N. (1990). *J. Gen. Virol.*, **71**, 1153–62.
Rodriguez-Couzino, N., Esteban, L.M. & Esteban, R. (1991). *J. Biol. Chem.*, **266**, 12772–8.
Rozanov, M.N., Morozov, S. Yu. & Skryabin, K.G. (1990). *Virus Genes*, **3**, 373–9.
Shapira, R., Choi, G.H. & Nuss, D.L. (1991). *EMBO J.*, **10**, 731–9.
Snijder, E.J., den Boon, J.A., Bredenbeek, P.J., Horzinek, M.C., Rijnbrand, R. & Spaan, W.J.M. (1990). *Nucl. Acids Res.*, **18**, 4535–42.
Snijder, E.J., den Boon, J.A., Horzinek, M.C. & Spaan, W.J.M. (1991). *Virology*, **180**, 448–52.
Steinhauer, D. & Holland, J.J. (1987). *Annu. Rev. Biochem.*, **41**, 409–33.
Strauss, J.H. & Strauss, E.G. (1988). *Annu. Rev. Microbiol.*, **42**, 657–83.
Strauss, E.G., Strauss, J.H. & Levine, A.J. (1991). In *Fundamental Virology*, 2nd edn B.N. Fields, D.M. Knipe (eds), Raven Press, New York, pp. 167–190.
Rossmann, M.G., Arnold, E., Erickson, J.W., Frankenberger, E.A., Griffith, J.P., Hecht, H.-J., Johnson, J.E., Kamer, G., Luo, M., Moser, A.G., Ruecker, R.R., Sherry, B. & Vriend, G. (1985). *Nature*, **317**, 145–53.
Rupasov, V.V., Morozov, S.Yu., Kanyuka, K.V. & Zavriev, S.K. (1989). *J. Gen. Virol.*, **70**, 1861–9.
Zimmern, D. (1988). Evolution of RNA viruses. In *RNA Genetics II*, J.J. Holland, E. Domingo, P. Ahlquist (eds), CRC Press, Florida, pp. 211–240.

5
Origins of DNA viruses
DUNCAN J. McGEOCH AND ANDREW J. DAVISON

Introduction

Viruses are a class of genetic element, dependent on suitable host cells for their propagation. Thus, the evolutionary development of viruses must have been contingent on the prior evolution of potential host cells, and the genetic material of viruses must primordially have derived from cellular nucleic acids. Many virus coded proteins, particularly enzymes, have characteristic sequence similarities to non-viral proteins, and this is taken to indicate that many viral genes have cellular genes as progenitors. Having made these almost axiomatic assertions, it is by no means easy to fill in further strategic detail on virus origins. We would like to know what kind of pre-viral elements gave rise to early viruses; what forces acted in the genesis and evolution of viruses; at what epochs in the evolution of cellular organisms viruses appeared; and to what degree contemporary viruses have radiated from common progenitors. This chapter aims to examine some aspects of these questions, but it will quickly become apparent just how limited is our capability in obtaining firm or detailed answers.

Two general classes of data on viruses are applicable to considering details of viral origins, evolution and relatedness. The first includes what we know of basic phenotypes in terms of virion structure (including gross structure of the genome) and replicative cycle; 'higher' aspects of phenotype, namely properties relating to pathogenesis and disease, are less useful for these purposes. The second class consists of direct information on genotypes, that is, determined DNA sequences and their interpretations in terms of gene organization and encoded protein sequences. Detailed information on basic viral phenotypes has been accumulating for several decades, and most of current viral taxonomy is based on this class of data. Access to large amounts of sequence data is relatively recent, and the process of incorporating the fundamental genotypic data of DNA sequences into taxonomic schemes is an ongoing research effort; the views of virus relationships which are emerging have greater detail and solidity, although the earlier classifications for the most part have remained valid.

We would like to be able to use nucleic acid and protein sequence comparisons to establish relationships between viral and non-viral genes, and to reconstruct events in viral evolution. However, these goals look to be only partly achievable. Although the amounts of viral and other sequence data now available are huge by expectations of 10 or 15 years ago, they remain sufficiently sparse that they still limit comparative analyses severely. This restriction is steadily being alleviated, but two more fundamental impediments remain. These are, first, the fact that viral genomes have probably explored truly enormous 'evolutionary spaces' during their history, in terms of local mutation and of larger scale transformations and, secondly, the lack of a good, general means of assigning time scales to viral evolution. Together, these factors constitute an apparently insurmountable obstacle to any real knowledge of early viral evolution. After this early confession of limitation, the remainder of our chapter is aimed at adding detail to these assertions.

Nature of viral progenitors

A possibility for the origins of viruses discussed some years ago was that they might have evolved by simplification (or 'regression') of single-cell organisms (Luria & Darnell, 1967). There are convincing non-viral examples of such processes, for instance with mycoplasmas. As our detailed knowledge of viruses and their genes has increased, this model has come to seem profoundly unattractive; even the most elaborate viruses (such as poxviruses or herpesviruses) do not resemble simplified cells in any critical aspect of phenotype or genotype. On the other hand, clear similarities in organization and functions between DNA virus genomes and genetic elements such as plasmids can now be seen, and it is attractive to view viruses as developing from a past that includes this class of genetic entity. Such thinking appears reasonable enough, but it is important to be aware of its limitations: it does not generate testable hypotheses and, more subtly, it applies common elements of our current scientific consciousness in the very remote and unfamiliar milieu of early evolution. In any case, we treat viruses in this chapter as having evolved from small, non-viral genetic elements, and wish this to be taken as plausible but not rigorously derived device.

We regard DNA viruses as genetic elements which have three basic, characteristic properties: first, they encode mechanisms to enable their replication within a suitable host cell; secondly, they can exist as packaged forms, virions, which typically are genetically inactive; and thirdly, the virions are transmissible to other cells, most commonly (or, most familiarly) by way of an extracellular phase. The first of these attributes is, of course, shared with other genetic elements such as plasmids and transposons. The second is close to constituting the defining characteristic of the concept 'virus' but this is less clear-cut than we at first thought, since a class of genetic element can be imagined which directs its packaging but remains intracellular and non-transmissible. Yeast retrotransposons (with RNA genomes) might be a real instance of this hypothetical class. The point is that, having decided to view viruses as evolved from obligately intracellular progenitors, we then must attempt to allow for all conceivable classes of such elements rather than just familiar ones, and we consider there is no logical basis for excluding such an intracellular packaged element; it need not represent merely a regressed virus. Justifications for the evolution of such a creature could certainly be attempted. The property of viruses which separates them best from other genetic elements is thus the transmissibility of the packaged genome. This definition relates to evolutionary origins, and need not be troubled by the existence of partially disabled forms of contemporary viruses which do not meet all our criteria. It should be noted that viruses are not unique among genetic elements in being transmissible: cell-cell transfer is a prominent aspect of the biology of bacterial plasmids but is mediated by a distinct type of mechanism.

We thus see viruses as having developed from intracellular and unpackaged progenitors by way of intermediate forms. The evolution of such forms could have been complex and extended, but is now hidden. Another aspect to attempting to clear our minds of prejudice concerns the usual view of viruses as disease agents. Applying this outlook to early evolution would treat viruses solely as selfish elements. We feel that this view must be to a large extent justified, but that it is possible that early viruses, like plasmids and transposons, could have had significant positive roles in cellular evolution as agents for dissemination of sequences from the host cell's genome.

Origins of viral genes

In this section we treat the sub-problem of the nature of processes by which the gene sets of viruses have been generated. Many virus genes have evidently resulted from recombinational capture of a non-viral gene followed by minor or major adaptation to the viral role. We suppose that the primary source of genes was the host cell's genome; genes could reach a viral genome either directly from the cellular genome or indirectly via other genetic elements, and transfer could involve only DNA sequences or could be via reverse transcription of messenger RNA.

In discussing relationships between viral and non-viral genes, it is useful to distinguish between small DNA viral genomes (less than 10 kbp) and large genomes (say, greater than 100 kbp), with some of intermediate size such as adenovirus (35 kbp). In all the large DNA virus genomes which have been extensively sequenced, many encoded protein sequences have been found to be similar to sequences of non-viral families of proteins, in such a characteristic way that evolutionary relatedness is clear. Most examples concern virally encoded enzymes. For instance, the replicative DNA polymerases of adenoviruses, baculo-viruses, herpesviruses and poxviruses (and also of certain phages) all have amino acid sequence motifs closely related to motifs conserved in a family of eukaryotic cellular DNA polymerases (Ito & Braithwaite, 1991). Studies on herpesviruses (Chapter 20) and poxviruses (Chapter 18) have been particularly prominent in uncovering such homologues. However, only the herpesvirus group has yet been well enough studied to show that viral genes with non-viral homologues fall into two distinct classes: some belong to the 'core set' of genes which are common to all sequenced mammalian herpesviruses, while others occur only sporadically. The core set includes genes responsible for central aspects of virus replication and virion structure. These findings are interpreted as indicating that the core genes with cellular homologues were present in an ancient progenitor of contemporary herpesviruses, and derived originally from a non-viral source. The sporadically occurring genes probably represent more recent acquisitions from the host genome or from another virus or genetic element, and are presumed to supply functions of value to a virus in a particular niche, but to be of less fundamental importance for virus replication.

It should be emphasized that these relationships are all seen via comparisons of encoded protein sequences rather than of the genomic

DNA sequences. The DNA sequences of viral genes with non-viral homologues resemble other genes of the same virus in attributes such as nucleotide and dinucleotide frequencies, and differ from their cellular counterparts; all have thus probably been in the viral genome for a long time. We are not aware of any exceptions to this generalization within the scope of this chapter, which excludes acute transforming retroviruses. The occurrence of multigene families is widespread in large viral genomes, and the processes of gene duplication and divergence which presumably generate these families are, in effect, an intramolecular version of external gene capture.

Small DNA viruses have a limited genetic content, and there is only one possible non-viral homology to register for this class: polyomaviruses, papillomaviruses, parvoviruses and geminiviruses all encode proteins with helicase activity (the better studied examples are known to have additional functions). All these proteins possess the very widespread Walker NTP-binding motifs, and have been assigned by Gorbalenya and Koonin (1989) to those authors' 'putative helicase superfamily 3', which contains one cellular protein, a protease of *Escherichia coli*; this group also includes some RNA viral proteins.

Among large DNA viruses, some of the proteins with cellular homologues are virion constituents; for instance, certain enzymes packaged in poxvirus and herpesvirus virions, and minor surface membrane proteins in some herpesviruses. However, in no case that we are aware of, for either large or small DNA viruses, has an unambiguous cellular homologue been identified for any major virion structural protein, including proteins of icosahedral shells, virion core proteins, surface membrane proteins and other types in complex virions. This leaves a most unsatisfactory information void at the very core of our consideration of the genesis of virus genes.

The lack of detectable cellular homologues for virion structural proteins could have several roots. For any given viral protein, it might simply be that the protein sequence database available for comparison is not adequate, but we feel that this surely cannot represent the whole explanation. Our view is that at least a proportion of viral proteins must have had cellular progenitors, but that during evolutionary development as virus components these have mutated to the point that their cellular antecedents are no longer discernible in amino acid sequence similarities. We see two aspects to such mutation. First, the architectural requirements of virion construction, perhaps unique in cellular terms, might have necessitated early large changes in order to optimize functionality; this possibility is essentially untestable. Secondly, we know that homologous proteins can indeed diverge beyond any amino acid sequence similarity while retaining recognizable similarity in three-dimensional structure. The best evidence for this comes from well-studied protein families such as the globins (Bashford, Chothia & Lesk., 1987) and it is also directly apparent from comparisons of viral protein families, for instance, of some herpesvirus structural proteins, that extensive divergence can take place in structural components of virions (McGeoch, 1989). At the end of this discussion, none the less, it does remain somewhat unsatisfactory that very stable sequence features, such as the invariant patterns of cysteine residues in otherwise strongly divergent membrane glycoproteins, have not identified cellular homologues of major virion proteins.

In principle, protein relationships can be much more sensitively detected by comparisons of three-dimensional structure than by sequence comparisons, but such structural information is harder to obtain and databases are more limited. An intriguing structural similarity, perhaps indicative of truly distant evolutionary connections, has recently been found in the coat proteins of three disparate groups of viruses (but not in a cellular protein, yet), namely the β-barrel component of small positive strand RNA viruses, parvoviruses and adenoviruses (Roberts *et al.*, 1986; Tsao *et al.*, 1991). Finally, we note that, outside the scope of this chapter, a major structural protein of Sindbis virus is reported to be related structurally to the chymotrypsin family of proteinases (Choi *et al.*, 1991).

There is another, quite distinct pathway for development of virus genes, namely genesis *de novo* of coding function in a previously non-coding genomic sequence or reading frame (Chapter 6; Keese & Gibbs, 1992). The clearest examples of this are in overlapping, out-of-frame, coding regions; at least in the more extreme of these, where one coding region is wholly or largely within the bounds of the other, it seems necessary to propose that one gene has arisen *in situ*. There are viral examples of these forms from φX174 to herpesviruses. Outside this special arrangement, it looks essentially impossible to infer what proportion of the virus genes that lack visible homologues may have originated *de novo*. Our provisional view is that genesis *de novo* could well be a common process, especially for smaller genes and in large DNA viruses.

Scope of viral evolutionary processes

Comparisons of DNA sequences among members of virus families show that a variety of mutational processes have occurred during divergence of genomes from a common progenitor. Very extensive local mutations (nucleotide substitutions and small additions/deletions) are universally seen for well-studied small DNA virus families (such as papovaviruses, papillomaviruses and parvoviruses) and for those large DNA viruses where comparative data are available (notable herpesviruses, and also poxviruses). For presently recognized families of small DNA viruses, the types of changes observed really stop at this local level, and the impression is that these have evolved into terminal forms which cannot readily be altered (but a qualification must be added: extensive change would render family connections unrecognizable).

Among large DNA viruses, the best studied in sequence terms are the herpesviruses, with seven genome sequences now published, as well as many partial sequences (Chapter 20). Conservation of the core set of genes among mammalian herpesviruses shows that they have indeed diverged from a common ancestor, with three major lineages discernible. During this divergence, the genomes have undergone several distinct types of profound change. Nucleotide substitution has been very extensive and has involved biasing mechanisms such that a large range in overall nucleotide compositions now exists. The linear order of genes has been scrambled by large scale rearrangements of blocks of genes. Distinct patterns of large repeat elements have developed. Outside the core set, the gene complements have changed substantially, with acquisition of additional genes and development of multigene families probably playing major roles, and genome sizes range from 125 kbp to 230 kbp. The available evidence for other families of large DNA viruses suggests that the herpesvirus picture is typical (McGeoch, 1992).

The recent analysis of the DNA sequence of the fish herpesvirus, channel catfish virus (CCV), yielded the unexpected result that its gene set showed no amino acid sequence similarities that might indicate a common origin with the mammalian herpesviruses, although phenotypically CCV is a typical herpesvirus (Davison, 1992). There are two possible interpretations of this finding: first, that CCV is not related phylogenetically to mammalian herpesviruses; and secondly, that it is related but is divergent to an unprecedented degree. In view of the detailed phenotypic similarities, in particular the common possession of a characteristic and complex virion morphology, we prefer the latter view. This would imply that the great differences previously noted among mammalian herpesviral genomes trace only the latest part of the evolution of this one sub-group of herpesviruses, and further that we cannot expect to use sequence comparisons, even of whole genomes, to follow herpesvirus evolution back to a pre-herpesviral or pre-viral ancestor.

We regard the CCV case as important for illustrating what may happen in long distance comparisons of viral genomes. Of course, it is weakened by the tentative nature of our interpretation. If this is accepted, it implies that there is a limit to how far into the evolutionary past of viruses we can probe by means of sequence comparisons. For viruses with less characteristic virion morphologies or smaller genomes, we can readily imagine divergence beyond any recognizable similarities, genotypic or phenotypic. These considerations are the basis for our assertion at the start of this chapter that virus genomes may have explored enormous evolutionary spaces, and for our belief that detailed knowledge on early forms of viruses or previral forms may be unobtainable in principle. These ideas imply that virus families now apparently unrelated could have had a common origin.

Time scales in virus evolution

A basic limitation of the study of processes of viral evolution is that, in general, it has not yet been feasible to identify a quantitative time scale. There are exceptions to this in cases of rapid nucleotide substitution, but the times so defined are for relatively recent events only. For most viruses there is no equivalent of a fossil record to help study prehistory; an exception, but outside the scope of this chapter, is that retroviral genomes integrated into germline DNA of the host take on low mutation rates characteristic of the host genome, and can be examined genetically and evolutionarily as part of the host genome.

Some estimate of order of magnitude for longer time scales in virus evolution may be obtained by the principle that where a group of viruses is found in a range of diverged hosts then a common viral progenitor may have existed in populations of the host progenitor predating radiation towards the present day hosts; that is, contemporary viruses have evolved with their hosts. This is certainly a plausible and useful approach, but it is not universally applicable: other pathways for virus speciation certainly exist, notably transfer between host species and divergence within a single host species. None the less, building up estimates of relative divergences among viruses from sequence data, and superimposing on this structure time scales of host evolution, might eventually give consistent estimates for the dates of divergence of lineages in virus families; this possibility remains a hope for the future.

Many features of virus families argue that their origins are of great antiquity – the very broad spectra of hosts, the great diversity within virus families, the probable large extent of viral evolutionary processes. We think it entirely possible that virus families could have originated during an early, cellular phase of the development of life on our planet, as distinct from subsequent organismal phases.

The subject of this chapter is speculative both inherently and from present lack of data. We have, none the less, kept chains of suppositions as short as possible. Beyond the point now reached, there lie topics yet more free of evidence, such as detail of possible primordial relationships among virus families and even descent of DNA viruses from the RNA World, that we judge should properly be left in the imagination.

Acknowledgements

We thank L.J.E. Kattenhorn for extensive help in preparing the text.

References

Bashford, D., Chothia, C. & Lesk, A.M. (1987). *J. Mol. Biol.*, **196**, 199–216.
Choi, H.-K., Tong, L., Minor, W., Dumas, P., Boege, U., Rossmann, M.G. & Wengler, G. (1991). *Nature*, **354**, 37–43.
Davison, A.J. (1992). *Virology*, **186**, 9–14.
Gorbalenya, A.E. & Koonin, E.V. (1989). *Nucl. Acids Res.*, **17**, 8413–40.
Ito, J. & Braithwaite, D.K. (1991). *Nucl. Acids Res.*, **19**, 4045–7.
Keese, P.K. & Gibbs, A. (1992). *Proc. Natl. Acad. Sci. USA*, **89**, 9489–93.
Luria, S.E. & Darnell, J.E., Jr (1967). In *General Virology*, 2nd edn. John Wiley and Sons, Inc, New York, London, pp. 439–454.
McGeoch, D.J. (1989). *Ann. Rev. Microbiol.* **43**, 235–65.
McGeoch, D.J. (1992). *Seminars Virol*, **3**, 399–408.
Roberts, M.M., White, J.L., Grütter, M.G. & Burnett, R.M. (1986). *Science*, **232**, 1148–51.
Tsao, J., Chapman, M.S., Agbandje, M., Keller, W., Smith, K., Wu, H., Luo, M., Smith, T.J., Rossmann, M.G., Compans, R.W. & Parrish, C.R. (1991). *Science*, **251**, 1456–64.

6

In search of the origins of viral genes
ADRIAN GIBBS AND PAUL K. KEESE

Introduction

Over the past half century, viruses have been classified and identified by such characters as the shape, size and serological specificity of their virions, and also by biological characters such as host and vector type. These features placed viruses into seemingly sensible genera and other taxa, but the evolutionary relationships of the genera (and higher taxa), and the processes that produced them, were unknown. However, knowledge of viral genomic sequences is starting to reveal the wealth of genetic processes involved in viral evolution, and the relative importance of each (Fig. 6.1). Sequence information has confirmed most of the earlier groupings, but also revealed a much greater shared genealogy than expected.

Taxonomic comparisons show that the genes of closely related viruses, such as isolates of a single species, usually only differ from one another by point mutations and, occasionally, by insertions or deletions of one or a few nucleotides. The different species of each viral genus differ by a greater number of changes of the same type. Thus mutation is a primary evolutionary driving force for viruses just as it is for all other organisms.

Comparisons of the genes of viruses placed in separate genera show that some are related, some are not. For example, all viral RNA replicase genes encode proteins that share sequence motifs, and hence may be related (Poch *et al.*, 1989). Likewise, the virion proteins of most viruses with isometric virions of 25–30 nm diameter have a characteristic eight-stranded β-barrel structure (Rossmann & Johnson, 1989) the only known exception being those of Qβ levivirus, which has virion proteins with a β-sheet/α-helical hook structure (Valegaord & Liljas, 1990). Different viral genera mostly have different combinations of the shared

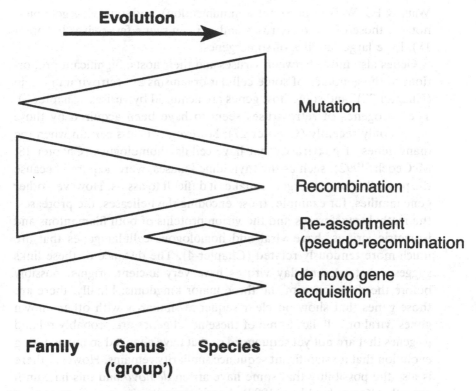

Evolution

Mutation

Recombination

Re-assortment
(pseudo-recombination

de novo gene
acquisition

**Family   Genus   Species
          ('group')**

Fig. 6.1. The apparent relative importance of some sources of genetic novelty
that influence virus evolution.

genes, and taxonomies calculated separately for each gene are often not
congruent (Chapters 4 and 24). The simplest explanation of this is that
the progenitors of existing viral genera acquired such genes from more
than one ancestral virus by recombination (Gibbs, 1987; Steinhauer &
Holland, 1987; Strauss & Strauss, 1988; Zimmern, 1982).

Recombination has resulted in widely divergent viral groups having
related genes. Perhaps the most surprising example of this is shown by
the clear sequence similarity of a virion glycoprotein of baculoviruses
and one from orthoacariviruses, even though these viruses seem, in
other respects, to be quite unrelated, and have genomes of dsDNA
and negative-sense ssRNA, respectively (Chapter 28). Recombination
has also produced duplicated viral genes, for example, bovine ephem-
eral fever and Adelaide River rhabdoviruses encode a non-virion
glycoprotein related to their virion glycoprotein (Walker *et al.*, 1992; Y.

Wang & P.J. Walker, personal communication), and some viral genomes, notably those of the poxviruses and African swine fever virus (Chapter 18), have large families of some genes.

Genes also move between viruses and their hosts. Significant proportions of the genomes of some cellular organisms are retroviral in origin (Chapter 27), and some host genes are acquired by viruses (Chapter 7). The oncogenes of retroviruses seem to have been acquired by those viruses only recently (Chapter 27); however, it is less certain when the many genes of poxviruses that have cellular homologues (Chapter 18; McGeoch, 1992), such as the thymidine kinases, were acquired because the possibility of convergence makes it difficult to assess. However, other gene families, for example, those encoding the helicases, the proteases, the methyl transferases and the virion proteins of both filamentous and isometric virions, have viral and homologous cellular genes that are much more tenuously related (Chapter 4). The distance of these links suggests that present day viruses have very ancient origins, possibly before the divergence of the three major kingdoms. Finally, there are those genes that show no clear sequence similarity with other known genes, viral or cellular. Some of these novel genes are probably related to genes that are not yet sequenced or that have changed so much during evolution that no significant sequence similarity remains. However there is also the possibility that some have arisen *de novo* and this has, until recently (Keese & Gibbs, 1992), been ignored.

Novel genes might be generated either by synthesizing entirely new nucleotide sequences, or alternatively by using the unused reading frames of existing genes, a process first mooted by Grasse (1977), who called it 'overprinting'. It would be virtually impossible to prove that a novel gene arose by the first mechanism; however proof that the latter mechanism, overprinting, does occur can be obtained from studies of functional overlapping genes, such as those which are found in most types of viral genomes, because taxonomic analysis of such linked genes can show which is the original gene that has been overprinted, and which is the novel gene. It is, however, more difficult to prove overprinting if the original gene has been copied, and the original and the new (frame-shifted) gene have evolved separately, but we have some evidence that this may occur in viruses. A gene of the latter type is perhaps best distinguished by calling it a 'palimpsest gene'; a palimpsest is 'a vellum, parchment or brass, the original writing on which has been effaced to make room for a second' (Greek *palimpsestos*; *palin*, again, and *psao*, rub smooth).

Overlapping genes

Overlapping genes were first found when the genome of φX174 microvirus was sequenced. Some of its genes behaved anomalously in mutation experiments and, when sequenced, were found to be in different reading frames of the same part of the genome (Barrell, Air & Hutchison, 1976; Shaw *et al.*, 1978). Overlapping genes have subsequently been found in about half of all viral genera, not only those with DNA genomes, but also those with RNA genomes, including viruses of bacteria and eukaryotes, both animals and plants.

The genomes of microviruses, φX174 (Sanger *et al.*, 1977), G4 (Godson et al., 1978) and S13 (Lau & Spencer, 1985), have two fully overlapping gene pairs (Fig. 6.2). Gene B is entirely within gene A, and E within gene D; the products of genes A and B are involved in genome replication, that of gene D is a scaffolding protein, and that of gene E is a lysis protein. In the φX174 genome, there is an excess of thymidine (G 23.3%, A 23.9%, C 21.5%, T 31.2%), especially in the third positions of most genes, which are probably the original genes (Sanger *et al.*, 1977). Genes B and D also have the same pattern of excess thymidine, whereas the codons of gene E and of gene A, where it overlaps B, are in reading frames one nucleotide 3′ of those of gene D and B, respectively; therefore the excess of thymidine in the third positions of genes D and B is in the second codon positions of genes E and A, where it overlaps B. Thus it is likely that genes B and D are the original genes, and A, where it overlaps B, and E have arisen more recently. There are other overlaps, and in one part of the φχ174 genome all three reading frames are used, but the thymidine bias indicates which gene or gene portion is the original gene. This inference is confirmed by studies of the genome of the related G4 phage, which differs from that of φX174 in an average of 33.1% of its bases (Godson *et al.*, 1978), for although the G4 genome has a smaller T content, except in its overlapping genes, the biases in the overlapping genes are the same. As evolutionary change is likely to occur more slowly in a genomic region that is encoding a pair of overlapping genes than in a comparable part encoding a single non-overlapping gene, the greater content of thymidine in G4's overlapping genes suggests that thymidine has been lost from the G4 genome since its divergence from φX174, rather than that the latter has gained thymidine.

It is, however, easier to establish which gene is the overprinted one and which one is novel, when the sequences of several relatives of the genes are known, as this allows the phylogenetic histories of the genes

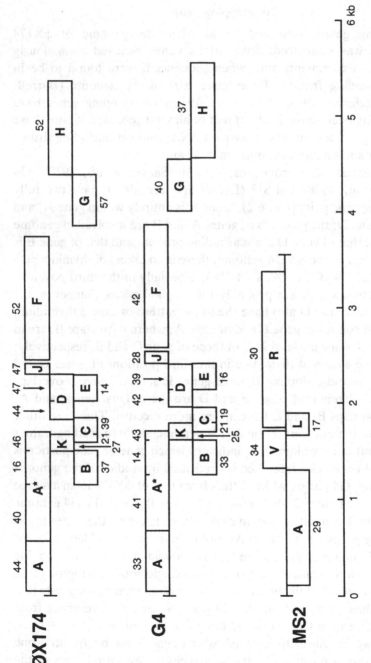

Fig. 6.2. Genome maps of φX174 and G4 microviruses and, on the same scale, that of MS2 levivirus. The proteins encoded by genes A, B and C of the microviruses are involved in DNA replication, that of gene D in capsid assembly, that of gene E is a lysin and those of genes F, G, H and J are capsid proteins. The A gene of the leviviruses encodes a maturation protein, that of V the virion protein, that of L is a lysin and the R gene encodes one subunit of the viral polymerase. The figures indicate the percentages of thymidine in the third codon positions of the microvirus genes, and of cytosine in the levivirus genes.

to be inferred, and the relative time of appearance of each gene member identified. Three clear examples of genes of this sort are found in the overlapping genes of tymoviruses, luteoviruses and lentiviruses.

The overlapping gene of tymoviruses

The genome of turnip yellow mosaic tymovirus (TYMV), like that of all tymoviruses, is single-stranded messenger-sense RNA of about 6300 nucleotides (Keese, Mackenzie & Gibbs, 1989). Its largest open-reading frame (ORF) is about 5400 nucleotides long and encodes a replicase (RP) which includes the characteristic GDD motif. The 5′-terminal third of this gene also encodes an out-of-frame overlapping gene (OP) (Fig. 6.3a), that is present in all tymoviruses, and that always begins from the AUG which is closest to the 5′-terminus and separated by four nucleotides from that of the RP ORF. It encodes a protein of unknown function that is required for normal replication (Weiland & Dreher, 1989).

Tymovirus RPs have at least three distinct regions; the N-terminal third shows clear sequence similarity to the RNA methyltransferases of tobacco mosaic and its relatives (Ding, Keese & Gibbs, 1990), especially the potexviruses, likewise their middle and C-terminal third are clearly homologous to the nucleotide-binding and replicase domains of the same viruses (Srifah, 1991). Separate taxonomies of these three RP regions are almost identical in topology, indicating that the three parts of the RP gene of this group of viruses have co-evolved without recombination.

Comparisons of the RPs of all viruses that have been sequenced show that, for all its three parts, the tymoviruses classify among the Sindbis-like viruses (Srifah, 1991). Within this group, the potexviruses, carlaviruses and closteroviruses are 'sister groups' of the tymoviruses, whereas the remainder, including the alphaviruses, bromoviruses, cucumoviruses and tobamoviruses, are 'outgroups'. However, among the Sindbis-like viruses only the tymoviruses have an OP gene overlapping the 5′-terminus of the RP gene (Fig. 6.4a). The simplest explanation of these facts is that the RP gene is the original one and the OP arose by overprinting after the tymoviruses and their sister groups diverged from a common ancestor. It is less likely that this ORF was present earlier but subsequently lost in all Sindbis-like viruses except the tymoviruses.

Nucleotide biases in the RP/OP overlap region also support the conclusion that the RP is the primitive gene. Tymovirus genomes have

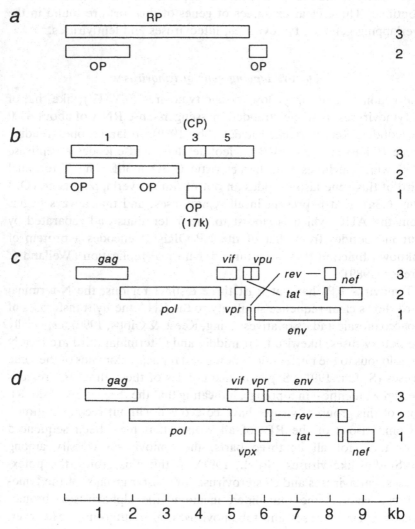

Fig. 6.3. Viral genome organizations. a) TYMV; RP, replicase protein; OP, overlapping protein; VP, virion protein. b) PLRV; CP, coat protein; 17K, ORF encoding 17K M_r polypeptide. c) HIV 1 and 2; *tat*, transactivator; *rev*, regulator of expression of virion proteins. (Redrawn from Keese & Gibbs, 1992.)

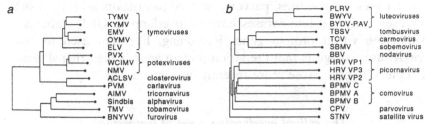

Fig. 6.4. Dendrograms showing the relationships of: a) the methyltransferase-like domains of the RPs of 14 Sindbis-like viruses. KYMV, kennedya yellow mosaic tymovirus; EMV, eggplant mosaic tymovirus; OYMV, ononis yellow mosaic tymovirus; ELV, erysimum latent tymovirus; PVX potato X potexvirus; WCIMV, white clover mosaic potexvirus; NMV, narcissus mosaic potexvirus; ACLSV, apple chlorotic leaf spot closterovirus; PVM, potato M carlavirus; AIMV, alfalfa mosaic alphamovirus; TMV, tobacco mosaic tobamovirus; BNYVV, beet necrotic yellow vein furovirus. b) the eight β-strands in the β-barrel domain of 15 virion proteins. BWYV, beet western yellows luteovirus; BYDV-PAV, PAV strain of barley yellow dwarf luteovirus; TBSV, tomato bushy stunt tombusvirus; TCV, turnip crinkle carmovirus; SBMV, southern bean mosaic sobemovirus; BBV, black beetle nodavirus; HRV, human rhinovirus; BPMV, bean pod mottle comovirus; CPV, canine parvovirus; STNV satellite of tobacco necrosis virus. (Redrawn from Keese & Gibbs, 1992.)

a large cytosine content, especially in the third codon position of the RP ORF (Keese *et al.*, 1989). This bias is found in all parts of the RP, but not in the OP, where cytosine is most common in the first codon position. This bias also incidentally decreases the chance occurrence of stop codons (UAA, UAG, and UGA) in all reading frames and, as a result, increases the chance occurrence and length of ORFs.

The VpG gene of luteoviruses

Another overlapping viral gene whose taxonomy is well known is found in luteoviruses. The virion protein genes of luteoviruses, such as potato leafroll virus (PLRV), all contain an embedded ORF encoding a protein of about 17K M_r (Fig. 6.3b) (Martin *et al.*, 1990). Antibodies to this protein react with the 5' genome-linked protein (R.R. Martin, personal communication). Sequence analyses of the luteovirus virion proteins shows their close similarity to the β-barrel domain of other viral virion proteins found in small isometric virions (Martin *et al.*, 1990). Comparisons show that the luteovirus virion proteins form a monophyletic group most closely related to those of the carmoviruses, sobemoviruses and tombusviruses, and more distantly to those of comoviruses, nodaviruses, parvoviruses and picornaviruses (Fig. 6.4b); however, only the luteoviruses have an overlapping ORF embedded within the virion protein gene. Reasoning, like that used for the tymoviruses, suggests that the virion protein gene is the original gene, and the 17K gene arose more recently by overprinting.

The spliced overlapping genes of lentiviruses

The genomes of lentiviruses, such as human immunodeficiency viruses 1 and 2 (HIV 1 and 2), also have several overlapping ORFs (Fig. 6.3c), expressed mostly by alternative splicing (Schwartz *et al.*, 1990). The 3'-terminal coding exons of the regulatory genes *rev* (Sodroski *et al.*, 1986) and *tat* (Arya *et al.*, 1985) overlap and also overlap the *env* gene, so that, in one region, all three reading frames are expressed. Although the *env* gene of HIV 1 and 2 are related to the *env* genes of all other retroviruses (McClure *et al.*, 1988) (Fig. 6.4c), it is only the lentiviruses that have the *tat* and *rev* genes, and so it is most likely that the *env* gene is ancestral and the *tat* and *rev* genes arose later, only in lentiviruses, by overprinting.

The HIV 1 genome, but not the HIV 2 genome, also has an extra gene, *vpu*, that overlaps the *env* gene (Fig. 6.3c) (Cohen *et al.*, 1988). Similarly, *vpx*, which overlaps the *vif* and *vpr* genes, is found only in the HIV 2 genome (Hu, Heyden & Ratner, 1989). As *vpu* and *vpx* are not found in HIV 2 and 1, respectively, or in other retroviruses, then they too probably arose *de novo* more recently than the *tat* and *rev* genes.

It is possible that functional overlapping ORFs originate in viral genomes, such as lentiviruses, because an ancestral nucleotide sequence with a single expressed ORF fortuitously expresses a second functional but dispensable protein, such as *vpu* or *vpx*, which becomes essential as it evolves and becomes adaptively integrated into the genome. This could be the way that the 3'-terminal coding exon of *rev*, which is embedded in the *env* gene (Fig. 6.3c), has become essential for HIV infectivity (Malim *et al.*, 1989).

Palimpsest genes

There are many viruses, such as some of the flaviviruses and some of the 'picornavirus supergroup' (Chapter 4; King, Lomonosoff & Ryan, 1991), that have monopartite genomes that are translated into single large polyproteins before being hydrolysed into their constituent proteins. Consequently, such viruses cannot obtain genetic novelty by expressing overlapping ORFs. However, some of their genes, like the overlapping genes of other viruses, have no known homologues and are very variable (Fig. 6.5). It is possible that some of these are palimpsest genes, analogous to overlapping genes, but resulting from gene duplication and then frameshifting.

When the genomes of such viruses are examined for 'RNY' (puRine-aNy base-pYrimidine) bias (Shepherd, 1981), the novel genes are found to have a consistently large RNY signal in the non-reading frame that has, as its first nucleotide, the third nucleotide of the ORF (i.e. reading frame 3 or −1), by contrast, the shared ancestral genes of those viruses do not have a large RNY signal in their non-reading frames. Fig. 6.6

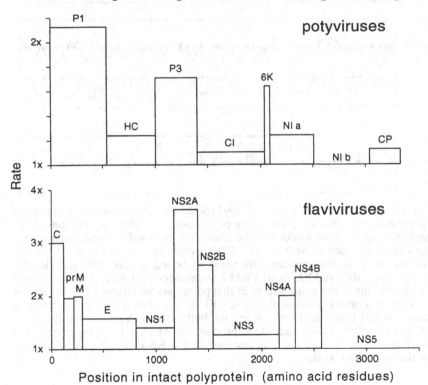

Fig. 6.5. The relative rate of change of different proteins of the potyviruses and flaviviruses. The rates are expressed as mean sequence identities relative to that of the protein with the smallest interspecific differences; NI b for potyviruses and NS5 for flaviviruses. Data for five potyvirus comparisons is from Yeh *et al.* (1992), and for 55 flavivirus comparisons from Blok *et al.* (1992).

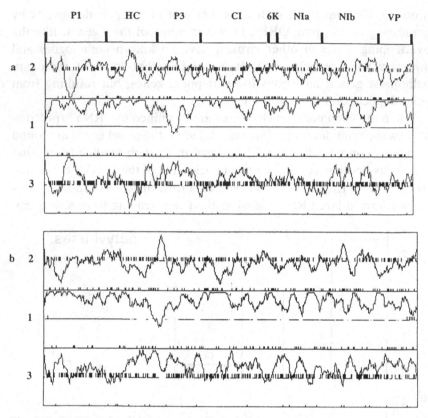

Fig. 6.6. RNY analyses (Shepherd, 1981) of the genomic sequences of a) papaya ringspot and b) tobacco vein mottling potyviruses. The analyses were done using Staden's algorithm networked by the Australian National Genomic Information Service, University of Sydney. The RNY signals in each of the three reading frames (RFs) of the sequence are given as boxed graphs; RF1 encodes the polyprotein, RF2 and RF3 start 1 and 2 nucleotides 3′ of RF1. A dot is marked on the midline of the graph which, at that point, has the largest RNY signal; this appears as a nearly continuous line in RF1. Start and stop codons are shown as short vertical lines; the former along the bottom of each graph, and the latter along its midline. Labels indicate approximately which parts of the polyprotein yield the 'mature' proteins. The decrease in the RNY signal in the P3 region of the main ORF is clear.

is a RNY analysis of the genome of two potyviruses and clearly shows the large RNY signal in the P3 protein. This phenomenon is also shown by the NS2 and NS4 proteins of the flaviviruses and their relatives (see Chapter 19). The 'virtual amino acid sequences' translated from the non-reading frames of these unusual genes have been compared

with the amino acid sequences of the proteins encoded by the genes
of these viruses (Gibbs & Keese, unpublished observations), and it is
found that they align best with the proteins immediately N-terminal to
them in the polyprotein (P3 to HC in the potyviruses, and the combined
NS4 reading frame −1 with the NS3 protein in the flaviviruses).
However, the similarities are only 2–4 standard deviations from the
mean similarity scores of random sequences of the same composition.
We are, at present, testing more sensitive methods for making these
comparisons.

Cellular genes

Interestingly, there are examples (Keese & Gibbs, 1992) of cellular genes
that have arisen by overprinting. Furthermore, Fukuchi and Otsuka
(1992) claim to have detected similarities between the sequences of
enzyme genes and 'virtual amino acid sequences' of the sense strand and
of the antisense strands of ribosomal and transfer RNA genes and their
flanking regions (i.e. amino acid sequences translated from non-reading
frames). They conclude that 'metabolic pathways have been developed
by the chance assembly of enzyme proteins generated from the sense
and antisense strands of pre-existing genes . . .'.

It is likely that overprinting is a common source of new genes in all
organisms, indeed of the 182 ORFs greater than 300 nucleotides in
length in yeast chromosome III, 17 overlap to a greater or lesser extent
(Oliver *et al.*, 1992). The expression of novel ORFs from pre-existing
nucleotide sequences seems to have involved both coding regions, such
as the thyroid steroid hormone receptor (Issemann & Green, 1990),
and non-coding regions, such as the mitochondrial intron of yeast that
now encodes the optional intron 2 of the *cob* gene (Lazowska, Jacq &
Slonimski, 1980).

Discussion

All nucleotide sequences have alternative ORFs that are potentially
functional; however, nucleotide frequency biases increase the chances
that longer ORFs occur (Casino *et al.*, 1981; Keese *et al.*, 1989;
Sharp, 1985), and changes in bias during evolution may produce new
overprinting possibilities.

Biases in nucleotide usage are often most pronounced in the third
codon position of ORFs, as this position is most redundant in the

coding sequences. When, however, alternative reading frames of such biased genes are used, the bias then occurs in the first or second codon positions of the new ORF, and the encoded protein may have an over-representation of amino acids with similar chemical or physical properties. This is because similar amino acids often have codons with the same first or second nucleotide. For example, the two codons for histidine and four of the six for arginine have cytosine in their first positions as do four for leucine; thus the overlapping protein of tymoviruses is a very basic and hydrophobic protein. Similarly, codons with thymidine in their second position mostly encode hydrophobic amino acids; thus over a fifth of the E protein of microviruses is leucine, twice that of their other proteins. So, in general, when genes with biased compositions are overprinted, the novel proteins are dominated by amino acids that fall into a row or column of the genetic code table (as it is usually arranged), and hence will have polarized chemical or physical properties. This suggests that genes produced by overprinting may first produce proteins whose usefulness results from their composition, and only later, after further evolution, will they acquire functions determined by their sequence and structure.

A range of mechanisms allow genes to be overprinted and novel genes to be expressed (Keese & Gibbs, 1992). These, and other mechanisms, may allow chance exploitation of alternative longer ORFs to produce significant functional novelty from any part of the genome including non-coding gene sequences, pseudogenes or duplicated genes. The outcome could be a large genetic change in the organism, and hence a large change in its phenotype. This is one potential source, at least, of the saltatory changes that contribute to the 'punctuated equilibrium' theory of evolutionary change (Eldridge & Gould, 1972). Consequently, the exclusive presence of the OP gene in all tymoviruses, and the 17K gene in all luteovirus genomes, not only indicates the relative times of origins of these genes but also suggests that these genes may have been fundamentally responsible for the establishment and success of these virus groups.

The presence of overlapping genes may also give some broad clues on earliest stages of viral evolution. For example, an overlapping gene has been found in the genome of f2 and MS2 leviviruses (Fig. 6.2; Atkins *et al.*, 1979; Beremand & Blumenthal, 1979; Model, Webster & Zinder, 1979), and others of this genus of male-specific phages. This overlapping gene encodes a lysis protein, and overlaps the 3' end of the coat protein gene and the 5' end of the replicase gene, together with the short untranslated intergenic region (Fig. 6.2). The base compositions of the different codon positions of the A protein, coat protein and replicase genes of the leviviruses are similar and unlike that of the lysis gene, which suggests that the lysis gene is the novel one. Incidentally, levivirus genomes have an excess of cytosine, although not as extreme as those of tymoviruses; however, the second codon position of the lysin gene is the third of the genes it overlaps, and thus the amino acid bias of the levivirus lysins is different from that of the tymovirus OPs. Beremand and Blumenthal (1979) noted that the amino acid sequence of the f2 lysis protein shows some relatedness to the lysis protein of ϕX174 microvirus. However, we have made detailed comparisons and have found that this similarity is no more than would be expected by chance, and, furthermore, that the lysin aligns best with different parts of the ϕX174 and G4 lysins; as one would expect of genes that have arisen *de novo* by overprinting unrelated genes. None the less, given that host lysis is an important step in the life cycle of many phages, including the microviruses and leviviruses, it is interesting that two unrelated phages of *E.coli* acquired lysis genes by overprinting older genes. This could indicate that both these phage genera acquired their older genes from plasmids (Chapter 16), or from phages that, like the inoviruses (i.e. fd and M13), leave and enter host cells without lysing them. Another possibility is that the earlier hosts of the ancestors of micro- and leviviruses had no cell wall or one of a different type, indeed the fact that two unrelated phages of *E. coli* acquired lysins supports this possibility.

References

Arya, S.K., Guo, C., Josephs, S.F. & Wong-Staal, F. (1985). *Science*, **229**, 69–73.

Atkins, J.F., Steitz, J.A., Anderson, C.W. & Model, P. (1979). *Cell*, **18**, 247–56.

Barrell, B.G., Air, G.M. & Hutchison III, C.A. (1976). *Nature*, **264**, 34–41.

Beremand, M.N. & Blumenthal, T. (1979). *Cell*, **18**, 257–66.

Blok, J., McWilliam, S.M., Butler, H.C., Gibbs, A.J., Weiller, G., Herring, B.L., Hemsley, A.C., Aaskov, J.G., Yoksan, S. & Bhamarapravati, N. (1992). *Virology*, **187**, 573–90.

Casino, A., Cipollaro, M., Guerrini, A.M., Mastrocinque, G., Spena, A. & Scarlato, V. (1981). *Nucl. Acids Res.*, **9**, 1499–517.

Cohen, E.A., Terwilliger, E.F., Sodroski, J.G. & Haseltine, W.A. (1988). *Nature*, **334**, 532–4.

Ding, S.-W., Keese, P. & Gibbs, A. (1990). *J. Gen. Virol*, **71**, 925–31.

Eldridge, N. & Gould, S.J. (1972). In *Models in Paleobiology*, T.J.M. Schopf (ed.). Freeman, San Fransisco, pp. 82–115.

Fukuchi, S. & Otsuka, J. (1992). *J. Theor. Biol*, **158**, 271–91.

Gibbs, A. (1987). *J. Cell Sci.* (Suppl.), **7**, 319–37.
Godson, G.N., Barrell, B.G., Staden, R & Fiddes, J.C. (1978). *Nature*, **276**, 236–47.
Grasse, P.-P. (1977) *Evolution of Living Organisms*. Academic Press, New York, p. 297.
Hu, W, Heyden, N. V. and Ratner, L. (1989). *Virology*, **173**, 624–30.
Issemann, I. & Green, S. (1990). *Nature*, **347**, 645–50.
Keese, P.K. & Gibbs, A.J. (1992). *Proc. Natl. Acad. Sci. USA*, **89**, 9489–93.
Keese, P., Mackenzie, A. & Gibbs, A. (1989). *Virology*, **172**, 536–46.
King, A.M.Q., Lomonosoff, G.P. & Ryan, M.D. (1991). *Sem. Virol*, **2**, 11–18.
Lau, P.C.K. & Spencer, J.H. (1985). *Gene*, **40**, 273–84.
Lazowska J., Jacq & Slonimski, p.p. (1980). *Cell*, **22**, 333–41.
McClure, M.A., Johnson, M.S., Feng, D.-F. & Doolittle, R.F. (1988). *Proc. Natl. Acad. Sci. USA*, **85**, 2469–73.
McGeoch, D.J. (1992). *Sem. Virol.*, **3**, 399–408.
Malim, M.H., Böhnlein, S, Hauber, J. & Cullen, B.R. (1989). *Cell*, **58**, 205–14.
Martin, R.R., Keese, P.K., Young, M.J., Waterhouse, P.M. & Gerlach, W.L. (1990). *Ann. Rev. Phytopath.* **28**, 341–63.
Model, P., Webster, R.E. & Zinder, N.D. (1979). *Cell*, **18**, 235–46.
Oliver et al. (1992). *Nature*, **357**, 38–46.
Poch, O., Sauvaget, I., Delarue, M. & Tordo, N. (1989). *EMBO J.*, **8**, 3867–74.
Rossmann, M.G. & Johnson, J.E. (1989). *Ann. Rev. Biochem.*, **58**, 533–73.
Sanger, F., Air, G.M., Barrell, B.G., Brown, N.L., Coulson, A.R., Fiddes, J.C., Hutchison III, C.A., Slocombe, P.M. & Smith, M. (1977) *Nature*, **265**, 687–95.
Schwartz, S., Felber, B.K., Benko, D.M., Fenyö, E.-M. & Pavlakis, G.N. (1990). *J. Virol.*, **64**, 2519–29.
Sharp, P.M. (1985). *Nucl. Acids Res.*, **13**, 1389–97.
Shaw, D.C., Walker, J.E., Northrop, F.D. Barrell, B.G., Godson, G.N. & Fiddes, J.C. (1978). *Nature*, **272**, 510–15.
Shepherd, J.C.W. (1981). *Proc. Natl. Acad. Sci. USA*, **78**, 1596–600.
Sodroski, J., Goh, W.C., Rosen, C., Dayton, A., Terwilliger, E. & Haseltine, W. (1986). *Nature*, **321**, 412–17.
Srifah, P. (1991). The molecular biology of erysimum latent tymovirus. PhD thesis, Australian National University.
Steinhauer, D.A. & Holland, J.J. (1987). *Ann. Rev. Microbiol.*, **41**, 409–33.
Strauss, J.H. & Strauss, E.G. (1988). *Ann. Rev. Microbiol.*, **42**, 657–83.
Valegaord, K. & Liljas, L. (1990). *Nature*, **345**, 36–41.
Walker, P.J., Byrne, K.A., Riding, G.A., Cowley, J.A., Wang, Y. & McWilliam, S. (1992). *Virology*, **191**, 49–61.
Weiland, J.J. & Dreher, T.W. (1989). *Proc. Natl. Acad. Sci. USA*, **17**, 4675–87.
Yeh, S.-D., Jan, F.-J., Chiang, C.-H, Doong, T.-J. Chen, M.-C. Chung, P.-H. & Bau, H.-J. (1992). *J. Gen. Virol.*, **73**, 2531–41.
Zimmern, D. (1982). *Trends Biochem.Sci.*, **7**, 205–7.

7

Cellular sequences in viral genomes
GREGOR MEYERS, NORBERT TAUTZ,
AND HEINZ-JÜRGEN THEIL

Introduction

Recombination of viral and host cellular nucleic acids probably plays a major role in the evolution of viruses. By this means, a virus is able to exploit a large gene pool and to acquire new properties very rapidly. The presence of cellular genes or gene fragments within the genomes of retroviruses (*v*-oncogenes) is well established. For other viruses, the cellular origin of certain genes has been suggested. Direct evidence is, however, difficult to obtain, probably because elaborate mutational changes of the cellular sequences occurred in the viral genomes. For some non-retroviruses, however, sequences homologous to cellular genes have been identified. We will discuss examples of these very recent recombination events with emphasis on bovine viral diarrhoea virus (BVDV).

Cellular sequences in the genome of BVDV

Identification of host cell-derived insertions

BVDV is a member of the genus pestivirus within the family Flaviviridae (Wengler, 1991). The pestiviral genome consists of single-stranded RNA with positive polarity of about 12 kilobases (kb) (Collett *et al.*, 1988; Meyers, Rümenapf & Thiel, 1989*a*). The first third of this RNA codes for an autoprotease and four structural proteins, while the 3' part of the genome encodes the other non-structural proteins (Collett *et al.*, 1991; Thiel *et al.*, 1991). The genome of BVDV, like that of all flaviviruses, is translated into a single polyprotein which is co- and post-translationally cleaved to give rise to mature viral proteins. While cellular proteases are probably responsible for processing of the three structural glyco-proteins, most or all of the cleavages in the carboxy-terminal part of

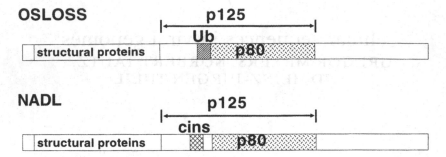

Fig. 7.1. Genomic location of the cellular inserts in the RNA genomes of the BVDV strains Osloss and NADL. The regions coding for the viral structural proteins and for p125/p80 are indicated. Ub: insertion coding for ubiquitin-like protein; cins: sequence homologous to cellular mRNA of unknown function.

the polyprotein are hydrolysed by a viral protease with a molecular weight of 125 kilodalton (p125) (Wiskerchen & Collett, 1991). In the pestiviral genome, the p125-coding region is located 3' to the structural protein region.

Comparison of the genomic sequences of two strains of BVDV, Osloss (Renard, Dino & Martial, 1987) and NADL (Collett *et al.*, 1988), with that from another pestivirus, hog cholera virus (HCV) (Meyers *et al.*, 1989*a*), resulted in identification of strain-specific insertions in the BVDV RNAs (Meyers, Rümenapf & Thiel, 1989*b*, 1990). The additional 228 or 270 nucleotides (strains Osloss and NADL, respectively) exhibit no significant homology to other pestiviral sequences, and are both located within the p125-coding region of the genomes (Fig. 7.1). The Osloss insertion codes for 76 amino acids which, in a data bank search, were found to be highly homologous to a well-known cellular protein, namely ubiquitin. The ubiquitin amino acid sequence is conserved for all animals analysed so far. The ubiquitin-like sequence from the BVDV Osloss strain exhibits two amino acid exchanges with respect to the animal ubiquitin sequence (Fig. 7.2) (Meyers *et al.*, 1989*b*, 1990).

Comparison of the inserted element in the BVDV NADL genome (termed cins) with the sequences available in different data banks did not result in identification of a cellular counterpart. However, an mRNA homologous to the NADL insertion was detected on a Northern blot with RNA from bovine cells (Meyers *et al.*, 1990). After cloning and sequencing, a cellular cDNA fragment was identified containing a region with a nucleotide sequence almost identical to the inserted element (Fig. 7.3). Only two nucleotide exchanges were found, one of which results in

```
BVDV, OSLOSS:    N L E H L G W I L K M Q I F V K T L T G K T I T L E V E P S D T
BVDV, NADL:      N L E H L G W I L R
Ubiquitin:                         M Q I F V K T L T G K T I T L E V E P S D T


BVDV, OSLOSS:    I E N V K A K I Q D K E G I P P D Q Q R L I F A G K Q L E D G R

Ubiquitin:       I E N V K A K I Q D K E G I P P D Q Q R L I F A G K Q L E D G R


BVDV, OSLOSS:    [S] L S D Y N I Q K E S T L H L V L R L R G [S] G P A V C K K I
BVDV, NADL:                                                  G P A V C K K I
Ubiquitin:       [T] L S D Y N I Q K E S T L H L V L R L R G [G]
```

Fig. 7.2. Amino acid sequence comparison of parts of the polyproteins encoded by the genomes of BVDV Osloss and NADL and the conserved animal ubiquitin. Amino acids differing between the animal ubiquitin and the protein encoded by the Osloss insertion are marked by boxes.

```
NADL:    agggt ATGTGCAGCCGATGCCAGGGAAAGCATAGGAGGTTTGAAATGGACCGGGAACCT    5046
               M  C  S  R  C  Q  G  K  H  R  R  F  E  M  D  R  E  P
cins:    cgatg ATGTGCAGCCGATGCCAGGGAAAGCATAGGAGGTTTGAAATGGACCGGGAACCT

  5047   AAGAGTGCCAGATACTGTGCTGAGTGTAATAGGCTGCATCCTGCTGAGGAAGGTGACTTT    5106
          K  S  A  R  Y  C  A  E  C  N  R  L  H  P  A  E  E  G  D  F
         AAGAGTGCCAGATACTGTGCTGAGTGTAATAGGCTGCATCCTGCTGAGGAAGGTGACTTT
                                                   ▼
  5107   TGGGCAGAGTCGAGCATGTTGGGCCTCAAAATCACCTACTTTGCGCTGATGGATGGAAAG    5166
          W  A  E  S  S  M  L  G  L  K  I  T  Y  F  A  L  M  D  G  K
         TGGGCAGAGTCAAGCATGTTGGGCCTCAAAATCACCTACTTTGCGCTGATGGATGGAAAG
                      ▲
  5167   GTGTATGATATCACAGAGTGGGCTGGATGCCAGCGTGTGGGAATCTCCCCAGATACCCAC    5226
          V  Y  D  I  T  E  W  A  G  C  Q  R  V  G  I  S  P  D  T  H
         GTGTATGATATCACAGAGTGGGCTGGATGCCAGCGTGTGGGAATCTCCCCAGATACCCAC
                                                   ▼
  5227   AGAGTCCCTTGTCACATCTCATTTGGTTCACGGATG cctttcaggcaggaa            5283
          R  V  P  C  H  I  S  F  G  S  R  M
                      Y
         AGAGTCCCTTATCACATCTCATTTGGTTCACGGATG ccaggcaccagtggg
                      ▲
```

Fig. 7.3. Comparison of the sequences from the region of the BVDV NADL genome containing the insertion and part of a bovine cell-derived cDNA clone. For the region identified as insertion in the viral genome, the nucleotide sequence is given in capital letters and the deduced amino acid sequence is shown. Nucleotide sequence differences between viral and cellular sequence are marked by triangles. Numbers refer to the sequence published for BVDV NADL (Collett *et al.*, 1988).

a change of the encoded amino acid sequence. According to our present knowledge, the cins insertion represents an internal fragment of a bovine mRNA coding for a protein which has not yet been identified.

The findings outlined above indicate that the genomes of BVDV Osloss and NADL are probably products of recombination between viral and cellular nucleic acids. Most likely these recombination events took place between RNA molecules, since reverse transcription has not been shown for any flavivirus. Because of the extreme conservation of the cellular inserts, the recombination probably occurred recently.

Biological significance of the cellular insertions: a model

Two biotypes of BVDV can easily be distinguished in tissue culture: non-cytopathogenic BVDV (non-cp-BVDV) replicates without damaging its host cells, whereas cytopathogenic BVDV (cp-BVDV) is cytopathic for the target cell. As a molecular marker for cp-BVDV, expression of a protein of 80 kD (p80) has been reported (Purchio, Larson & Collett, 1984; Donis & Dubovi, 1987a,b). p80 is structurally and antigenically closely related to the above-mentioned viral protease p125 and therefore was believed to result from proteolytic processing of p125.

Both BVDV biotypes play a central role in development of mucosal disease (MD), the most severe disease resulting from BVDV infections. The first step in pathogenesis of MD is the infection of a pregnant cow with non-cp-BVDV, which results in birth of a persistently infected calf immunotolerant against the respective BVDV strain. Such animals can develop MD, usually at an age of 6 to 18 months. Interestingly, not only non-cp-BVDV but also cp-BVDV can be isolated from these animals (Brownlie, Clarke & Howard, 1984; Bolin *et al.*, 1985). With regard to one animal, the two viruses are called a 'BVDV pair'. Because of the extreme antigenic similarity between the two viruses of a BVDV pair, it has been proposed that the cytopathogenic virus is a mutant of the non-cp-BVDV strain (Fig. 7.4) (Pocock *et al.*, 1987; Corapi, Donis & Dubovi, 1988).

The presence of cellular insertions close to the predicted processing site of p125 in the genomes of cp-BVDV lets us propose the following working hypotheses: (i) One possible mutation changing non-cp-BVDV into cp-BVDV is a recombination between cellular and viral RNA which leads to integration of additional sequences into the viral p125 gene. (ii) The inserted cellular element within p125 is responsible for processing of the protein thereby allowing expression of p80; the presence of p80

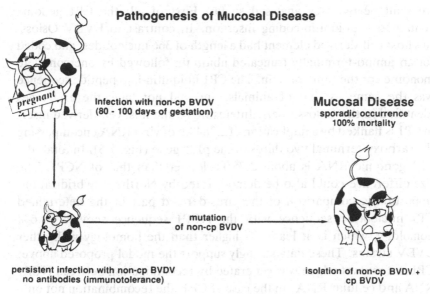

Pathogenesis of Mucosal Disease

Infection with non-cp BVDV
(80 - 100 days of gestation)

Mucosal Disease
sporadic occurrence
100% mortality

mutation
of non-cp BVDV

persistent infection with non-cp BVDV
no antibodies (immunotolerance)

isolation of non-cp BVDV +
cp BVDV

Fig. 7.4.

in infected cells correlates with the cytopathogenic phenotype. (iii) Fatal mucosal disease results from generation of such a cytopathogenic mutant in the persistently infected animal. Apparently, very many replication cycles are required to generate a cytopathogenic mutant.

Verification of the working hypotheses

Up to this point, the conclusions were based on a comparison between the genomes of two laboratory cp-BVDV strains and one hog cholera virus (HCV) strain; the analysis of a non-cp-BVDV strain was missing. It was therefore appropriate to investigate cp and non-cp virus from a defined BVDV pair. According to the hypotheses outlined above, one would expect a very close relationship between the cp and non-cp virus and a genomic change in the cytopathogenic isolate only. We therefore started cDNA cloning and sequencing of the genomes of BVDV NCP1 and BVDV CP1, a pair of viruses which had been isolated from one animal with MD. Like the HCV sequence the genome of the non-cytopathogenic virus NCP1 was found to contain no insert or other discontinuity in the p125 gene (Fig. 7.5). The cytopathogenic virus CP1, however, had a complicated genome organization different from all pestiviruses analysed so far. First of all, the CP1 genome contained a ubiquitin-coding insertion. In contrast to BVDV Osloss, this host cell-derived element had a length of 366 nucleotides, and coded for an amino-terminally truncated ubiquitin followed by one complete monomer of the same protein. The CP1 ubiquitin-like peptide sequence was the same as that of animals, and did not have the exchanges identified in the Osloss insert. Interestingly, the host cell-derived insert in CP1 is flanked by a duplication of 2.384 kb of viral RNA encompassing the carboxy-terminal two thirds of the p125 gene (Fig. 7.5). In total, the CP1 genomic RNA is about 2.75 kb longer than that of NCP1. The size difference could also be demonstrated by Northern hybridization. Importantly, comparison of the virus-derived part of the determined CP1 nucleotide sequence with the NCP1 sequence showed 99.6% homology, which is at least 9% higher than the homology with other BVDV strains. These data strongly support the model proposed above: CP1 is a mutant which was generated by recombination between NCP1 RNA and cellular RNA. In the case of CP1, the recombination not only resulted in integration of the cellular ubiquitin-coding sequence but also

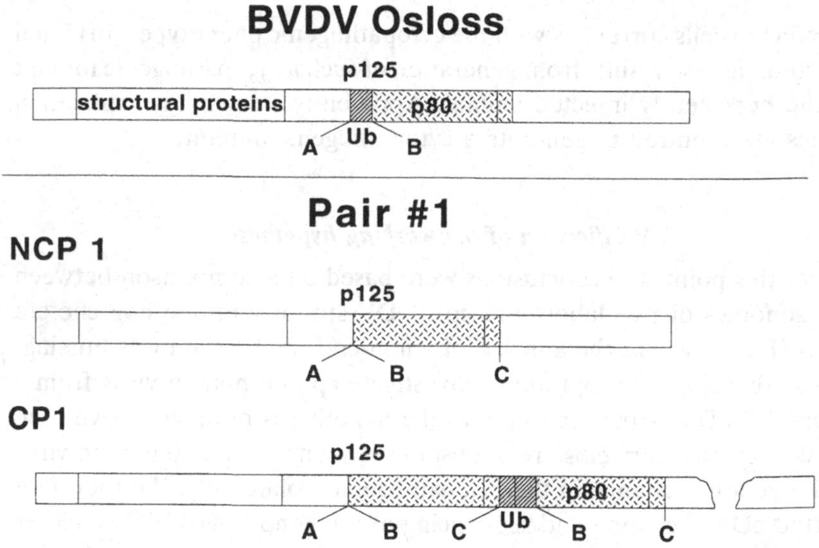

Fig. 7.5. Genome organization of BVDV strains Osloss, NCP1 and CP1. Ubiquitin-coding regions are denoted 'Ub' and the location of the sequences coding for p125/p80 is indicated. 'A' and 'B' mark the residues preceding and following the ubiquitin insertion in the BVDV Osloss genome. 'C' represents the last nucleotide of the region duplicated in the RNA from BVDV CP1.

in duplication of viral RNA. Protein analyses revealed that processing of p125 as observed for BVDV Osloss and NADL does not occur in cells infected with CP1. Instead, p80 is encoded by the duplicated sequence following the ubiquitin insertion in the CP1 RNA, whereas the genomic region preceding the cellular element codes for p125.

The possible recombination mechanism

Reverse transcription has not been reported for any flavivirus. Accordingly, the recombination resulting in integration of cellular sequences into the genomes of cp-BVDV probably happened between RNA molecules. RNA recombination has been described for different viruses, i.e. picornaviruses, coronaviruses and alphaviruses. A so-called 'copy choice' mechanism has been proposed as the molecular basis for such reactions. This involves a template switch of the viral RNA polymerase during replication. Most RNA recombinations observed so far represent homologous recombinations between two molecules of viral RNA and can be explained by a single template switch (Lai *et al.*, 1985; Kirkegaard & Baltimore, 1986; Makino *et al.*, 1986). For the integration of host cellular sequences into the BVDV genome, however, two switches are required. The cellular mRNAs involved in these reactions are of positive polarity. As the same orientation was identified for the insertions within the BVDV genomic RNA, the proposed template switches occurred during synthesis of the viral genomic negative strand. The finding of more than one ubiquitin gene monomer inserted into the BVDV CP1 genome strongly suggests a polyubiquitin-coding mRNA as cellular recombination partner.

Comparison of the sequences flanking the cellular inserts in the genomes of BVDV Osloss and CP1 revealed that at the 3' ends both ubiquitin-coding elements are located at exactly the same genomic position. This finding could indicate that the recombination is site specific.

The genomes of other cp-BVDV strains

Analysis of additional BVDV genomic RNAs by Northern hybridization and classical or PCR-based cDNA cloning revealed the presence of cellular sequences for a variety of cytopathogenic viruses but not for any non-cytopathogenic virus (Table 7.1). Most of the so far identified insertions code for ubiquitin (Table 7.1); the insertions of some strains

G. Meyers, N. Tautz and H.-J. Theil

Table 7.1. *Analysis of BVDV strains by Northern hybridization*

	Type of insertion	Number	Genome size	
I. Lab strains only cp BVDV				
Osloss	Ubiquitin	1	12.5 kb	
NADL	cINS	1	(as described for	
Others (Danmark, Singer, Oregon)	?	3	pestiviruses)	
II. BVDV Pairs			12.5 kb	>13.5 kb
A. cp BVDV	Ubiquitin	7	1	6
	cINS	1	1	–
	?	8	4	4
B. non-cp BVDV	Insertions so far not identified	16	16	–

Source: Meyers *et al.*, 1991 and unpublished observations. Qi *et al.*, 1992. De Moerlooze *et al.*, 1990

are flanked by duplications of viral RNA as described above for BVDV CP1 whereas others are similar to BVDV Osloss with the cellular sequence inserted into the p125 gene (Qi *et al.*, 1992, Tautz *et al.*, 1993). Remarkably, in all analysed isolates, the location of the 3' ends of the ubiquitin insertions is the same as found for Osloss and CP1. The reason for this is probably functional fitness as ubiquitin represents an almost universal intracellular protease cleavage signal suitable for generation of p80. Moreover, ubiquitin-coding mRNAs are abundant within eukaryotic cells.

After Northern hybridization, two more cp-BVDV strains have been found that contain the cins-insertion which was first identified in the BVDV NADL genome. Investigation of the size of these inserted elements, their homology to the cellular counterpart and their localization within the viral genome is in progress. The finding of more than one BVDV strain containing cins sequences is surprising if one considers the low copy number of the respective mRNAs, at least in tissue culture cells used for propagation. Again, functional aspects are probably responsible for the presence of this sequence in the viral genome. A very important question for future work will therefore be which function the cellular protein has, and how the truncated form found within the viral polyprotein is able to promote cleavage of p125 resulting in generation of p80.

For a variety of cp-BVDV strains, the genetic basis for the cytopathogenicity is still unknown. Out of 16 cp viruses, 8 belonging to BVDV pairs do not contain a ubiquitin-coding element or a cins insert. In 4 of these cases the viral genomic RNA is much larger than the one of the corresponding non-cp virus. It therefore seems likely that these viruses have also been generated by recombination. For two of these isolates duplication and rearrangement of only viral sequences was found (Meyers *et al.*, 1992) and one had a deletion (Tautz *et al.*, 1994). It might turn out that integration of cellular sequences is just one of several possible recombination processes which change non-cp-BVDV into cp-BVDV.

Cellular sequences in the genomes of two other viruses

For influenza viruses, proteolytic cleavage of the haemagglutinin glycoprotein (HA) by host cellular proteases is known to be essential for production of infectious virus. During adaptation of a non-pathogenic influenza virus to chicken cells which are not permissive for HA cleavage, Khatchikian, Orlich and Rott (1989) isolated a variant able to form plaques on non-permissive cells in the absence of trypsin. The respective virus was highly pathogenic for chicken. Protein analyses revealed that HA cleavage happened in all cell types tested.

For investigation of the genetic basis of the observed HA cleavage, the HA genes of both the original virus and the variant were sequenced (Khatchikian *et al.*, 1989). A major change, consisting of an insertion of 54 nucleotides, was identified. After sequence comparison studies, the inserted element was found to be complementary to a region within 28S ribosomal RNA of the host. Only one out of 54 nucleotides differed from part of a sequence complementary to human 28S rRNA sequence, which is probably very close to the avian gene sequence. As the inserted element is present within the HA mRNA in opposite orientation, template switching of the viral polymerase during positive strand synthesis was proposed as mechanism for the recombination reaction. In contrast to the BVDV system, the two recombining RNAs show sequence complementarity at the first cross-over position. The variant HA sequence contains a 14 nucleotide palindrome at the 5′ border of the insertion.

Within the HA protein, the inserted sequence is located only one amino acid upstream of the position where HA cleavage occurs in permissive cells. Insertion of 18 additional amino acids into this region of the protein may change the conformation of the cleavage site in a way which renders it susceptible to cellular proteases leading to HA processing in previously non-permissive cells. Accordingly, the generation of an infectious and highly pathogenic influenza virus during the adaptation process is most likely the consequence of a recombination between viral and cellular RNA. Whether the short insertions identified at the cleavage sites of naturally occurring pathogenic avian influenza viruses also result from such recombination reactions cannot be decided at the moment.

Guarino (1990) apparently detected a host cellular sequence in the genome of the insect virus *Autographa californica* nuclear polyhedrosis virus (AcMNPV), a member of the family Baculoviridae. Interestingly, the sequence codes for a ubiquitin-like protein, termed *v*-ubi. At the level of nucleotides, the *v*-ubi gene is 66% homologous to an insect cellular ubiquitin gene. The amino acid sequence is 76% identical to the animal ubiquitin sequence. Importantly, 10 out of 18 observed amino acid exchanges are conservative and functionally important residues have not mutated (Guarino, 1990). The clear homology suggests the host cellular origin of this baculoviral sequence. As the similarity is much less than that of the BVDV ubiquitin, the recombination integrating the cellular sequence into the baculovirus was probably a much more ancient event.

Determination of the nucleotide sequence upstream of the *v*-ubi gene revealed the presence of a typical baculovirus late promoter with two putative transcription start sites. In transcriptional analyses expression of *v*-ubi during the late phase of the viral life cycle could be demonstrated (Guarino, 1990). The function of *v*-ubi has not yet been elucidated. However, the similarity of the sequences of the baculovirus and animal ubiquitin suggest they may have similar functions. Besides its well-known role in intracellular proteolysis, cellular ubiquitin functions as a molecular chaperone in ribosome assembly (Finley, Bartel & Varshavsky, 1989). It might turn out that *v*-ubi has a similar function for the incorporation of viral proteins into virions. This hypothesis fits nicely with the expression of *v*-ubi during the late phase of viral replication.

Conclusions

Integration of cellular sequences into the genomes of viruses with either DNA or RNA genomes has to be regarded as important for the evolution of viruses. BVDV represents an especially interesting example for such a recombination, as the correlation between recombination and the outbreak of a naturally occurring fatal disease gives the opportunity to study a very specific evolutionary step almost immediately afterwards. As outlined above for influenza virus, and for baculovirus, there are other cases of clearly host cell-derived sequences in viral genomes. These two examples have been chosen since they are related to the BVDV system either by the functional aspect of protease cleavage site generation and pathogenicity change or by the integration of ubiquitin-coding sequences. Employment of the ubiquitin system for different steps in replication might turn out to be a common feature for a variety of viruses. In addition to the presence of genes coding for ubiquitin or ubiquitin-like proteins in the genomes of BVDV and baculovirus, respectively, recent publications report on the presence of ubiquitinated proteins in the virions of different viruses (Dunigan *et al.*, 1988; Hazelwood & Zaitlin, 1990). Moreover, the genome of African swine fever virus codes for a polypeptide homologous to ubiquitin-conjugating proteins, and this was shown to have the same enzymatic function (Hingamp *et al.*, 1992; Rodriguez, Sales & Vinuela, 1992). The ubiquitin system could represent another example of a cellular system which viruses not only learned to employ but found so advantageous that it was worthwhile to own part of it.

References

Bolin, S.R., McClurkin, A.W., Cutlip, R.C. & Coria, M.F. (1985). *Am. J. Vet. Res.*, **46**, 573–6.

Brownlie, J., Clarke, M.C. & Howard, C.J. (1984). *Vet. Rec.*, **114**, 535–536.

Collett, M.S., Larson, R., Gold, C., Strinck, D., Anderson, D.K. & Purchio, A.F. (1988). *Virology*, **165**, 191–9.

Collett, M.S. Wiskerchen, M.A., Welniak, E. & Belzer, S.K. (1991). *Arch. Virol. (Suppl. 3)*, 19–27.

Corapi, W.V., Donis, R.O. & Dubovi, E.J. (1988). *J. Virol.*, **62**, 2823–7.

De Moerlooze, L., Desport, M., Renard, A., Lecomte, C., Brownlie, J. & Martial, J. A. (1990). *Virology*, **177**, 812–15.

Donis, R.O. & Dubovi, E.J. (1987*a*). *J. Gen. Virol.*, **68**, 1597–605.

Donis, R.O. & Dubovi, E.J. (1987*b*). *J. Gen. Virol.*, **68**, 1607–16.

Dunigan, D., Dietzgen, R.G., Schoelz, J.E. & Zaitlin, M. (1988). *Virology*, **165**, 310–12.

Finley, D., Bartel, B. & Varshavsky, A. (1989). *Nature*, **338**, 394–401.

Guarino, L.A. (1990). *Proc. Natl. Acad. Sci. USA*, **87**, 409–13.

Hazelwood, D. & Zaitlin, M. (1990). *Virology*, **177**, 352–6.

Hingamp, P.M., Arnold, J.E., Mayer, R.J. & Dixon, L.K. (1992). *EMBO J.*, **11**(1), 361–6.

Khatchikian, D., Orlich, M. & Rott, R. (1989). *Nature*, **340**, 156–7.

Kirkegaard, K. & Baltimore, D. (1986). *Cell*, **47**, 433–43.

Lai, M.M.C., Baric, R.S., Makino, S., Deck, J.G., Egbert, J., Leibowitz, J.L. & Stohlmann, S.A. (1985). *J. Virol.*, **56**, 449–56.
Makino, S., Keck, J.G., Stohlmann, S.A. & Lai, M.M.C. (1986). *J. Virol.*, **57**, 729–37.
Meyers, G., Rümenapf, T. & Thiel, H.-J. (1989*a*). *Virology*, **171**, 555–67.
Meyers, G., Rümenapf, T. & Thiel, H.-J. (1989*b*). *Nature*, **341**, 491.
Meyers, G., Rümenapf, T. & Thiel, H.-J. (1990). In *New Aspects of Positive Strand RNA Viruses* (M.A. Brinton and F.X. Heinz, Eds.), American Society for Microbiology, Washington DC, pp. 25–29.
Meyers, G., Tautz, N., Dubovi, E.J. & Thiel. H.-J. (1991). *Virology*, **180**, 602–16.
Meyers, G., Tautz, N., Stark, R., Brownlie, J., Dubovi, E.J., Collett, M.S. & Thiel H.-J. (1992). *Virology*, **191**, 368–86.
Pocock, D.H., Howard, C.J., Clarke, M.C. & Brownlie, J. (1987). *Arch. Virol.*, **94**, 43–53.
Purchio, A.F., Larson, R. & Collett, M.S. (1984). *J. Virol.*, **50**, 666–9.
Qi, F., Ridpath, J.F., Lewis, T., Bolin, S.R. & Berry, E.S. (1992). *Virology*, **189**, 285–92.
Renard, A., Dino, D. & Martial, J. (1987). European Patent Application number 86870095. 6. Publication number 0208672, 14 January.
Rodriguez, J.M., Salas, M.L. & Vinuela, E. (1992). *Virology*, **186**, 40–52.
Tautz, N., Meyers, G. & Thiel, H.-J. (1993). *Virology*, **197**, 74–85.
Tautz, N., Thiel, H.-J., Dubovi, E.J. & Meyers, G. (1993). *J. Virol.*, **68**, 3289–97.
Thiel, H.-J., Stark, R., Weiland, E., Rumenapf, T. & Meyers, G. (1991). *J. Virol.*, **65**, 4705–12.
Wengler, G. (1991). Family Flaviviridae. In: *Classification and nomenclature of viruses*. Fifth report of the international committee on taxonomy of viruses Francki, R.I.B., Fauquet, C.M., Knudson, D.L. and Brown, F., (eds.). Springer-Verlag, Wien.
Wiskerchen, M.A., Belzer, S.K. & Collett, M.S. (1991). *J. Virol.*, **65**, 4508–14.
Wiskerchen, M.A. & Collett, M.S. (1991). *Virology*, **184**, 341–50.

Part III
Sources of virus variation

8
Molecular mechanisms of point mutations in RNA viruses

BERTHA-CECILIA RAMÍREZ, PASCALE
BARBIER, KARIN SÉRON, ANNE-LISE HAENNI
AND FRANÇOISE BERNARDI

Introduction

Analyses of populations of viruses with RNA genomes find a large amount of variability that results from re-assortment of genome segments of those viruses that have multipartite genomes, recombination events, point mutations and small deletions or insertions. The reverse transcriptases (RT) and the RNA-dependent RNA polymerases (RdRp) cause the last three types of change. These polymerases are error-prone owing to their intrinsic low level of fidelity and to their lack of correction mechanisms. In addition, host factors can also be responsible for certain mutations.

This chapter deals with point mutations; indeed, point mutations represent the main source of diversity occurring during replication of the genome of RNA viruses. It attempts to briefly survey 1) the genetic variation in virus populations, 2) the methods used to determine mutation rates and frequencies, 3) the estimates of such mutation rates and frequencies obtained *in vivo* and of error rates obtained *in vitro* for the most studied viruses, and 4) our present knowledge concerning the mechanisms of point mutations. It does not deal with recombination and re-assortment; the reader is referred to other chapters of this book for information on these topics.

Genetic variation in virus populations

Most RNA virus infections correspond to heterogeneous virus populations that can be referred to as quasi-species; this term describes complex distributions of replicating molecules subject to mutation and competitive selection (Domingo *et al.*, 1992). If RNA virus populations are maintained at constant selective pressures, genomic variability is high, but a stable distribution is reached that usually undergoes minimal changes (Domingo et al., 1978). When selective pressures change, rapid evolution of the viral population ensues by selection of the fittest (Rowlands et al., 1980), reaching another distribution.

Variation in virus populations has been documented in the following two situations.

1. It has been studied by comparing isolates sampled from different geographical locations or sampled in the course of time during epidemics or chronic infections, or by comparing cDNA clones derived from one isolate; the diversity observed in the latter case is referred to as microheterogeneity.

In this situation, mutation rates can be estimated from the comparison of isolates if the divergence between these isolates results from cumulative mutations over time. This is valid only in those cases where the co-ancestry of isolates can be established. For a certain number of viruses, such as influenza C virus and to some extent for influenza B virus (Yamashita et al., 1988), a phylogenetic analysis of isolates sampled over the years points to the existence of multiple co-circulating strains. In contrast, other viruses such as influenza A virus are characterized by a single evolutionary lineage, where isolates composed of a single dominant virus derive from one another.

Because of the rapid evolution of RNA genomes, recent work has aimed at characterizing the short-term change in RNA virus populations from a single individual during persistent infection. Extremely high rates of accumulation of mutations (3×10^{-2} to 3×10^{-3} per nucleotide per year) were observed in hepatitis delta virus (HDV) sampled over 27 months from a chronically infected patient (Lee et al., 1992). The various clones were obtained by PCR amplification after reverse transcription, and sequenced. From the sequence microheterogeneity observed, it seems clear that the three samples sequenced during the course of time are such that they could not have evolved from one another by the sole process of cumulative point mutations. Thus, genetic drift or selection of minor variants present in the original virus population is likely to have occurred, eliminating certain virus lineages. The values reported may thus represent an overestimation, since the original virus population may already have contained some of the mutations detected in the later samples.

Similarly, high rates of accumulation of mutations per site per year were found in influenza A virus (3.5×10^{-3} to 24×10^{-3} for the nucleoprotein gene, and 5.8×10^{-3} to 17×10^{-3} for the haemagglutinin gene) sampled eight times over a 10-month period in an immunodeficient patient (Rocha *et al.*, 1991). Although the choice of an immunodeficient patient theoretically eliminated the possibility of selection of virus variants by the immune system of the host, the samples sequenced did not derive from one another by progressive accumulation of mutations. Consequently, the values reported are also overestimates as in the case of the HDV discussed above. The microheterogeneity or quasi-species nature of some RNA virus populations thus renders estimation of mutation rates difficult, because of abrupt shifts in the equilibrium distribution of these populations.

Yet chronic infection with an RNA virus does not result necessarily in extensive genetic variation. Indeed, in lymphocytic choriomeningitis virus (Ahmed *et al.*, 1991) a U→C transition leading to a $Phe_{260} \rightarrow$ Leu change (predominantly in viruses isolated from spleens but not from the central nervous system) in the viral glycoprotein correlates with organ-specific selection.

Examples of high genetic stability have also been documented in two plant tobamoviruses, pepper mild mottle virus (PMMV) (Rodríguez-Cerezo, Moya & García-Arenal, 1989) and tobacco mild green mosaic virus (Rodríguez-Cerezo *et al.*, 1991), even during epidemic episodes in the case of PMMV.

Consequently, the pattern of variation observed in nature is the outcome of the mutational process, together with factors such as functional constraints and selection imposed by the host or by the vector.

2. Variation in virus populations has also been studied by comparing isolates of a viral clone recovered after replication in cell culture. In this situation, selection pressure is greatly reduced in so far as experimental conditions can be made to minimize competition between variants and selection of variants. Such conditions are reached when infection is initiated using a single plaque and is performed at low multiplicity of infection (m.o.i.). Estimates of the error rates of viral replicases have thus been obtained by investigating the variation in a virus population generated by one passage in cell culture.

Methods used to determine mutation rates and frequencies

The following methods have been used to estimate mutation rates and frequencies *in vivo* (methods 1–5; Table 8.1) or *in vitro* (methods 2, and 6–8; Table 8.2).

1. Direct sequencing of mutants (DS).

2. RNase T1 fingerprinting (T1). This method was devised by Steinhauer and Holland (1986) to quantify error frequencies during VSV replication. It permits the analysis of a region of the viral genome defined or tagged by hybridization with an oligodeoxyribonucleotide. It is limited to detecting mutations at the level of G residues; consequently, estimates must be extrapolated to the three other nucleotides. Sequencing is required to determine the nature of the nucleotide replacing the G residue.

3. Screening for monoclonal antibody (Mab) resistant mutants (Smith & Inglis, 1987). Using this method, large viral populations can be screened, and the variation in mutation rates between different viral genes or between virus clones can be investigated. However, a proportion of mutants defective in one viral function can be overlooked if they are competed out by the wild-type virus or other mutants, so that this method may tend to lead to an underestimate (Holland *et al.*, 1989).

4. Guanidine resistance (G) (Pincus *et al.*, 1986). In this method, guanidine-resistant variants are first selected. Altered proteins produced by these mutants are detected by electrofocussing and tryptic peptide fingerprinting. The RNA region encoding the altered protein is then sequenced to locate the site of mutation.

5. Denaturing gradient gel electrophoresis (DGGE) (Leider *et al.*, 1988). This method rests on the expected melting map of double-stranded RNAs (or double-stranded DNAs) based on the total nucleotide sequence of the viral genome. Heteroduplexes between wild-type and mutant molecules are thermodynamically less stable than perfect duplexes and hence will yield a different melting map. This is visualized by DGGE in which the concentration of denaturant varies linearly through the gel either perpendicularly or parallel to the direction of electrophoresis. In such gels, a shift in the migration profile of the duplex reflects the presence of a mutation in the RNA region examined. The sensitivity of the method is such that single-point mutations are accurately detected.

The following methods have been applied to both RNA and DNA templates; they have in common an *in vitro* primer elongation step. In methods 7 and 8, this step is followed by an *in vivo* screening or selection method.

6. Kinetic constants (V_{max}, K_m) can be determined for nucleotide misincorporation by a gel assay (GA) that makes it possible to quantify the amount of elongated primer in various conditions such as defined

nucleotide pools, and to calculate the relative misincorporation frequency (incorrect versus correct nucleotide). By this procedure, a detailed analysis of transitions and transversions can be obtained. With this method, it is also possible to determine the ability of the polymerase to elongate a 3′ mismatched primer (Boosalis *et al.*, 1987).

7. The rate of forward mutations (F) can be measured using the lacZ complementation test; after DNA synthesis of a well-defined region of lacZ carried by M13 and transformation in bacteria, plaques can be screened for mutants (Kunkel, 1985; Kunkel & Alexander, 1986). This method has been used to study error rates caused by RTs (Roberts, Bebenek & Kunkel, 1988; Roberts *et al.*, 1989).

8. The product synthesized on a single-stranded bacteriophage template can be introduced into bacteria to select for reversion (R) of a termination codon which allows survival of the phage (Weymouth & Loeb, 1978). This method has been adapted for RNA templates (Hübner *et al.*, 1992). In this *in vivo* selection procedure, results are biased, since viability of the phages is required.

Mutation rates and frequencies

Mutation rates are defined as the proportion of misincorporation events occurring during nucleic acid synthesis, expressed as substitutions per nucleotide per round of template copying, and mutation frequencies as the proportion of mutants in a population, and is expressed as substitutions per nucleotide (Domingo & Holland, 1992).

Mutation rates and frequencies in vivo

Table 8.1 summarizes the mutation rates and frequencies estimated *in vivo*. Specific points of this table are discussed below.

Direct sequencing without cDNA synthesis was used by Parvin *et al.* (1986) to estimate mutation rates in influenza A virus and poliovirus type 1 by comparing for each virus more than 100 viral clones over a 1000 nucleotide stretch. These viral clones were derived from one plaque obtained by low multiplicity passage (m.o.i. of 0.2). This study required sequencing of over 90 000 nucleotides; it yielded estimates for the mutation rate of influenza A virus of 0.15×10^{-4} (resulting in seven substitutions after five replication cycles) and of poliovirus of $<0.02 \times 10^{-4}$. These estimates turned out to be low

Table 8.1. In vivo *mutation rates and frequencies in RNA viruses*

Virus	Method	Value ×10⁴	Reference
Poliovirus type 1 (Mahoney strain)	DS	<0.02	Parvin *et al.*, 1986
Poliovirus type 1 (Mahoney strain)	G	2.1	de la Torre *et al.*, 1990
Poliovirus type 1 (Mahoney strain)	G	2–9	Holland *et al.*, 1990
Poliovirus type 1 (Mahoney strain)	T1	20–41	Ward & Flanegan, 1992
VSV (Indiana strain)	T1	2.2	Steinhauer & Holland, 1986
VSV (Indiana strain)	T1	1.1–18	Steinhauer *et al.*, 1989
VSV (Indiana strain)	Mab	1.1	Holland *et al.*, 1989
VSV (Indiana strain)	Mab	0.7–1.0	Holland *et al.*, 1990
Influenza A virus (WSN/33)	DS	0.15	Parvin *et al.*, 1986
Influenza A virus (Victoria strain)	Mab	0.06–0.3	Valcárcel & Ortín, 1989
Influenza A virus (Victoria strain)	Mab	0.02–0.42	Suárez *et al.*, 1992
RSV	DGGE	1.4	Leider *et al.*, 1988

Methods: T1, RNase T1 fingerprinting; G, guanidine-resistance; DS, direct sequencing; Mab, *mar* mutants; DGGE, denaturing gradient gel electrophoresis. *Abbreviations of viruses*: VSV, vesicular stomatitis virus; RSV, Rous sarcoma virus.

in the case of poliovirus compared to those obtained by other methods.

Recently, higher estimates (20×10^{-4} to 41×10^{-4}, at eight sites) were obtained by RNAse T1 oligonucleotide fingerprinting for poliovirus type 1 (Ward & Flanegan, 1992). In this report, variants were harvested in bulk, so that infectivity of individual viruses was not required. In such experiments, encapsidation is, however, mandatory, since the RNA genomes are extracted from virions. Interestingly, the estimates obtained were consistent over the eight genomic sites probed, irrespective of whether variable or constant regions were considered. In a VSV population issued from one virus clone, estimates of mutation frequencies during replication were 2.2×10^{-4} substitutions per base.

These estimates were consistent with an estimate of 7×10^{-4} obtained *in vitro* using the replicase purified from the virion (Table 8.2) (Steinhauer & Holland, 1986). Later estimates using the same RNAse T1 method, several clones of VSV, and focussing on several genomic locations, were in the same range of 1.1×10^{-4} to 18×10^{-4} (Steinhauer, de la Torre & Holland, 1989).

Estimates of the error frequency of viral replicases based on the number of mutants obtained by repeatedly plating virus clones on Mab- or replication inhibitor-containing plates, are valid only if the new resistant phenotypes result from single point mutations; this is checked by sequencing, but occasional second-site mutations cannot be avoided. For Mab resistance experiments, a good correlation between amino acid and/or nucleotide substitution rates and frequency of neutralization-resistant phenotypes can only be found if a panel of Mabs is used. A statistical determination of the mutation rate obtained from the number of resistant plaques appearing in a population of influenza A virus clones selected on Mab-containing plates yielded estimates of 0.02×10^{-4} to 0.42×10^{-4}, depending on the Mab, i.e. the region of the genome probed (Suárez, Valcárcel & Ortin, 1992). In earlier estimates, a wide range of values was reported; this is partly due to phenotypic mixing, i.e. encapsidation by a mixture of wild-type and mutant capsid proteins, that are or are not recognized by each particular Mab (Valcárcel and Ortín, 1989).

Estimates of 2.1×10^{-4} for poliovirus type 1 based on the number of revertants to guanidine resistance (de la Torre, Wimmer & Holland, 1990; Holland *et al.*, 1990) were far lower than those obtained by RNAse T1 fingerprinting (Ward & Flanegan, 1992). This is probably due not only to the difference in methods used, but also to differences in the region of the viral genome examined, and to the fact that the guanidine resistance method only scores for viable mutants.

Error rates in vitro

Polymerase fidelity estimates with purified enzymes are expressed as error rates per detectable nucleotide polymerized per round of synthesis (Eckert & Kunkel, 1991).

As opposed to RdRps that synthesize RNA exclusively from RNA templates, RTs synthesize DNA first from an RNA template, and then from the newly produced DNA template, yielding double-stranded DNA. It is at the level of the action of the RT that the majority

Table 8.2. In vitro *error rates for RNA virus polymerases*

Virus	Method	Value ×10⁴	Reference
AMV	R	1.1	Preston *et al.*, 1988
	R	0.4	Roberts *et al.*, 1988
	F	0.6	Roberts *et al.*, 1988
	R	0.6	Roberts *et al.*, 1989
	GA	4.9	Ricchetti & Buc, 1990
HIV-1	R	2.5	Preston *et al.*, 1988
	R	0.5	Roberts *et al.*, 1988
	F	5.7[a]	Roberts *et al.*, 1988
	GA	1.9	Ricchetti & Buc, 1990
MoMuLV	F	0.3	Roberts *et al.*, 1988
	GA	10	Ricchetti & Buc, 1990
VSV	T1	7	Steinhauer & Holland, 1986

Note: [a] average value recalculated from original data in corresponding reference. *Methods*: F, forward mutation; GA, gel assay; R, reversion. *Abbreviations of viruses*: AMV, avian myeloblastosis virus; HIV-1, human immunodeficiency virus type-1; MoMuLV, Moloney murine leukaemia virus. All other indications are as in Table 8.1.

of mutations occur in retroviruses (Coffin, 1992), rather than during subsequent replication steps involving the cellular DNA polymerase (which is 1×10^6 times more accurate) and RNA polymerase.

Polymerase error rate data are particularly abundant with regard to RTs which have been purified and are commercially available from different sources. On the other hand, the only RdRp for which error rates have been determined *in vitro* stems from VSV, a virus with a negative-stranded RNA genome. Indeed, since such viruses encapsidate their RdRp, a convenient source of the enzyme complex is available. On the contrary, viruses with a positive-stranded RNA genome do not encapsidate their RdRp, and studies on the error rates *in vitro* of such viruses have been hampered by the lack of active RdRp.

DNA has been the template of choice in most studies. No significant difference was found in the errors rates of HIV-1 and AMV RT when RNA or DNA was used as a template for *in vitro* replication estimated by the gel assay method (Yu & Goodman, 1992) or when HIV-1, HIV-2 and MoMuLV were similarly compared (Bakhanashvili & Hizi, 1992). On the other hand, differences in error rates attributable to the DNA or

RNA template used with the HIV-1 RT have been reported (Hübner *et al.*, 1992; Boyer, Bebenek & Kunkel, 1992).

Using the gel assay, the detailed analysis of transition and transversion frequencies at well defined sites on the DNA template shows that transitions occur at a higher frequency than transversions, and that the substitution pattern depends on the RT studied and on the template sequence (Weber & Grosse, 1989; Ricchetti & Buc, 1990; Yu & Goodman, 1992). An extreme case of such a situation occurs in the G→A mutations of HIV-1 within GpA dinucleotides (Vartanian *et al.*, 1991). Using the same gel assay, it has been shown that an important feature of RTs is their capacity to elongate primers with a 3' mismatch (Roberts *et al.*, 1988; Perrino *et al.*, 1989; Ricchetti & Buc, 1990; Yu & Goodman, 1992).

Table 8.2 summarizes the error rates observed *in vitro* using various purified RTs. It also indicates the corresponding results obtained with the VSV RdRp.

Molecular mechanisms of point mutation

A fundamental step in nucleic acid polymerization is pairing between the base in the template and the base to be incorporated into the growing chain by the polymerase. The difference in free energy $\Delta\Delta G$ between correct and incorrect base-pairs in water is of the order of 1 to 2 kcal/mol which would lead to a misincorporation frequency of 10^{-1} to 10^{-2}; these values are those observed in the case of non-enzymatic template copying (Orgel, 1992). Polymerases are able to further discriminate between correct and incorrect base-pairs (Loeb & Kunkel, 1982; Echols & Goodman, 1991; Kunkel, 1992). This ability is reflected in the K_m value of the polymerization kinetics which may increase by two to three orders of magnitude for incorrect base-pairs, bringing error rates down to $10^{-3} - 10^{-4}$. For a given mismatch, the K_m values differ from one polymerase to another resulting in the observed varieties of misincorporation displayed by the different polymerases (Ricchetti & Buc, 1990). Even though kinetic amplification is the basis for misincorporation, the mechanism of discrimination between correct and incorrect base-pairs is not known. One possible mechanism proposed by Petruska, Sowers and Goodman (1986) and Petruska *et al.* (1988) suggests that DNA polymerases can amplify the value of $\Delta\Delta G$ *in vitro* by exclusion of water at the binding cleft.

Incorrect base-pairing during replication can also result from the shift of the base from the normal configuration to rare tautomeric forms either in the template or in the incoming nucleotide, making mutations unavoidable. Major tautomeric forms lead mostly to transitions. Topal and Fresco (1976) suggested that rare tautomer mispairing would dictate transition frequencies at a level approximating 10^{-4} to 10^{-5} per site; this value would thus represent the lowest limit of fidelity. Transversions were estimated to be less frequent than transitions, because in addition to rare tautomeric forms, they also require anti→syn isomerization that can occur only at the level of the incoming nucleotide but not at the level of the template. Their frequencies would be 10 and 500 times lower than the corresponding transition frequencies for G and A residues, respectively.

Other elements may also contribute to misincorporation frequencies, such as velocity of polymerization (Ricchetti & Buc, 1990), processivity (Kunkel, 1992), nucleotide pools and metal ions (Holland, de la Torre & Steinhauer, 1992).

Other possible sources of errors arising during DNA synthesis come from template-primer strand misalignments that can lead to deletions or additions if slippage occurs, to frameshifts if misincorporation is followed by template-primer strand rearrangements, or to a single base substitution if, after slippage, correct nucleotide incorporation and template-primer strand rearrangements occur. The latter mechanism has been referred to as dislocation mutagenesis (Kunkel, 1992). Extensive base substitutions in RNA viruses are designated hypermutations, and when these are predominantly of one type (monotonous nucleotide substitutions) and are clustered in specific regions of the viral genomes, the term 'biased hypermutation' is used. In the course of studies of frequent passages of HIV-1 to various cell lines, a predominance of G→A hypermutations within GpA dinucleotides was observed (Vartanian *et al.*, 1991). The authors have proposed that dislocation mutagenesis might explain this biased hypermutation. Clustering of G→A substitutions in GpA dinucleotides has also been reported in a single spleen necrosis provirus (Pathak & Temin, 1990) and in clones of simian immunodeficiency virus (Johnson *et al.*, 1991).

A related yet distinct situation exists in the satellite virus of tobacco mosaic virus (STMV). Analyses of STMV clones (Kurath, Rey & Dodds, 1992, see also Chapter 26) reveal a predominance (18 of the 29 single base differences characterized) of G→A substitutions. As opposed to the examples provided by the animal viruses just mentioned, the

G→A transitions in the STMV population are essentially scattered over the entire genome. Moreover, because the G→A substitutions are not surrounded by common context features, another mechanism than dislocation mutagenesis has been proposed. If during RNA replication, a U rather than a C pairs with a G in the template strand (G:U base-pairs do occur at low frequencies) then during the following round of replication a common A:U base-pair would result, and as a consequence an A would be positioned where a G had existed in the original template RNA.

In the previous examples, a viral polymerase is involved in promoting hypermutation. A very different situation is encountered in measles virus (MV) (Cattaneo *et al.*, 1988) and in a defective interfering (DI) particle of VSV (O'Hara *et al.*, 1984). In MV grown in the cell line IP-3-Ca, there is a predominant and constant type of mutation in a specific region of the haemagglutination protein. Within a stretch of 333 nucleotides, a cluster of 20 A residues is mutated to G, resulting in 16 amino acid changes (Cattaneo *et al.*, 1989). In the case of VSV, one of the sequenced DI particles contains a cluster of 14 A residues that is mutated to G residues (O'Hara *et al.*, 1984). A possible model that could explain this biased hypermutation has been proposed (Bass *et al.*, 1989). It entails the participation of a host factor. A cellular unwindase activity can covalently alter adenosine residues contained in double-stranded RNAs into inosine. It has been proposed that this double-stranded cellular unwindase activity could be responsible for the A→G mutations observed in MV (Cattaneo *et al.*, 1989). According to the model proposed, in double-stranded RNA regions arising during viral RNA replication, the unwinding activity converts the adenosine residues into inosine residues that in the next round of RNA synthesis direct the incorporation of cytosine. Ultimately, an A→G mutation results in the viral RNA, and in the opposite positive strand, a U→C mutation ensues. It is conceivable that a similar mechanism could give rise to the A→G mutations of the VSV DI particle.

Conclusions

Replication of viral RNA genomes by RTs and RdRps results in high levels of mutations, in most cases substitutions but also small insertions and deletions, leading to the production of a heterogeneous viral population termed quasi-species. High mutation rates are, however, a prerequisite for the survival of viruses, since they constitute the basis of virus adaptation to the environment.

Even though mutation rates of the viral RdRps are in the same range for most viruses (as reflected also by error rates obtained *in vitro* using the VSV RdRp), with RTs showing the highest mutation and error rates, the level and distribution of mutations in the natural population are different from one virus to another, reflecting among others, selection for fitness or evasion from host immune response.

Acknowledgements

We are most grateful to Thomas A. Kunkel for his useful comments and constructive suggestions concerning the text, and to Esteban Domingo for providing us with manuscripts prior to publication. B.-C.R. and K.S. are indebted to the Ministère des Affaires Etrangères and to the Ministère de la Recherche et de la Technologie respectively, for fellowships. This study was supported in part by the Rockefeller Foundation, the Appel d'Offre: 'Virologie Fondamentale' of the Ministère de l'Education Nationale de la Jeunesse et des Sports, and the Ligue Nationale Française Contre le Cancer. The Institut Jacques Monod is an 'Institut Mixte, CNRS-Université Paris 7'.

References

Ahmed, R., Hahn, C.S., Somasundaram, T., Villarete, L., Matloubian, M. & Strauss, J.H. (1991). *J. Virol.*, **65**, 4242–7.

Bakhanashvili, M. & Hizi, A. (1992). *Biochemistry*, **31**, 9393–8.

Bass, B.L., Weintraub, H., Cattaneo, R. & Billeter, M.A. (1989). *Cell*, **56**, 331.

Boosalis, M.S., Petruska, J. & Goodman, M.F. (1987). *J. Biol. Chem.*, **262**, 14689–96.

Boyer, J.C., Bebenek, K. & Kunkel, T.A. (1992). *Proc. Natl. Acad. Sci. USA*, **89**, 6919–23.

Cattaneo, R., Schmid, A., Eschle, D., Baczko, K., ter Meulen, V. & Billeter, M.A. (1988). *Cell*, **55**, 255–65.

Cattaneo, R., Schmid, A., Spielhofer, P., Kaelin, K., Baczko, K., ter Meulen, V., Pardowitz, J., Flanagan, S., Rima, B.T., Udem, S.A. & Billeter, M.A. (1989). *Virology*, **173**, 415–25.

Coffin, J.M. (1992). *Curr. Top. Microbiol. Immunol.*, **176**, 143–64.

de la Torre, J.C., Wimmer, E. & Holland, J.J. (1990). *J. Virol.*, **64**, 664–71.

Domingo, E., Sabo, D., Taniguchi, T. & Weissmann, C. (1978). *Cell*, **13**, 735–44.

Domingo, E. & Holland, J.J. (1994). In *Evolutionary Biology of Viruses*. S.S. Morse (ed.). Raven Press, New York, 161–84.

Domingo, E., Escarmis, C., Martinez, M.A., Martinez-Salas, E. & Mateu, M.G. (1992). *Curr. Top. Microbiol. Immunol.*, **176**, 33–47.

Echols, H. & Goodman, M.F. (1991). *Ann. Rev. Biochem.*, **60**, 477–511.

Eckert, K.A. & Kunkel, T.A. (1991). *PCR Methods and Applications*, **1**, 17–24.

Holland, J.J., Domingo, E., de la Torre, J.C. & Steinhauer, D.A. (1990). *J. Virol.* **64**, 3960–62.
Holland, J.J., de la Torre, J.C., Steinhauer, D.A., Clarke, D., Duarte, E. & Domingo, E. (1989). *J. Virol.*, **63**, 5030–6.
Holland, J.J., de la Torre, J.C. & Steinhauer, D.A. (1992). *Curr. Top. Microbiol. Immunol.*, **176**, 1–20.
Hübner, A., Kruhoffer, M., Grosse, F. & Krauss, G. (1992). *J. Mol. Biol.*, **223**, 595–600.
Johnson, P.R., Hamm, T.E., Goldstein, S., Kitov, S. & Hirsch, V.M. (1991). *Virology*, **185**, 217–28.
Kunkel, T.A. (1985). *J. Biol. Chem.*, **260**, 5787–96.
Kunkel, T.A. (1992). *J. Biol. Chem.*, **267**, 18251–4.
Kunkel, T.A. & Alexander, P.S. (1986). *J. Biol. Chem.*, **261**, 160–6.
Kurath, G., Rey, M.E.C. & Dodds, J.A. (1992). *Virology*, **189**, 233–44.
Lee, C.-M., Bih, F.-Y., Chao, Y.-C., Govindarajan, S. & Lai, M.M.C. (1992). *Virology*, **188**, 265–73.
Leider, J.M., Palese, P. & Smith, F.I. (1988). *J. Virol.*, **62**, 3084–91.
Loeb, L.A. & Kunkel, T.A. (1982). *Ann. Rev. Biochem.*, **52**, 429–57.
O'Hara, P.J., Nichol, S.T., Horodyski, F.M. & Holland, J.J. (1984). *Cell*, **36**, 915–24.
Orgel, L.E. (1992). *Nature*, **358**, 203–9.
Parvin, J.D., Moscona, A., Pan, W.T., Leider, J.M. & Palese, P. (1986). *J. Virol.*, **59**, 377–83.
Pathak, V.K. & Temin, H.M. (1990). *Proc. Natl. Acad. Sci. USA*, **87**, 6019–23.
Perrino, F.W., Preston, B.D., Sandell, L.L. & Loeb, L.A. (1989). *Proc. Natl. Acad. Sci. USA*, **86**, 8343–7.
Petruska, J., Goodman, M.F., Boosalis, M.S., Sowers, L.C., Cheong, C. & Tinoco, I. (1988). *Proc. Natl. Acad. Sci. USA*, **85**, 6252–6.
Petruska, J., Sowers, L.C. & Goodman, M.F. (1986). *Proc. Natl. Acad. Sci. USA*, **83**, 1559–62.
Pincus, S.E., Diamond, D.C., Emini, E.A. & Wimmer, E. (1986). *J. Virol.*, **57**, 638–46.
Preston, B.D., Poiesz, B.J. & Loeb, L.A. (1988). *Science*, **242**, 1168–73.
Ricchetti, M. & Buc, H. (1990). *EMBO J.*, **9**, 1583–93.
Roberts, J.D., Bebenek, K. & Kunkel, T.A. (1988). *Science*, **242**, 1171–3.
Roberts, J.D., Preston, B.D., Johnston, L.A., Soni, A., Loeb, L.A. & Kunkel, T. A. (1989). *Mol. Cell. Biol.*, **9**, 469–76.
Rocha, E., Cox, N.J., Black, R.A., Harmon, M.W., Harrison, C.J. & Kendal, A. P. (1991). *J. Virol.*, **65**, 2340–50.
Rodríguez-Cerezo, E., Elena, S.F., Moya, A. & García-Arenal, F. (1991). *J. Mol. Evol.*, **32**, 328–32.
Rodríguez-Cerezo, E., Moya, A. & García-Arenal, F. (1989). *J. Virol.*, **63**, 2198–203.
Rowlands, D., Grabau, E., Spindler, K., Jones, C., Semler, B. & Holland, J. (1980). *Cell*, **19**, 871–80.
Smith, D.B. & Inglis, S.C. (1987). *J. Gen Virol.*, **68**, 2729–40.
Steinhauer, D.A., de la Torre, J.C. & Holland, J.J. (1989). *J. Virol.*, **63**, 2063–71.
Steinhauer, D.A. & Holland, J.J. (1986). *J. Virol.*, **57**, 219–28.
Suárez, P., Valcárcel, J. & Ortín, J. (1992). *J. Virol.*, **66**, 2491–4.
Topal, M.D. & Fresco, J.R. (1976). *Nature*, **263**, 285–9.

Valcárcel, J. & Ortín, J. (1989). *J. Virol*, **63**, 4107–9.
Vartanian, J.P., Meyerhans, A., Åsjö, B. & Wain-Hobson, S. (1991). *J. Virol.*, **65**, 1779–88.
Ward, C.D. & Flanegan, J.B. (1992). *J. Virol.*, **66**, 3784–3793.
Weber, J. & Grosse, F. (1989). *Nucl. Acids Res.*, **17**, 1379–93.
Weymouth, L.A. & Loeb, L.A. (1978). *Proc. Natl. Acad. Sci. USA*, **75**, 1924–8.
Yamashita, M., Krystal, M., Fitch, W.M. & Palese, P. (1988). *Virology*, **163**, 112–22.
Yu, H. & Goodman, M.F. (1992). *J. Biol. Chem.*, **267**, 10888–96.

9

Recombination and its evolutionary effect on viruses with RNA genomes

MICHAEL M.C. LAI

Sequence mutation is the most common mechanism for the evolution of RNA viruses. However, major changes of viral genotype often involve exchanges of RNA segments between viruses, as exemplified by influenza virus evolution. For viruses with non-segmented RNA genomes, genetic exchange was previously thought to be rare. In recent years, however, sequence analyses of viral RNA revealed that many RNA viruses have apparently evolved from other viruses by exchange or rearrangement of genome sequences. Furthermore, experimental systems have been established for several viruses, in which recombination could be demonstrated between two viral RNAs. Thus, genetic recombination appears to be more important in RNA virus evolution than was previously realized.

Mechanistically, RNA recombination is similar to the generation of defective-interfering (DI) RNA, as both involve polymerase jumping during RNA synthesis. Many RNA viruses generate DI RNA and, thus, can potentially undergo genetic recombination. However, recombination has not been commonly observed. Therefore, there may be genetic restriction on RNA recombination or selective disadvantage for recombinant viruses.

RNA recombination in experimental systems with animal viruses

Evidence for RNA recombination was first obtained in classical genetic studies involving polioviruses (family *Picornaviridae*). When cells were co-infected with virus resistant to the antibodies in horse or cattle serum and virus resistant to guanidine, viruses resistant to both antibody and guanidine could be isolated at a higher frequency than by spontaneous mutation, suggesting that recombination may have occurred (Hirst, 1962; Ledinko, 1963). Similar observations were made on foot-and-mouth disease virus (FMDV) (Pringle, 1965). However, attempts to demonstrate genetic recombination in other RNA viruses and phages failed (Pfefferkorn, 1977; Horiuchi, 1975). RNA recombination was eventually established as a *bona fide* genetic phenomenon by the biochemical analysis of possible recombinant FMDVs and the presence of sequences derived from two different viruses was demonstrated (King *et al.*, 1982).

Recombination of polioviruses and FMDV occurs not only between viruses of the same serotype (intratypic recombination), but also between viruses of different serotypes (intertypic recombination). Recombination frequency was less for intertypic recombination (Kirkegaard & Baltimore, 1986), suggesting that recombination depends on sequence homology between the two RNAs. The smallest amount of sequence homology required for recombination to occur is not known. For instance, polioviruses and coxsakieviruses are genetically compatible despite sequence differences, as artificially engineered recombinants between the two can replicate (Semler, Johnson & Tracy, 1986). However, natural recombinants between the two have not yet been reported. Recombination in picornaviruses is strictly homologous, i.e. cross-overs occur at homologous sites between the two parental RNAs. The recombination frequency of picornaviruses has been estimated to range from 2% (Cooper, 1968, 1977) to 10–20% for the entire genome (King, 1988). Recombination can occur virtually anywhere in the genome; however, there are recombination hot spots, which correspond to certain secondary structures in the genomic RNA (Romanova *et al.*, 1986; Tolskaya *et al.*, 1987).

The second virus family, species of which have been shown to recombine, is the *Coronaviridae*. Like picornaviruses, murine coronavirus (mouse hepatitis virus, MHV) very readily undergoes homologous recombination in tissue culture. The recombination frequency has been estimated to be as much as 25% for the entire genome (Baric *et al.*, 1990). This high recombination frequency reflects the unusually large size (31 kb) of the coronavirus RNA genome. There are apparent recombination 'hot spots' (Banner, Keck & Lai, 1990). However, in a study designed to examine the mechanism of recombination as it occurs in the cells, without resorting to selection of recombinant viruses, it was shown that recombination occurred randomly between two selection markers; only after subsequent virus passages did recombination 'hot spots' appear (Banner & Lai, 1991). Thus, recombination 'hot spots' probably result from selection of certain types of recombinants. This finding, plus the observation that recombinants became the predominant virus species under certain culture conditions even in the absence of specific selection pressure (Makino *et al.*, 1986), suggests that recombination is an evolutionary strategy of coronaviruses. Recombination also occurs in animal infections (Keck *et al.*, 1988). In addition, homologous and non-homologous recombinations have been shown to occur between DI RNA and helper viral RNA during the experimental and natural evolution of DI RNA (Van der Most *et al.*, 1992; Furuya *et al.*, 1993). Recombination also could occur between viral RNAs and non-replicating viral RNA fragments (Koetzner *et al.*, 1992; Liao & Lai, 1992).

Viruses of a third animal virus family, the *Togaviridae*, have been shown to undergo RNA recombination. Recombination occurred between two defective Sindbis alphaviruses when cotransfected into cells (Weiss & Schlesinger, 1991). Surprisingly, most of the recombinations occurred between two non-homologous sites in the parental RNAs, even though the two RNAs shared long stretches of homologous sequence. A large number of gene rearrangements also accompanied these recombination events, termed 'aberrant homologous recombination' (Lai, 1992). It is very likely that multiple rounds of recombination and subsequent selection were involved to yield these recombinant RNAs. This has been the only type of recombination observed for togaviruses; and several classical genetic studies aimed at detecting homologous recombinants between these viruses have failed (Burge & Pfefferkorn, 1966; Atkins *et al.*, 1974).

A related study is relevant to the discussion of RNA recombination in virus evolution. Sindbis virus has a monopartite RNA genome. It was demonstrated that this RNA could be divided into two segments, to each of which was added a replication signal, and co-transfection of the two RNAs led to the production of infectious virus particles with a bipartite genome (Geigenmüller-Gnirke *et al.*, 1991). This study suggested the potential of RNA recombination in the evolution of RNA viruses from a segmented RNA virus to a non-segmented RNA virus, or vice versa.

A similar type of recombination occurred between influenza virus genome segments, when they were cotransfected (Bergman, García-Sastre & Palese, 1992). The recombinants obtained, like those of Sindbis virus, had multiple gene rearrangements, and the cross-over sites were not at homologous sites of the two parental genomes. The biological relevance of this type of recombination is not clear. However, an influenza virus DI RNA has been shown to consist of sequences derived by recombination (or polymerase jumping) between different

RNA segments (Fields & Winter, 1982). This was the first report of recombination for a negative-stranded RNA virus.

Recently, flock house nodavirus has also been shown to recombine non-homologously (Li & Ball, 1993).

RNA recombination in experimental systems with plant viruses

The first plant virus shown to undergo RNA recombination was brome mosaic virus (BMV), a bromovirus. This virus has a genome comprising three RNA segments that have the same sequences at the 3'-ends. When a defective RNA segment of BMV was cointroduced with the other two RNAs into plants or protoplasts, wild-type RNA was generated by recombination in the shared sequence between these RNA segments (Bujarski & Kaesberg, 1986; Rao, Sullivan & Hall, 1990). Homologous recombination was the rule; however, in some studies, most of the recombinants had resulted from cross-overs at nonhomologous sites and had extensive sequence rearrangements (Bujarski & Dzianott, 1991). Sequence analysis of recombinants and their parental RNAs suggests that recombination took place at sites where potential hybrid formation between two parental RNAs could occur (Bujarski & Dzianott, 1991; Nagy & Bujarski, 1992). Another bromovirus, cowpea chlorotic mottle virus (CCMV), also can recombine in a manner similar to this (Allison, Thompson & Ahlquist, 1990).

Turnip crinkle virus (TCV), a carmovirus, also recombines. This virus contains several satellite RNAs, some of which (e.g. satellite RNA C) have sequence elements derived from both the helper virus RNA and from another satellite RNA (Simon & Howell, 1986), suggesting that they were derived by recombination during natural viral infection. When a defective satellite RNA was transfected with a helper virus RNA into protoplasts, recombination occurred between the defective RNA and another satellite RNA in the virus (Cascone *et al.*, 1990). Recombination took place in the homologous region between both RNAs, but not at precisely the homologous sites, resulting in recombinant RNAs containing redundant nucleotides (Cascone *et al.*, 1990). The sequence requirement for this recombination has been determined (Cascone, Haydar & Simon, 1993); the presence of a specific stem-and-loop structure at the acceptor RNA site is required for recombination. In some recombinant RNAs, the cross-over sites appear to correspond to the replicase recognition sequence (Cascone *et al.*, 1990).

Another plant RNA virus shown to recombine is alfalfa mosaic virus (AMV). Sequence analysis of some AMV RNAs suggests that they were derived by RNA recombination in nature (Huisman *et al.*, 1989). This virus has a tripartite genome. When one of its segments was introduced into transgenic plants expressing the other two RNA segments, recombination occurred between these RNA segments (Van der Kuyl, Neeleman & Bol, 1991).

Recently tomato bushy stunt and cucumber necrosis tombusviruses have been shown to recombine *in planta* (White & Morris, 1994).

Evidence of recombination in nature

Animal viruses

Several animal viruses have been demonstrated to undergo recombination during natural viral infections. First of all, poliovirus vaccine strains frequently recombine between each other in the gastrointestinal tracts of vaccinees (Kew & Nottay, 1984; Minor *et al.*, 1986). A child who had received oral poliovirus vaccines yielded recombinant polioviruses, particularly intertypic recombinants between type 2 and 3 vaccine strains, in stools as early as eight days after vaccination and new variants were found thereafter (Minor *et al.*, 1986). These recombinants replaced the original type 3 virus as the dominant virus strain, suggesting that the recombinants had selective advantages. Recombinant polioviruses have also been isolated during outbreaks of poliomyelitis (Rico-Hesse *et al.*, 1987; Kew *et al.*, 1990). Secondly, many field isolates of avian infectious bronchitis virus (IBV), a coronavirus, are likely to be recombinants between different IBV strains, as different parts of their spike protein genes and other genomic regions have different degrees of homology to different wild-type and vaccine strains (Kusters *et al.*, 1989; Cavanagh & Davis, 1988; Cavanagh *et al.*, 1992). Thirdly, Western equine encephalitis virus (WEEV) is homologous to Eastern equine encephalitis virus (EEEV), a New World monkey alphavirus; however, the 3' end of its genomic sequence is more closely related to that of Sindbis virus, an Old World monkey virus (Hahn *et al.*, 1988) than to that of EEEV. Thus, WEEV is probably a recombinant of Sindbis virus and EEEV or their ancestors. Recombination, therefore, is a way in which coronaviruses and alphaviruses evolve, although no homologous recombination has been detected among alphaviruses in tissue culture.

Recombination between unrelated viruses also has contributed to the natural evolution of viruses. Several coronaviruses contain a haemagglutinin-esterase gene, which is evolutionarily very close to that of influenza C virus (Luytjes *et al.*, 1988). This gene might have been derived by non-homologous recombination between coronaviruses and influenza C virus, as these two viruses are not closely related (one has a positive-strand RNA genome and the other a negative-strand RNA). As this gene is only present in some coronaviruses (Lai, 1990), this recombination probably occurred fairly recently in virus evolution.

Plant viruses

There are several documented examples of RNA recombination in the natural evolution of plant viruses, as suggested from the sequence analyses of viral RNAs. For example, there has been recombination between different virus strains of tobacco rattle virus, which is a tobravirus (Robinson *et al.*, 1987; Angenent *et al.*, 1989; Goulden *et al.*, 1991). Some of the RNA segments of BMV contain sequences related to CCMV, suggesting the occurrence of recombination between bromoviruses (Allison *et al.*, 1990). Also, recombination has clearly contributed to the generation of satellite RNAs of turnip crinkle virus (Simon & Howell, 1986) and some alfalfa mosaic virus RNA segments (Huisman *et al.*, 1989).

Recombination in the speciation of RNA viruses

Many animal and plant viruses contain structural elements and functional domains in their gene products which have remote sequence homology to each other, suggesting that they were evolutionarily related, and yet their genome structures are significantly different. For instance, brome mosaic virus, tobacco mosaic virus (TMV) and Sindbis virus have sequence domains in the non-structural protein genes which are related to each other, but BMV has three RNA segments whereas TMV has only one (Goldbach, 1987). Sindbis virus also has a single genomic RNA, which is, however, significantly bigger than that of TMV and contains several additional genes. Thus, during the process of evolution of these viruses, RNA recombination and rearrangement must have occurred. A similar genetic relationship can be seen between enterovirus, potyvirus, comovirus and nepovirus although they have different genomic structures and different host ranges (Strauss, Strauss & Levine, 1990). Thus, a high degree of RNA recombination and rearrangement has occurred during the evolution of plant and animal viruses.

Another type of RNA recombination is that which results in gene rearrangements, deletions or duplications that are found in some of the viral genomes. For example, coronaviruses of different animal species have significant variations in their gene arrangement: the gene encoding the viral matrix protein occurs in different parts of the genome in different coronaviruses (Lai, 1990). This could be the result of homologous intra- or intermolecular recombination, as the regions flanking each gene of coronavirus RNA share consensus intergenic sequences; thus, each gene can be viewed as a gene cassette, which is moved by recombination between the consensus intergenic sequences. This type of gene rearrangement has also been observed in the evolution of toroviruses (e.g. Berne virus) from coronaviruses, or vice versa (Snijder *et al.*, 1991). Deletions of genes in viral RNA genomes have been noted in several plant RNA viruses during the course of evolution (Bouzoubaa *et al.*, 1991; Hilllman, Carrington & Morris, 1987; Shirako & Brakke, 1984). Gene deletion can be considered to be the result of non-homologous recombination.

Non-homologous recombination between viral and cellular RNAs

Although most of the genetic recombination seen in RNA viruses has involved homologous or aberrant homologous recombination between related viral RNAs (Lai, 1992), recombination between unrelated RNA species, particularly between viral and cellular RNAs, occasionally occurs. The most striking example is bovine viral diarrhea virus (BVDV), a pestivirus. Many of the cytopathogenic strains of BVDV, which cause mucosal disease in cows, have various cellular sequences inserted in the non-structural protein-coding region of the genome (Collett *et al.*, 1988). Surprisingly, most of these cellular sequences are ubiquitin genes (Meyers *et al.*, 1991; Chapter 7). The length of the ubiquitin gene inserted and its site of insertion are variable, but often in the same region of the genome. Although no homology is apparent in the viral and ubiquitin sequence, this is clearly a favoured recombination event. However, other cellular sequences of unknown nature have also been detected in some recombinant BVDV strains (Meyers *et al.*, 1991; Collett *et al.*, 1988). The derivation of ubiquitin sequences from cellular sequences has been unequivocally established, as cytopathogenic strains have emerged from non-cytopathogenic strains during viral infections in cows (Meyers *et al.*, 1991; Corapi, Donis & Dubovi, 1988). This recombination invariably leads to increased cytopathogenicity of the viruses, the mechanism of which is not clear. Non-homologous recombination has also been reported to occur between the haemagglutinin gene of an influenza virus and 28S ribosomal RNA (Khatchikian, Orlich & Rott, 1989). This recombinant was more pathogenic than the wild-type virus. Other examples of recombination between viral and cellular RNAs have been found in an isolate of potato leafroll virus, which had a sequence of approximately 100 nucleotides similar to the sequence of the tobacco chloroplast gene at the 5'-end of its viral RNA (Mayo & Jolly, 1991), beet yellows virus, which has heat shock protein-like sequences in its RNA genome (Agranovsky *et al.*, 1991), and a poliovirus mutant which had part of the 28S-ribosomal RNA gene in its 3C gene (Charini *et al.*, 1994). In addition, many functional domains of viral gene products, such as proteases, helicases, polymerases and nucleotide-binding proteins of RNA viruses have different degrees of sequence similarity to the corresponding sequences of cellular proteins (Chapter 4). Although the evolutionary convergence of sequences cannot be rigorously ruled out, the most plausible explanation is the occurrence of recombination between viral and cellular RNAs during early viral evolution. Thus, non-homologous recombination is clearly involved in the natural evolution of viruses.

The selection of recombinant RNA viruses

Recombination clearly plays a role in the evolution of RNA viruses. However, recombination has been detected only in a few viruses so far. As RNA recombination occurs by a copy-choice mechanism (Kirkegaard & Baltimore, 1986; Lai, 1992), the ability of RNA viruses to recombine probably reflects the ability of viral RNA polymerase to dissociate from the template RNA, carry the nascent RNA transcripts to a different RNA template and continue transcriptional elongation; thus, the ability of a virus to recombine probably is inversely proportional to the processivity of its own RNA polymerase. As most RNA viruses can generate DI RNAs, which are derived by polymerase jumping, similar to the mechanism of recombination, it can be assumed that recombination should occur in most RNA viruses. The failure to detect recombination, particularly homologous recombination, in many RNA viruses suggests that recombination may impose selective disadvantages under some circumstances, while providing advantages under other conditions. This possibility is congruent with the fact that recombinant viruses rapidly dominate virus populations under some culture conditions (Makino *et al.*, 1986), but that recombinant RNAs in the virions are less heterogenous than those in the cells, and recombinant viruses became more homogenous after serial passage (Banner *et al.*, 1990; Banner & Lai, 1991).

Selective advantages of RNA recombination

Substituted proof-reading of the polymerase errors

RNA viruses in general have very high mutation rates, as a result of high error frequencies of RNA polymerases (Steinhauer, de la Torre & Holland, 1989; Ward & Flanegan, 1992; Ward, Stokes & Flanegan, 1988). These errors are not normally repaired as RNA polymerases, in general, do not have a proof-reading activity. Genetic complementation partially ameliorates this problem. However, some viral gene products cannot be complemented because they act *in cis*. Furthermore, to ensure the long-term genetic stability of the RNA genomes, those genetic errors have to be corrected. Re-assortment of RNA segments can bypass the need for proof-reading by generating functional RNA molecules from a pool of RNA segments. RNA recombination probably serves the same purpose. This concept is consistent with the finding that a coronavirus has the largest RNA genome and the highest recombination frequency, as this RNA is expected to accumulate mutations at a faster rate if one assumes that viral RNA polymerases have the same error frequencies; thus, the evolutionary pressure to recombine will be greatest for coronaviruses. However, recombination does not, *a priori*, generate functional RNA molecules free of polymerase errors, as the error sites do not necessarily cause recombination; nevertheless, recombination can generate diverse RNA molecules from which a functional RNA is selected. Given the proper selection pressure provided by nature, the RNA molecules with best fitness should emerge.

Escape from the 'Muller's ratchet'

During virus evolution, when the effects of mutation are, on average, deleterious and the mutation rate is high, the mutation-free individuals may become rare and may be lost by genetic drift (Muller, 1964; Felsenstein, 1974). This loss may become irreversible if no genetic recombination or re-assortment takes place. This mutational effect, called 'Muller's ratchet', which results in the gradual decline of the fitness of an entire population, has been demonstrated to operate in some RNA virus populations (Chao, 1990). The backward and compensatory mutations alone, without recombination or re-assortment, cannot reverse Muller's ratchet. Genetic exchange by RNA recombination or re-assortment thus provides an evolutionary advantage for RNA viruses by enabling them to overcome the effect of Muller's ratchet.

Rapid evolution of viruses in response to environmental pressure

Reassortment of segmented RNA genomes clearly demonstrates the power of genetic exchange in the evolution of viruses. For instance, re-assortment of influenza virus genomic segments allows the virus to change drastically its antigenic properties and escape the immune response of the host. This could not be achieved by single nucleotide mutations. This evolutionary ability also applies to RNA recombination. In poliovirus vaccinees, recombinants between strains in vaccines emerge and are rapidly selected in the gastrointestinal tract after vaccination (Minor et al., 1986). This selection may be in response to immune pressure or to the growth conditions in cells lining the gastrointestinal tract. Similarly, many field isolates of avian infectious bronchitis virus appear to be recombinants between different IBV strains (Kusters et al., 1989). As most of these recombinations occur within the spike protein gene, which encodes the immunodominant protein of the virus, recombination probably reflects immune selection.

Repair of a deletion or insertion which imparts growth disadvantage to the virus

Deletions or insertions of RNA sequences, which result in the disruption of viral gene functions, cannot be repaired by simple mutations; however, they can conceivably be removed by exchange with the wild-type sequences via RNA recombination. If the recombinant RNA has selective advantages, it will be rapidly selected. This has been clearly demonstrated experimentally in the recombination of defective-interfering RNA of several animal viruses, such as coronaviruses (Van der Most et al., 1992), and plant viruses, such as BMV (Bujarski & Kaesberg, 1986) or TCV (Cascone et al., 1990), with variants that have deletions in their genomes. More recently, it has also been shown that a deletion in the Qβ phage genome could be 'reversed' by recombination with a separate mRNA (Palasingam & Shaklee, 1992).

Alteration of the biological properties of the virus

The recombination of bovine viral diarrhea virus with cellular RNA, particularly from the ubiquitin gene, is correlated with increased pathogenicity of the virus (Meyers *et al.*, 1991), as was the recombination between influenza virus RNA and ribosomal RNA (Khatchikian *et al.*, 1989). These properties may aid the spread of viruses.

Expansion of the viral host range and generation of new virus species

As demonstrated by the sequence relationship of various animal and plant viruses, recombination has played a role in the evolution of viruses crossing the boundaries between various animal and plant species. The emergence of western equine encephalitis virus by recombination between Sindbis virus and eastern equine encephalitis virus is an example of natural virus evolution by recombination.

Selective disadvantages of RNA recombination

Recombination has only been found in nature in a limited set of viruses, although the ability to recombine has been demonstrated in a larger number of RNA viruses experimentally. Recombination may be restricted in nature by selection against the recombinants. For instance, although polioviruses recombine frequently in poliovirus vaccinees, and recombinant virus strains can outgrow the parental poliovirus vaccine strains (Minor *et al.*, 1986), the recombinant viruses do not replace the original three serotypes of polioviruses in nature. Thus, recombinant viruses seem to have some selective advantages but only under certain conditions. It is most likely that recombinants have selective disadvantages resulting from structural or functional incompatibility of their proteins. Various viral proteins may require perfect sequence and structural compatibility to interact with each other for optimum virus growth, whereas recombinant viruses may have RNA sequences or proteins whose slightly different structures prevent their proper interaction and thus reduce the fitness of the viruses. Thus, during the generation and evolution of WEEV, which was derived by recombination between Sindbis virus and EEEV, the structural proteins of this virus evolved so that all now resemble those of Sindbis virus (Hahn *et al.*, 1988), and are structurally or functionally compatible. It is likely that viruses have stringent sequence and structural requirements so that most recombinants are less fit than their parents. In a study on recombination of coronaviruses, it was shown that recombination initially occurred randomly; however, after several passages, the range of recombinants became limited (Banner & Lai, 1991). Another study on brome mosaic virus recombination also indicated that only some recombinants survive (Nagy & Bujarski, 1992). Thus, the viruses in which no recombinants have been detected may have very inflexible requirements, and recombinants may be unable to survive.

Epilogue

Genetic recombination clearly plays a role in the evolution of RNA viruses. The extent to which RNA recombination occurs in viruses and the conditions which favour the selection of recombinants remain unclear. The potential for recombination appears to be innate to most RNA viruses; with the sequencing of an increasing number of these viruses, additional evidence of genetic recombination is likely to emerge. Recombination is also an important issue to consider in the use of live attenuated virus vaccines. Recombination between vaccine strains and other viruses may yield viruses with unexpected properties. These issues will require careful future study.

References

Agranovsky, A.A., Boyko, V.P., Karasev, A.V., Koonin, E.V. & Dolja, V.V. (1991). *J. Mol. Biol.*, **217**, 603–10.

Allison, R.F., Thompson, C. & Ahlquist, P. (1990). *Proc. Natl. Acad. Sci. USA*, **87**, 1820–4.

Angenent, G.C., Posthumus, E., Brederode, F.T. & Bol, J.F. (1989). *Virology*, **171**, 271–4.

Atkins, G.J., Johnston, M.D., Westmacott, L.M. & Burke, D.C. (1974). *J. Gen. Virol.*, **25**, 381–90.

Banner, L.R., Keck, J.G. & Lai, M.M.C. (1990). *Virology*, **175**, 548–55.

Banner, L.R. & Lai, M.M.C. (1991). *Virology*, **185**, 441–5.

Baric, R.S., Fu, K., Schaad, M.C. & Stohlman, S.A. (1990). *Virology*, **177**, 646–56.

Bergmann, M., García-Sastre, A. & Palese, P. (1992). *J. Virol.*, **66**, 7576–80.

Bouzoubaa, S., Niesbach-Klosgen, U., Jupin, I, Guilley, H., Richards, K. & Jonard, G. (1991). *J. Gen Virol.*, **72**, 259–266.

Bujarski, J.J. & Dzianott, A.M. (1991). *J. Virol.*, **65**, 4153–9.

Bujarski, J.J. & Kaesberg, P. (1986). *Nature*, **321**, 528–31.

Burge, B.W. & Pfefferkorn, E.R. (1966). *Virology*, **30**, 214–23.

Cascone, P.J., Carpenter, C.D., Li, X.H. & Simon, A.E. (1990). *EMBO J.*, **9**, 1709–15.

Cascone, P.J., Haydar, T.F. & Simon, A.E. (1993). *Science*, **260**, 801–5.

Cavanagh, D. & Davis, P.J. (1988). *J. Gen. Virol.*, **69**, 621–9.

Cavanagh, D., Davis, P.J. & Cook, J.K.A. (1992). *Avian Pathol.* **21**, 401–8.

Chao, L. (1990). *Nature*, **348**, 454–5.
Charim, W.A., Todd, S., Gutman, G.A. & Semler, B.L. (1994). *J. Virol.* **68**, 6547–52.
Collett, M.S., Larson, R., Gold, C., Strick, D., Anderson, D.K. & Purchio, A.F. (1988). *Virology*, **165**, 191–9.
Cooper, P.D. (1968). *Virology*, **35**, 584–96.
Cooper, P.D. (1977). In *Comprehensive Virology*, H. Fraenkel-Conrat and R.R. Wagner (eds). Plenum Press, New York, Volume 9, pp. 133–208.
Corapi, W.V., Donis, R.O. & Dubovi, E.J. (1988). *J. Virol.*, **62**, 2823–7.
Felsenstein, J. (1974). *Genetics*, **78**, 737–56.
Fields, S. & Winter, G. (1982). *Cell*, **28**, 303–13.
Furuya, T., MacNaughton, T.B., La Monica, N. & Lai, M.M.C. (1993). *Virology*, **194**, 408–13.
Geigenmüller-Gnirke, U., Weiss, B., Wright, R. & Schlesinger, S. (1991). *Proc. Natl. Acad. Sci. USA*, **88**, 3253–7.
Goldbach, R.W. (1987). *Microbiol. Sci.*, **4**, 197–202.
Goulden, M.G., Lomonossoff, G.P., Wood, K.R. & Davies, J.W. (1991). *J. Gen. Virol.*, **72**, 1751–4.
Hahn, C.S., Lustig, S., Strauss, E.G. & Strauss, J.H. (1988). *Proc. Natl. Acad. Sci. USA*, **85**, 5997–6001.
Hillman, B.I., Carrington, J.C. & Morris, T.J. (1987). *Cell*, **51**, 427–33.
Hirst, G.K. (1962). *Cold Spring Harbor Symp. Quant. Biol.*, **27**, 303–9.
Horiuchi, K. (1975). In *RNA Phages*, N. Zinder (ed). Cold Spring Harbor Laboratory, Cold Spring Harbor, NY, pp. 29–50.
Huisman, M.J., Cornelissen. B.J.C., Groenendijk, C.F.M., Bol, J.F. & van Vloten-Doting, L. (1989). *Virology*, **171**, 409–16.
Keck, J.G., Matsushima, G.K., Makino, S., Fleming, J.O., Vannier, D.M., Stohlman, S.A. & Lai, M.M.C. (1988). *J. Virol.*, **62**, 1810–13.
Kew, O.M. & Nottay, B.K. (1984). In *Modern Approaches to Vaccines: Molecular and Chemical Basis of Virus Virulence and Immunogenicity*, R.M. Chanock and R. A. Lerner (eds). Cold Spring Harbor Laboratory, Cold Spring Harbor, NY, pp. 357–62.
Kew, O.M., Pallansch, M.A., Nottay, B.K., Rico-Hesse, R., De, L. & Yang, C.-F. (1990). In *New Concepts of Positive-Strand RNA Viruses*, M.A. Brinton and F.X. Heinz (eds). American Society of Microbiology, Washington, DC, pp. 357–65.
Khatchikian, D., Orlich, M. & Rott, R. (1989). *Nature*, **340**, 156–7.
King, A.M.Q. (1988). In *RNA Genetics*, E. Domingo, J.J. Holland and P. Ahlquist (eds). CRC Press, Inc., Boca Raton, Florida, pp. 149–65.
King, A.M.Q., McCahon, D., Slade, W.R. & Newman, J.W.I. (1982). *Cell*, **29**, 921–8.
Kirkegaard, K. & Baltimore, D. (1986). *Cell*, **47**, 433–43.
Koetzner, C.A., Parker, M.M., Ricard, C.S., Sturman, L.S. & Masters, P.S. (1992). *J. Virol.*, **66**, 1841–8.
Kusters, J.G., Niesters, H.G.M., Lenstra, J.A., Horzinek, M.C. & van der Zeijst, B. A.M. (1989). *Virology*, **169**, 217–21.
Lai, M.M.C. (1990). *Ann. Rev. Microbiol.* **44**, 303–33.
Lai, M.M.C. (1992). *Microbiol. Rev.*, **56**, 61–79.
Ledinko, N. (1963). *Virology*, **20**, 107–19.
Li, Y & Ball, L.A. (1993). *J. Virol.*, **67**, 3854–60.
Liao, C.-L. & Lai, M.M.C. (1992). *J. Virol.*, **66.**, 6117–24.

Luytjes, W., Bredenbeek, P.J., Noten, A.F.H., Horzinek, M.C. & Spaan,
    W.J.M. (1988). *Virology*, **166**, 415–22.
Makino, S., Keck, J.G., Stohlman, S.A. & Lai, M.M.C. (1986). *J. Virol.*,
    57, 729–37.
Mayo, M.A. & Jolly, C.A. (1991). *J. Gen. Virol.*, **72**, 2591–5.
Meyers, G., Tautz, N., Dubovi, E.J. & Thiel, H.-J. (1991). *Virology*,
    180, 602–16.
Minor, P.D., John, A., Ferguson. M. & Icenogle, J.P. (1986). *J. Gen. Virol.*,
    67, 693–706.
Muller, H.J. (1964). *Mut. Res.*, **1**, 2–9.
Nagy, P.D. & Bujarski, J.J. (1992). *J. Virol.*, **66**, 6824–8.
Palasingam, K. & Shaklee, P.N. (1992). *J. Virol.*, **66**, 2435–42.
Pfefferkorn, E.R. (1977). In *Comprehensive Virology*, H. Fraenkel-Conrat
    and R.R. Wagner (eds). Plenum Publishing Co., New York, Volume 9,
    pp. 209–89.
Pringle, C.R. (1965). *Virology*, **25**, 48–54.
Rao, A.L.N., Sullivan, B.P. & Hall, T.C. (1990). *J. Gen. Virol.*, **71**, 1403–7.
Rico-Hesse, R., Pallansch, M.A., Nottay, B.K. & Kew, O.M. (1987).
    Virology, **160**, 311–22.
Robinson, D.J., Hamilton, W.D.O., Harrison, B.D. & Baulcombe, D.C.
    (1987). *J. Gen. Virol*, **68**, 2551–61.
Romanova, L.I., Blinov, V.M., Tolskaya, E.A., Viktorova, E.G.,
    Kolesnikova, M.S., Guseva, E.A. & Agol, V.I. (1986). *Virology*,
    155, 202–13.
Semler, B.L., Johnson, V.H. & Tracy, S. (1986). *Proc. Natl. Acad. Sci.
    USA*, **83**, 1777–81.
Shirako, Y. & Brakke, M.K. (1984). *J. Gen. Virol.*, **65**, 855–8.
Simon, A.E. & Howell, S.H. (1986). *EMBO J.*, **5**, 3423–8.
Snijder, E.J., den Boon, J.A., Horzinek, M.C. & Spaan, W.J.M. (1991).
    Virology, **180**, 448–52.
Steinhauer, D.A., de la Torre, J.C. & Holland, J.J. (1989). *J. Virol.*,
    63, 2063–71.
Strauss, E.G., Strauss, J.H. & Levine, A.J. (1990). In *Virology*, B. Fields
    and D.M. Knipe (eds). Raven Press, New York, 2nd edn, pp. 167–90.
Tolskaya, E.A., Romanova, L.I., Blinov, V.M., Viktorova, E.G., Sinyakov,
    A.N., Kolesnikova, M.S. & Agol, V.I. (1987). *Virology*, **161**, 54–61.
Van der Kuyl, A.C., Neeleman, L. & Bol, J.F. (1991). *Virology*, **183**, 731–8.
Van der Most, R.G., Heijnen, L., Spaan, W.J.M. & de Groot, R.J. (1992).
    Nucl. Acids Res., **20**, 3375–81.
Ward, C.D. & Flanegan, J.B. (1992). *J. Virol.*, **62**, 3784–93.
Ward, C.D., Stokes, M.A. & Flanegan, J.B. (1988). *J. Virol.*, **62**, 558–62.
Weiss, B.G. & Schlesinger, S. (1991). *J. Virol.*, **65**, 4017–25.
White, K.A. & Morris, T.J. (1994). *Proc. Natl. Acad. Sci. USA* **91**, 3642–6.

Part IV

Molecular interactions of viruses
and their hosts

10

Viruses as ligands of eukaryotic cell surface molecules

THOMAS L. LENTZ

Introduction

A virus receptor is defined as a component of the cell surface that contains a domain which is structurally and conformationally complementary with a domain on the viral attachment protein (VAP) and forms a stable complex with the VAP. Association of the VAP with the receptor is followed by a biological response which may include internalization and replication of the virus, cytopathic changes in the cell, activation of second messenger systems, inhibition of host protein synthesis, or down regulation of receptors. The interaction with the host cell receptor is the initial stage in the viral infectious cycle. Besides being necessary for entrance of the virus into the cell, the receptor plays a major role in determining cell and species tropisms. Binding of virions to host cell receptors places them in intimate physical contact with the cell surface, and sets the stage for crossing the membrane barrier. Viruses then enter the cell either by direct fusion with the plasma membrane or by adsorptive- or receptor-mediated endocytosis. Many viruses have been shown to enter by both mechanisms, although some may preferentially utilize one of the two pathways. With either process, the viral genome gains entrance to the cytoplasm without lysing the cell membrane, which would be fatal to the cell. Paulson (1985) and Marsh and Helenius (1989) reviewed data on the entry of enveloped and non-enveloped viruses into cells.

Virus receptors

Any normal constituent comprising or associated with the cell membrane of the host cell is potentially a virus receptor. These surface components include phospholipids, glycolipids, and integral membrane glycoproteins, such as enzymes, ion channels, ion pumps, transporters, adhesion molecules, major histocompatibility complex (MHC) antigens, immunoglobulins, and receptors for substances such as neurotransmitters, peptides, growth factors, and hormones. Thus, viruses have opportunistically usurped and used as receptors normal surface molecules serving otherwise useful physiological functions. Molecules that have been shown or hypothesized to serve as virus receptors are listed in Table 10.1.

Carbohydrates, lipids, and proteins have been found to act as virus receptors. Many viruses bind to the sialyloligosaccharides of glycoproteins and glycolipids (gangliosides). These include encephalomyocarditis virus, murine cytomegalovirus, orthomyxoviruses (influenza viruses), paramyxoviruses (Sendai virus, Newcastle disease virus), polyoma virus, rabies virus, reoviruses, and rotaviruses. Vesicular stomatitis virus (Indiana) recognizes phosphatidylserine and phosphatidylinositol. Most other viruses attach to integral membrane proteins present on the cell surface. Many of these proteins are members of the immunoglobulin superfamily (White & Littman, 1989). These include: the CD4 molecule which serves as the receptor for human immunodeficiency virus (HIV)-1, HIV-2, and simian immunodeficiency virus (SIV); IgM, suggested to be a receptor for murine leukaemia virus; intercellular adhesion molecule-1 (ICAM-1) which is the rhinovirus receptor; the receptor for poliovirus; the receptor for mouse hepatitis virus; and MHC molecules which may act as receptors for several viruses, including human adenoviruses, human cytomegalovirus, lactate dehydrogenase-elevating virus, Semliki Forest virus, simian virus 40, and visna virus. Other proposed virus receptors function normally as receptors for hormones or neurotransmitters. These include epidermal growth factor receptor for vaccinia virus, β adrenergic receptor for reovirus 3, the acetylcholine receptor for rabies virus, interleukin-2 receptor for human T-cell leukaemia (HTLV)-1 virus, and fibroblast growth factor receptor for herpes simplex virus. Foot-and-mouth disease virus utilizes members of the integrin family of adhesion receptors.

Viral attachment protein (VAP)

The glycoproteins of enveloped viruses and the capsid proteins of non-enveloped viruses form the VAP. In order to construct a molecular model of the virus–receptor interface, the binding domain on the VAP must also be identified. Several approaches are used to identify binding sites on VAPs. Comparison of the amino acid sequence of the VAP with

Table 10.1. *Host cell receptors for viruses*

Virus	Host cell receptor
Epstein–Barr virus	C3d receptor CR2 (CD21) of B lymphocyte (Fingeroth et al., 1984)
Foot-and-mouth disease virus	Integrins (adhesion proteins) (Fox et al., 1989)
Gibbon ape leukaemia virus	679 amino acid protein, possibly a phosphate transporter (O'Hara et al., 1990)
Hepatitis B virus	Hepatocyte receptor for polymerized serum albumin via albumin (Machida et al., 1984)
Herpes simplex virus type 1	Heparan sulphate (WuDunn & Spear, 1989)
	Fibroblast growth factor receptor (Kaner et al., 1990)
Human adenovirus	Class I HLA MHC molecule (Chatterjee & Maizel, 1984)
Human cytomegalovirus	Class I HLA MHC molecule via β2-microglobulin (Grundy et al., 1987)
Human immunodeficiency virus (HIV-1)	CD4 molecule of T lymphocyte[a] (Dalgleish et al., 1984; Klatzmann et al., 1984; McDougal et al., 1986)
Human rhinovirus	Intercellular adhesion molecule-1 (Greve et al., 1989; Staunton et al., 1989; Tomassini et al., 1989)
Human T-cell leukaemia virus (HTLV-1)	Class I HLA MHC molecule (Clarke, Gelmann & Reitz, 1983)
Influenza virus	Interleukin 2 receptor (Kohtz et al., 1988; Lando et al., 1983)
	Sialyloligosaccharides[b] (Paulson, 1985; Paulson, Sadler & Hill, 1979)
Lactate dehydrogenase elevating virus	Class II Ia MHC molecule of macrophage (Inada & Mims, 1984)
Mouse hepatitis virus	Member of carcinoembryonic antigen family (Williams, Jiàng & Holmes, 1991)
Murine leukaemia viruses	Lymphoma cell surface IgM (McGrath et al., 1987) 622 amino acid protein, possibly an amino acid transporter (Albritton et al., 1989)
Poliovirus	Member of immunoglobulin superfamily (Mendelsohn et al., 1989)
Rabies virus	Acetylcholine receptor (Lentz et al., 1982)
	Sialylated gangliosides (Superti et al., 1986)
Radiation leukaemia virus	T cell receptor-L3T4 molecule complex (O'Neill et al., 1987)
Reovirus 3	β Adrenergic receptor (Co et al., 1985)

Table 10.1. (cont.)

Virus	Host cell receptor
Semliki Forest virus	Class I HLA and H-2 MHC molecules (Helenius et al., 1978)
Simian virus 40	Class I MHC molecule (Attwood & Norkin, 1989)
Vaccinia virus	Epidermal growth factor receptor (Eppstein et al., 1985)
Vesicular stomatitis virus (Indiana)	Phosphatidylserine (Mastromarino et al., 1987; Schlegel et al., 1983) Phosphatidylinositol (Mastromarino et al., 1987)
Visna virus	Ovine class II MHC molecule (Dalziel et al., 1991)

Note: [a] Other viruses binding to CD4 are HIV-2 and SIV.
[b] Other viruses reported to bind to sialic acid include encephalomyocarditis virus, hepatitis B virus, human rhinovirus 87, murine cytomegalovirus, Newcastle disease virus, polyomavirus, reovirus 3, rotaviruses, and Sendai virus.

sequenced vertebrate proteins has, in some cases, revealed structural similarities with natural ligands of receptors. Anti-VAP antibodies that inhibit attachment can be used to map domains of the VAP involved in binding. Determination of the three-dimensional structure of virus particles has revealed exposed projections or depressions on the virus surface that could function as attachment sites. Finally, the effects of site-directed mutagenesis of the putative VAP binding site on attachment and infectivity can yield direct evidence on structure–function relationships providing the mutations do not alter the conformation of the VAP.

Molecular mimicry of ligands by the VAP

The ability of the virus to bind to a host cell receptor may be conferred by a domain on the VAP that resembles in amino acid sequence or conformation the normal ligand of the cellular receptor. Many regions of viruses and other pathogens share amino acid similarities with normal host proteins, a phenomenon termed molecular mimicry (Damian, 1987). Molecular mimicry can cause autoimmune disease as a result of antibodies against the virus reacting with similar epitopes on the host-cell proteins (Oldstone, 1987). Mimicry of the binding domain of a ligand by the VAP may confer the ability to bind to the host cell receptor for the ligand.

Examples of similarities between portions of VAPs and physiological ligands of receptors are shown in Table 10.2. In some cases, these regions of the VAP may mediate binding to a host cell receptor. For example, portions of the gp350 of Epstein–Barr virus show structural similarities with complement fragment C3d (Nemerow *et al.*, 1987). These regions of gp350 could mediate binding of Epstein–Barr virus to the C3d receptor, CR2 (CD21), of B lymphocytes. Similarly, foot-and-mouth disease virus contains an RGD sequence characteristic of proteins such as fibronectin which bind to integrins (Fox *et al.*, 1989). A segment of the rabies virus glycoprotein bears an amino acid similarity to snake venom curaremimetic neurotoxins which bind with high affinity to the nicotinic acetylcholine receptor (Lentz *et al.*, 1984).

Biological effects of the VAP-receptor interaction

Similarities between the VAP and normal ligands (Table 10.2), besides mediating attachment followed by entry of virus into the cell, may also be responsible for certain biological effects of viruses independent of

Table 10.2. *Structural similarities between viral attachment proteins and ligands of receptors*

Adenovirus glycoprotein (102–116)	**DI**TMYMSK**QYK**LW**PP**
Immunoglobulin Mκ chain, human (83–97) (Chatterjee & Maizel, 1984)	**DI**ATYYCQ**QYN**NW**PP**
Epstein–Barr virus gp350 (372–378), (21–29)	**TPSGC**EN **EDPG**--FF**NVE**
Complement C3d fragment (1006–1012),(1221–1231) (Nemerow *et al.*, 1987)	**TPSGC**GE **EDPG**KQLY**NVE**
Epstein–Barr virus BCRFI protein (41–67)	**DFKGYLGCQALSEMIQFYL**EE**VMPQAE**
Cytokine synthesis inhibitory factor (interleukin-10), mouse (46–72) (Moore *et al.*, 1990)	**DFKGYLGCQALSEMIQFYL**VE**VMPQAE**
Foot-and-mouth disease virus VP1 (145–147) (Fox *et al.*, 1989); coxsackie-virus A9 VP1 (Chang *et al.*, 1989)	**RGD**
Fibronectin, fibrinogen, type 1 collagen, and proteins binding to integrins	**RGD**
Hepatitis B virus preS1 env protein (21–47)	PL**GFFP**DH**Q**L**D**PAFGANSNN-PDW**DFNP**
Immunoglobulin A α1 constant region (28–55) (Neurath & Strick, 1990)	VQ**GFFP**QQPLSVTWSESGEGVTARD**FPP**
Human cytomegalovirus H301 protein (261–286)	**DGTF**HQ--GCY**V**-AIFCN**QNYTC**R**VTH**
Class I MHC molecule α-1 domain, human (239–264) (Beck & Barrell, 1988)	**DGTF**QKWAAVV**V**-PSG-E**QRYTC**H**VQH**
Human immunodeficiency virus-1 gp120 (193–200)	**A**ST**TTNYT**
Vasoactive intestinal polypeptide (4–11) (Ruff *et al.*, 1987)	**A**VFT**DNYT**

Table 10.2. (*cont.*)

Human immunodeficiency virus-1 gp120 (61–88)	**GASDAKAYDTEVHKVWATHAGVPTDPNP**
Immunoglobulin γ2 heavy chain, human (84–111) (Maddon *et al.*, 1986)	**GNVDHKPSNTKVDKTVERKC GVECPPCP**
Human immunodeficiency virus-1 gp120 (245–273)	**VQ**CTHG**IR**PVVSTQ-L**LL**NGS**LAEEE**VV**IR**
Neuroleukin (410–438) (Lee *et al.*, 1987)	**VQ**TQ**HPIRK**GLHHK**ILL**ANF**LA-QTEALMR**
Human T-cell leukaemia virus-1 gp (246–253)	**SVPSSSST**
Interleukin 2, human (20–27) (Kohtz *et al.*, 1988)	**SAPTSSST**
Rabies virus glycoprotein (189–199)	**CDIFTNSRGKR**
Naja naja naja (India) toxin b (30–40) (Lentz *et al.*, 1984)	**CDGFCSSRGKR**
Vaccinia virus VGF protein (69–80)	**CRCSHGYTGIRC**
Epidermal growth factor, human (31–42) (Eppstein *et al.*, 1985)	**CNCVVGYIGERC**

internalization. In some viral infections, the glycoprotein or coat protein is synthesized in excess and is secreted or shed extracellularly. The proteins may also dissociate from virus particles. Biological effects exerted by viruses during the course of infection include toxic effects on cells, activation or inactivation of the immune response through several mechanisms, activation of second messenger systems, suppression or induction of protein synthesis and secretion, and mitogenesis. Vaccinia virus encodes a polypeptide (VGF) that is structurally similar to epidermal growth factor and transforming growth factor α (Brown *et al.*, 1985; Reisner, 1985; King *et al.*, 1986). VGF has been shown to stimulate tyrosine kinase activity of epidermal growth factor receptors (King *et al.*, 1986) and to stimulate mitogenesis (Twardzck *et al.*, 1985). The receptor-mediated leukaemogenesis hypothesis proposes that murine leukaemia viruses induce proliferation of lymphocytes by binding to antigen-specific receptors complementary to virus envelope gene products (McGrath, Tamura & Weissman, 1987). Similarly, gp55 of Friend spleen focus-forming virus binds directly to the erythropoietin receptor experimentally expressed in non-erythroid cells and triggers their proliferation. It was suggested that the first step in leukaemogenesis induced by Friend virus is mimicry of erythropoietin by gp55 and stimulation of proliferation of erythroid cells (Li *et al.*, 1990).

HIV has been reported to interfere with the neurotrophic factors neuroleukin (Lee, Ho & Gurney, 1987) and vasoactive intestinal polypeptide (Brenneman *et al.*, 1988) by virtue of regions of the gp120 that bear an amino acid similarity to these factors. Thus, some of the neurological effects and dementia observed in patients with acquired immune deficiency syndrome (AIDS) could be the result of inhibition of normal neurotrophic factors. Synthetic peptides of a region of rabies virus glycoprotein that is structurally similar to snake venom curare-mimetic neurotoxins (Lentz *et al.*, 1984) act as competitive antagonists of the nicotinic acetylcholine receptor and inhibit carbachol-induced ^{22}Na flux into BC3H-1 cells (Donnelly-Roberts & Lentz, 1989). Thus, soluble glycoprotein or proteolytic fragments of the glycoprotein could be responsible for some of the bizarre behavioral manifestations of rabies.

Some similarities between the VAP and host cell proteins may aid the virus in evading host immune or defence responses. For example, Epstein–Barr virus protein BCRFI exhibits a sequence similarity to cytokine synthesis inhibitory factor (interleukin-10) and blocks the production of interferon-γ, thereby interfering with host defences (Hsu *et al.*, 1990).

Evolutionary origins

The most frequently used receptors for viruses appear to be sialyloligosaccharides of glycoproteins and glycolipids, and members of the immunoglobulin superfamily. The latter includes immunoglobulins, MHC molecules, some CD molecules, ICAMs, T cell receptors, lymphocyte function-related antigens, and some growth factor receptors (Springer, 1990). Together, the two groups comprise the major portion of cell surface molecules. They are widely distributed among cell types, although particular molecules may be restricted to certain cell types. Glycoproteins and adhesion molecules of the immunoglobulin superfamily occur throughout the animal kingdom. Thus, viruses have had ample opportunity to evolve mechanisms for interacting with these molecules.

The mechanisms that led to the ability to bind cell surface molecules and the structural similarities between portions of VAPs and normal ligands for host cell receptors are not clear. It would be expected that within a large group of protein sequences, some similarities are simply the result of chance. Other sequence similarities may be the result of divergence from a common ancestor. In this case, the virus may have acquired from the host during the course of infection genetic material (DNA or messenger RNA) coding the binding domain of a normal functional ligand. The VAP could also acquire the binding domain of another virus due to genetic recombination between two viruses infecting the same cell.

Viruses could also develop binding domains similar to that of a ligand and capable of binding to a host cell receptor through evolutionary convergence due to common functions. Mutations that enhance binding might have survival value. In the case of both convergent and divergent evolution, the acquisition of ability to bind to a new host cell receptor might confer a selective advantage by providing another mechanism for binding and entry into cells, the ability to bind to a new, more hospitable cell or host, or suppression of defence mechanisms.

Conservation of binding domains

It seems likely that the binding domain on the VAP will be conserved among strains of virus that maintain the same tropism. Mutations of binding domains which result in inability to attach to cells would be lethal to the virus. A single mutation (Trp 432) in the region of gp120 of HIV-1 considered to interact with CD4 blocks binding to CD4 (Cordonnier *et al.*, 1989). Mutations of other regions of coat proteins not directly involved in binding may also affect tropisms and attachment to receptors by altering the conformation of the protein. It is also likely that the binding domains on the receptor are conserved if these domains are binding sites of normal ligands.

The receptor-binding sites of human rhinovirus 14 (Rossmann *et al.*, 1985), of Mengo virus (Luo *et al.*, 1987), and of the haemagglutinin glycoprotein of influenza virus (Weis *et al.*, 1988) lie within a canyon, pit, or pocket on the virus surface. A similar depression is present on the surface of poliovirus (Hogle, Chow & Filman, 1985). Determination of the three-dimensional structure of the influenza haemagglutinin complexed with sialyllactose reveals the sialic acid filling the haemagglutinin pocket (Weis *et al.*, 1988). The amino acids lining these depressions are highly conserved whereas the surrounding surface residues are variable and could represent the binding sites of neutralizing antibodies. According to the canyon hypothesis, these viruses can accept mutations in the hyper-variable antibody binding sites thereby escaping the immune response while at the same time maintaining a constant receptor binding site which is physically inaccessible to antibodies (Rossmann & Palmenberg, 1988).

Inhibition of the virus-receptor interaction

The interaction of the VAP with a host cell receptor represents a target to which agents that prevent infection by blocking binding can be directed (Lentz, 1988, 1990). Targeting the attachment process may be a particularly useful strategy for attacking viruses that evade immune clearance by frequent mutation. As discussed above, binding domains are likely to be conserved so that agents directed against these regions should be effective against different strains of virus. Two general classes of agents should theoretically prevent viral entry into cells by blocking attachment of virus particles to host cell receptors (Fig. 10.1). The first group of agents resembles in structure or conformation the binding domain on the VAP and should competitively inhibit binding of the VAP to the cellular receptor. Substances mimicking the binding domain of the VAP are (1) antibodies against the binding site of the receptor (2) anti-idiotypic antibodies against antibodies to the binding domain of the VAP, (3) natural ligands of the binding site of the receptor, and (4) designed ligand-mimics. The latter include synthetic peptides of the VAP or designed drugs that mimic the VAP. The second group of agents that should prevent attachment of virus to host cell receptors are agents that resemble the binding domain on the receptor to which the VAP binds. These agents will bind to the VAP and prevent attachment to the host cell receptor. This group of agents includes (1) antibodies against the binding domain of the ligand, (2) anti-idiotypic antibodies against antibodies to the binding site on the receptor, (3) the receptor itself and, (4) receptor-mimics. Receptor-mimics are designed receptor binding domains prepared synthetically or by genetic engineering.

All of the classes of agents listed above have been shown to prevent infection of cells *in vitro* by certain viruses (Lentz, 1990). As examples of ligand-mimics, synthetic peptides of the binding domains of the VAP inhibited binding of vaccinia virus (Eppstein *et al.*, 1985), hepatitis B virus (Neurath *et al.*, 1988), foot-and-mouth disease virus (Fox *et al.*, 1989), and Epstein–Barr virus (Nemerow *et al.*, 1989) to their receptors. Fibroblast growth factor residues 103 to 120 inhibited herpes simplex virus type 1 uptake by cultured cells (Kaner *et al.*, 1990). A potential

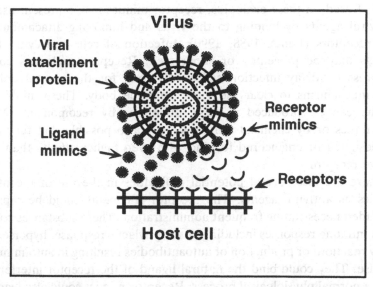

Fig. 10.1. Viruses attach to host cell surfaces as a result of binding of a viral attachment protein to a receptor on the surface of the cell. Substances with the potential for blocking the attachment step are those that mimic the binding domain on the attachment protein (ligand mimics) and those that mimic the attachment site on the receptor (receptor mimics).

difficulty with synthetic peptides as competitors of virus attachment is that a single linear sequence may not be effective if the binding site consists of several discontinuous segments brought into proximity in the native, folded structure of the protein. In addition, since these agents are ligands of the receptors, they may have adverse or toxic effects.

In the case of receptor-mimics, agents that contain the binding site of the receptor have been shown to have anti-viral activity. Sialylated gangliosides have been proposed to comprise part of a rabies receptor complex (Superti *et al.*, 1986), and it was shown that gangliosides were effective in inhibiting rabies virus infection of cultured cells (Conti *et al.*, 1988). Heparin inhibits attachment of herpes simplex virus to cells (Vaheri, 1964), presumably by blocking the interaction of virus with heparan sulfate on the cell surface (WuDunn & Spear, 1989). Monovalent sialosides and sialyloligosaccharide sequences were effective in inhibiting influenza virus adsorption to erythrocytes (Pritchett *et al.*, 1987). Synthetic peptides of CD4 residues 25–58 of CD4 (Jameson *et al.*, 1988) and a benzylated peptide of residues 81–92 (Lifson *et al.*, 1988) specifically inhibited HIV-1 induced cell fusion and infection *in vitro*. Based on these examples, receptor-mimics may represent useful antiviral agents by binding to the virus and inhibiting attachment to cell receptors (Lentz, 1988, 1990). Infection of cells and cytopathic effects may be prevented or suppressed. Receptor analogues could suppress or delay infection sufficiently long for the normal defensive mechanisms to clear the virus from the body. These inhibitory agents can be produced in large quantities by recombinant DNA techniques or by chemical synthesis. It is also possible that they can be designed or engineered to bind with even higher affinity than the native receptor.

There are also several potential difficulties in the use of receptor-mimics as antiviral agents. The receptor fragments could be rapidly degraded necessitating frequent administration. These substances could elicit immune responses including an anaphylactic response, hypersensitivity reaction, or production of autoantibodies resulting in autoimmune disease. They could bind the natural ligand of the receptor interfering with a normal physiological process. Receptor-mimics could also lead to the selection of mutants that are resistant to neutralization by the receptor analogue or that can utilize other receptors. As discussed above, it could be expected that the receptor binding domain of the VAP is highly conserved. However, it has been shown that infectious poliovirus mutants resistant to neutralization with soluble poliovirus receptor can be isolated (Kaplan, Peters & Racaniello, 1990). Finally, soluble CD4, which blocks HIV-1 infection, significantly enhances infection of a human T cell line by SIV (Allan, Strauss & Buck, 1990). However, it may be possible to design receptor analogs that avoid the adverse effects.

Acknowledgements

The author's research is supported by grants from the National Institutes of Health (NS 21896) and the National Science Foundation (BNS 88–05780).

References

Albritton, L.M., Tseng, L., Scadden, D. & Cunningham, J.M. (1989). *Cell*, **57**, 659–66.
Allan, J.S., Strauss, J. & Buck, D.W. (1990). *Science*, **247**, 1084–8.
Attwood, W.J. & Norkin, L.C. (1989). *J. Virol.*, **63**, 4474–7.
Beck, S. & Barrell, B.G. (1988). *Nature*, **331**, 269–72.
Brenneman, D.E., Westbrook, G.L., Fitzgerald, S.P., Ennist, D.L., Elkins, K.L., Ruff, M.R. & Pert, C.D. (1988). *Nature*, **335**, 639–42.
Brown, J.P., Twardzik, D.R., Marquardt, H. & Todaro, G.J. (1985). *Nature*, **313**, 491–2.
Chang, K.H., Auvinen, P., Hyypiä, T. & Stanway, G. (1989). *J. Gen. Virol.*, **70**, 3269–80.

Chatterjee, D. & Maizel Jr., J.V. (1984). *Proc. Natl. Acad. Sci. USA*, **81**, 6039–43.
Clarke, M.F., Gelmann, E.P. & Reitz Jr., M.S. (1983). *Nature*, **305**, 60–2.
Co, M.S., Gaulton, G.N., Tominaga, A., Homcy, C.J., Fields, B.N. & Greene, M.I. (1985). *Proc. Natl. Acad. Sci. USA*, **82**, 5315–18.
Conti, C., Hauttecoeur, B., Morelec, M.J., Bizzini, B., Orsi, N. & Tsiang, H. (1988). *Arch. Virol.*, **98**, 73–86.
Cordonnier, A., Montagnier, L. & Emerman, M. (1989). *Nature*, **340**, 571–4.
Dalgleish, A.G., Beverley, P.C.L., Clapham, P.R., Crawford, D.H., Greaves, M.F. & Weiss, R.A. (1984). *Nature*, **312**, 763–7.
Dalziel, R.G., Hopkins, J., Watt, N.J., Dutia, B.M., Clarke, H.A. & McConnell, I. (1991). *J. Gen. Virol.*, **72**, 1905–11.
Damian, R.T. (1987). *Parasitol. Today*, **3**, 263–6.
Donnelly-Roberts, D.L. & Lentz, T.L. (1989). *Peptide Res.*, **2**, 221–6.
Eppstein, D.A., Marsh, Y.V., Schreiber, A.B., Newman, S.R., Todaro, G.J. & Nestor Jr., J.J. (1985). *Nature*, **318**, 663–5.
Fingeroth, J.D., Weis, J.J., Tedder, T.F., Strominger, J.L., Biro, P.A. & Fearon, D.T. (1984). *Proc. Natl. Acad. Sci. USA*, **81**, 4510–14.
Fox, G., Parry, N.R., Barnett, P.V., McGinn, B., Rowlands, D.J. & Brown, F. (1989). *J. Gen. Virol.*, **70**, 625–37.
Greve, J.M., Davis, G., Meyer, A.M., Forte, C.P., Yost, S.C., Marlor, C.W., Kamarck, M.E. & McClelland, A. (1989). *Cell*, **56**, 839–47.
Grundy, J.E., McKeating, J.A., Ward, P.J., Sanderson, A.R. & Griffiths, P.D. (1987). *J. Gen. Virol.*, **68**, 793–803.
Helenius, A., Morein, B., Fries, E., Simons, K., Robinson, P., Schirrmacher, V., Terhorst, C. & Strominger, J.L. (1978). *Proc. Natl. Acad. Sci. USA*, **75**, 3846–50.
Hogle, J.M., Chow, M. & Filman, D.J. (1985). *Science*, **229**, 1358–65.
Hsu, D.-H., Malefyt, R.d.W., Fiorentino, D.F., Dang, M.-N., Vieira, P., deVries, J., Spits, H., Mosmann, T.R. & Moore, K.W. (1990). *Science*, **250**, 830–2.
Inada, T. & Mims, C.A. (1984). *Nature*, **309**, 59–61.
Jameson, B.A., Rao, P.E., Kong, L.I., Hahn, B.H., Shaw, G.M., Hood, L.E. & Kent, S.B.H. (1988). *Science*, **240**, 1335–9.
Kaner, R.J., Baird, A., Mansukhani, A., Basilico, C., Summers, B.D., Florkiewicz, R.Z. & Hajjar, D.P. (1990). *Science*, **248**, 1410–13.
Kaplan, G., Peters, D. & Racaniello, V.R. (1990). *Science*, **250**, 1596–9.
King, C.S., Cooper, J.A., Moss, B. & Twardzck, D.R. (1986). *Mol. Cell. Biol.* **6**, 332–6.
Klatzmann, D., Champagne, E., Chamaret, S., Gruest, J., Guetard, D., Hercend, T., Gluckman, J.-C. & Montagnier, L. (1984). *Nature*, **312**, 767–8.
Kohtz, D.S., Altman, A., Kohtz, J.D. & Puszkin, S. (1988). *J. Virol.*, **62**, 659–62.
Lando, Z., Sarin, P., Megson, M., Greene, W.C., Waldman, T.A., Gallo, R.C. & Broder, S. (1983). *Nature*, **305**, 733–6.
Lee, M.R., Ho, D.D. & Gurney, M.E. (1987). *Science*, **237**, 1047–51.
Lentz, T.L. (1988). *Trends Pharmacol. Sci.*, **9**, 247–52.
Lentz, T.L. (1990). *J. Gen. Virol.*, **71**, 751–66.
Lentz, T.L., Burrage, T.G., Smith, A.L., Crick, J. & Tignor, G.H. (1982). *Science*, **215**, 182–4.

Lentz, T.L., Wilson, P.T., Hawrot, E. & Speicher, D.W. (1984). *Science*, **226**, 847–8.

Li, J.-P., D'Andrea, A.D., Lodish, H.F. & Baltimore, D. (1990). *Nature*, **343**, 762–4.

Lifson, J.D., Hwang, K.M., Nara, P.L., Fraser, B., Padgett, M., Dunlop, N.M. & Eiden, L.E. (1988). *Science*, **241**, 712–16.

Luo, M., Vriend, G., Kamer, G., Minor, I., Arnold, E., Rossmann, M.G., Boege, U., Scraba, D.G., Duke, G.M. & Palmenberg, A.C. (1987). *Science*, **235**, 182–91.

McDougal, J.S., Kennedy, M.S., Sligh, J.M., Cort, S.P., Mawle, A. & Nicholson, J.K.A. (1986). *Science*, **231**, 382–5.

McGrath, M.S., Tamura, G. & Weissman, I.L. (1987). *J. Mol. Cell. Immunol.*, **3**, 227–42.

Machida, A., Kishimoto, S., Ohnuma, H., Baba, K., Ito, Y., Miyamoto, H., Funatsu, G., Oda, K., Usuda, S., Togami, S., Nakamura, T., Miyakawa, Y. & Mayumi, M. (1984). *Gastroenterology*, **86**, 910–18.

Maddon, P.J., Dalgleish, A.G., McDougal, J.S., Clapham, P.R., Weiss, R.A. & Axel, R. (1986). *Cell*, **47**, 333–48.

Marsh, M. & Helenius, A. (1989). *Adv. Virus Res.*, **36**, 107–51.

Mastromarino, P., Conti, C., Goldoni, P., Houttecoeur, B. & Orsi, N. (1987). *J. Gen. Virol.*, **68**, 2359–69.

Mendelsohn, C.L., Wimmer, E. & Racaniello, V.R. (1989). *Cell*, **56**, 855–65.

Moore, K.W., Vieira, P., Fiorentino, D.F., Trounstine, M.L., Khan, T.A. & Mosmann, T.R. (1990). *Science*, **248**, 1230–4.

Nemerow, G.R., Houghten, R.A., Moore, M.D. & Cooper, N.R. (1989). *Cell*, **56**, 369–77.

Nemerow, G.R., Mold, C., Schwend, V.K., Tollefson, V. & Cooper, N.R. (1987). *J. Virol.*, **61**, 1416–20.

Neurath, A.R., Kent, S.B.H., Strick, N. & Parker, K. (1988). *Ann. Inst. Pasteur/Virol*, **139**, 13–38.

Neurath, A.R. & Strick, N. (1990). *Virology*, **178**, 631–4.

O'Hara, B., Johann, S.V., Klinger, H.P., Blair, D.G., Rubinson, H., Dunn, K.J., Sass, P., Vitek, S.M. & Robins, T. (1990). *Cell Growth Different.* **1**, 119–27.

O'Neill, H.C., McGrath, M.S., Allison, J.P. & Weissmann, I.L. (1987). *Cell*, **49**, 143–51.

Oldstone, M.B.A. (1987). *Cell*, **50**, 819–20.

Paulson, J.C. (1985). In *The Receptors*, Vol. 2, M. Conn (ed.) Academic Press, Orlando pp. 131–219.

Paulson, J.C., Sadler, J.E. & Hill, R.L. (1979). *J. Biol. Chem.*, **254**, 2120–4.

Pritchett, T.J., Brossmer, R., Rose, U. & Paulson, J.C. (1987). *Virology*, **160**, 502–6.

Reisner, A.H. (1985). *Nature*, **313**, 801–3.

Rossmann, M.G., Arnold, E., Erickson, J.W., Frankenberger, E.A., Griffith, J.P., Hecht, H.-J., Johnson, J.E., Kamer, G., Luo, M., Mosser, A.G., Rueckert, R. R., Sherry, B. & Vriend, G. (1985). *Nature*, **317**, 145–53.

Rossmann, M.G. & Palmenberg, A.C. (1988). *Virology*, **164**, 373–82.

Ruff, M.R., Martin, B.M., Ginns, E.I., Farrar, W.L. & Pert, C.B. (1987). *FEBS Lett.*, **211**, 17–22.

Schlegel, R., Tralka, T.S., Willingham, M.C. & Pastan, I. (1983). *Cell*, **32**, 639–46.

Springer, T.A. (1990). *Nature*, **346**, 425–34.
Staunton, D.E., Merluzzi, V.J., Rothlein, R., Barton, R., Marlin, S.D. &
  Springer, T.A. (1989). *Cell*, **56**, 849–53.
Superti, F., Hauttecoeur, B., Morelec, M.-J., Goldoni, P., Bizzini, B. &
  Tsiang, H. (1986). *J. Gen. Virol.*, **67**, 47–56.
Tomassini, J.E., Graham, D., DeWitt, C.M., Lineberger, D.W., Rodkey,
  J.A. & Colonno, R.J. (1989). *Proc. Natl. Acad. Sci. USA*, **86**, 4907–11.
Twardzck, D.R., Brown, J.P., Ranchalis, J.E., Todaro, G.J. & Moss, B.
  (1985). *Proc. Natl. Acad. Sci. USA*, **82**, 5300–4.
Vaheri, A. (1964). *Acta Pathol. Microbiol. Scand. Suppl.*, **171**, 1–98.
Weis, W., Brown, J.H., Cusack, S., Paulson, J.C., Skehel, J.J. & Wiley,
  D.C. (1988). *Nature*, **333**, 426–31.
White, J.M. & Littman, D.R. (1989). *Cell*, **56**, 725–8.
Williams, R.K., Jiàng, G.-S. & Holmes, K.V. (1991). *Proc. Natl. Acad. Sci.
  USA*, **88**, 5533–6.
WuDunn, D. & Spear, P.G. (1989). *J. Virol.*, **63**, 52–8.

11

The influence of immunity on virus evolution

CHARLES R.M. BANGHAM

Introduction

The vertebrate immune system contains both broadly reactive and highly specific mechanisms to detect and destroy foreign proteins and nucleic acids. The specific arms of the immune system, antibodies and T cells, change extremely rapidly in response to the appearance of a new antigen, while the non-specific mechanisms of antiviral defence, principally interferons (IFN) and natural killer (NK) cells, are relatively invariant.

To escape destruction by the immune system, viruses use many tactics and strategies that allow them to survive in the individual, at least long enough to be transmitted to another host. The immune system therefore exerts a powerful selection pressure on a virus in one individual, a pressure that can sometimes be demonstrated experimentally. However, the extent to which the resulting variation in the virus is reflected in evolutionary changes in the virus population is largely a matter of inference. Recent advances in our knowledge of how the immune system, particularly the T cell, recognizes its target antigen, and the ability to sequence quickly many variants of viral genomes, should soon lead to a better understanding of the importance of immunity in affecting viral evolution.

I shall consider here the mechanisms used by viruses of vertebrates to avoid the immune system. There are three main categories of mechanism: i) changes in antigenic structure, allowing immediate escape from immune recognition; ii) changes in gene composition of the virus, usually allowing escape from non-specific viral destruction, and iii) changes in viral behaviour, such as its tissue tropism or the time-course of infection. The main message I wish to convey is this: although each method the virus uses to avoid the immune response gives it only a small, relative advantage in replication or survival over other viruses, such an advantage can quickly lead to the emergence of a new variant virus, because of rapid replication.

Components of the antiviral immune system

Non-specific mechanisms

The important parts of the immune defence against viruses which are not specific to the antigens coded by the virus are i) natural killer cells and ii) interferons. NK cells are found in the large granular lymphocyte fraction, which makes up about 15% of peripheral blood lymphocytes in the human. They spontaneously kill many tumour cells and virus-infected cells; this killing, unlike T-cell mediated killing, does not involve the T-cell antigen receptor, and does not require the sharing of major histocompatibility (MHC) antigens between NK and target cells. They arise early in infection (approximately 1–5 days), before cytotoxic T lymphocytes (CTL) (5–10 days) and well before antibody (7–10 days onwards in a primary infection). Because these cells are functionally defined and have no unique cell-surface marker, and because the molecular basis of target cell recognition by NK cells is unknown, it has been hard to demonstrate their importance experimentally. However, they are believed to be important in the early stages of a virus infection.

Interferons constitute a family of heterogeneous proteins, which inhibit viral replication and cell proliferation. IFNα (leukocyte interferon), and IFNβ (fibroblast interferon) are produced by the respective cell on infection with viruses; RNA viruses as a rule are better inducers of IFN than are DNA viruses (except poxviruses). IFNγ, which is unrelated in sequence to IFNα and β, is produced by activated T cells of both the helper (CD4+) and cytotoxic (CD8+) classes, and by some NK cells. Like NK cells, IFNs are probably of greatest importance early in viral infections.

Interferons have a remarkable range of actions, including induction of MHC expression, activation of macrophages and NK cells, proliferation of B cells, increased expression of high-affinity Fc receptors for IgG on certain cells and of IL-2 receptor on T cells.

At the molecular level, interferons induce two main antiviral enzymes: a protein kinase and 2′, 5′A synthetase. When activated by dsRNA, these enzymes lead, respectively, to inhibition of translation initiation, and mRNA degradation. These two paths are primarily responsible for the antiviral actions of the interferons, although their relative importance varies with the cell type and virus species.

Antigen-specific components of antiviral immune system: antibodies and T cells

Until the 1970s it was believed that antibodies were the main specific defence of the organism against a viral infection. However, it is now established that, in most cases, T cells are necessary for protection against a virus, particularly in eliminating a virus from the body.

Antibody

The binding site, the variable region, of an immunoglobulin molecule recognizes an area of about six amino acids on its target protein. These six amino acid residues may not be contiguous in the primary protein sequence, since the antibody binds to an intact, folded protein; most antibody epitopes are therefore said to be 'conformational'.

The gene encoding a functional antibody light chain consists of variable (V), junctional (J) and constant (C) segments; the Ig heavy chain also contains a diversity (D) segment between the V and J segments. During the development of a B cell, the genomic DNA in the Ig locus is rearranged (spliced) to form a full-length gene containing one each of the V, (D), J and C gene segments. The enormous diversity of antibody specificity is generated by three mechanisms: i) the multiplicity of genes coding each segment of the chains (V, D, J, C), which in principle can be joined in any combination; ii) sequence diversity at the junctions of V/D and D/J segments, and iii) somatic hypermutation, in which the nucleotide mutation rate in the V region gene rises by three to four orders of magnitude for a short time during B cell differentiation. It is this hypermutation that is largely responsible for the increase in affinity of antibodies for their target antigens during the maturation of the immune response. The diversity of binding site is further increased because any light chain can then be joined to any heavy chain. The net effect of these mechanisms is to generate a repertoire of perhaps 10^{16} antibody sequences in the mouse (Davis & Bjorkman, 1988).

Antibodies may alone be sufficient to neutralize a virus *in vivo*, for example, by blocking adsorption of the virion to its cellular receptor, or it may act together with cells, for example, to promote ingestion of the virus by phagocytic cells (a process called opsonization), by including lysis of an infected cell by killer cells (known as antibody-dependent, cell-mediated cytotoxicity or ADCC) or by complement. The relative importance of these mechanisms of killing *in vivo* is not known, but *in vitro* neutralization is usually a poor correlate of protection *in vivo*: a measure of total antibody titre is a better predictor of protection.

T cells

T cells are broadly divided into cytotoxic T lymphocytes (CTL), characterized by the surface marker CD8, and helper T cells, characterized by the CD4 marker. CTL are the primary agents of the immune system that kill virus-infected cells; helper T cells are responsible for secretion of the lymphokines that activate both CTL and the non-antigen-specific cells in the antiviral inflammatory response, principally NK cells and macrophages. Two classes of helper T cells are found, Th1 and Th2, distinguished by the range of lymphokines they secrete.

While antibodies recognize intact viral proteins, T cells recognize short fragments of proteins, produced by the degradation of intact proteins in the cytosol. These short peptide fragments, 9 to 12 amino acid residues long, associate with MHC molecules, i.e. the HLA system in humans, and are transported to the cell surface. It is the peptide-MHC complex that is specifically recognized by the T cell antigen receptor. Only those peptides which can bind MHC molecules can therefore be recognized by T cells, and so the T cell response is said to be 'MHC-restricted'. CD4[1] (helper) T cells recognize peptide-class II MHC complexes, and are said to be class II restricted, while CTL (CD8+) recognize peptide-class I MHC complexes. It has recently become clear that certain residues in the antigenic peptide fragment are particularly important in determining whether the peptide will bind a given MHC molecule, particularly class I MHC molecules: as a result, it is now beginning to be possible to predict the epitopes in a viral antigen that will be recognized by cytotoxic (class I MHC-restricted) T cells.

The T-cell receptor (TCR) for antigen, which recognizes the peptide-MHC complex, is closely related to the Ig molecule in both structure and gene organization. Like the Ig molecule, the occurrence of multiple germline gene segments and sequence diversity at the segment junctions results in an enormous diversity of mature TCR structures: the potential repertoire in the mouse is probably about 10^{17} different TCR molecules (Davis & Bjorkman, 1988). Unlike antibodies, however, the TCR does not vary further by somatic hypermutation after gene rearrangement.

Binding of the TCR to its specific target of viral peptide plus MHC leads to the activation of the T cell. As a result, helper T cells are induced to release lymphokines; CTL are induced to kill their target cells, using three apparently distinct pathways (Ostergard & Clark, 1989; Male *et al.*, 1991). Certain CTL and helper T cells also release IFNγ.

Effects of the immune system on viral evolution

Introduction

The virus must avoid or combat this formidable combination of specific and non-specific defences in order to survive. The selection forces exerted by the immune system have therefore, and not surprisingly, played a central part in shaping the genome of each vertebrate virus. Certain points must be made here before we consider the consequences of these selection forces.

 (i) It is uneconomical for a virus to invest more than a certain proportion of its genome in immune evasion, as it would sacrifice speed of replication and so be outgrown.

 (ii) It does not invariably benefit a virus to attenuate its virulence in order to co-exist with its host population. Host populations can co-exist stably with highly virulent parasites (May & Anderson, 1983).

(iii) The means available to a virus to escape immune destruction depend, to a large extent, on its genome type. DNA viruses, with their large genomes, can co-opt host cell genes during evolution, which are useful in avoiding or inhibiting the immune response. RNA viruses, whose genome size is limited by their higher mutation rate (Eigen, 1971), cannot afford the burden of extra genes, and must adopt other methods.

(iv) Evolution of a virus during the infection of an individual must be sharply distinguished from the evolution of the virus in the population as a whole. Many mutations in a virus that allow it to persist in an individual, place it at a strong selective disadvantage in the population, since they reduce infectivity or replication rate. RNA viruses, in particular, generate an extraordinary diversity of sequences in a persistent infection *in vivo*, but only a small proportion of these sequences may be transmitted to a subsequent host.

 (v) In infected individuals, the functional requirements of viral proteins limit the sequence variation that can be tolerated by the virus.

(vi) Since antibodies recognize intact proteins, antibody-mediated immune selection pressure is directed mainly against surface proteins of the virus. However, CTL recognize fragments of degraded viral proteins from inside the cell: they can therefore exert a selection pressure on any viral antigen. In fact, the most powerful CTL-stimulating antigens usually are not surface glycoproteins but 'internal' viral proteins (Townsend & Bodmer, 1989).

To evade immune recognition, a virus may change its antigenic structure, its gene composition (i.e. by acquiring new genes) or its behaviour. These strategies, which are not mutually exclusive, will be illustrated below with examples from different virus species.

Viral immune escape by changes in antigenic structure

Escape from antibody

The most obvious way for a virus to escape specific immune recognition is for a mutation to occur in the region of the protein, the epitope, recognized by antibodies or T cells. The sequence of many virus surface proteins, which are most exposed to antibody, is highly variable within and between viral isolates, and it is naturally inferred that this variation is the result of antibody-mediated immune selection pressure. It is relatively straightforward to demonstrate such pressure *in vitro*, by means of monoclonal antibodies that select neutralization escape mutants of the virus, but evidence for *in vivo* selection by antibody is much harder to come by.

In influenza A virus infection in humans (Webster *et al.*, 1982; Chapter 34), new strains of the virus arise by point mutations (genetic drift) or reassortment of the genome segments between strains (genetic shift), and may cause epidemics when the human population lacks antibody that recognizes the haemagglutinin (HA), and to a lesser extent the neuraminidase (NA), of the new strain. For this reason, when the H1N1 subtype of influenza A recurred in 1977, those highly susceptible to influenza infection were people born since the replacement of H1N1 strains by H2N2 in 1957. The strong evidence that these genetic changes are the result of antibody-mediated selection comes from two observations: i) the very close correlation between sequence changes and antibody reactivity (antigenic changes) in the virus, and ii) the fact that a high proportion of sequence changes in epidemic strains of influenza viruses are found in the sites on HA (Wilson, Skehel & Wiley, 1981) and

NA (Colman, Varghese & Laver, 1983) that are known to be important antibody epitopes.

Infection with influenza A virus is perhaps the only clear-cut instance in which the antibody is known to select novel virus subtypes in the human population, and so influences directly the evolution of the virus. Interestingly, influenza B virus, which evolves much more slowly than influenza A, appears to behave quite differently: about 30% of observed nucleotide changes in influenza B were found to be amino acid coding changes, compared with 50% in influenza A, suggesting that antibody-mediated selection is much less important in the evolution of influenza B virus (Air *et al.*, 1990).

An often-cited example of antibody-mediated virus selection *in vivo* is in infection of horses with the lentivirus, equine infectious anaemia virus (EIAV). In the first few months after infection, there are episodes of anaemia and viraemia; serum taken after one such episode can neutralize the virus from preceding episodes, but not from subsequent ones, suggesting that antibody selects against the earlier virus variants. But studies by Carpenter *et al.* (1990) showed that specific neutralizing antibody does not cause the emergence of EIAV variants early after infection, and they concluded that cell-mediated immunity may be more important in limiting the infection than is antibody. A crucial observation is that the antibody did not eliminate *in vivo* the EIAV variant that it was found to neutralize *in vitro*. It therefore appears that variation in antibody specificity is the effect rather than the cause of the variation in EIAV sequence.

Escape from cytotoxic T lymphocytes

Cytotoxic T lymphocytes might in theory exert a stronger selection force on viruses than antibody, because they can recognize fragments of both internal viral proteins and surface membrance proteins. However, the number of CTL epitopes in a viral protein which can be recognized by one individual is limited by the class I MHC antigens (see above). Also, CTL can act only after infection of the cell, and cannot kill free virus. Less work has been done on virus selection mediated by CTL than by antibody, largely because of the greater experimental problems in isolating and handling T cell lines and clones.

Lymphocytic choriomeningitis virus (LCMV)

The first example of selection of virus variants *in vivo* by CTL was demonstrated by Pircher *et al.* (1990). Transgenic mice, which expressed a T-cell receptor specific to the LCMV glycoprotein (GP) antigen on some 80% of their T cells, were acutely infected with the WE strain of LCMV. The transgenic mice failed to eliminate the virus within 15 days of infection, and sequence analysis of the surviving virus showed that point mutations had been selected in the known H-2D^b-restricted CTL epitope in LCMV GP. Synthetic peptides containing these sequences failed to be recognized on target cells by the transgenic CTL, which still recognized the original WE sequence. This was clear evidence of specific selection of LCMV sequences by the CTL. As Pircher and colleagues pointed out, such selection is not seen in non-transgenic mice. This is probably because, in non-transgenic mice, CTL specific to other LCMV epitopes rapidly arise and eliminate the virus; in the TCR-transgenic mice these CTL either do not arise, or arise later in infection. More recently, similar selection of CTL escape mutants of LCMV has been demonstrated *in vitro* (Aebischer *et al.*, 1991).

HIV

Recently, we have obtained evidence that cytotoxic T lymphocytes exert a significant selection pressure in a natural infection with HIV (Phillips *et al.*, 1991). In a longitudinal study of HIV seropositive patients, sequence variation in and near known CTL epitopes in the *gag* protein of HIV, recognized by the T cells in association with HLA-B8, was found to be much more frequent in those patients who carried the HLA-B8 molecule than in HIV-infected control patients who did not. These results suggest that mutations occurred in the HLA-B8-restricted, *gag*-specific CTL. This conclusion was strengthened by the demonstration (Phillips *et al.*, 1991) that some of the 'new' peptide epitope sequences that arose in the HLA-B8-positive patients failed to be recognized *in vitro* by the patients' own CTL, at a time when earlier peptide sequences were still recognized. Accumulation of such unrecognized epitopes could give the virus a strong survival advantage.

However, three points must be made: i) the evidence for CTL-mediated selection remains circumstantial, although in our view it is compelling; actual replication-competent CTL escape mutants of HIV have not been demonstrated; ii) *gag* proteins with the previous peptide sequences appear to persist, in spite of their recognition by CTL: this recalls the studies of Carpenter *et al.* (1991) on persistent infections with EIAV, in which virus variants persisted in the face of an antibody response that neutralized the virus *in vitro*; iii) we do not yet know whether HIV with a putative CTL escape mutant epitope sequence is more likely to be transmitted to another individual, and so the consequences of this apparent CTL-mediated selection for evolution of HIV in the population have yet to be determined.

Recent work by M.A. Nowak *et al.* (submitted for publication) suggests that a viral variant with an epitope that is recognized by the immune response can persist *in vivo* because an immune escape mutation shifts the dominant selection force (the immune response) to another epitope.

Hepatitis B virus (HBV)

Ehata *et al.* (1992) found a cluster of amino acid changes in the HBV core antigen in 15 of 20 patients with chronic liver disease, but no coding changes were found in 10 asymptomatic HBV carriers. The authors concluded that these mutations may be the result of CTL-mediated selection; but the CTL activity was not assayed in this study.

In a separate study, a strong association has been found in HBV-infected patients between severe liver disease and amino acid change in helper T-cell antibody epitopes in the HBV core antigen (Carman *et al.*, 1995). However, it again remains to be demonstrated whether this sequence change affects the efficiency of HBV recognition by T cells or antibodies.

Epstein-Barr virus

Campos-Lima *et al.* (1993) found a predominance of a variant of EBV in Papua-New Guinea that contained an amino acid substitution known to abolish recognition of a CTL epitope by HLA-A11-restricted CTL. The authors suggested that this variant predominated because of the high gene frequency of HLA-A11 in the indigenous population.

A short peptide from a viral sequence that contains a CTL escape mutation can partly inhibit the recognition of the wild-type peptide by CTL *in vitro*. However, the importance of this 'peptide antagonism' in the persistence of viruses *in vivo* is not known (Bertoletti *et al.*, 1994; Klenerman *et al.*, 1994).

Viral immune evasion by acquisition of new genes

The genome size of RNA viruses is limited by their mutation rate (Eigen, 1971), since as a rule RNA virus polymerases lack a proof-reading function. The greater replicative fidelity of DNA viruses allows them a much greater range of genome size (poxviruses encode about 100 polypeptides in the virus particle) and they can therefore acquire new genes, during evolution, which confer a growth or survival advantage on the virus. Of course, interference with immune function may not be the only function of these genes, but in many cases the gene clearly imparts an advantage to the virus. Several such examples of this have now been described: nearly all are found in DNA viruses (but cf. influenza virus in section below).

Expression of Fc receptors for immunoglobulin

Several human herpesviruses induce an IgG Fc receptor to be expressed on the surface of infected cells, including herpes simplex virus (HSV; Baucke & Spear, 1979), cytomegalovirus (CMV; Keller, Peitchel & Goldman, 1976), and varicella-zoster virus (Ogata & Shigata, 1979). The precise nature of the advantage conferred on the virus is unknown, perhaps the receptors reduce the exposure of viral proteins to circulating antibodies.

Interference with complement function

The vaccinia virus complement control protein VCP, which has sequence similarity to eukaryotic proteins that control complement function, interferes with both the classical and alternative paths of complement activation, and increases vaccinia virus virulence *in vivo* (Isaacs, Kotwal & Moss, 1992). Two glycoproteins from HSV, gC-1 and gC-2, bind to the third component of complement and protect the virus against complement-mediated neutralization (McNearney *et al.*, 1987). Epstein–Barr virus contains a factor(s) which acts with a plasma enzyme, factor I, to degrade the C3b and C4b components of complement (Mold *et al.*, 1988).

Disruption of the actions of interferon

Vaccinia virus contains a gene with sequence similarity to the eukaryotic initiation factor 2α (eIF-2α), which is normally phosphorylated by the IFN-induced protein kinase, and so inhibits translation initiation (see interferons, above). The vaccinia virus eIF-2α homologue was found to confer resistance to α and β IFNs on the vaccinia virus (Beattie, Tartaglia & Paoletti, 1991). Several viruses inhibit the action of the protein kinase itself: adenoviruses (Kitajewski *et al.*, 1986; O'Malley *et al.*, 1986), influenza viruses (Katze *et al.*, 1988), and reoviruses (Imani & Jacobs, 1988).

Interference with class I MHC expression

Reduced surface expression of class I MHC molecules would clearly reduce the efficiency of surveillance by cytotoxic T cells, and confer a strong survival advantage on the virus. The best characterized example of this is the inhibition of class I MHC expression by adenoviruses; in the case of adenovirus type 2 this has been shown to be mediated by a viral protein, E3/19K, which specifically binds intracellular class I molecules and prevents their transport to the cell surface (Burgert & Kvist, 1985). In adenovirus type 12, a gene in the E1 region of the genome is responsible for inhibiting class I MHC expression (Bernards *et al.*, 1983; Schrier *et al.*, 1983).

Human cytomegalovirus produces a protein from the UL18 gene with sequence similarity to MHC class I proteins. The UL18 gene product binds β_2-microglobulin, which is required for normal class I MHC expression; this is apparently one mechanism by which human CMV infection leads to inhibition of HLA expression on the cell surface (Browne *et al.*, 1990). Other, as yet unidentified, genes of CMV expressed in the early phase of viral infection selectively prevent the presentation of a major CTL antigen of CMV to T cells (Del Val *et al.*, 1989).

Serpins: interference with degradation of viral antigens

The B13R and B24R genes of vaccinia virus encode two proteins with *ser*ine *p*rotease *in*hibitory activity ('serpins'). Disruption of these genes led to an increased antibody response to another protein expressed by a recombinant vaccinia virus (Zhou *et al.*, 1990). The same proteins might be involved in inhibiting the 'processing' (degradation) of viral proteins into the short peptides that can be recognized when bound to class I MHC antigens (Townsend *et al.*, 1988). Serpins may also inhibit the inflammatory response by interfering with the production of chemo-attractant substances (Palumbo *et al.*, 1989).

Interference with cytokine activity other than interferons

Smith and Chan (1991) showed that vaccinia virus also encodes two genes with sequence similarity to the IL-1 receptor gene. These authors suggested that the products of these genes may adsorb IL-1 and so curb an effective immune response, but this awaits experimental demonstration. Epstein–Barr virus has been shown to encode a gene, BCRF1, with homology to the gene of cytokine synthesis inhibitory factor IL-10, with which it shares certain properties (Vieira *et al.*, 1991); the BCRF1 product probably confers an advantage to the virus by inhibiting production of IFNγ.

Natural killer cell escape

Some evidence for the emergence *in vitro* of Pichinde virus variants that escape destruction by natural killer cells was obtained by Vargas-Cortes *et al.* (1992); whether such variants arise *in vivo* is not known.

Viral immune evasion by change in behaviour of the virus

The evidence for this third category of effects of the immune system on viral evolution is entirely inferred, and the evolutionary pathways have not been demonstrated. However, the presumption that the lifestyle or behaviour of the virus concerned leads to avoidance of immune destruction is in many cases very strong.

Several viruses have evolved to replicate or persist in the central nervous system, where they are relatively protected from the immune system (Sedgwick & Dorries, 1991), for example, rabies virus, herpes simplex virus, the polyomavirus JC and, less commonly, measles virus.

Other viruses have developed a life cycle which involves prolonged latency, during which the virus is inapparent to the immune system. Certain herpesviruses latently infect their hosts for life but are periodically reactivated. Retroviruses integrate into the host cell genome and in most cases presumably rely on this latency to avoid elimination by the host; however, the human retroviruses HIV-1 (Walker *et al.*, 1987) and HTLV-I (Jacobson *et al.*, 1990; Parker *et al.*, 1992, 1994) persistently express at least some genes in the face of an active antibody and CTL response.

Certain viruses have a transient or long-lasting immunosuppressive effect which has not been attributed to individual viral genes as described above. For example, von Pirquet showed in 1908 that acute infection with measles virus inhibits the tuberculin skin sensitivity reaction (Norrby & Oxman, 1990). The mechanisms by which HIV suppresses the immune response are not understood (Miedema, Tersmette & van Lier, 1990), but may include inhibition of CTL function, or lysis of infected CD4$^+$ T cells (Siliciano *et al.*, 1988).

Conclusions

(i) The emergence of a viral antigenic escape mutant *in vivo* does not usually lead to the complete replacement of the pre-existing sequence. This is because the survival advantage of the new variant is relative, not absolute, and the balance between old and new variants can only be understood in terms of population dynamics; at present the population dynamics of viruses *in vivo* are not understood.

(ii) There is good evidence of both antibody-mediated and CTL-mediated selection of viral sequences in single infected individuals. However, the extent to which these selection forces result in evolution of the virus in the population remains largely unexplored. The strongest evidence to date exists for influenza A virus.

(iii) The wide range of mechanisms used by viruses, particularly DNA viruses, to avoid the immune response, clearly shows the strength of immune selection forces in viral evolution.

Acknowledgements

I thank my colleagues S. Daenke, A. McMichael, S. Niewiesk and R. Phillips for comments on the manuscript, and the Wellcome Trust for financial support.

References

Aebischer, T., Moskophidis, D., Rohrer, U.H., Zinkernagel, R.M. & Hengartner, H. (1991). *Proc. Natl. Acad. Sci. USA*, **88**, 11047–51.

Air, G.M., Gibbs, A.J., Laver W.G. & Webster, R.G. (1990). *Proc. Natl. Acad. Sci. USA*, **87**, 3884–8.

Bangham, C.R.M. & McMichael, A.J. (1989). In *T Cells*, M. Feldmann, J. Lamb and M.J. Owen (eds) J. Wiley, New York, Chapter 14.

Baucke, R.B. & Spear, P.G. (1979). *J. Virol.*, **32**, 779–89.

Beattie, E., Tartaglia, J. & Paoletti, E. (1991). *Virology*, **183**, 419–22.

Bernards, R., Schrier, P.I., Houweling A., Bos, J.L., van der Eb, A.J., Zisjltra, M., & Melief, C.J.M. (1983). *Nature*, **305**, 776–9.

Bertoletti, A., Sette, A., Chisari, F.V., Penna, A., Levrero, M., De Carli, M., Fiaccadori, F. & Ferrari, C. (1994). *Nature*, **369**, 407–11.

Browne, H., Smith, G., Beck, S. & Minson, A. (1990). *Nature*, **347**, 770–2.

Burgert, H.G. & Kvist, S. (1985). *Cell*, **41**, 987–97.

Campos-Lima, P.-O. de, Gavioli, R., Zhang, Q.-J., Wallace, L., Dolcetti, R., Rowe, M., Rickinson, A. & Masucci, M. (1993). *Science* **260**, 98–100.

Carman, W., Thursz, M., Hadziyannis, S., McIntyre, G., Colman, K., Gioustozi, A., Fattovich, G., Alberti, A. & Thomas, H.C. (1995). *J. Viral Hepatitis* (in press).

Carpenter, S., Evans, L.H., Sevoian, M. & Chesebro, B. (1990). Chapter 6, in: *Applied Virology Research, vol.2: Virus Variability, Epidemiology, and Control.* E. Kurstak, R.G. Marusyk and F.A. Murphy (eds), Plenum Medical Book Co., New York/London.

Colman, P.M., Varghese, J.N. & Laver, W.G. (1983). *Nature*, **303**, 41–4.

Davis, M.M. & Bjorkman, P.J. (1988). *Nature*, **334**, 395–402.

Del Val M., Munch, K., Reddehase, M.J. & Koszinowski, U.H. (1989). *Cell*, **58**, 305–15.

Ehata, T., Omata, M., Yokosuka, O, Hosoda, K. & Ohto, M. (1992). *J. Clin. Invest.*, **89**, 332–8.

Eigen, M. (1971). *Naturwissenschaften*, **58**, 465–523.

Imani, F. & Jacobs, B.L. (1988). *Proc. Natl. Acad. Sci. USA*, **85**, 7887–91.

Isaacs, S.N., Kotwal, G.S. & Moss, B. (1992). *Proc. Natl. Acad. Sci. USA*, **89**, 628–32.

Jacobson, S., Shida, H., McFarlin, D.E., Fauci, A.S. & Koenig, S. (1990). *Nature*, **348**, 245–8.

Joklik, W.F. (1990) in *Fields Virology* 2nd edn, B.N. Fields, D.M. Knipe, R.M. Chanock, M.S. Hirsch, J.L. Melnick, T.P. Monath and B. Roizman (eds), Raven Press, New York, Chapter 16.

Katze, M.G., Tomita, J., Black T., Krug, R.M., Safer B. & Hovanessian, A. (1988). *J. Virol.*, **62**, 3710–17.

Keller, R., Peitchel R. & Goldman, J.N. (1976). *J. Immunol*, **116**, 772–7.

Kitajewski, J., Schneider, R.U., Safer, B., Munemitsu, S.M., Samuel, C.E., Thimmappaya, B. & Shenk, T. (1986). *Cell*, **45**, 195–200.

Klenerman, P., Rowland-Jones, S., McAdam, S., Edwards, J., Daenke, S., Lalloo, D., Koppe, B., Rosenberg, W., Boyd, D., Edwards, A., Giangrande, P., Phillips, R.E. & McMichael, A.J. (1994). *Nature*, **369**, 403–7.

McNearney, T.A. Odell C. Holers V.M., Spear, P.G. & Atkinson, J.P. (1987). *J. Exp. Med.*, **166**, 1525–35.

Male, D., Champion, B., Cooke, A. & Owen, M. (1991). *Advanced Immunology* 2nd edn, Gower Medical Publ. New York/London.

May, R.M. and Anderson, R.M. (1983). *Proc. Roy. Soc. Lond. (B)* **219**, 281–313.

Miedema, F., Tersmette, M. & van Lier, R.A.W. (1990). *Immunol. Today*, **11**, 293–7.

Mold, C., Bradt, B.M., Nemerow, G.R. & Cooper, N.R. (1988). *J. Exp. Med.*, **168**, 949–69.

Norrby, E. & Oxman, M.N. (1990). Measles virus, in: Fields, B.N., Knipe, D.M., Chanock, R.M., Hirsch, M.S., Melnick, J.L., Monath, T.P. and Roizman, B. (eds). *Virology*, 2nd edn, Chap.37, Vol.1, pp. 1013–1044. Raven Press, New York

O'Malley, R.P., Marino, T.M., Siekerka, J. & Mathews, M. (1986). *Cell*, **44**, 391–400.

Ogata, M. & Shigata, S. (1979). *Infect. Immun.* **26**, 770–4.

Ostergard, H.L. & Clark, W.R. (1989). *J. Immunol.*, **143**, 2120–6.

Palumbo, G.J., Pickup, D.J., Fredrickson, T.N., McIntye, L.J. & Buller, R.M.L. (1989). *Virology*, **172**, 262–73.

Parker, C.E., Daenke, S.D., Nightingale, S. & Bangham, C.R.M. (1992). *Virology*, **188**, 628–36.

Parker, C.E., Nightingale, S., Taylor, G.P., Weber, J. & Bangham, C.R.M. (1994). *J. Virol.*, **68**, 2860–8.

Phillips, R.E., Rowland-Jones, S., Nixon, D.F., Gotch, F.M., Edwards, J.P.

Ogunlesi, A.O., Elvin, J.G., Rothbard, J.A., Bangham, C.R.M., Rizza, C.R. & McMichael, A.J. (1991). *Nature*, **354**, 453–9.
Pircher, H., Moskophidis, D., Rohrer, U., Burki, K., Hengartner, H. & Zinkernagel, R.M. (1990). *Nature*, **346**, 629–33.
Schrier, P.I., Bernards, R., Vaessen, R.T.M.J., Houweiling, A. & van der Eb, A.J. (1983). *Nature*, **305**, 771–5.
Sedgwick, J.D. & Dorries, R. (1991). *Semin. Neurosci.* **3**, 93–100.
Siliciano, R.F., Lawton, T., Knall, C., Karr, R.W., Berman, P., Gregory, T. & Reinherz, E.C. (1988). *Cell*, **54**, 561–75.
Smith, G.L. & Chan, Y.S. (1991). *J. Gen. Virol.*, **72**, 511–18.
Townsend, A., Bastin, J., Gould, K., Brownlee, G., Andrew, M., Coupar, B., Boyle, D., Chan, S. & Smith, G. (1988). *J. Exp. Med.*, **168**, 1211–24.
Townsend, A.R.M. & Bodmer, H. (1989). *Ann. Rev. Immunol.*, **7**, 601–24.
Vargas-Cortes, M., O'Donnell, C.L., Maciaszek, J.W. & Welsh, R.M. (1992). *J. Virol.*, **66**, 2532–5.
Vieira, P., de, Waal-Malefyt R., Dang, M.-N. *et al.* (1991). *Proc. Natl. Acad. Sci. USA*, **88**, 1172–6.
Walker, B.D., Chakrabarti, S., Moss, B., Paradis, T.J., Flynn, T., Durno, A.G., Blumberg, R.S., Kaplan, J.C., Hirsch, M.S. & Schooley, R.T. (1987). *Nature*, **328**, 345–8.
Webster, R.G., Laver, W., Vair, G.M. & Schild, G.C. (1982). *Nature*, **296**, 115–21.
Wilson, I.A., Skehel, J.J. & Wiley, D.C. (1981). *Nature*, **289**, 366–73.
Zhou, J., Crawford, L., McLean, L., Sun, X.-Y., Almond, N. & Smith, G.L. (1990). *J. Gen. Virol.*, **71**, 2185–90.

Further reading

Immunoglobulin and T-cell receptor; NK cells, Male *et al.* (1991); *Cytotoxic T lymphocytes function*, Bangham and McMichael (1989); *Recognition of target cell*, Townsend and Bodmer (1989); *Interferons*, Joklik (1990).

12

Effect of variation within an HIV-1 envelope region containing neutralizing epitopes and virulence determinants

JAAP GOUDSMIT

Introduction

The third variable domain (V3) of the external envelope glycoprotein (gp120) of the human immunodeficiency virus type 1 (HIV-1) contains determinants of virus cytopathicity, cell tropism and virus infectivity (Cheng-Mayer, Shioda & Levy, 1991; Chesebro *et al.*, 1991; Freed & Risser, 1991; Hwang *et al.*, 1991; Ivanoff *et al.*, 1991; Takeuchi *et al.*, 1991; de Jong *et al.*, 1992; Fouchier *et al.*, 1992). Antibodies binding to the virus V3 region block cell-free virus infectivity and cell-to-cell spread of virus (Goudsmit *et al.*, 1988; Palker *et al.*, 1988; Rusche *et al.*, 1988). Recent evidence suggests that the ability to induce V3-specific antibodies may be pivotal to vaccine efficacy. Berman *et al.* (1990) and Girard *et al.* (1991) have demonstrated that chimpanzees can be protected by active immunization from infection by virions. Subsequently, Girard and colleagues have shown that the vaccination protocol used previously (Girard *et al.*, 1991) protected against challenge with infected cells (M. Girard, personal communication). Both Berman and Girard suggested that antibodies to the virus V3 region contributed to protection, a notion that was corroborated by the fact that monoclonal antibodies to the V3 domain protected chimpanzees against infection by HIV-1 virions (Emini *et al.*, 1992). Variation of the HIV-1 envelope protein and of the V3 domain in particular, may control HIV-1 pathogenesis and vaccine efficacy by modulating virus virulence and sensitivity to neutralizing antibodies.

Two major forms of variation are observed during the course of infection: one general and one V3-specific. Host-to-host transmission has a major impact on one form of variation, but hardly one on the other, resulting in very important principles of V3 diversification in the population. The consequences of these principles of V3 variation for virus phenotype and vaccine strategies will be discussed.

Characteristics of V3 variation in HIV-1 DNA of peripheral blood mononuclear cells and in HIV-1 RNA in serum

To avoid *in vitro* selection, most of the available sequence information has been obtained by polymerase chain reaction (PCR) amplified sequences from uncultured material. Thus far, however, only a single report by Simmonds *et al.* (1991) compares sequences from uncultured peripheral blood mononuclear cells (PBMCs) and plasma. Sequentially sampled sequences from one patient showed that new variants initially appeared in plasma and later became dominant in the PBMCs. A similar study was done by Wolfs *et al.* (1992*a*) and included genomic RNA sequences from sequential samples from four chronically infected children from whom corresponding DNA sequences have also been published (Wolfs *et al.*, 1990). Two samples were taken from each child at different intervals (21–45 months) and varying times (44–96 months) after infection. The four children were from a larger group of children who were infected in 1981 with HIV-1 by one or more transfusions of plasma from a single donor (Wolfs *et al.*, 1990). For each sample, RNA nucleotide consensus sequences, made by assigning to each position the most frequently occurring nucleotide, were always more like the corresponding DNA consensus sequence than any of the other samples (Fig. 12.1). Thus, although consensus sequences made from at least four individual short sequences (180 nucleotides) may not represent the master sequence of the quasi-species, they appear to be close to the most prevalent sequences in the population. In addition, within one sample, complexity of the RNA and DNA populations of all samples is the same, especially following host-to-host transmission of HIV-1. During the period that HIV-1 antibodies appear, and up to 6 months thereafter, the population of HIV-1 DNA in cells and HIV-1 RNA in serum is virtually monoclonal. All of these datasets, including those of Simmonds *et al.* (1991), contain nucleic acid sequences encoding the envelope protein. The observed early monoclonality may primarily be due to immune selection playing a major role in HIV envelope variation, but not necessarily in variation of proteins that are not (or less) involved in the process of infectivity and virus spread.

It is likely that the intracellular HIV-1 DNA population contains as many different kinds of genomes (defectives, unintegrated forms, etc)

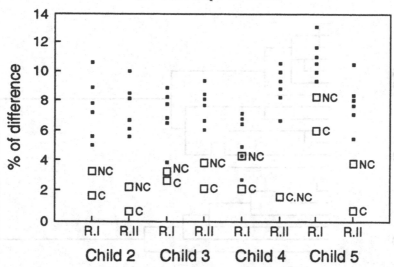

Fig. 12.1. Nucleotide differences between consensus RNA and DNA sequences. Consensus sequences were made by assigning to each position the most frequently found nucleotide. The RNA consensus sequence of a child (R.I, early RNA consensus sequence; R.II, late RNA consensus sequence) is compared to the DNA consensus sequence of that same child (squares) at the corresponding timepoint (C) and at the non-corresponding timepoint (NC) and to DNA consensus sequences of the other children (dots). Differences are expressed in percentages.

as the extracellular HIV-1 RNA (non-viable genomes, etc). All of this information is lost when one analyses genomes of HIV-1 isolates. Most of the biological and immune-functional data are obtained from intact replicating viruses. Therefore, the next issue is the bias that is introduced by virus isolation.

In vitro *selection of particular genotypes:*
impact on assessment of V3 variation

In vitro culture of HIV has been shown to select a subset of variants from the pool of variants present *in vivo* (Meyerhans *et al.*, 1989; Delassus, Cheynier & Wain-Hobson, 1991; Martins *et al.*, 1991; Vartanian *et al.*, 1991; Kusumi *et al.*, 1992). These tissue-culture isolates are probably the variants that grow best in transformed human cell lines or in primary peripheral blood mononuclear cells.

It is of importance to know, first, how quickly does co-cultivation with either PBMCs or cell-lines select for a certain V3 sequence, and is that

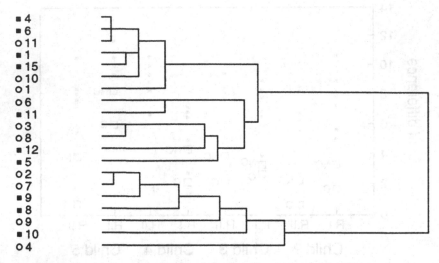

Fig. 12.2. A dendrogram showing the relationships of 20 HIV-1 sequences
(276 nucleotides) from uncultured virus isolates calculated from their nucleotide
differences. Sequences were derived from genomic RNA from plasma (○) or
proviral PBMCs (■). Type A is top cluster, type B is bottom cluster (Wolfs
et al., 1991).

V3 sequence associated with increased *in vitro* replication? Secondly, are
similar selective forces operational *in vivo* and in which phase of HIV
infection?

To answer the first question Wolfs *et al.* (1992*b*) separated blood from
a participant of the Amsterdam Cohort Studies and obtained by PCR
initially 10 molecular clones of 276 bp in length that included the V3
region, and were either derived from genomic RNA in plasma or
proviral DNA from PBMC's. Though no apparent clustering of PBMC
or plasma sequences was observed, indicating that also in this sample (as
has been indicated above) the V3 populations were indistinguishable, we
found that sequences were clearly divided into two groups (Fig. 12.2)
designated type A and type B.

To examine whether types A and B were both sequence variants that
originated from diversification of a single virus isolate or whether they
belonged to two different virus isolates (as the result of re-infection),
Wolfs and colleagues determined the similarity of the V3 flanking regions
of both type A and type B. Fig. 12.3 shows again a cluster analysis based
on nucleotide distances between the V3 flanking regions of the two
types. The strict separation into two groups disappeared and sequences

of both types were dispersed in the tree. Therefore, it appeared that, in spite of the large difference in their V3 regions, types A and B originated from one virus isolate.

Subsequently, the effect of culture was studied by determining the sequences of multiple isolates after cocultivating a patient's plasma and PBMC with donor PBLs and the MT-2 T-cell line. All 20 sequenced clones were of type A. This selection occurred within 96 hours of cocultivation with PBLs. To examine the role of the V3 domain in this selection process, three V3 variants (2 of type A and one of type B) were cloned into an infectious molecular HIV-1 clone (de Jong *et al.*, 1992). Following transfection into a CD4+ T-cell line all three clones produced virus progeny. The infectious clone with the type A V3 domain produced more virions than the infectious clone with the type B V3 domain, suggesting that the V3 envelope domain directly affected the selection process of its coding region.

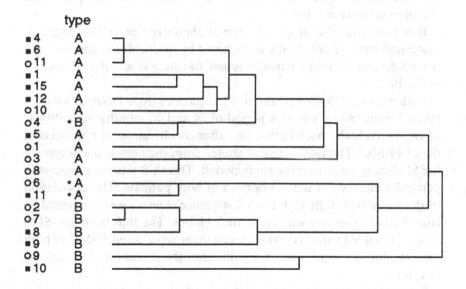

Fig. 12.3. A dendrogram showing the relationships of 20 sequences of uncultured virus isolates calculated from the nucleotide differences between V3 flanking sequences. Nucleotides coding for the V3 region were not included for comparison. The sequence type to which the sequences belong is indicated. An asterisk indicates sequences not included in the original groups. Sequences were derived from genomic RNA from plasma (o) or proviral PBMCs (■).

In vivo *effect of phenotype-associated V3 mutations*

How does this selection process affect viral replication *in vivo*? This question was addressed by Kuiken *et al.* (1992) studying the evolution of the V3 envelope domain in proviral sequences and isolates of HIV-1 during changes of their biological phenotype.

Isolates of viruses that form syncytia in donor PBMCs grow rapidly and infect transformed T-cell lines, and soon cause AIDS (Cheng-Mayer *et al.*, 1988; Tersmette *et al.*, 1988, 1989). Such viruses are designated SI (syncytium-inducing) viruses. Primary isolates showing the non-SI phenotype exclusively contain clones with the non-SI phenotype. In contrast, SI isolates nearly always constitute a mixed population of clones with and without SI capacity (Tersmette & Miedema, 1990; Schuitemaker *et al.*, 1992). Apparently, the phenotype of the primary isolate is determined by the phenotype of the most virulent clone present. It seems that non-SI monocytotropic clones are responsible for the persistence of HIV-1 infection. Progression of disease is associated with a selective increase of T-cell tropic, non-monocytotropic clones (Schuitemaker *et al.*, 1992).

It is clear that the SI (T-cell tropic) phenotype of HIV-1 strains is associated with, if not totally determined by, particular residues within the V3 domain of that particular isolate (de Jong *et al.*, 1992; Fouchier *et al.*, 1992).

Kuiken *et al.* (1992) generated V3 sequences from DNA of samples taken 3 months apart over a period of 24 and 30 months from PBMC of two individuals, both before and after cocultivation with uninfected donor PBMC. The isolated virus shifted from the non-SI phenotype to the SI phenotype during the study period. This shift was associated with particular changes in the V3 domain of both patients. The association of the phenotype shift with the V3 sequence changes was confirmed by construction of viruses with chimeric V3 loops. The shift from non-SI to SI-associated V3 variants was also seen in the uncultured PBMC of both patients, but not until 3 and 9 months after the detection of SI variants in culture.

To further investigate the evolution of the SI phenotype, multi-dimensional scaling can be done on the matrix of Hamming distances between consensus sequences of a set of samples (Kuiken *et al.*, 1992). This multivariate analysis method can be used to represent the distances between a set of sequences (N) in fewer than N−1 dimensions. If one dimension is sufficient, the sequences can be represented as lying on a straight line. If the relationships between the sequences are very heterogeneous because many mutations have been required in particular lineages and there is no unidirectional evolutionary line, more than one dimension is needed to accurately represent the Hamming distance matrix. Such an analysis is shown in Fig. 12.4. This figure shows the V3 sequence shift of variants coinciding with the non-SI (C1–4) to SI (C5–8) phenotypic shift. In addition, it shows that the V3 sequence population in the uncultured material (V3–8) moves in sequence space in the same direction as the V3 sequence population in the cultured material, however, with a lag time of about 9 months. The uncultured sample 6 (V6) was the main cause of deviation from a one-dimensional representation. This pattern looks like a 'fitness landscape'. Several fitness peaks (concentrations of similar variants) are formed by viruses that are almost as fit as the 'master' virus. Variants in the fitness lowlands between the peaks generate fewer offspring and exist only at low frequency. In crossing the lowlands, it may be advantageous for a variant to take a longer route, if that enables it to avoid a particularly crippling mutation. If this concept, put forward by Kuiken, is correct,

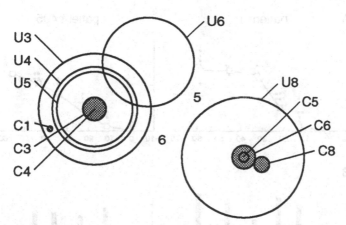

Fig. 12.4. Two-dimensional classification of the changes in the sequences isolated from patient H479, calculated by multidimensional scaling from a matrix of Hamming distances between nucleotide consensus sequences. Distances between the circle centres represent these Hamming distances. Thus, the difference between cultured samples 3 (C3) and 6 (C6) is 6 nucleotides, while between uncultured samples 5 and 6 it is 3 nucleotides. The radius of each circle is proportional to the mean Hamming distance between the clones and the consensus sequence of each sample. The circles were added to the figure manually. Hatched circles represent the sequences from cultured material (C) and open circles those from uncultured patient material (U).

then neither the calculation of a minimum-length phylogenetic tree nor the computation of evolutionary rates for the V3 domain are justified.

Still, fixation of particular biologically significant mutations within the V3 domain may help interpreting the evolution of the V3 domain during the course of infection.

Evolution of V3 mutations during the course of infection

A general increase in genomic heterogeneity is observed during persistently high neutralizing and V3-specific antibody titres in genomic RNA (Wolfs *et al.*, 1991). Fig. 12.5A illustrates this for two patients, one (patient 1) that remained asymptomatic during the observation period of 60 months and the other (patient 495) that developed AIDS. As can be observed the continuous increase in genomic diversity over a period of years resulting in an increasing genomic distance from the initial genome population appears to be independent of both the rapidity of disease progression and the virion concentration (Wolfs *et al.*, 1991). In panel B of Fig. 12.5 two phenomena are clear. Shortly after virus

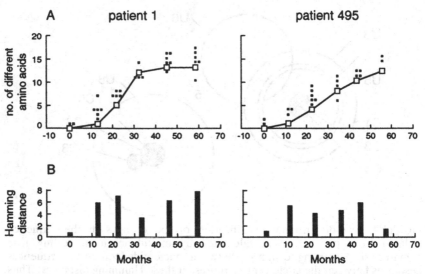

Fig. 12.5 A. Number of deduced amino acid differences in clonal sequences (dots) and consensus sequences (open squares) at different times after the first sample was taken. A total of 92 amino acids (position 270–363 of the HXB2 envelope gp120, Los Alamos 1990) for each isolate was compared. B. Amino acid diversity of the population of sequences at each time of sampling; the diversity is expressed as the mean Hamming distance.

transmission, around the time of antibody seroconversion, the HIV-1 V3 RNA population is homogeneous, both the extracellular HIV-1 RNA as well as the intracellular HIV-1 DNA. Another observation also seen both in HIV-1 RNA and HIV-1 DNA populations is a decrease in the Hamming distances later in the course of infection and related to disease development (Fig. 12.5B, patient 495). However, it is still unresolved whether the increase of homogeneity later in infection is caused by virus phenotype, use of drugs like AZT or other pathogenic factors. Nowak *et al.* (1991) explained this observation as the existence of an antigenicity threshold that is passed at a certain time due to overwhelming of the immune system by a myriad of HIV variants.

The second major form of genomic variation occurs in a more discontinuous fashion. This concerns mutations that are fixed in all molecular clones within a very limited timespan of a few months. Two to four years after the acquisition of HIV-1 infection an amino acid-changing mutation occurs at position 308 within the V3 loop. This change directly affects the binding of patient's antibody both in specificity and affinity to V3 peptides that are distinct for position 308. From these data one may conclude that naturally occurring mutations within HIV-1 V3 genomic RNA lead to antigenic variation dependent on this one particular amino acid substitution.

A second set of fixed mutations in the V3 region occurs approximately 2 years prior to disease development and is associated with the emergence of SI viruses. These changes at positions 306, 320 and 324 (de Jong *et al.*, 1992; Fouchier *et al.*, 1992; Kuiken *et al.*, 1992) may also have impact on the antigenicity of the V3 domain although this is not certain.

In summary, during the course of HIV-1 infection and despite continuous genomic diversification, two distinct events occur: mutations are fixed at position 308, and these alter the antigenicity of HIV-1, and subsequently mutations are fixed at positions 306 and 320 and these influence the biological phenotype of the virus.

HIV-1 V3 RNA diversification following transmission between humans

Wolfs *et al.* (1992c) studied seven presumed donor-recipient pairs following sexual (6/7) or parental (1/7) transmission. Again, HIV-1 RNA was studied. In this study the first RNA-positive serum sample was studied of each recipient as well as the serum sample of the virus source, identified by epidemiological history and taken within three months of the recipient's seroconversion. Again, the sequence populations of the recipients were, without exception, homogeneous, while the sequence populations of the donors showed varying degrees of heterogeneity. In addition the consensus sequence of the recipient was closest to that of the likely donor among the 5 donor-recipient pairs, including two pairs that were epidemiologically linked beyond doubt. From these data it may be concluded that when HIV-1 is transmitted, a limited set of genomes carries the infection to the new host, and does not often lead to a shift from the consensus sequence population.

The fact that HIV-1 genomic RNA populations are homogeneous shortly after infection when antibodies first circulate suggests that antigenic diversification of HIV-1 can be studied during an epidemic.

The V3 sequences of isolates obtained from sera of individuals who acquired HIV-1 within the Netherlands in 1981/1982, in 1985 and in 1990 have been determined. Preliminary analysis of these results indicate that diversification of V3 sequences occurred during the last ten years of the AIDS epidemic. When changes at position 308, shown to be important for antigenicity, were studied, histidine in particular predominated at that position in 1981/1982 (9/10), while in 1990 4/10 viruses had histidine, 4/10 had proline, 1/10 had asparagine and 1/10 arginine, illustrating not so much an influenza-like sequential drift or shift, but a diversification.

The crucial phenotypic consequence for the human population: how to cope with HIV-1 V3 variation?

Previously Nowak *et al.* (1991) postulated that an antigenic threshold exists above which the patient's, often compromised, immune system is unable to cope with the number of HIV variants infecting or affecting CD4+ cells. I would like to propose a similar model for the spread of HIV-1 in the population. The argument stems in particular from the observation that a dominant strain, usually the non-SI phenotype, is transmitted whatever the genomic heterogeneity of the virus in the donor (Nowak *et al.*, 1991; Wolfs *et al.*, 1991). Perhaps the spread of V3 mutants in the population depends on the frequency of random transfers during individual HIV-1 infections. Although V3 heterogeneity increases continuously during an infection, particular mutations related to antigenicity or the biological phenotype of HIV-1 appear at fixed times during the asymptomatic phase of individual infections. If a vaccine has to protect against multivalent V3 antigens, the antigenic variability of the virus in the seropositive population may exceed the 'antigenic threshold' of the vaccinated population and result in the spread of infection to the seronegatives at risk.

An apparent consequence of this idea might be that vaccines should be given simultaneously to seronegatives to protect them against infection and to seropositives to decrease the antigenic variability of the viruses in them at a time when they can still immunologically respond. This moment is best before the switch in amino acids at residue 308 in the V3 domain has occurred, and when the broadly neutralizing (low affinity) antibodies are still absent. In practical terms this means that vaccination has to occur among seropositives within the first year of infection. In high incidence areas this could be accomplished by vaccinating, every six months, the complete population regardless of their HIV-1 antibody status.

The vaccine should be composed of antigens that decrease the variability in general through the induction of broad immunity, as well as of antigens that prevent the residue 308 antigenic switch (and maybe also the switch at phenotype-related positions) from occurring. The caveat, of course, of the use of a multivalent V3-specific set of antigens, is that escape mutants will be selected. The experience, however, with Visna virus in sheep is that the presence of broadly neutralizing antibody rarely yields resistant mutants (Narayan, Griffin & Chase, 1977; Narayan *et al.*, 1981).

The advantage of such a vaccine strategy, which also may be useful for other persistent virus infections, is that the whole population of an AIDS high-incidence area could be vaccinated without the necessity of costly antibody pre-screening to ensure that only the seronegatives are vaccinated. Repeated vaccination of the whole population at short intervals (half yearly?) is required for this strategy to work. Using V3-derived antigens may have the additional benefit that early immunization of this kind may prevent or delay AIDS development because of the link between HIV-1 envelope variability (Nowak *et al.*, 1991), V3-encoded protective epitopes and V3 encoded virus virulence determinants (de Jong *et al.*, 1992; Fouchier *et al.*, 1992; Kuiken *et al.*, 1992).

References

Berman, P.W., Gregory, T.J., Riddle, L., Nakamura, G.R., Champe, M.A., Porter, J.P., Wurm, F.M., Hershberg, R.D., Cobb, E.K. & Eichberg, J.W. (1990). *Nature*, **345**, 622–5.

Cheng-Mayer, C., Seto, D., Tateno, M. & Levy, J.A. (1988). *Science*, **240**, 80–2.

Cheng-Mayer, C., Shioda, T. & Levy, J. (1991). *J. Virol.*, **65**, 6931–41.

Chesebro, B., Nishio, J., Perryman, S., Cann, A., O'Brien, W., Chen, I.S.Y. & Wehrly, K. (1991). *J. Virol.*, **65**, 5782–9.

de Jong, J.J., Goudsmit, J., Keulen, W., Klaver, B., Krone, W., Tersmette, M. & de Ronde, A. (1992). *J. Virol.*, **66**, 757–65.

Delassus, S., Cheynier, R. & Wain-Hobson, S. (1991). *J. Virol.*, **65**, (1), 225–31.

Emini, E.A., Schleif, W.A., Nunberg, J.H., Conley, A.J., Eda, Y., Tokiyoshi, S., Putney, S.D., Matsushita, S., Cobb, K.E., Jett, C.M., Eichberg, J.W. & Murthy, K.K. (1992). *Nature*, **355**, 728–30.

Fouchier, R.A.M., Groenink, M., Kootstra, N.A., Tersmette, M., Huisman, H. G., Miedema, F. & Schuitemaker, H. (1992). *J. Virol.*, **66**, 3183–7.

Freed, E.O. & Risser, R. (1991). *AIDS Res. Human Retrov.*, **7**, 807–11.

Girard, M., Kieny, M., Pinter, A., Barré-Sinoussi, F., Nara, P., Kolbe, H., Kusumi, K., Chaput, A., Reinhart, T., Muchmore, E., Ronco, J., Kaczorek, M., Gomard, E., Gluckman, J. & Fultz, P.N. (1991). *Proc. Natl. Acad. Sci. USA*, **88**, 542–6.

Goudsmit, J., Debouck, C., Meloen, R.H., Smit, L., Bakker, M., Asher, D.M., Wolff, A.V., Gibbs, C.J., Jr. & Gajdusek, D.C. (1988). *Proc. Natl. Acad. Sci. USA*, **85**, 4478–82.

Hwang, S.S., Boyle, T.J., Lyerly, H.K. & Cullen, B.R. (1991). *Science*, **253**, 71–4.

Ivanoff, L.A., Looney, D.J., McDanal, C., Morris, J.F., Wong-Staal, F., Langlois, A.J., Petteway, S.R., Jr. & Matthews, T.J. (1991). *AIDS Res. Human Retrov.* **7**, 595–603.

Kuiken, C.L., Jong de, J.J., Baan, E., Keulen, W., Tersmette, M. & Goudsmit, J. (1992). *J. Virol.*, **66**, 4622–7.

Kusumi, K., Conway, B., Cunningham, S., Berson, A., Evans, C., Iversen, A.K. N., Colvin, D., Gallo, M.V., Coutre, S., Shpaer, E.G., Faulkner, D.V., Ronde de, A., Volkman, S., Williams, C., Hirsch, M.S. & Mullins, J.I. (1992). *J. Virol.*, **66**, 875–85.

Martins, L.P., Chenciner, N., Ajö, B., Meyerhans, A. & Wain-Hobson, S. (1991). *J. Virol.*, **65**, 4502–7.

Meyerhans, A., Cheynier, R., Albert, J., Seth, M., Kwok, S., Sninsky, J., Morfeldt-Manson, L., Ajö, B. & Wain-Hobson, S. (1989). *Cell*, **58**, 901–10.

Narayan, O., Clements, J.E., Griffin, D.E. & Wolinsky, J.S. (1981). *Infec. Immunol. 323*: 1045–1050.

Narayan, O., Griffin, D.E. & Chase, J. (1977). *Science*, **197**, 376–8.

Nowak, M.A., Anderson, R.M., McLean, A.R., Wolfs, T.F.W., Goudsmit, J. & May, R.M. (1991). *Science*, **254**, 963–9.

Palker, T.J., Clark, M.E., Langlois, A.J., Matthews, T.J., Weinhold, K.J., Randall, R.R., Bolognesi, D.P. & Haynes, B.F. (1988). *Proc. Natl. Acad. Sci. USA*, **85**, 1932–6.

Rusche, J.R., Javaherian, K., McDanal, C., Petro, J., Lynn, D.L., Grimaila, R., Langlois, A., Gallo, R.C., Arthur, L.O., Fischinger, P.J., Bolognesi, D.P., Putney, S.D. & Matthews, T.J. (1988). *Proc. Natl. Acad. Sci. USA*, **85**, 3198–202.

Schuitemaker, H., Koot, M., Kootstra, N.A., Dercksen, M.W., Goede de, R.E.Y., Steenwijk van, R.P., Lange, J.M.A., Eeftink Schattenkerk, J.K.M., Miedema, F. & Tersmette, M. (1992). *J. Virol.*, **66**, 1354–60.

Simmonds, P., Zhang, L.Q., McOmish, F., Balfe, P., Ludlam, C.A. & Leigh Brown, A.J. (1991). *J. Virol.*, **6511**, 6266–76.

Takeuchi, Y., Akutsu, M., Murayama, K., Shimizu, N. & Hoshino, H. (1991). *J. Virol.*, **65**, 1710–18.
Tersmette, M., Goede de, R.E.Y., Al, B.J.M., Winkel, I.N., Coutinho, R.A., Cuypers, H. Th. M., Huisman, J.G. & Miedema, F. (1988). *J. Virol.*, **62**, 2026–32.
Tersmette, M., Lange, J.M.A., Goede de, R.E.Y., Wolf de, F., Eeftinck Schattenkerk, J.K.M., Schellekens, P.T.H., Coutinho, R.A., Huisman, J.G., Goudsmit, J. & Miedema, F. (1989). *Lancet*, **i**, 983–5.
Tersmette, M. & Miedema, F. (1990). *AIDS 4* (suppl 1), S57–66.
Vartanian, J.-P., Meyerhans, A., Ajö, B. & Wain-Hobson, S. (1991). *J. Virol.*, **65**, 1779–88.
Wolfs, T.F.W., Jong de, J., Berg van den, H., Tijnagel, J.M.G. H, Krone, W.J.A. & Goudsmit, J. (1990). *Proc. Natl. Acad. Sci. USA*, **87**, 9938–42.
Wolfs, T.F.W., Zwart, G., Bakker, M., Valk, M., Kuiken, C.L. & Goudsmit, J. (1991). *Virology*, **185**, 195–205.
Wolfs, T.F.W., Clement, M., Bakker, M., Nerg van den, H. & Goudsmit, J. (1992a). (*in preparation*)
Wolfs, T.F.W., Jong de, J.J., Boucher, C.A.B., Hartman, S., Schipper, P.J., Keulen, W., Nara, P. & Goudsmit, J. (1992b). (*in preparation*)
Wolfs, T.F.W., Zwart, G., Bakker, M. & Goudsmit, J. (1992c). *Virology*, **189**, 103–10.

Part V

Viruses, hosts and populations

13

Quasi-species: the concept and the word

ESTABAN DOMINGO, JOHN HOLLAND,
CHRISTOF BIEBRICHER AND MANFRED EIGEN

Introduction

A type of dynamic genetic organization of a population in which individual genomes have a fleeting existence, and usually differ in one or more positions from the consensus or average sequence of the population, is known as a quasi-species. Quasi-species have been found among viruses with RNA genomes and originate as a result of the high mutation rates during RNA replication and reverse transcription, as well as in the competition among continuously arising variant genomes. Because the concept of the quasi-species is of fundamental importance to understand RNA viruses, in this brief chapter we address the origins of the idea, summarize its biological relevance by referring the reader to other articles and reviews, and discuss the adequacy of the term quasi-species.

Historical background: the origins

The quasi-species concept had two origins: one theoretical and the other experimental.

The first theoretical treatment of a system involving regular generation of error copies by replication with limited fidelity was by one of the authors (Eigen, 1971) in considering earliest life forms on earth. This study was stimulated by Sol Spiegelman's experiments on *in vitro* replication of Qβ RNA, and a 'breakfast discussion' with Francis Crick. 'Self-instructive' behaviour was distinguished from 'general autocatalytic' behaviour, the former being required for 'template' activity. A quality factor was defined which determines the fraction of copying processes that leads to an exact copy of the template. The 'master copy' will produce 'error copies' which will occur with a certain probability distribution. The term 'comet tail' of error copies was used but 'quasi-species' and 'mutant spectrum' were introduced later (Eigen & Schuster, 1977, 1978).

The quasi-species mutant spectrum is at a 'selection equilibrium' (Eigen, 1971; Eigen & Schuster, 1977, 1978), a concept of paramount importance for RNA virology. This equilibrium is usually metastable; it will collapse when a more advantageous mutant appears, and be replaced by a new selection equilibrium. As discussed below, repeated fluctuations in population equilibrium can determine to a large extent the dynamics of RNA virus populations.

The origin of self-organization during early life, one of the main aims of Eigen's treatment, represented a link between the principles of Darwinian evolution and classical information theory. It was recognized that there is a maximum information content which can be preserved reproducibly for a given average copying fidelity (or 'recognition parameter', q). Error rates less than about 10^{-2} per base incorporated would be difficult to reach in the absence of catalytic activities (Inoue & Orgel, 1983). Eigen also recognized that the treatment of evolution as a *deterministic* process (that is, the concept that whenever a mutant with a selective advantage occurs, it will *inevitably* outgrow the former mutant distribution) has two limitations: (1) The elementary process of mutagenesis during replication is inherently *non-deterministic*. It is an example of an elementary event subject to quantum-mechanical uncertainty. (2) The growth and competition process is subject to statistical fluctuations which may modify the result anticipated for any deterministic theory.

A self-instructive 'catalytic hypercycle' during early life on earth may ideally combine the coding potential of nucleic acids with the catalytic capabilities of proteins. However, a theoretical model is worth only as much as its capacity for experimental testing; a general theory is valuable to the extent that it guides such work and defines clear and reproducible conditions for comparative studies. (Eigen, 1971). For a recent review of the theoretical quasi-species concept see Eigen, McCaskill and Schuster (1989a, b).

The second birth of the quasi-species concept was experimental. It was initiated by the observation by C. Weissmann and co-workers of rapid reversion, at a frequency of about 10^{-4} per single base site per round of copying, of an extracistronic mutant of phage Qβ generated by *in vitro* site-directed mutagenesis (Domingo, Flavell & Weissmann, 1976; Batschelet, Domingo & Weissmann, 1976). Genetic heterogeneity and a selective disadvantage of randomly sampled mutants were documented in phage populations. 'The genome of Qβ phage cannot be described as a defined unique structure, but rather as a weighted average of a large number of different individual sequences'. (Domingo *et al.*, 1978). An interesting encounter took place when C. Weissmann presented these experimental results to M. Eigen and colleagues at the annual Klosters meeting of the Max Planck Institut in 1978. It was the beginning of a fruitful interaction between theoretical biophysics and experimental virology.

Several observations that had been made in previous decades had strongly suggested that RNA virus pathogens were very variable in their biological behaviour. However, it was only when methods of molecular analysis of RNA genomes became available that the extent of RNA virus variation was quantified. Work with several animal and plant viruses, as well as with other RNA genetic elements (viroids, satellites, retrotransposons) steadily supported the generality of high mutation rates, population equilibrium, and potential for rapid evolution of RNA genomes (Holland *et al.*, 1982). Equally important for the understanding and testing of the quasi-species concept is the work with a model system, where short-chained RNA species are replicated by Qβ replicase (Haruna *et al.*, 1963; Mills, Peterson & Spiegelman, 1967). Using this system, accurate selection and mutation values of mutants of an RNA population can be derived from kinetic parameters, and used to precisely predict the evolutionary behaviour (Biebricher, 1983, 1986, 1987; Biebricher, Eigen & Gardiner, 1985, 1991).

Experimental evidence of quasi-species

Evidence for the quasi-species structure of RNA genomes is based on:

(1) Determinations of mutation rates (number of misincorporation events per nucleotide site and per round of copying) and mutant frequencies (the proportion of a given mutant genome in a virus population) by genetic and biochemical methods. They include quantifications *in vivo*, and *in vitro* using cell-free enzyme preparations.

(2) Sequence comparisons of the genomes that compose a viral population (natural isolates; sequential isolates from infected humans, animals and plants; plaque-purified viruses; and generation of variants in infections initiated by biological or chemical viral clones).

(3) Measurements of relative selection rate values (the relative population change of a type in the total population) or the gain of individual genomes composing a viral population. Often relative fitness values have

been defined as the selection success of a mutant in the population in relation to that of a master sequence per generation time, or in relation to a reference virus strain upon completion of an infectious cycle. However, fitness values are difficult to determine in viral populations because 'generations' in viral populations are undefined. Some authors have defined a passage or an infectious cycle as a generation, but this view is not justified because new mutations arise during a single replication round. The definition difficulty is further aggravated by the fact that the numbers of viral genome copies in infected cells vary with time. [See Eigen and Biebricher (1988) for a discussion of the meaning of selection values and fitness landscape].

Since it is not possible in this brief chapter to quote the relevant primary publications concerning the experimental evidence of quasi-species, the reader is referred to the following review articles: Holland et al. (1982); Eigen and Biebricher (1988); Domingo et al. (1985); Steinhauer and Holland (1987); Domingo , Holland and Ahlquist (1988); Temin (1989); Eigen & Nieselt-Struwe (1990); Holland (1992); Domingo (1992); Domingo and Holland (1992, 1994); Nowak (1992). Two recent observations must be added. At least two viruses, vesicular stomatitis virus and poliovirus, replicate close to the maximum error rate compatible with maintaining their genetic information (Holland et al., 1990). Poliovirus, however, seems to accept preferentially changes in the third codon position (Eigen & Nieselt-Struwe, 1990). Most characteristic of the quasi-species is its error threshold, i.e. a critical mutation rate above which the information represented by the selected distribution becomes unstable. Closely below the error threshold, the quasi-species shows optimal evolutionary behaviour. It produces a maximum amount of mutants at the periphery of its mutant spectrum that may provide selective advantage or may allow for escape from deleterious environmental alterations. The critical mutation rate is inversely proportional to the nucleotide length of the selected sequence and depends also on the relative fitness values of the mutant spectrum. For RNA viruses (comprising around 10^4 nucleotides) typical error threshold values have been found between 10^{-4} and 10^{-5} (Holland et al., 1990, 1992). The second observation is that repeated plaque-to-plaque passage of phage $\phi6$ or vesicular stomatitis virus leads to viral populations with decreased competitive fitness (Chao, 1990; Duarte et al., 1992). That is, Muller's ratchet operates in RNA genome populations, as expected from the fact that randomly sampled genomes from a clonal viral population will tend, on average, to show a certain decrease in fitness as compared with the average population from which they were isolated. This is, again, a prediction of the quasi-species organization of RNA viruses, since genomes replicating and competing in a defined environment will be rated according to their relative fitness, and the most abundant (or 'master') sequence is often a minority in the population.

Physical, chemical and biological meaning of quasi-species

Quasi-species has a physical, a chemical and a biological definition. Physically, quasi-species can be regarded as a cloud in sequence space (Eigen & Biebricher, 1988). This cloud has a defined population structure, dependent on distance and fitness values of mutants relative to the wild type represented by the consensus sequence. Mathematically, the whole distribution refers to a maximum eigen value and thereby it behaves as if it were (i.e. 'quasi') a single species. Chemically, quasi-species is a rated distribution of related, but non-identical RNA (or DNA) sequences, the definition most familiar to virologists. Biologically, quasi-species is the target of selection. It seems natural to think of individual virus genomes (or individual viral genome sequences) as being the objects of environmental selection. However, the true target of selection is the quasi-species ensemble. Individual RNA genomes (or individual RNA genome sequences) may have only fleeting existence, and their evolution is heavily influenced by the mutant spectrum of variants that surround them (Swetina & Schuster, 1982). Deliberate seeding of very small numbers of highly fit variants into much larger quasi-species populations of lower average fitness does not always ensure that they will rise to dominance, nor even survive; that is, quasi-species swarms may suppress variants of superior fitness unless they are present above a critical threshold frequency (de la Torre & Holland, 1990).

Effects of complex quasi-species populations on RNA virus selection

Even within a clonal population (or subpopulation) selection necessarily must act on the parental genome sequence together with the complex swarm of related variants which it quickly (and inevitably) produces and from which it cannot escape. In virology, it is no more meaningful to discuss the environmental fitness of an individual RNA virus genome sequence than it is in statistical mechanics to refer to the temperature of a single gas molecule. In statistical mechanics, temperature is representative of the *average* kinetic energy, distributed among the degrees of freedom, of all gas molecules in a closed system at equilibrium. Likewise, in RNA virology, fitness in a defined microenvironment depends upon *average* values for the complex genome population. When large quasi-species swarms of RNA virus genomes are replicating and competing in the limited micro-environment of a host tissue site (or cell culture flask), the behaviour of single genomes or single defined genome sequences can have little significance. However, when single genomes temporarily escape their mutant swarms during transmission of one or a few infectious particles to new host (or to a distant site in the same host), such genetic bottlenecks or founder events might sometimes have profound evolutionary consequences. A detailed theoretical discussion is given in Eigen *et al.* (1989*a,b*). Realistic quasi-species are far from being simple symmetrical distributions of mutants with fitness values that decrease monotonously towards the periphery. They rather include many neutral or nearly neutral mutants causing quite bizarre distributions in a rugged mutant frequency landscape involving far - reaching protrusions. Theory can easily cope with such distributions (if known, e.g. by experiments) revealing quite unexpected evolutionary behaviour as indeed is found in viral systems.

Replication and transmission of large populations of RNA viruses

As outlined above, in a large quasi-species population of RNA viruses, there should be one or more (and often many) master sequences which represent the most fit genomes present in that population and in that defined environment. Such master sequences may exist only transiently and evolve rapidly or they may remain in equilibrium and show evolutionary stasis for long periods of time. This will depend strongly on environmental factors and on distances in sequence space required to evolve variants of greater (or at least equal) fitness.

Rapid evolution of master and consensus sequence is frequently observed for RNA viruses. For example, foot-and-mouth disease virus, poliovirus and HIV-1 regularly and rapidly change their master sequences and consensus sequences during replication and transmission in their natural host. In contrast, some plant viruses, eastern equine encephalitis virus in all of North America, and avian influenza A virus genes in their wild avian host can exhibit remarkable evolutionary stasis (see individual chapters in Holland, 1992 and relevant chapters in this volume). Clearly, their master sequences and consensus sequences are able to maintain population equilibrium for years or decades under appropriate conditions. When environmental selective conditions change, such evolutionary stasis might be replaced by rapid evolution. Note that the avian influenza A genes which exhibit prolonged equilibrium in their avian hosts can undergo rapid evolution in humans.

Repetitive transmissions of large populations under unchanging environmental conditions might be expected to cause repetitive selection of the most-fit variant genomes which arise by mutation from these inocula (i.e. master sequences and their swarms of related variants). Following their selection, relative evolutionary stasis (relative population equilibrium) might be expected because highly-fit genomes become well established and are transmitted with high probability. Moreover, fluctuations of environment greatly increase the probability of finding the fittest quasi-species.

Multiple biological implications

Quasi-species are endowed with interesting, not easily appreciated, biological potentialities. On the one hand, at any one time and replicative niche, quasi-species are often well-organized distributions of replicons. On the other hand, such distributions are exquisitely sensitive to environmental changes and, also, frequently perturbed by random sampling events. Consideration of four parameters (average number of mutations per infectious genome found in viral populations, virus populations numbers, genome length, and the number of mutations likely to be required for any particular phenotypic change) render the quasi-species structure of RNA genomes extremely relevant, as discussed in detail previously (Domingo *et al.*, 1985). If phenotypic changes well known to virologists (such as changes in tropism, virulence, resistance to antibodies or to antiviral agents, etc.) required a number of mutations orders of magnitude above the number of mutations found in representative individuals from mutant spectra, then the relevance of such a genetic organization for viral behaviour would be questionable. However, numerous examples show that one or a few mutations suffice to cause significant phenotypic variations. Stochastic fluctuations in genomic distributions are likely to be an important adaptive strategy, relevant to viral pathogenesis (Domingo & Holland, 1992, 1994; Eigen, 1992).

Mutants able to establish persistence rather than killing their host cells, variants with expanded host range, variants resistant to antiviral agents or able to overcome an immune response, may each arise with a finite probability in the ever evolving mutant spectra of quasi-species swarms of RNA viruses. Several recent reviews have discussed additional implications of quasi-species in relation to control of viral disease (Domingo *et al.*, 1985, 1988; Domingo & Holland, 1992).

Is the term quasi-species appropriate?

Few virologists would now deny the concept conveyed by the term quasi-species, although some would still argue that it cannot be applied to all (or most) RNA viruses but only to a few representatives (typically, the influenza viruses and retroviruses). However, improved techniques for genome analysis are increasingly showing that most RNA viruses, even those manifesting relative antigenic stability in nature, are best described as viral quasi-species. The term 'quasi-species', however, has not been fully supported as a descriptor of the extremely heterogeneous nature of RNA genome populations. Alternative terms that have been proposed are 'swarms' (Temin, 1989), 'heterospecies' or 'heteropopulations' (G. Kurath, personal communication), 'clone', etc. Nevertheless, we prefer 'quasi-species' since the alternative proposed terms do not convey the concept of microheterogeneity *along* with that of genetic relatedness (or common ancestry).

An analysis of the word 'quasi-species' may help to dispel some misunderstandings about its meaning. 'Quasi' is a widely used prefix (in mathematics, physics, law, music, etc.) to indicate 'almost', 'seemingly', 'nearly', etc. We do not believe that this widely used prefix should be the source of any ambiguity. However, we understand that the term 'species' is ambiguous. It could either refer to 'biological species' or to 'molecular species' and this ambiguity is obvious in some articles. When Eigen and Schuster (1977) described the physical meaning of quasi-species they stated: 'In biology, a species is a class of individuals characterized by a certain phenotypic behaviour. On the genotypic level, the individuals of a given species may differ somewhat, but, nevertheless, all species are represented by DNA-chains of a very uniform structure. What distinguishes them individually is the very sequence of their nucleotides. In dealing now with such molecules, being the replicative units, we just use these differences of their sequences in order to define the (molecular) species'. Nowak and Schuster (1989) in the introduction to their interesting paper on error threshold referred also to the notion of species in biology. Whether we consider the conventional view of species as a class of biological objects or we regard species as units of behaviour or evolutionary entities, this controversial concept (see, for example, Mayr, 1988) is highly problematic when applied to viruses. Bonds provided by sexual reproduction (and to a large extent by geographical isolation) are often non-existent. As evolutionary entities, viruses being molecular parasites, cannot evolve independently of their hosts. Recent commendable efforts (Van Regenmortel, Maniloff & Calisher, 1991) led to the following definition which obviates some of the above difficulties: 'A virus species is a polythetic class of viruses that constitutes a replicating lineage and occupies a particular ecological niche'. The replicating lineage may also include occasional horizontal gene transfers, as RNA recombination has been found in several virus groups (Lai, 1992). Certainly, viruses can be grouped as polythetic classes, albeit with rather indeterminate spectra of properties even within the confines of a single, clonal population, as discussed in previous sections. Also, even taking an ecological niche in the sense of spatial habitat (rather than in the broader sense of a functional role of a biological entity in a community) it would be difficult to state in precise terms the ecological niche of many viruses. Perhaps an alternative, more general definition would be to consider species as an ensemble that occupies a coherent part of the sequence space which is continuously populated for prolonged periods of time and under a wide variety of environmental conditions.

Thus, the term quasi-species should not be used in relation to the concept of biological species, since the latter cannot be accurately delimited, at least for viruses. The term quasi-species can be used, however, in contradistinction to 'molecular species', to represent a defined ensemble of related, non-identical genomic sequences. The term emphasizes that, due to their error-prone replication, *even clones* of RNA genomes are not homogeneous collections of sequences, although they may behave as if they were. Its use seems justified because, even though genetic variability is a universal feature of all living forms, there are important differences in the space–time coordinates in which variations occur and their effects are felt. For RNA viruses, error-prone replication is inherent to their very existence as molecular parasites. For higher organisms, evolution must be inferred by comparing individuals which have diverged over aeons, and most of their normal biochemical activities proceed according to an essentially fixed programme. Quasi-species define the dynamics of systems in which error copies are continuously being generated even as they complete a single life cycle within a span of hours.

Acknowledgements

We are indebted to many colleagues who have contributed to the quasi-species concept. Work in Madrid is supported by CICYT, EC and Fundación R. Areces, in La Jolla by NIH grant AI 14627.

References

Batschelet, E., Domingo, E. & Weissmann, C. (1976). *Gene*, **1**, 27–32.
Biebricher, C.K. (1983). In *Evolutionary Biology*, Vol. 16, M.K. Hechet, B. Wallace, and G.T. Prance (Eds.) Plenum, New York, pp. 1–52.
Biebricher, C.K. (1986). *Chemica Scripta*, **26B**, 51–7.
Biebricher, C.K. (1987). *Cold Spring Harbor Symp. Quant. Biol.*, **52**, 299–306.
Biebricher, C.K., Eigen, M. & Gardiner, W.C. (1985). *Biochemistry*, **24**, 6550–60.
Biebricher, C.K., Eigen, M. & Gardiner, W.C. (1991). In *Biologically Inspired Physics*, Peliti, L. (ed.) Plenum Press, New York, pp. 317–337.
Chao, L. (1990). *Nature*, **348**, 454–5.
De la Torre, J.C. & Holland, J.J. (1990). *J. Virol.*, **64**, 6278–81.
Domingo, E. (1992). *Current Opinion in Genetics and Development*, **2**, 61–3.
Domingo, E., Flavell, R.A. & Weissmann, C. (1976). *Gene*, **1**, 3–25.
Domingo, E. & Holland, J.J. (1992). In *Genetic Engineering, Principles and Methods*, Vol. 14, J.K. Setlow (ed.) Plenum, New York, pp. 13–32.
Domingo, E. & Holland, J.J. (1994). In *Evolutionary Biology of Viruses*, S.S. Morse (ed.), Raven Press, New York, pp. 161–84.
Domingo, E., Holland, J.J. & Ahlquist, P. (1988). *RNA Genetics*, CRC Press Inc., Boca Raton, Florida.
Domingo, E., Martínez-Salas, E., Sobrino, F., de la Torre, J.C., Portela, A., Ortín, J., López-Galíndez, C., Pérez-Breña, P., Villanueva, N., Nájera, R., VandePol, S., Steinhauer, D., De Polo, N. & Holland, J.J. (1985). *Gene*, **40**, 1–8.
Domingo, E., Sabo, D., Taniguchi, T. & Weissmann, C. (1978). *Cell*, **13**, 735–44.
Duarte, E., Clarke, D., Moya, A., Domingo, E. & Holland, J.J. (1992). *Proc. Natl. Acad. Sci. USA*, **89**, 6015–19.
Eigen, M. (1971). *Naturwiss enschaft*, **58**, 465–523.
Eigen, M. (1993). *Sci. Am.*, **269**, 32–9.
Eigen, M. & Biebricher, C.K. (1988). *In RNA Genetics Vol. 3: Variability of RNA Genomes*, E. Domingo, J.J. Holland and P. Ahlquist (eds.) CRC Press, Inc., Boca Ratón, Florida, pp. 211–245.
Eigen, M., McCaskill, J. & Schuster, P. (1989a). In *Advances in Chemical Physics*, Vol. 75, I. Prigogine and S.A. Rice (eds.), Wiley, New York, pp. 149–263.
Eigen, M., McCaskill, J. & Schuster, P. (1989b). *J. Phys. Chem.*, **92**, 6881–91.
Eigen, M. & Nieselt-Struwe, K. (1990). *AIDS*, 4, S85–93.
Eigen, M. & Schuster, P. (1977). *Naturwissenschaft*, **64**, 541–65.
Eigen, M. & Schuster, P. (1978). *Naturwissenschaft*, **65**, 341–69.
Haruna, I., Nozu, K., Ohtaka, Y. & Spiegelman, S. (1963). *Proc. Natl. Acad. Sci. USA*, **50**, 905–11.
Holland, J.J. (1992). *Curr. Top. Microbiol. Immunol.*, Vol. 176, Springer-Verlag, Berlin.

Holland, J.J., Domingo, E., de la Torre, J.C. & Steinhauer, D.A. (1990). *J. Virol.*, **64**, 3960–2.

Holland, J., Spindler, K., Horodyski, F., Grabau, E., Nichol, S. & VandePol, S. (1992). *Science*, **215**, 1577–85.

Inoue, T. & Orgel, L.E. (1983). *Science*, **219**, 859–61.

Lai, M.M.C. (1992). *Curr. Top. Microbiol. Immunol.*, **176**, 21–32.

Mayr, E. (1988). *Towards a New Philosophy of Biology. Observations of an Evolutionist.* Harvard University Press, Cambridge, Massachusetts.

Mills, D.R., Peterson, R.L. & Spiegelman, S. (1967). *Proc. Natl. Acad. Sci. USA*, **58**, 217–24.

Nowak, M.A. (1992). *Trends in Ecology and Evolution*, **7**, 118–21.

Nowak, M. & Schuster, P. (1989). *J. Theor. Biol.*, **137**, 375–95.

Steinhauer, D.A. & Holland, J.J. (1987). *Ann. Rev. Microbiol.*, **41**, 409–33.

Swetina, J. & Schuster, P. (1982). *Biophys. Chem.*, **16**, 329–45.

Temin, H. (1989). *J. AIDS*, **2**, 1–9.

Van Regenmortel, M.H.V., Maniloff, J. & Calisher, C. (1991). *Arch. Virol.*, **120**, 313–14.

14

The co-evolutionary dynamics of viruses and their hosts

ROBERT M. MAY

Introduction

This paper aims to give an overview of the dynamics and genetics of interactions between populations of viruses and their hosts. I emphasize that 'successful' parasites are those individuals who leave the most offspring, and that this does not necessarily mean evolving toward being harmless to the host (even though this may be the best for the viral population as a whole); many evolutionary trajectories are possible, depending on the constraints and trade-offs among viral virulence, transmissibility, and the costs of host resistance. These general ideas are illustrated by the particular cases of myxoma virus and Australian rabbits, and of HIV-1 and human hosts (where the role of 'escape mutants' of the virus, and of a possible 'diversity threshold' beyond which the immune system can no longer control the virus, are emphasized).

Viruses as 'microparasites'

Viruses and other infectious agents that afflict humans and other animals are conventionally classified along taxonomic lines. In discussing the ecology or evolution of host/parasite associations, however, it is often more useful to make distinctions on the basis of the population biology of the interaction.

Microparasites (*sensu* Anderson & May, 1979) are those which have direct reproduction, usually at very high rates, within the host. They characteristically have small sizes and short generation times; the duration of infection is typically very short relative to the expected lifespan of the host, and hosts that recover from infection usually acquire immunity against re-infection for some time, often for life. The result is that most microparasitic infections are of a transient nature in individual hosts. Most viruses fall

192

broadly into the microparasite category, along with most bacterial parasites and, more equivocally, many protozoan and fungal parasites.

To analyse the interaction between populations of such microparasites and their hosts, we may divide the host population into relatively few classes of individuals: susceptible, infectious, recovered-and-immune. Our operational definition of a microparasite is an infectious agent whose population biology can, to a sensible first approximation, be described by some such compartmental model.

A variety of other kinds of parasites, whose life cycles and evolution are intimately entwined with those of one or more host species, can be distinguished. *Macroparasites* (*sensu* Anderson & May, 1979), broadly, parasitic helminths and arthropods, may be thought of as those having no direct reproduction within the host. They are typically larger and longer-lived than microparasites, with generation times that often are an appreciable fraction of the host lifespan. Lasting immune responses are rarely elicited, so that macroparasitic infections are characteristically persistent, with hosts being continually reinfected. The various factors that characterize host/macroparasite associations (egg output per female parasite, pathogenic effects on the host, parasite death rates, and so on), can all depend on the number of parasites in a given host, and mathematical models must take account of the details of the way the population of macroparasites is distributed among the hosts. *Parasitoids* are arthropods (usually dipteran or hymenopteran species) that lay their eggs in or on the larvae or pupae of other insects; the parasitoid offspring kill their host. This lifestyle accounts for as much as 10% or more of all metazoan species. Dobson (1982) has given a useful summary of some of the relations among life history characteristics of microparasites, macroparasites, parasitoids and conventionally defined predators. This summary would serve as a point of departure for a wider discussion of similarities and differences among various categories of host/parasite associations, if space permitted. Interspecific and intraspecific brood parasitism is yet another category of animal host/parasite association, with themes running parallel to those below. A brief guide to the literature on co-evolution in the systems mentioned in this paragraph is given toward the end of this chapter.

Dynamics of host/microparasite associations

Suppose we have a population of hosts where generations overlap completely, so that population change is a continuous process. The

basic model for the interaction between such a host population and a
virus or other microparasitic agent takes the familiar form of differential
equations, describing the rate of change in the number or density of
susceptible, infected-and-infectious, and recovered-and-immune hosts,
$X(t)$, $Y(t)$, $Z(t)$, respectively:

$$dX/dt = B(X,Y,Z,) - \mu X - \beta XY + \gamma Z \tag{1}$$
$$dY/dt = \beta XY - (\mu+\nu+\alpha)Y \tag{2}$$
$$dZ/dt = \nu Y - (\mu+\gamma)Z \tag{3}$$

Here $B(X,Y,Z)$ represents the rate at which new hosts are born; μ
is the per capita death rate; ν and α are the rates at which hosts
move out of the infectious category by recovering or by dying from
the infection, respectively; and γ is the rate at which immunity is
lost ($\gamma = 0$ if immunity is lifelong). A key feature is the non-linear
term βXY, which in this simplest model describes the rate at which
new infections appear (proportional both to the number susceptible,
X, and the number infectious, Y; β is a rate parameter characterizing
transmission efficiency). The total host population is $N = X+Y+Z$.

   The population dynamics of these and related equations have been
thoroughly studied (Anderson & May, 1991). The details depend on
the assumptions made about density-dependence in the birth rate, B,
but broadly the microparasite can regulate an otherwise exponentially
growing population to a stable equilibrium value (either monotonically
or via damped oscillations), provided the disease-induced death rate or
virulence, α, is large enough. The equations also describe the epidemics
that arise when infection is introduced into a susceptible population.
Analogous equations describe the regulatory effects of microparasites in
host populations with discrete, non-overlapping generations (as found in
many temperate-zone insects); here the 'regulated' state may be chaotic
fluctuations in host abundance (May, 1985).

Genetics of host/microparasite associations

The previous section dealt with epidemiology, with no reference to
genetics. We now sketch work on the co-evolutionary genetics of
host/microparasite associations that makes essentially no reference
to the epidemiology. The next section will draw these two strands
together.

   Following the early work of Mode (1958), Day (1974) and Van der
Plank (1975), there has been much work on the population genetics of

'gene-for-gene' interactions between hosts and pathogens. These studies assume specific associations between individual genotypes of hosts and corresponding genotypes of pathogens. The work is mainly directed toward pathogens of plants, where there are documented instances of such gene-for-gene associations; for a detailed and critical review, see Barrett (1985). Constant values are assigned to the fitnesses of each host genotype when attacked by a parasite of a specified genotype, and the ensuing net fitnesses of the various host genotypes are weighted sums over the appropriate fitnesses (weighted according to the relative abundances of the parasite genotypes). Similar calculations give the fitnesses of the various parasite genotypes. The result is a system in which the host fitnesses depend on the relative gene frequencies within the parasite population, and parasite fitnesses on host gene frequencies. The simplicity of these assumptions is, however, such that threshold and other important density-dependent effects associated with epidemiological processes are neglected.

These studies of gene-for-gene associations between hosts and microparasites suggest that polymorphisms in the gene frequencies of both hosts and pathogens can easily arise and be maintained. The polymorphisms may be stable, or cyclic, or even chaotic. Levin (1983), however, emphasizes that many of the simpler models can be seen to be neutrally stable, although this fact is often obscured by round-off errors and by the proliferation of parameters in numerical studies.

The essential mechanisms responsible for maintaining polymorphisms in these models, whatever the dynamical details, are the interplay between parasite virulence and the costs of host resistance.

Co-evolution of host/microparasite associations

What we need are studies that take the earlier gene-for-gene framework, and combine it with fitness functions computed from epidemiological analyses that pay full respect to the non-linear nature of the transmission process. The pioneering study here is by Gillespie (1975) who, however, only considered the statics and not the dynamics of his model. Studies of the full dynamics of such combined genetic and epidemiological models reveal interesting biological and mathematical features (Hamilton, 1980; May & Anderson, 1983, 1990; Beck, 1984; Seger, 1988; Seger & Hamilton, 1988; Levin *et al.*, 1982); for a recent review, see Anderson and May (1991).

Most of these studies are for the conventional metaphor of one locus with two alleles. The studies typically show that gene-for-gene associations between hosts and microparasites promote genetic diversity, giving rise to polymorphisms that may be stable or varying from generation to generation in cyclic or even chaotic fashion. The reason is straightforward: once the epidemiological dynamics are recognized, rarer alleles of hosts or parasites are seen often to be favoured (essentially because rare host genotypes are typically less susceptible to currently-successful parasite genotypes, and rare parasite genotypes typically find few hosts resistant to them), which leads to polymorphisms that may be static or shifting, depending on the details.

Going beyond single-locus studies, Seger (1988; see also Seger & Hamilton, 1988) has used computer simulations to explore the co-evolutionary properties of a model in which two loci in the host determine a strain-specific defence against parasites. He finds that intermediate rates of recombination tend to produce cyclically varying polymorphisms (while extreme recombination rates do not). Seger's model suggests that parasites can, in effect, generate a fluctuating environment, which can favour the evolution of intermediate rates of recombination in the host population.

In short, it has long been recognized that the frequency-dependent and density-dependent selective effects that act reciprocally between animal host populations and viruses or other microparasites are likely to create and maintain genetic polymorphisms (Haldane, 1949). But essentially all earlier work, both theoretical and empirical, has implicitly assumed that such polymorphisms will be at some steady level. It now appears that host/microparasite interactions will often produce polymorphisms that vary cyclically or chaotically. These may even have wider implications for the evolution of sex (Hamilton, 1980; Seger & Hamilton, 1988). All this has obvious and important implications for empirical studies of such polymorphisms, both in the laboratory and in the field.

Evolution of virulence

It is commonly asserted that 'well-adapted' or 'successful' parasites will have evolved to be relatively harmless to their hosts. While it is possible to assemble data that seem to support this view (Allison, 1982; Holmes, 1982; for a more equivocal look at the data, see May & Anderson, 1983), the theoretical arguments advanced to support it are blatantly group-selectionist: it is supposedly in the interest of the population of parasites not to harm its population of hosts too much. More careful appraisal makes it clear that the co-evolutionary trajectories of hosts and microparasites depend on the detailed interplay among the transmission and virulence of the parasite and the costs of host resistance. If virulence were entirely unconnected with other factors, then indeed the evolutionary interests of both hosts and parasites would, independently, be best served by avirulence. But transmissibility, virulence, and resistance costs are rarely unconnected, which makes things more complex.

Some feeling for what is involved can be gained by returning to Eqns. 1–3, and focussing only on the parasite. The intrinsic fitness (Fisher's 'net reproductive value') of a microparasite is measured by the number of new infections produced, on average, by each infected host in a population that is almost entirely susceptible. Looking at Eqn. 2 (in this limit when $X \simeq N$), we see that this net reproductive value, R_0, is given by

$$R_0 = \beta N / (\alpha + \mu + v) \tag{4}$$

This result can be understood intuitively. Each infected individual produces new infections at a rate βN per unit time, and does so, on average, for a time that is the reciprocal of the sum of the rates of moving out of the infected class (by dying from disease, α, or other causes, μ, or by recovering, v), namely $1/(\alpha+\mu+v)$. As Eqn. 2 makes clear, the microparasite can invade and maintain itself only if $R_0 > 1$, and the larger R_0, the higher the parasites's reproductive capacity or fitness.

If recovery rates (v) and transmission (βN) were uncoupled from virulence (α), then clearly the microparasite maximizes R_0 by being avirulent ($\alpha \rightarrow 0$). But most of the nasty effects that parasites have on hosts are connected with producing transmission stages, so that both transmission efficiency (β) and recovery rates (v) will in general be explicitly connected with α. Once this is recognized, it is obvious that the value of α that maximizes R_0, the optimal degree of virulence, depends on the details of the functional relations among βN, v, and α. Evolution could favour decreasing virulence, increasing virulence, or convergence to some intermediate level. There is no *a priori* way of knowing. The details matter.

It must be emphasized that the argument in the preceding two paragraphs focusses exclusively (and in a frankly oversimplified way) on the evolutionary pressures experienced by parasite individuals. It is not a co-evolutionary argument. Even so, the argument in general, and

Eqn. 4 in particular, makes it clear that the extent to which parasites will tend to evolve toward avirulence depends on the degree to which transmissibility and recovery rates are linked with harmful effects on host physiology or behaviour. Unfortunately, virtually no information is available about these kinds of linkages. Yet, without such information, studies of the evolution of virulence are doomed to sterile abstraction.

Myxoma virus and Australian rabbits

There is one example for which sufficient data are available to make a very rough assessment of the interrelationships in Eqn 4, and thence to say something about the likely evolution of avirulence in a specific instance. This example is the Australian rabbit/myxoma virus system, following the introduction of the virus into Australia in 1950. This case study has been fully discussed elsewhere (Anderson & May, 1982; for a summary see Anderson & May, 1991, pp.649–652). Suffice it to say that a theoretical analysis based on Eqn. 4 (with relations among α, v, and βN roughly estimated from data) suggests myxoma evolved from the highly virulent strain that was originally introduced, to a strain of intermediate virulence. Compared with these strains of intermediate virulence, those with too high a virulence kill rabbits too fast, whereas those with too low a virulence allow the rabbits to recover too fast; the parasite's intrinsic reproductive rate, R_0 is maximized at intermediate values of α (Fig. 14.1 (A) and 14.1 (B)). This theoretical conclusion is not the story commonly told in introductory texts, which usually tell of ever-diminishing virulence, but it matches the facts (Fig. 14.2(a) and 14.2(b)).

To our knowledge, no other studies provide enough information to infer the shape of functional relationships among virulence, transmissibility, and recovery rates. But there are certainly instances in which empirical evidence indicates that transmissibility and damage to the host are so entangled that it is difficult, and probably impossible, for the parasite to evolve toward harmlessness. Moore (1984), for example, has reviewed many examples where parasites with indirect life cycles modify the behaviour of their vertebrate or invertebrate hosts in such a way as to facilitate transmission to the next stage in the parasite's life cycle, even though this behaviour increases host mortality.

Many invertebrate hosts have relatively short lives anyway, and such pressures toward avirulence as do exist are correspondingly weak. Thus,

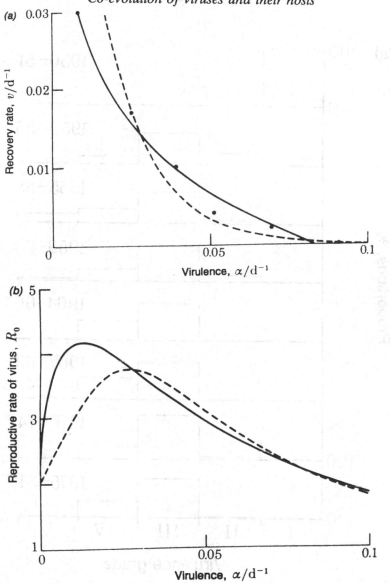

Fig. 14.1.(*a*) This figure shows the empirical relationship between virulence, α, and recovery rate, v, for various strains of myxoma virus in wild populations of rabbits in Australia. The full circles are the observed values, and the two functional fits are as described in detail in Anderson and May (1982). (*b*) The relationship between the basic reproductive rate, R_0, of Eq. (4) and virulence, α, for the various grades of myxoma virus in wild populations of rabbits in Australia, for the two functional relations between v and α displayed in Fig. 14.1(*a*). The disease-free death rate is $\mu = 0.011$ per day, and βN is arbitrarily set constant at 0.2 per day (after Anderson & May, 1982).

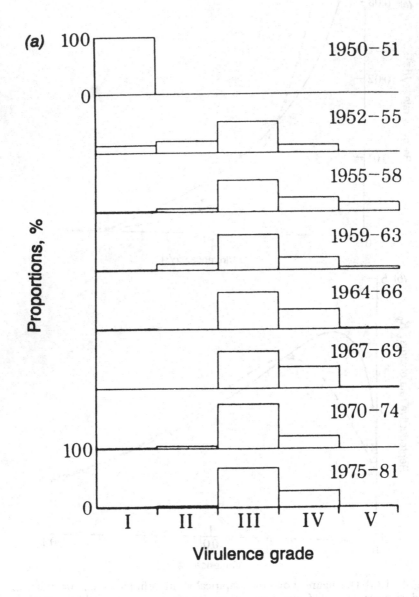

Fig. 14.2.(a) This figure shows the proportions in which various grades of myxoma virus have been found in wild populations of rabbits in Australia, at different times from 1951 to 1981. Data are those compiled by Fenner (1983), and come from various sources.

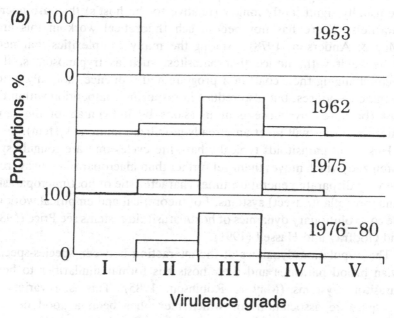

Fig. 14.2. (*b*) The proportions of the various grades of myxoma virus recovered from wild populations of rabbits in Britain are shown, for different times from 1953 to 1980. The data again come from a variety of sources, as summarized by Fenner (1983).

for example, many microsporidian protozoans kill their insect hosts fairly quickly and, in doing so, effectively turn them into masses of transmission stages.

To summarize, both theory and some empirical evidence suggest that the co-evolution between hosts and microparasites can follow many evolutionary paths, depending on the relations among parasite transmissibility and virulence, and host cost of resistance.

Other 'host/parasite' systems

At the start of this chapter, I mentioned other associations that could be gathered under the umbrella of 'animal hosts and parasites'. Lacking space to pursue these in detail, I provide here a very brief guide to the literature on co-evolutionary aspects of such associations (with the warning that much of the work really deals with evolution of one partner, or non-interactively with both, rather than with genuine coevolution).

For macroparasites, as defined earlier, the trade-offs among transmissibility, virulence, and cost of host resistance are as for microparasites, but further complicated by the facts that most hosts harbour many worms or ectoparasitic arthropods and that macroparasite lifetimes are usually significantly longer (relative to the host's) than are microparasite's. There has not been much theoretical work in this area (May & Anderson, 1978). Among the many complexities that need to be dealt with, notice that parasites, such as trypanosomes, that keep 'changing their coat' in a programmed sequence, not only evade immune responses, but also reduce intraspecific competition within the host (because any subsequent invasions by individuals of the same parasite species will meet an already-mobilized immune system).

Hosts and parasitoids typically have life cycles that are roughly synchronized, which moves them yet further than macroparasites away from the very disparate generation times characteristic of host/microparasite (and most plant/insect) systems. For theoretical and empirical work on the co-evolutionary dynamics of host/parasitoid systems see Price (1980) and Godfray and Hassell (1991).

The population dynamics of the interaction between species-specific avian brood parasites and their hosts has formal similarities to host/parasitoid systems (May & Robinson, 1985). This is a variety of host/parasite association for which there has been a good deal of work combining field observations with qualitative models (for recent reviews, see Davies and Brooke (1988) and Rothstein and Mason (1986)), and a limited amount of analysis of the population dynamics, but little connection either between the two approaches or with other host/parasite systems (and very little on population genetics).

Finally, intraspecific brood parasitism, among birds and other animals, poses some of the basic host/parasite co-evolutionary questions in sharp form (Andersson, 1984; Moller & Petrie, 1991). Theoretical work that blends game theory with population dynamics suggests that once the habit of brood parasitism arises in a population, it is likely to spread, in turn eliciting some kind of vigilant response. The result is likely to be a population in which different genotypes within the population play, as it were, different strategies or in which individual birds play mixed strategies. In either event, this kind of behavioural polymorphism can, in principle, drive cycles in the overall abundance of the population (May, Nee & Watts, 1991; Nee & May, 1993). If this theoretical possibility turns out to be manifested by real animals, it will be a surprising example where the coevolution of host and parasite behaviours within a single population leads to cyclic changes in population density.

HIV and the dynamics of the immune system

I end this chapter by discussing a rather different example of co-evolution between a virus and its host. Specifically, I outline a mathematical model that may explain why, in most individuals, infection with the human immunodeficiency virus (HIV-1 or HIV-2) leads to acquired immunodeficiency syndrome (AIDS) after a long and variable incubation period, despite the fact that the immune system appears to deal successfully with the initial infection. Here we are looking not at the co-evolution between a virus and the populations of hosts, but rather at the micro-evolutionary interaction between a virus and the population of immune cells within an individual host. If the ideas sketched here, and developed more fully elsewhere (Nowak *et al.*, 1990, 1991; Nowak & May, 1991), are correct in essentials, then AIDS disease can be thought of as resulting from co-evolution between the HIV virus and the immune system.

Following infection with the HIV, there is typically a long, but variable, incubation period before the immune system undergoes the collapse characterized as AIDS. The incubation period averages more than 8–10 years in adults in developed countries, is somewhat shorter in older people (and possibly in adults in developing countries), and is much shorter in infected infants. The initial infection, however, gives the appearance of being coped with by the immune system. For a short but variable period following infection, a few weeks to a few months, virus is typically found in the blood (viraemia), and high levels of viral replication can be observed. Antibodies then appear in blood serum (seroconversion), after which it becomes difficult to isolate the virus; viral antigens are often undetectable after seroconversion has occurred. After the long and variable quiescent or latent interval, viraemia again appears as AIDS develops. These details of the natural history of HIV infection are reviewed more fully elsewhere (e.g. Nowak *et al.*, 1991).

The mechanism proposed by Nowak *et al.* to explain all this rests on three basic assumptions, each backed by experimental data.

First, as a retrovirus, HIV lacks error-correcting mechanisms during replication, and thus produces variant forms at a high rate; the error rate for HIV is about 10^{-4} per base, or about one misincorporation per genome per replication cycle. As a result, HIV continually produces 'escape mutants', which largely evade the surveillance of the already-mobilized immune system. Indeed, Nowak (1990) has calculated the theoretical error rate for HIV replication that would optimize the probability of producing escape mutants, and has shown it is roughly equal to the observed rate, which may be a coincidence.

Secondly, the immunological responses directed against HIV involve a specific response to individual strains or 'escape mutants' (subpopulations of CD4+ cells specifically directed toward immunological attack against that strain), along with a cross-reactive response that acts against all strains.

Thirdly, each immunologically distinct strain of HIV can infect and subsequently kill any CD4+ cell, regardless of its specificity to a particular viral mutant.

These three empirically based assumptions add up to a most peculiar kind of prey–predator system. New prey species keep appearing (assumption one); each requires a specific predator species to regulate its abundance (assumption two); yet any of the prey species can kill any of the predators (assumption three). For a more detailed and jargon-laden account of these assumptions, see Nowak *et al.* (1991).

A dynamical system possessing these properties is highly non-linear, and can exhibit very interesting behaviour. To begin with, if the original viral strain of HIV does not produce, on average, more than one escape mutant before being driven to very low levels by the immune responses (that is, if '$R_O < 1$', where R_O is the basic reproductive rate for producing escape mutants), the infection will be cleared. If, on the other hand, $R_O > 1$ for escape mutants, then over time more and more escape mutants will appear, each being regulated by appropriately specific immune responses. Thus the diversity of different strains of HIV, as measured, for example, by the ecologists' Simpson Index, will steadily increase. But in the very peculiar 'prey–predator' system enunciated above, there eventually appears a 'diversity threshold'; that is, once the number of distinct viral strains, n, exceeds some threshold value, n_c, the immune system can no longer control the virus population. Following the breaching of the diversity threshold, the immune system's population of CD4+ cells declines to very low levels, and viral abundance rises rapidly. In this above-threshold phase, the faster replicating strains of HIV predominate, and as a consequence viral diversity will appear to decrease when finite samples of the population are taken.

Some intuitive understanding can be obtained from a highly over-simplified mathematical model of the above-described system. Let $v_i(t)$ represent the abundance of viral strain i at time t, and $x_i(t)$ represent the abundance of the strain-specific immune response; $v = \Sigma_i v_i$ is the total abundance of HIV virus, and in this simplest case the cross-reactive immune responses are neglected. The dynamics of v_i and x_i can be described by the differential equations

$$dv_i/dt = v_i (r - px_i) \qquad (5)$$
$$dx_i/dt = kv_i - uvx_i \qquad (6)$$

Here, r is the replication rate of HIV, p is the rate at which virus is killed on contact with the strain-specific immune cells, kv_i is the rate at which viral strain i stimulates a strain-specific immune response, and u is the rate at which virus (of any strain) kills any immune cell on contact. The initial conditions are that $v_i(0)$ is specified for the initially infecting viral strain 1, whereas all subsequent v_i ($i > 1$) are produced by mutations, at some rate that depends on the magnitude of the total viral population, $v(t)$; $x_i(0) = 0$ for all i. It can easily be seen that all viral strains will be controlled by the immune response (all dv_i/dt asymptotically negative), provided

$$pk\, v_i > ruv \qquad (7)$$

for all i. If $pk/ru \gg 1$, this can be satisfied so long as the number of strains or escape mutants, n, is not too big. But eventually, for $n > n_c \equiv pk/ru$, it will be impossible to satisfy Eqn. (7) for all i. At this point, the diversity threshold is exceeded, the viral population runs away, and AIDS develops. Clearly, this mathematical model is grossly too simple. Models where different viral strains have different values of r_i, k_i, p_i, and ui, and where a cross-reactive immune response and other realistic refinements are included, are discussed elsewhere (Nowak & May, 1991; Nowak, May & Sigmund, 1992), but Eqns. (5)–(7) make the essential elements clear.

Figures 14.3(*a*), 14.3(*b*), 14.3(*c*), 14.3(*d*) illustrate these nonlinear properties of the possible coevoluntary interactions between HIV and the immune system (using a somewhat more complicated and more realistic version of Eqns. (5), (6); see Nowak *et al.*, 1991). Specifically, Fig. 14.3(*a*) shows the total virus population (*v*) and Fig. 14.3(*b*) shows the total CD4+ cell abundance ($x = \Sigma_i\, x_i$), as functions of time, for a more realistic model than that above (Nowak *et al.*, 1991). Note that the HIV abundance behaves as typically observed, with an initial peak, a long period of very low levels, and a final runaway. Fig. 14.3(*c*) teases apart the underlying details, showing the abundance, over time, of each of the 40 different strains of HIV that are added together in Fig. 14.3(*a*); notice how the fastest replicating strains predominate once the immune system no longer regulates the virus. Finally, Fig. 14.3(*d*) shows how viral diversity (as measured by Simpson's Index, $D = [\Sigma \; (vi/v)^2]^{-1}$) changes over time in the model: D increases as more and more escape mutants arise during the regulated phase, but D appears to fall as the fastest-replicating viral strains predominate in the runaway phase once the diversity threshold is exceeded.

Fig. 14.4 illustrates how the element of chance associated with the production of the first few escape mutants can influence the rate of progression to AIDS in different individuals. This figure gives 30 different versions of Fig. 14.3(a), showing total viral abundance, as a function of time, for 30 different simulations (all with the same parameters) of a more realistic version of Eqns. (5) and (6). The simulations correspond to each initially-infecting viral strain producing, on average, 1.5 escape mutants over its lifetime in the host. It can be seen that, in the simulations, some hosts effectively clear the infection before any escape mutants are produced, and so never progress to develop AIDS. In other hosts, escape mutants appear early, and the incubation period for developing AIDS is significantly shorter than the average. The figure thus mimics the significant variability in lengths of incubation periods found in AIDS cases.

In summary, the system described above has the property that the immune system can control the initial infection. But the continual generation of escape mutants gives rise to an evolving dynamical system, whose peculiar non-linear features result in a viral diversity threshold, above which the immune system is overwhelmed.

Whether or not these ideas about the evolutionary interplay between HIV virus and the immune system turn out to be correct, they point toward a widening of the theme of this chapter. Earlier in the chapter, I discussed the evolutionary dynamics of interactions between populations of viruses (and other parasites) and populations of hosts. Now we see that there may be some interesting parallels in the evolutionary dynamics of interactions between viruses and the immune systems of individual hosts.

Discussion: co-evolution and 'emerging viruses'

The appearance of HIV/AIDS, in a world where, at least to the inhabitants of developed countries, mortal epidemics of infectious diseases were thought to be things of the past, has led to many questions being asked about the 'emergence' of new viruses. Pointing not only to HIV, but also to more transiently-appearing viruses like Marburg or Rift Valley

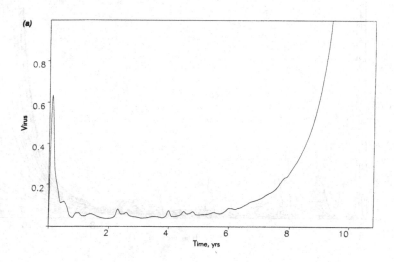

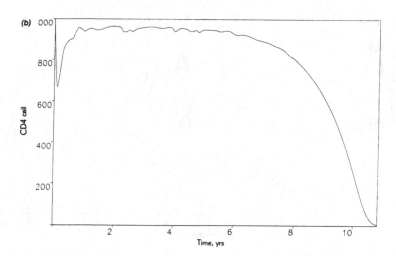

Fig. 14.3 (*a*) and (*b*). See p. 209 for caption.

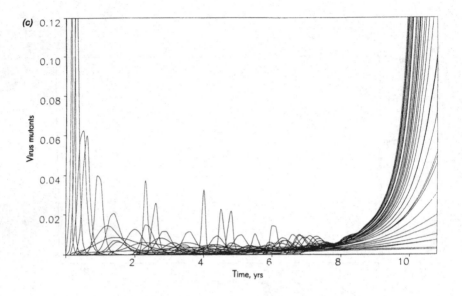

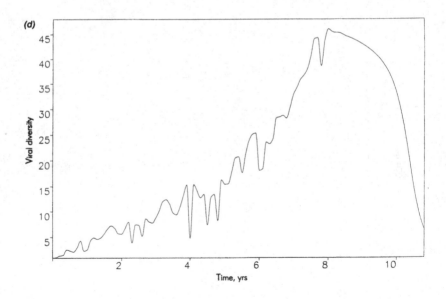

Fig. 14.3 (*c*) and (*d*).

Fever, some researchers have called for a global network of monitoring centres, to give early warning of future viral emergences.

The molecular and other techniques that can be deployed quickly to elucidate the structure and function of such 'emergent' viruses are awe-inspiring. Some of the accompanying evolutionary speculation, however, can be generously described as naive (for instance, the notion that we should expect new viruses to emerge in the tropics because biological diversity is greatest there, while superficially plausible, does not stand up to close analysis).

This chapter has aimed to sketch an analytic framework for examining the evolutionary and population dynamics of viruses and their hosts. Central to the discussion is the concept of R_0, the basic reproductive rate of the virus within a particular group of hosts, at a particular place and time. If $R_0 > 1$, a virus that appears in a host population by mutation, or by introduction from another place, or from another

Fig. 14.3. Figs 14.3(a–d) illustrate the dynamical properties of a more realistic version of the model described by eqns. (5) and (6) for the interplay between HIV and the immune system. The model, which is described more fully elsewhere (Nowak & May, 1992), includes CD4+ cell dynamics, strain-specific and cross-reactive immune responses, and viral strains whose replication rates and other parameters differ one from another (so that Eqn. (5), for example, becomes $dv_i/dt = v_i [r_i + r_iy - s_iz - p_ix_i]$). Fig. 14.3($a$) shows total virus density, v (in arbitrary units), as a function of time; $v(t)$ shows an initial peak of viraemia, followed by a long period with low abundance of virus, and a final increase in the AIDS phase. Fig. 14.3(b) shows the CD4+ cell population, x (in arbitrary units), as a function of time; $x(t)$ increases slightly during the asymptomatic phase, and decreases rapidly as the virus population increases to high levels in the AIDS phase. Fig. 14.3(c) gives a picture of the co-evolution and co-existence of many different strains of HIV over the course of the infection in a single patient. Specifically, the figure shows the abundance, v_i, of each of 40 strains ($i = 1,2, \ldots, 40$), as a function of time. It can be seen that, initially, the strains attain fairly high levels of abundance, which may cause the clinical symptoms sometimes observed during primary HIV infection. The subsequently emerging escape mutants are suppressed faster, as a result of the cross-reactive immune responses that are now mobilized. Different viral strains grow to different levels, according to their growth rates. In this simulation, the accumulation of viral diversity breaches the diversity threshold after about 7 years. In the final phase the fastest growing strains dominate the viral population. Fig. 14.3(d) shows the diversity of the community of different strains of HIV, as conventionally measured by the ecologists' Simpson's Index, $D = [\Sigma (v_i/v)^2]^{-1}$, as a function of time. The diversity displays a one-humped pattern, with a maximum just before the virus escapes the immune system's control. The antigenic diversity increases as long as the immune response is selecting for escape mutants, but declines as faster replicating strains predominate once the diversity threshold is crossed. For explicit details of Fig. 14.3, see Nowak and May (1992).

R.M. May

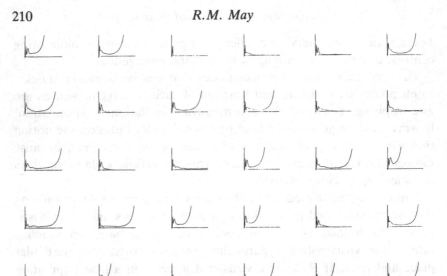

Fig. 14.4. Simulations of 30 HIV infections, in 30 different individuals. The underlying model is as in Fig. 14.3, and all 30 simulations use the same parameter values. The only difference from patient to patient is the seed for the random numbers involved in generating escape mutants. Notice that the outcome of the infection and the rate of progression towards AIDS are extremely variable, even though each patient has the same immunological parameters and the viral populations have the same replication and mutation rates. Some patients develop AIDS shortly after the initial viraemia, while others are still asymptomatic after a long time and in some instances the immune system has wiped out the virus. The length of the incubation period in a given individual depends on the stochastic emergence of new virus variants. Thus the basic ideas outlined here can explain both the length and the high degree of variability in the asymptomatic period. For explicit details and parameter values, see Nowak & May (1992).

host species, will tend to persist and spread within the population. If R_0 < 1, such a newly- appearing virus may harm hosts in a decaying chain, but will not maintain itself within the population.

We have seen, in deliberately oversimplified examples, how R_0 depends on both biological and behavioural/social factors (for example, for HIV, R_0 depends on the biological details of the interactions between HIV and immune cells, which influence infectiousness, and it also depends on patterns of sexual behaviour, which affect transmission and can vary from group to group and from place to place). It follows that behavioural changes within the host population, or environmental changes that affect transmission, can move R_O from below unity to above it (or conversely), thus triggering the emergence (or disappearance) of the virus within the host population.

The criterion $R_0 > 1$ can equivalently be expressed in terms of a threshold density of hosts (or, *mutatis mutandis*, a threshold rate of acquiring new sexual partners, or of sharing needles). Most viral infections of humans have fairly high threshold densities. Measles, for example, requires populations of around 200 to 300 thousand if it is to remain endemic. Thus most viral and bacterial infections of childhood have probably been with us only for 10 thousand years or less (since the agricultural revolution led to human aggregations of above-threshold size), which is a blink of the eye on an evolutionary time scale. Viral emergence is no new thing and, as emphasized in their different ways by Haldane (1949), McNeill (1976) and others, much of human history can be interpreted as the interplay among viral and bacterial diseases, threshold population sizes, and patterns in the movement of groups of humans.

These, and other aspects of the dynamics of host–virus co-evolution, are reviewed more fully elsewhere (Anderson & May, 1991, Chapter 23). My chapter here has sought to outline some of the main points. The essential message is that the dynamic interplay between viral and host populations is necessarily a non-linear one, and as such has many properties that defy simple intuition.

References

Allison, A.C. (1982). In *Population Biology of Infectious Diseases*. R.M. Anderson and R.M. May (eds). Springer-Verlag, New York, pp. 245–268.

Anderson, R.M. & May, R.M. (1979). *Nature*, **280**, 361–7.

Anderson, R.M. & May, R.M. (1982). *Parasitology*, **85**, 411–26.

Anderson, R.M. & May, R.M. (1991). *Infectious Diseases and Control*. Oxford University Press, Oxford.

Andersson, M. (1984). In *Producers and Scroungers*. C.J. Barnard (ed). Chapman Hall, London, pp. 195–228.

Barrett, J. (1985). In *Ecology and Genetics of Host–Parasite Interactions*. D. Rollinson and R.M. Anderson (eds). Academic Press, London, pp. 215–225.

Beck, K. (1984). *J.Math. Biol.*, **19**: 63–78.

Davies, N.B. & Brooke, M.L. (1988). *Animal Behav.* **36**, 262–284.

Day, P.R. (1974). *Genetics of Host Parasite Interactions*. W.H. Freeman, San Francisco.

Dobson, A.P. (1982). In *Population Biology of Infectious Diseases*. R.M. Anderson and R.M. May (eds). Springer-Verlag, New York, p.5.

Fenner, F. (1983). *Proc. Roy. Soc. B*, **218**, 259–5.

Gillespie, J.H. (1975). *Ecology*, **56**, 493–5.

Godfray, H.C.J. & Hassell, M.P. (1991). In *Parasite–Host Associations:*

212   R.M. May

Coexistence or Conflict? C.A. Toft, A. Aeschlimann and L. Bolis (eds).
    Oxford University Press, Oxford, pp. 131–147.
Haldane, J.B.S. (1949). *La Ricerca Sci. Suppl.* **19**, 68–76.
Hamilton, W.D. (1980). *Oikos*, **35**, 282–90.
Holmes, J.C. (1982). In R.M. Anderson & R.M. May (1982), *op. cit.*,
    pp. 37–51.
Levin, S.A. (1983). In *Coevolution*. M. Nitecki (ed). University of Chicago
    Press, Chicago, pp. 21–65.
Levin, B.R., Allison, A.C., Bremermann, H.J., Cavalli-Sforza, L.L., Clarke,
    B.C., Frentzel-Beyme, R., Hamilton, W.D., Levin, S.A., May, R.M.
    & Thieme, H.R. (1982). In *Population Biology of Infectious Diseases*.
    R.M. Anderson and R.M. May (eds). Springer-Verlag, New York,
    pp. 212–243.
McNeill, W.H. (1976). *Plagues and Peoples*. Doubleday, New York.
May, R.M. (1985). *Am. Natur.*, **125**, 573–84.
May, R.M. & Anderson, R.M. (1978). *J. Anim. Ecol.*, **47**, 249–67.
May, R.M. & Anderson, R.M. (1983). *Proc. Roy. Soc. B*, **219**, 281–313.
May, R.M. & Anderson, R.M. (1990). *Parasitology*, **100**, S89–101.
May R.M., Nee, S. & Watts, C. (1991). In *Proceedings of the 20th International Ornithological Congress*.
May, R.M. & Robinson, S.K. (1985). *Am. Natur.*, **126**, 475–94.
Mode, C.J. (1958). *Evolution*, **12**, 158–65.
Moller, A.P. & Petrie, M. (1991). In *Proceedings of the 20th International Ornithological Congress*.
Moore, J. (1984). *Am. Natur.*, **123**, 572–7.
Nee, S. & May, R.M. (1993). *J, Theor. Biol.* **161**, 95–109.
Nowak, M.A. (1990). *Nature*, **347**, 522.
Nowak, M.A., May, R.M. & Anderson, R.M. (1990). *AIDS* **4**, 1095–103.
Nowak, M.A. & May, R.M. (1991). *Math. Biosci*, **106**, 1–21.
Nowak, M.A. & May, R.M. (1992). *J. Theor. Biol.* **159**, 329–42.
Nowak, M.A. *et al.* (1991). *Science*, **254**, 963–9.
Price, P.W. (1980). *Evolutionary Biology of Parasites*. Princeton University
    Press, Princeton.
Rothstein, S.I. & Mason, P. (1986). *Evolution*, **40**, 1207–14.
Seger, J. (1988). *Phil Trans. Roy. Soc. B*, **319**, 541–55.
Seger, J. & Hamilton, W.D. (1988). In *The Evolution of Sex*. R.E. Michod
    and B.R. Levin (eds). Sinnauer, Sunderland, Mass., pp. 176–193.
Van der Plank, J.E. (1975). *Principles of Plant Infection*. Academic Press,
    New York.

15

Population genetics of viruses: an introduction
A. MOYA AND F. GARCÍA-ARENAL

Introduction

In the introduction to his classical textbook *Molecular Evolutionary Genetics* Nei (1987) says: 'in the study of evolution there are two major problems. One is to clarify the evolutionary histories of various organisms, and the other is to understand the mechanisms of evolution'. These two problems have been traditionally addressed by paleontologists (and taxonomists) and by population geneticists, respectively. The introduction of molecular techniques, first protein electrophoresis and sequencing, later restriction analysis and sequencing of nucleic acids, has removed the boundary between these two aspects of evolutionary studies. Molecular techniques have allowed the analysis of a large and representative sample of viral genomes, and make viruses especially suitable for molecular evolutionary studies. In fact, a large amount of information on viral evolution has been collected during the last 20 years.

As a perusal of the contents of this book will clearly show, virologists have been concerned mainly with the evolutionary histories of viral species or groups as analysed by different procedures of phylogenetic reconstruction. This may be related perhaps to the traditional interest of virologists (particularly animal virologists) in following the track of epidemic developments.

Phylogenetic analyses, in addition to clarifying evolutionary histories, can also serve other purposes, from the simplest graphic representation of genetic distance (Dopazo *et al.*, 1988; Air *et al.*, 1990) to the testing of different evolutionary models (Gojobori, Moriyama & Kimura, 1990; Fitch *et al.*, 1991). This approach can, and we dare say should, be complemented with a population genetics' view of viral evolution, ranging from the quantitative description of genetic variation within and between populations to ascertain what mechanisms are responsible for the observed genetic structure of these populations.

213

Population geneticists have developed sophisticated mathematical tools to analyse the mechanisms of population evolution, reviewed in some excellent recent books (Nei, 1987; Hartl & Clark, 1989; Weir, 1990). Most work has been done for sexual, diploid organisms, but with appropriate modifications these mathematical tools, and particularly those developed for haploid genomes such as bacteria (Milkman, 1973; Levin, 1981; Hartl & Dykhuizen, 1984; Selander, Beltrán & Smith, 1991) or cellular organelles (Birky, Fuerst & Maruyama, 1989; Birky, 1991), can be applied to the quantitative analysis of the genetic structure of viral populations and to test hypotheses on the mechanisms of viral evolution.

It is well known that current hypotheses on the mechanisms of evolution can be divided into two major groups. The first one is centred around the 'neutral theory of molecular evolution' (Maruyama & Kimura, 1980; Kimura, 1983), and states that the overwhelming majority of evolutionary changes at the molecular level are caused by random fixation of selectively neutral, or almost neutral, mutations under continuous mutation pressure. The second set could generically be grouped in the context of positive selection, and would integrate different but converging lines as the neo-Darwinian theory of evolution (Dobzhansky, 1970, see Fitch *et al.*, 1991, for a case relevant to viruses) or the quasi-species theory of molecular evolution (Chapter 13).

In this chapter we will limit our scope to introducing the reader to the analysis of the genetic variation of viral populations. We will also present some case studies illustrating different possible mechanisms for viral evolution.

The estimation of genetic variation within and between viral populations

The quantitative description of genetic variation in populations, and the study of the mechanisms responsible for the maintenance of variation, are central objectives of population genetics. The mathematical procedures for these analyses have been recently reviewed (Nei, 1987). In this section we will discuss how these procedures can be applied to the data provided by the experimental methods currently used to detect viral variants.

Genetic variation can be measured at different levels. The identification of allelic variants of polymorphic genes, and their frequencies in populations, allows the estimation of the population's gene diversity (the probability that two randomly chosen genes from a population are different) as well as of the genetic distance between populations (the extent of gene differences between them). Genetic polymorphism has been assessed largely by identifying electrophoretic variants of the proteins encoded by randomly sampled structural genes. Similar data of electrophoretic variants of viral encoded proteins, of peptide mapping or of immunological analyses of viral proteins can be analysed as allelic variants. This type of data, though, will rarely come from randomly sampled genes, as most reported work has been done with structural proteins.

Experimental procedures that will identify polymorphism of viral nucleic acids also yield data which can be used to analyse the genetic variation of populations. The simplest of these data, such as the identification of electrophoretic variants (Po *et al.*, 1987) or of dsRNA patterns (Dodds *et al.*, 1987) may be analysed just as allelic variants. Other experimental procedures permit an estimate to be made of the number of nucleotide differences per site (d_{ij}) between two equivalent nucleic acid sequences, *i* and *j*. Direct sequencing of the nucleic acid is an obvious source of information to estimate d_{ij}, as it allows us to determine the number of nucleotides that are different between two sequences (n_d) once the sequences have been properly aligned. When the number of differences between two sequences is small, d_{ij} can be estimated by n_d/n, *n* being the total number of nucleotides compared. If the number of differences is larger, this expression gives an underestimate of d_{ij}, and more complex estimates are used (e.g. methods of Jukes and Cantor, Kimura's two parameter, or Tajima and Nei), and these are based on different assumptions on the nature of mutational process. All these methods give similar estimates of d_{ij} for $\hat{d}_{ij} < 0.5$ ($\hat{d}_{ij}$ being the estimated value of d_{ij}).

Sequencing can be cumbersome and expensive, and for the rapid comparison of a large number of sequences other methods may be used that also allow the estimation of d_{ij}. Restriction analysis of viral DNA or of a DNA copy of the viral nucleic acid, is one such method. d_{ij} may be estimated from data on restriction site polymorphism or, more frequently, from data on restriction fragment polymorphism. It has been shown that d_{ij} can be estimated from the fraction of fragments shared between sequences *i* and *j*: if $F_{ij} = 2m_{ij}/(m_i + m_j)$ (where m_{ij} is the number of fragments shared by *i* and *j*, and m_i and m_j are the number of fragments of *i* and *j*, respectively), $\hat{d}_{ij} = -(2/r)\log_e G$ (Nei & Li, 1979). In this expression *r* is the length of the recognition sequence of the restriction enzyme used and G is estimated by interation from $G=[F(3-2G_1)]^{1/4}$. Variances for F_{ij} (Nei & Tajima, 1981) and for $\hat{d}_{ij}$ (González-Candelas, Elena & Moya, 1992) may be calculated.

A commonly used method to compare nucleic acids from RNA viruses is the characterization of the two-dimensional electrophoretic maps of oligonucleotides generated by complete digestion with a specific ribonuclease, frequently T1 ribonuclease (RNA fingerprinting). These fingerprints can be interpreted with the same theoretical treatment as restriction fragment polymorphism: Base-specific ribonucleases (RNases) can be assimilated to restriction enzymes with a recognition sequence of 1 nucleotide, and the expression above can be used to estimate d_{ij} making $r=1$ (Rodríguez-Cerezo et al., 1991). Another method widely used for the study of viral variability is the analysis of fragments generated by the cleavage by RNases of mismatches in heteroduplexes between the sequences to be analysed and RNA complementary to a reference sequence (ribonuclease protection assay) (Chapter 35). Although presumably data yielded by this method could be used to estimate d_{ij}, there is not yet enough theoretical background to derive it properly.

Once $\hat{d}_{ij}$ is known, nucleotide diversity in population k can be estimated by $\hat{D}_k=(n_k/n_k-1)\,\Sigma x_i x_j \hat{d}_{ij}$, where n_k is the number of sequences in the sample, and x_i and x_j are the sample frequencies of sequence type (haplotype) i and j, respectively. Nucleotide diversity is a concept equivalent to gene diversity.

When two populations, k and l, are to be compared, the net average number of nucleotide substitutions (genetic distance) between k and l, D_{kl}, is estimated by $\hat{D}_{kl}=\Sigma x_i x_j \hat{d}_{ij}-1/2\,(\hat{D}_k+\hat{D}_l)$, where $\hat{d}_{ij}$ is the nucleotide substitution value between the ith haplotype from population k and the jth haplotype from population l, and x_i and x_j the frequency of haplotypes i and j in populations k and l respectively.

Based on these parameters, population subdivision, population isolation, migration, bottleneck effects, and other phenomena related to the maintenance of genetic variation can be analysed.

Genetic structure and evolution of viral populations: some case studies

We will present here some examples of the analysis of genetic variation in viral populations using the above described methods. These case studies may illustrate the use of different types of experimental data, the approach to different biological issues, and how conclusions can be drawn that would not be obvious without this kind of analysis.

The genetic variation of field populations of tobacco mild green mosaic tobamovirus (TMGMV) was studied by Rodríguez-Cerezo *et al.* (1991). This virus naturally infects the perennial wild plant *Nicotiana glauca* Grah. By randomly sampling *N. glauca* plants along a 300 km transect in SE Spain, and by sampling plants in five of the discrete stands (only some tens of metres wide) in which this plant grows in that area, two sets of TMGMV isolates were obtained. The genomic RNAs of 73 isolates were analyzed by RNase T1 fingerprinting. The population was found to be composed of 51 haplotypes, as was expected for a virus with a RNA genome, and this indicated initially that the virus has great genetic variability. When, however, values for nucleotide diversities were calculated they were found to be very small: 0.0142, on the average, for isolates from each of the five stands; 0.0216 for isolates from the 300 km transect; and 0.0342 for a pool of Spanish and Australian isolates. Furthermore, these small values were maintained regardless of the distance between the sites from which the isolates came, indicating a high genetic stability for TMGMV (Rodríguez-Cerezo *et al.*, 1991). This extreme genetic stability was further analysed. From the data of within, and between, population divergence for the subpopulations from the five *N. glauca* stands (Table 15.1 (*a*)) no evidence for population subdivision was found using the procedure devised by Lynch and Crease (1990). It was also shown that this genetic structure cannot be explained exclusively in terms of the neutral theory of molecular evolution: assuming for TMGMV mutation rates as found for other RNA viruses, and based on the observed level of variation, an unrealistic estimate for the effective size of the viral population was obtained. A model of periodic, positive selection, could account for the small variability observed (Moya, Rodríguez-Cerezo & Garćia-Arenal, 1993), but other factors, such as bottleneck effects during transmission could also contribute to it.

The analysis of the temporal variation of a second tobamovirus, pepper mild mottle tobamovirus (PMMV) has given a similar view (Rodríguez-Cerezo, Moya & García-Arenal, 1989). Fingerprinting analysis of the genomic RNAs of 23 isolates of PMMV from glasshouse-grown peppers in SE Spain, representing epidemics from 1983 (the first recorded one) until 1987, again showed a large number of haplotypes (9) probably arising from a main one that accounts for 35% of the isolates: this main haplotype prevailed during the whole studied period and was not replaced by the new ones. Nucleotide diversities, calculated within, and between, populations, are, as for TMGMV, small (Table 15.1 (*b*)), and values were maintained regardless of the time elapsed between each of the populations compared. This shows the great stability of PMMV populations.

How general this high genetic stability is for plant viruses with RNA genomes we do not know, as we are not aware of similar work having

Table 15.1. *Nucleotide diversity within, and between, populations of different viruses*

(*a*) *Tobacco mild green mosaic tobamovirus*[a]

Population	E	S	P	R	C
E	0.0121				
S	0.0018	0.0146			
P	0.0023	0.0013	0.0191		
R	0.0012	0.0025	0.0024	0.0154	
C	0.0024	0.0040	0.0047	0.0035	0.0100

Note: [a] From Rodríguez-Cerezo *et al.* (1991).

(*b*) *Pepper mild mottle tobamovirus*[b]

Population	1983	1984	1985	1986	1987
1983	0.0026				
1984	0.0192	0.0322			
1985	0.0037	0.0201	0.0070		
1986	0.0141	0.0318	0.0151	0.0000	
1987	0.0059	0.0218	0.0073	0.0206	0.0071

Note: [b] Estimated from data in Rodríguez-Cerezo *et al.* (1989).

(*c*) *Vesicular stomatitis vesiculovirus*[c]

Population	US1982	US1983	US1984	US1985	Mx1984	Mx1985
US1982	0.0033					
US1983	0.0550	0.0059				
US1984	0.0495	0.0515	0.0011			
US1985	0.0590	0.0632	0.0190	0.0014		
Mx1984	0.0479	0.0503	0.0094	0.0096	0.0082	
Mx1985	0.0540	0.0565	0.0146	0.0057	0.0091	0.0064

Note: [c] Estimated from data in Nichol (1987).

been done with other viruses. A detailed analysis of the variability of 137 Spanish isolates of citrus tristeza closterovirus (CTV) from four different geographical areas as detected by dsRNA patterns from infected tissues (Guerri *et al.*, 1991) provides a further example of plant virus variation. When the published data are recalculated so that similar numbers of samples are compared for each area, it is found that genetic diversity is greatest (0.8146 ± 0.0490) in the area where CTV was first introduced (possibly around 1930), that is also the area of highest incidence (70%). Diversity is less in surrounding areas to which CTV has spread more recently (after 1957) and is less common (5–20%): values for the three areas are 0.4444 ± 0.0830; 0.1141 ± 0.0713; and 0.121 ± 0.0747. As for PMMV one haplotype out of ten, accounted for 75% of the isolates, and dominates the population. This type appears to be the one that has colonized the new areas: it accounts for 94% of the isolates in the areas of lowest incidence versus 38% of the isolates in the area of oldest colonization. On the other hand, new haplotypes seem not to disappear from the population, as shown by the increase in diversity of the population with time, a situation clearly different from that of PMMV. Still, considering the large time intervals involved, CTV populations seem also to be quite stable.

A different situation is found when the temporal variation of vesicular stomatitis vesiculovirus (VSV) is considered. Nichol (1987) has published RNase T1 fingerprints of the genomic RNAs of 41 isolates from epizootic epidemics among domestic animals in the western United States of America and in Mexico from 1982 to 1985; 22 haplotypes were found. From the published data, we have calculated nucleotide differences per site between haplotypes, as well as within, and between, population diversities for the 1982, 1983, 1984, and 1985 epizootics in the USA, and the 1984 and 1985 epizootics in Mexico (not enough data were available for 1982 and 1983 from Mexico). As for the tobamoviruses discussed above, within-populational diversities are very small (Table 15.1 (c)). For each epidemic, populations are very homogeneous. Values for between-population diversities are of the same order as for within-populations, or are considerably greater depending on the populations compared: the US1982 and US1983 populations differed notably from each other, and from the other four populations considered, but differences did not increase with time, and there is no evidence for a continuous evolution of the population. On the other hand, the populations from 1984 and 1985 are very homogeneous, both from the USA and from Mexico (possibly subsets of a single widespread population, although this should be further analysed). The data suggest that the genetic variation in the USA population results from the migration of haplotypes, most probably from Mexico as Nichol deduced from phylogenetic analysis. Our data support this conclusion for 1984 and 1985, but no such evidence exists for 1982 or 1983.

Temporal changes in the population structure of human immuno-deficiency lentivirus (HIV) from one individual were reported by Meyerhans *et al.* (1989). Variation was analysed by direct amplification by PCR of viral nucleic acids from the patient's peripheral blood mononuclear cells (PBMC) sampled during an eighteen month period (samples L1 to L4). Also, from each sample, HIV was isolated by culture in PBMC (samples V1 to V4). Twenty clones from the *tat* gene were sequenced for each of populations L1-L3 and V1-V4. The deduced amino acid sequences comprised 36 haplotypes, with different frequency distributions for each of the different populations. A detailed quantitative study of this data is in progress (Moya *et al.*, in preparation) but some preliminary results are presented here. Amino acid differences per site between sequences were estimated using the Jukes and Cantor method, and within, and between, population diversities were calculated (Table 15.2 (*a*)). In spite of the high number of haplotypes, and of fluctuations in haplotype composition of the population samples L1-L3 (Meyerhans *et al.*, 1989), the nucleotide diversities of these *in vivo* populations stayed constant with time. Also, the diversity among populations L1-L3 are much smaller than within populations, stressing the genetic stability of the populations in the host. On the other hand, within-population diversity values of the samples from the cultured cells are much greater, and they increase with time. For *in vitro* populations, values for between-population diversity are greater than for within-population diversity (Table 15.2 (*b*)). These data not only show that 'to culture is to disturb', as Meyerhans *et al.* (1989) had concluded, but also that selection of variants when culturing HIV may give a picture of genetic diversity for this virus that does not correspond to, and is greater than, that in the host. Also, it can be concluded that mechanisms, possibly negative selection, operate *in vivo* to maintain the genetic structure of the population.

The genetic variation of West Nile flavivirus (WNV) from Madagascar was analysed by typing antigenic variants with a panel of monoclonal antibodies (Morvan *et al.*, 1990). Data on 83 isolates from human beings, birds and mosquitoes, fitting into five antigenic types, were reported. As for CTV we have re-analysed the data to provide similarly sized,

Table 15.2. (a) *Nucleotide diversity within and between,* in vitro *(V1–V4) and* in vivo *(L1–L3) populations of human immunodeficiency lentivirus I*[a]

Population	V1	V2	V3	V4	L1	L2	L3
V1	0.0050						
V2	0.0003	0.0039					
V3	0.0368	0.0385	0.0164				
V4	0.0239	0.0252	0.0058	0.0231			
L1	0.0179	0.0188	0.0109	0.0025	0.0118		
L2	0.0071	0.0078	0.0166	0.0067	0.0030	0.0243	
L3	0.0126	0.0134	0.0131	0.0046	0.0017	0.0026	0.0180

Note: [a] Estimated from data in Meyerhans *et al.* (1989)

(b) *Ratios showing differences in the levels of genetic divergence between the above* in vitro *and* in vivo *populations of human immunodeficiency lentivirus I*

	In vivo	In vitro	$\dfrac{\text{In vitro}}{\text{In vivo}}$
a= Average within population diversity	0.0172	0.0121	0.7000
b= Average between population diversity	0.0027	0.0218	8.0700
b/a	0.1570	1.7948	11.4300

synchronous (for 1982) samples. Differences in genetic diversities were found for the populations of the three hosts, values being greater for mosquitoes (0.6015 ± 0.0804) than for human beings (0.3032 ± 0.1240) or for birds (0.2333 ± 0.1256). This may indicate that selection occurs when WNV isolates are transmitted to either human beings or birds from a common reservoir in mosquitoes, and this is not apparent from the heterogeneous original data.

The examples presented above show that, although data that can be used to test hypotheses and evolutionary models for viruses (such as the one for TMGMV) are scarce, there are some reports presenting data that can be analysed from a population genetics' viewpoint. Such analyses yield information on the genetic structure of viral populations, and this helps us understand virus evolution.

A complementary area of research would be the population genetics of viral hosts and vectors as they may affect important aspects of virus evolution through, for instance population isolation, migration, bottleneck effects, etc. We are not aware of any study that deals with population genetics of viruses and their hosts. In one of our laboratories research has started recently on the population genetics of *Rhopalosiphum padi* L., the aphid vector of the more important serotypes (PAV and RPV) of barley yellow dwarf luteovirus in Spain. Data from restriction analysis of mitochondrial DNA of aphid populations from four different regions in Spain show that only 28% of the total variability observed is due to differences among populations (Martínez *et al.*, 1992). This suggests that long-range migration of the aphids, possibly carrying viruses, does occur, an aspect to be considered when analysing the evolution of the vector-borne viruses.

Conclusions

In this short review we have attempted to show that a perspective of population genetics can significantly contribute to understanding the evolution of viruses. We want to stress again how infrequently has this kind of work been done with viruses, and how important it is if we are to get a full understanding of the mechanisms of viral evolution. In many studies of virus evolution populations are represented by just one individual isolate. Still, there is a small but significant number of published reports in which several individuals were sampled from each population, and these data may be analysed by the methods and from the point of view we have presented here. The examples discussed above do show that this kind of approach may bring strong support to conclusions drawn from other kind of analyses. What may be more important: they show that aspects of viral evolution not apparent from the data, or even opposed to what may be concluded from a simpler analysis, may be revealed from a population genetics approach.

We want to conclude by expressing our hope of an increase of the interest of virologists in the population genetics of viruses.

Acknowledgements

This work was in part supported by grants BIO89–0668–C03–03 and AGR90–0152, from CICYT, Spain, to A.M. and F.G.-A., respectively.

References

Air, G.M., Gibbs, A., Laver, W.G. & Webster, R.C. (1990). *Proc. Natl. Acad. Sci. USA*, **87**, 3884–8.
Birky, C.W. (1991). In *Evolution and the Molecular Level*, R.K. Selander, A.G. Clark and T.S. Whittam (eds.). Sinauer Sunderland, Massachusetts, pp. 112–134.

Birky, C.W., Fuerst, P. & Maruyama, T. (1989). *Genetics*, **121**, 613–27.
Dobzhansky, T. (1970). *Genetics of the Evolutionary Process*. Columbia University Press, New York.
Dodds, J.A., Jordan, R.L., Roistacher, C.N. & Jarupat, T. (1987). *Intervirology*, **27**, 177–88.
Dopazo, J., Sobrino, F., Palma, E.L., Domingo, E. & Moya, A. (1988). *Proc. Natl. Acad. Sci. USA*, **85**, 6811–15.
Fitch, W.M., Leiter, J.M.E., Li, X. & Palese, P. (1991). *Proc. Natl. Acad. Sci. USA* **88**, 4270–4.
Gojobori, T., Moriyama, E.N. & Kimura, M. (1990). *Proc. Natl. Acad. Sci. USA*, **87**, 10015–18.
González-Candelas, F., Elena, S.F. & Moya, A. (1992). Submitted.
Guerri, J., Moreno, P., Muñ, N. & Martínez, M.E. (1991). *Plant Path.* **40**, 38–44.
Hartl, D.L. & Clark, A.G. (1989). *Principles of Population Genetics*. Sinauer, Sunderland, Massachusetts.
Hartl, D.L. & Dykhuizen, D.E. (1984). *Ann. Rev. Genet.*, **18**, 31–68.
Kimura, M. (1983). *The Neutral Theory of Molecular Evolution*. Cambridge University Press, Cambridge, UK.
Levin, B.R. (1981). *Genetics*, **99**, 1–23.
Lynch, M. & Crease, T.J. (1990). *Mol. Biol. Evol.*, **7**, 377–94.
Martínez, D., Moya, A., Latorre, A. & Fereres, A. (1992). *Ann. Entomol. Soc. Am.* **85**, 241–6.
Maruyama, T. & Kimura, M. (1980). *Proc. Natl. Acad. Sci. USA*, **77**, 6710–14.
Meyerhans, A., Cheynier, R., Albert, J., Seth, M., Kwok, S., Sninsky, J., Morfeldt-Manson, L., Asjö, B. & Wain-Hobson, S. (1989). *Cell*, **58**, 901–10.
Milkman, R. (1973). *Science*, **182**, 1024–6.
Morvan, J., Besselaar, T., Fontenille, D. & Coulanges, P. (1990). *Res. Virol*, **141**, 667–76.
Moya, A., Rodríguez-Cerezo, E. & García-Arenal, F. (1993) *Mol. Biol. Evol.* **10**, 449–56.
Nei, M. (1987) *Molecular Evolutionary Genetics*. Columbia University Press, New York.
Nei, M. & Li, W.H. (1979). *Proc. Natl. Acad. Sci. USA*, **76**, 5269–73.
Nei, M. & Tajima, F. (1981). *Genetics*, **97**, 145–63.
Nichol, S.T. (1987). *J. Virol.*, **61**, 1029–36.
Po, T., Steger, G., Rosenbaum, W., Kaper, J. & Riesner, D. (1987). *Nucl. Acids Res.*, **15**, 5069–83.
Rodríguez-Cerezo, E., Moya, A., Elena, S.F. & García-Arenal, F. (1991). *J. Mol. Evol.*, **32**, 328–32.
Rodríguez-Cerezo, E., Moya, A. & García-Arenal, F. (1989). *J. Virol.*, **63**, 2198–203.
Selander, R.K., Beltrán, P. & Smith, N.H. (1991). In *Evolution and the Molecular Level*, R.K. Selander, A.G. Clark, and T.S. Whittman (eds). Sinauer, Sunnderland, Massachusetts, pp. 25–57.
Weir, B.S. (1990). *Genetic Data Analysis*. Sinauer, Sunderland, Massachusetts.

16

Origin and evolution of prokaryotes
E. STACKEBRANDT

Introduction

Genealogical relationships among prokaryotes have traditionally been defined by very diverse morphological, physiological and biochemical properties as well as by immunological comparisons of a few proteins. The main problems in interpreting the results of this deductive approach come from uncertainty about whether or not these characters are actually homologous. These genealogies can only be verified by inductive methods such as the analysis of palaeochemical data, the identification of fossils and the determination of the order of their appearance in time, and also by the biochemical investigation of the genes of ubiquitously distributed macromolecules that have conserved primary and secondary structure and functional constancy. In contrast to geological and fossil evidence, analyses of macromolecules offer an almost unlimited source of invaluable information about the genealogy of living organisms (Zuckerkandl & Pauling, 1965). In particular, estimation of the natural relationships among organisms presently alive can be by sequence analysis of ribosomal (r)RNAs and their genes, genes coding for certain enzymes (ATPases) and regulatory proteins (tuf-factors), and by nucleic acid hybridization studies. These methods are able to span the full range of relationships, from the hypothetical ancestor of life on earth to most recently evolved strains. In his superb reviews Woese (1987, 1991) discussed the rationale for using rRNA as a molecular clock in phylogenetic studies, rather than the traditional classification system which failed to delineate phylogenetic relationships, and discussed the power of sequences and nucleotide signatures to unravel phylogenetic trees.

'Nothing in biology makes sense except in the light of evolution'

With the publication of the first results of comparative analysis of small subunit rRNA sequences (Woese & Fox, 1977) biology entered a new era. Most exciting was the discovery of a 'third form of life', the archaebacteria. Their discovery proved the traditional view wrong that life was dichotomously organized into pro- and eukaryotes. Virtually all microbiological disciplines benefited from the new insight into the intra- and interrelationships of the two major prokaryotic groups. The finding that prokaryotes probably evolved more than 4.0 billion years ago explains their great genetic and phenotypic diversity. Prokaryotes dominated life on earth long before the first eukaryotes evolved and they have achieved this domination by producing the atmosphere of oxygen. Life of eukaryotes would have not been possible without prokaryotic symbionts that evolved into cytoplasmic organelles. These organelles, the mitochondria and chloroplasts, originated as members of the alpha subclass of proteobacteria and cyanobacteria, respectively, and have allowed the eukaryotes to thrive under aerobic conditions.

Following the discovery of the archaebacteria, systematists and chemotaxonomists extended molecular techniques with the goal of assessing the natural relationships of all forms of life. The small subunit rRNA of members of nearly all described genera of prokaryotes and representatives from all eukaryotic phyla have been analysed and the phylogenetic branching pattern has been used to determine groups of genetically coherent organisms. In a polyphasic approach to systematics, i.e. the consideration of both phylogenetic and phenotypic data, a natural system of relationship has now been established that combines both phylogenetic consistency and practicability (Stackebrandt, 1991). A logical consequence of this development is a system that 'will repair the damage that has been the unavoidable consequence of constructing taxonomic systems in ignorance of the likely course of microbial evolution' (Woese, Kandler & Wheelis, 1990). The most revolutionary step is the description of the domain, a taxon above the kingdom level, with Archaea (for the archaebacteria), Bacteria (for the 'typical' bacteria, formerly eubacteria) and Eucarya (for the eukaryotes) (Woese *et al.*, 1990).

Progress, however, is not restricted to systematics and the development of an improved classification system. There is hardly any field in microbiology that has not been influenced by the discovery of the archaeas and the arrangement of organisms according to genealogical relationships. The saying by Theodor Dobzhansky (1973) 'nothing in biology makes sense except in the light of evolution' was proven true by the increased international collaboration and the interest scientists from other biological disciplines have developed towards molecular phylogeny and an understanding of molecular evolution. Microbiologists working on morphologically and biochemically distinct groups of organisms, such as budding, prosthecate, chemolithotrophic and phototrophic organisms, realized that the failure to find more than one key character to define these individual groups was due to the large amount of genetic incoherency that actually occurs among its members.

Despite obvious progress, large gaps still exist in our understanding about the earliest, the progenote, stage of prokaryotes, the underlying mechanisms of evolution, the correlation between sequence divergence and time and the characteristics of the hypothetical common ancestors that occupy the branching points that lead to recent organisms. Further problems are introduced by the failure of present algorithms to distinguish between mutational changes of tachytelic (rapidly evolving) or bradytelic (slowly evolving) strains, and the inability of the available sophisticated algorithms to handle the large numbers of sequences in the dramatically increasing data base. Nevertheless, the finding that phylogenetic branching patterns derived from the analysis of certain proteins and nucleic acids are encouragingly well correlated, provides us with a basis for understanding the phylogeny of prokaryotes and for testing traditional concepts of bacterial evolution.

Phylogeny of prokaryotes

Several review articles have been published during the last ten years about the phylogeny of prokaryotes (Woese, 1987; Stackebrandt, 1991) that also reflect to a certain extent the evolution of the generation of phylogenetic trees. I will not try to outline all of this rapidly growing body of information, and it is beyond the scope of this chapter even to describe the most important findings. Instead, some problems in data analysis and certain major achievements will be indicated and for extra information the reader is referred to the reviews listed above.

Problems with the interpretation of phylogenetic trees

Phylogenetic trees of bacteria usually have long branches but only small internode distances. This arrangement not only causes problems in the determination of the order in which the ancestors of the main lines of descent evolved but the bush-like appearance of the tree makes it difficult to delineate higher taxa. This is obvious in the poor resolution of the main lines of descent within the supercluster of bacteria embracing spirochetes, green-sulphur bacteria, the bacteroides group, planctomycetes, chlamydiae, gram-positives, cyanobacteria and proteobacteria. A delineation between the deep branching members of these phyla would not be possible without the presence of certain taxon-specific features, both phenotypic and genotypic, such as signature nucleotides (Woese *et al.*, 1985; Woese, 1987). It can, however, be predicted that the branching pattern will become even more diffuse once the masses of sequences generated from gene libraries of DNA and/or RNA isolated directly from environmental samples are included in the database.

Another problem that relates to branching patterns is caused by the presence of rapidly (tachytelic) evolving sequences. Mycoplasmas, leptospiras, animal and fungal mitochondria and certain planctomycetes are prime examples. Standard treeing algorithms would not have discovered the relationships between the mycoplasmas and the bacilli. The planctomycetes are of special interest since they are pheno- and genotypically unique bacteria but neither their habitat nor their life cycle give any clues about the selective pressure that caused them to evolve in a tachytelic manner. When analysed together, members of all four genera actually form a coherent cluster, that branches next to the *Thermotoga* line (Liesack *et al.*, 1992). However, members of the genera *Planctomyces* and *Pirellula* group with *Chlamydia psittaci* within the main cluster of the domain Bacteria when the sequences of members of *Isosphaera* and *Gemmata* are excluded from the analysis. When analysed without members of *Planctomyces* and *Pirellula* members of the latter two genera not only form deep branches of the tree of Bacteria but their presence in the tree causes significant distortions of the branching order of the other main lines of descent. The switching of the phylogenetic position is probably caused by the varying degree of tachytely of planctomycete 16S rRNAs that makes them appear to have an ancient origin, whereas they are actually highly derived molecules, probably remotely related to the chlamydiae.

A third problem is encountered by 'sequence composition induced artifacts' (Woese *et al.*, 1991). This expression refers to the fact that the 16S rDNA of thermophilic bacteria evolved towards a higher DNA G+C base composition than their mesophilic relatives. This compositional bias results in artificially greater similarity values between thermophilic species. The most prominant examples are the clustering of *Thermotaga* with *Thermus* and *Thermomicrobium* (Weisburg, Giovannoni & Woese, 1988) and the clustering of *Archaeoglobus fulgidus* between the thermophilic sulfur-dependent archaeas and *Thermococcus*. Restriction of the analysis to transversions, or to sites where a single base comprises at least 50% of that position's total composition, the three thermophilic bacteria form three independent lines of descent. Archaeoglobus is found to be an offshoot of one of the methanogenic lines (Woese *et al.*, 1991), which is supported by its metabolic properties.

Phylogenetic problems also arise when an unbalanced selection of reference strains is used, and when the analysis is restricted to certain parts of an operon, where the information content is too small to reflect the evolution of the complete gene (Stackebrandt, Liesack & Witt, 1992).

The existence of phylogenetic branching patterns

The two prokaryotic domains

The distinctness of the three major lines of descent, originally detected by 16S rRNA cataloguing (Woese & Fox, 1977), was later confirmed by the analyses of complete rRNA, and their genes, and by analysis of the genes coding for the proteins involved in the regulation of translation and in energy yielding processes (Schleifer & Ludwig, 1989). The isolated position of the Archaea, considered to be an entity as unrelated to Eucarya and Bacteria as the latter two domains are related among themselves, was strengthened by the finding that certain genetic and epigenetic properties were unique to archaeas (e.g. the chemistry of cell envelopes and lipids, the subunit structure of the RNA polymerase, and the modification pattern of rRNA species). In addition, certain taxa of this domain exhibit specific properties absent in other organisms, such as the set of co-enzymes involved in methanogenesis, the purple membrane of *Halobacterium* and survival in hyperthermophilic conditions. As judged from the range of sequence similarity values of 16S rRNAs of members of the three domains, archaeas evolved at a slightly slower rate than bacteria (Fig. 16.1). The reasons are not known, but this effect may perhaps have been caused by the selection pressure and ecological forces of the unusual niches most of these organisms have occupied probably for several billion years.

Each of the two prokaryotic domains contain a number of more or less well separated sublines of descent, the phyla or divisions. According to Woese *et al.* (1990) two main lines have so far been defined within the domain Archaea (Fig. 16.1). The first, broad line contains three sublines of methanogenic organisms. Related to them are non-methanogenic taxa such as the wall-less thermoplasmas, the extreme halophilic and alkaliphilic forms and the peculiar thermophilic *Archaeoglobus fulgidus*, an organism capable of forming methane and reducing sulphate. The thermophilic *Thermococcus celer* seems to represent a sister group. The second line is defined by hyperthermophilic archaeas which grow optimally at temperatures up to 105 °C, and which mostly require elemental sulfur for optimal growth. The dichotomic phylogenetic structure of archaeas, the broad spectrum of niches occupied, and the varied patterns of metabolism, have been the basis for the elevation of these two major lines to the rank of kingdoms, Crenarchaeota and Euryarchaeota.

The phylogenetic structure of the domain Bacteria is much more complex in that more than ten main lines of descent have been identified so far, some of which, however, are only separated by small internode distances (Fig. 16.1(*a*)). The relationships within the individual lines are often unexpected, and support for these groupings from the sharing of properties other than similarities in conserved genes is rare. Furthermore, owing to the substantially longer history of the taxonomy of bacteria, as mentioned above, the uniqueness of archaeas was discovered only recently; the number of species is so much greater than those of archaeas that, for many groups of organisms, the branching pattern becomes difficult to interpret.

The branching pattern of Woese and colleagues has been challenged by Lake (1988). Using a rate-independent technique of evolutionary parsimony (Lake, 1987) in combination with the neighbourliness procedure (Fitch, 1981) and transversion parsimony (Fitch, 1977), five groups were recovered (Fig. 16.1(*b*)). These were the eubacteria, halobacteria, methanogens, eocytes (the sulphur-dependent and mostly thermophilic archaeas), and eukaryotes. Eubacteria, halobacteria and methanogens were separated from the eocytes and eukaryotes. The main point of criticism of Lake (1988) towards the concept of Woese is the difference in the substitution rate of rRNA genes of archaebacteria (about ten times slower in archaeas than in eubacteria and eukaryotes), which favour the formation of an independent archaebacterial line. The presence of archae-unique characters, as well as the rather large variety of bio-chemical markers in archaeas which range from being bacteria-specific

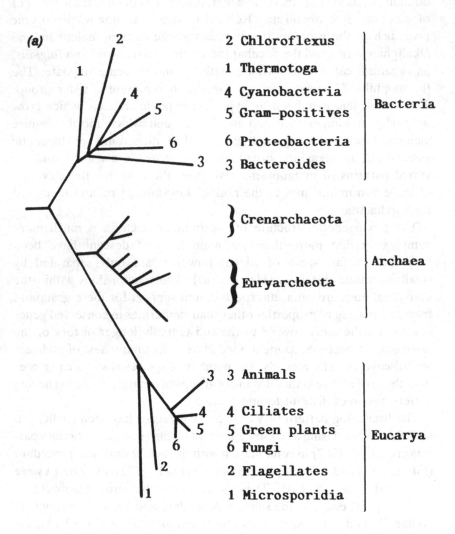

(a)

2	2 Chloroflexus
1	1 Thermotoga
4	4 Cyanobacteria
5	5 Gram–positives
6	6 Proteobacteria
3	3 Bacteroides

Bacteria

} Crenarchaeota
} Euryarcheota

Archaea

3	3 Animals
4	4 Ciliates
5	5 Green plants
6	6 Fungi
2	2 Flagellates
1	1 Microsporidia

Eucarya

Fig. 16.1(a). See opposite for caption.

to eukaryote-specific, support both kinds of interpretation. At present, however, most textbooks, have adopted the scheme of Woese.

The failure of the traditional classification system to reflect phylogeny

Comparison of the phylogenetic position of an organism relative to its neighbours, with its taxonomic position in the traditional classification system revealed the failure of the latter system to reflect natural relationships. Phenotypic properties used in the past to define higher taxa were shown to have little or no phylogenetic importance at all. The presence of a photosynthetic apparatus, and the chemical

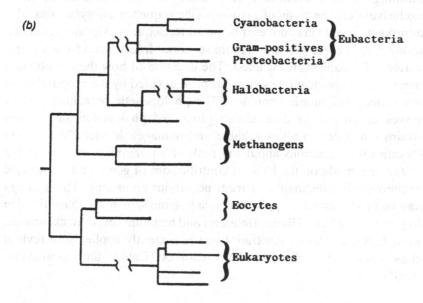

Fig. 16.1. Phylogenetic tree of the main lines of descent of organisms based on the analysis of the small ribosomal subunit RNA organisms. (*a*) Distance tree (De Soete, 1983) based on evolutionary distance values (Jukes & Cantor, 1969). Organisms are clustered into three domains (after Woese *et al.*, 1990). (*b*) Evolutionary parsimony tree (Lake, 1988). Organisms are clustered in five groups, separating the (eu)bacteria, halobacteria and methanogens from the eocytes and the eukaryotes.

composition of bacteriochlorophylls, were used to define orders and families. Similarly, gliding motility and budding mode of reproduction were considered key characters in the definition of higher taxa. In combination with other features these properties are now only used to circumscribe lower taxa. Phototrophic organisms are found in 5 of the 11 or so bacterial phyla, whereas gliding motility and budding are even more widely distributed. In many examples, groups of anaerobic and facultatively phototrophic bacteria are intermixed with organotrophic and chemolithotrophic relatives. This makes perfect sense from the viewpoint of the evolution of biochemical pathways, see below. Other features now recognized to be of little value in revealing phylogeny include morphology (except the spirochete shape), mycelium formation, spore formation, and biochemical and metabolic properties.

Towards a polyphasic approach to prokaryotic classification

Although it may eventually be possible to define taxa at all levels exclusively on the basis of taxon-specific sequence idiosyncrasies, this approach appears may not ever be possible. Sequence signatures defined today may become blurred as more sequence information from a larger variety of organisms is included. The question of how then to obtain a meaningful classification system has been decided by the proposal of an integrated, polyphasic approach. The phylogenetic branching pattern serves as an aid to determining groups of phylogenetically related strains. In order to allow reliable and manageable identification and classification, decisions about the rank of an organism in a hierarchic system are made on the basis of contributions of geno- and phenotypic properties that distinguish it from neighbouring groups. This strategy has been successfully applied to many groups of proteobacteria (De Ley and coworkers, Ghent, Belgium) and actinomycetes (Stackebrandt, Kiel, FRG; Brisbane, Australia) and is currently applied to a revised classification of bacilli and their relatives (Collins and co-workers, Reading, UK).

Traditional and modern ideas for the evolution of prokaryotes

The most widely accepted thesis for the origin and early evolution of life was expressed as early as 1924 (Oparin, 1924). This hypothesis and later modifications thereof assumed that life evolved in a soup of moderately concentrated organic molecules that originated through the action of ultraviolet light and electric discharges. In a series of events of increasing complexities (coacervate-proteinoids), a primitive type of organism evolved that could have resembled the present *Clostridium* type of organism, i.e. anaerobic and heterotrophic. From this stage on, newly evolved organisms faced an increasingly hostile environment as the exponentially proliferating species depleted the store of energy-rich compounds. As a consequence, biochemical pathways evolved, and this led step by step to the biosynthesis of essential monomers from their steadily depleted precursors. At the end of this development, photosynthesis and autotrophic fixation of carbon evolved which guaranteed continuation of life on earth by producing an unlimited supply of nutrients.

The deductive way of unravelling evolution

Considering the lack of alternative hypotheses for the origin of life (Woese, 1979), it is not surprising that an anaerobic and strictly fermentative bacterium has been considered to be the ancestor of life in the two main alternative proposals for the evolution of prokaryotes: the 'conversion hypothesis' (Broda, 1978) and the 'segregation hypothesis' (Margulis, 1970). These genealogical patterns were derived from deductive analyses of properties of recent organisms, for example, their biochemical pathways and metabolic properties. The fundamental differences between the two ideas concern the origin of anaerobic photosynthesis and the evolution of respiration. Since archaeas were not described at the time these hypotheses were outlined, their properties were considered to indicate that they were special metabolic types of the bacteria and consequently they were included as parts of the general outline of bacterial evolution.

According to Margulis (1970), the evolution of electron transport in anaerobic respirers preceded the evolution of the cyclic electron transport in anaerobic phototrophic organisms and aerobic respirers. Each of these biochemical properties was thought to have evolved only once during evolution, so that clusters of all recent fermentative, anaerobic respirers, phototrophic, chemoautotrophic and aerobic types were placed in separate lines of descent.

According to Broda (1978), phototrophic organisms evolved directly from the most ancient, fermentative organism. In a series of events, different types of phototrophs evolved in several lines of descent, some of which became extinct, while others still exist today. Reacting to changes in the oxygen level during time, each of the early phototrophic organisms gave rise to individual lines of descent in which metabolically different types evolved, e.g. organotrophic and lithotrophic forms. Consequently, anaerobes, phototrophes, anaerobic respirers and aerobic forms evolved independently from each other several times during evolution.

Comparison of the deductive and the inductive approach

Comparison of genealogical patterns derived from deductive and inductive methods by and large supports the 'conversion hypothesis'. As briefly indicated above, phototrophic bacteria are found in several lines of descent within Bacteria. Considering the complexity of the photosynthetic apparatus it is reasonable to assume that this system (photosystem I) evolved monophyletically with its variations developing independently under selective pressure in certain niches. As judged from the position of the oldest phototrophic organism in the tree, *Chloroflexus*, the photosynthetic apparatus was 'invented' early in the evolution of bacteria. Loss of the photosynthetic apparatus independently in certain members of all but two lines of descent (green sulphur bacteria, cyanobacteria), and changes in the function of the cylic electron chain to cope with the new nutritional condition, led to the evolution of novel metabolic types. The descendant of an anaerobic photosynthetic bacterium, predicted to be the ancestor of gram-positive bacteria by Broda has actually recently been discovered. *Heliobacterium chlorum*, representing a novel photosynthetic type, is phylogenetically related to clostridia and bacilli. Also, in accord with Broda, is the finding that several metabolic types such as fermentation, anaerobic respiration, chemoautotrophy and aerobic respiration did not evolved monophyletically, but originated from different ancestors in different main lines of descent. Phototrophic bacteria often share close relationships with chemoautotrophic and aerobic bacteria. Especially the relatedness between *Rhodopseudomonas palustris* and the nitrite oxidizing, strictly aerobic species *Nitrobacter winogradskii* shows most impressively how the respiration chain evolved from the photosynthetic electron chain. It should, however, be mentioned that, although Broda's scheme is, by and large, in accord with the branching pattern of sequences of evolutionary conserved macromolecules, significant differences occur in detail. In agreement with both deductively derived genealogies, anaerobes represent the most ancient forms, which are the organotrophic *Thermotogales*, i.e. Thermotoga, *Thermosiphon*, *Fervidobacterium*, and *Dictyoglomus* in the domain Bacteria, and the organotrophic and lithotrophic *Crenarchaeota, Thermococcus* and *Thermoproteus*, and *Pyrodictium*, respectively, in the domain Archaea. All the organisms associated with deep branches are extreme or even hyperthermophilic. Like anaerobiosis, thermophily may be seen as a second meaningful property to describe the environment and the hallmark of the phenotype of the ancestors of the two domains, if not that of the common ancestor, the progenote.

Correlation between the geological record and the branching pattern of the 16SrRNA

The fossil record

The fossil record in sedimentary rocks of the Precambrian is only sparse. The oldest fossils known are in the 3.8 Gy (giga years; 10^9 years before present) old Isua metasediments (Greenland), the 3.5 Gy old Waarawoona group (Australia) and the Swartkoppie formation (South Africa), but these types, like those from the Onverwacht series (South Africa) (3.35 Gy) or the Fig Tree series (Swaziland) (3.2 Gy), are less well preserved than those of the younger Soudan iron formation, (Minnesota) which are >2.7 Gy old. Rich microfossils have been found in the Gunflint chert in Ontario which is 2.0 Gy old. Microfossils resembling cyanobacteria in morphology have been described in the Fig Tree and Gunflint formations and in stromatolites from the Bulawayo rocks of South Africa which are 2.9 Gy old. Well-preserved species of cyanobacteria are detectable in stromatolites from Paradise Creek, Queensland, and these 1.6 Gy old formations are apparently devoid of eukaryotic structures. The first definite eukaryotic fossils were found in Beck Springs, California which is 1.2–1.4 Gy before present, and they are more abundant in the younger Bitter Springs rocks of central Australia which are 0.9–1.0 Gy before present.

Although certain morphological features of these fossils resemble those of cyanobacteria, no definite conclusions can be drawn about the biochemical properties of these organisms, such as whether they had an oxygen evolving photosynthetic apparatus. In the light of the poor predictive ability of morphology to indicate relationships, allocation of these fossils to the cyanobacteria or any other recent taxon is very speculative. Moreover, anaerobic phototrophic bacteria of the *Chloroflexus* type have been shown to form stromatolites as well. Considering that the ancestor of the *Chloroflexus* line of descent (harbouring

Thermomicrobium, Herpetosiphon and *Chloroflexus*) evolved earlier than that of the cyanobacteria, some of the ancient stromatolites could actually have been formed by the ancestors of the green non-sulfur bacteria. Nevertheless, although of very limited potential for any proof of ancient biochemical properties, the fossil record indicates that life existed as early as 3.2 Gy ago.

The palaeochemical record

More informative than the fossil record is the chemical analysis of inorganic molecules of geologically ancient sedimentary rocks (Fig. 16.2). The increased ratio of ^{12}C to ^{13}C in sedimentary organic matter is commonly considered to indicate the carbon-fixing reaction of photosynthesis or other autotrophic carbon-fixing mechanisms. In biological processes, the slightly lighter ^{12}C isotope is enriched as compared to the ^{13}C isotope. The discrimination factor for ^{13}C is −25%. This value has remained, by and large, constant over billions of years. Surprisingly, the ^{13}C value of material from the 3.8 Gy old rocks from the Archaean terrane of the Isua metasediments, Greenland, is −13%. This value is not as large as those in slightly younger rocks, but it has been interpreted as meaning that an originally more negative value has become clouded by metamorphic reconstitution of the rocks (Schidlowski, 1988). Isotope discrimination of ultralight carbon nevertheless indicates that carbon has been fixed autotrophically on earth for at least 3.8 Gy. Carbon reduction occurs during photosynthesis and chemoautotrophy, e.g. methanogenesis. Both kinds of energy yielding process have been found in members of ancient lines of descent, although not the most ancient lines, such as the green non-sulphur organism *Chloroflexus* (Bacteria) and the methanogenic species (Archaea) (Fig. 16.1). As shown by Schidlowski (1988) the origin of methanogenesis can be dated 2.7 Gy ago. Although a descendent of such a hypothetical ancestor has not as yet been discovered, it is likely that this type will be found once more extreme environmental niches have been surveyed. The distribution of CO_2 fixing mechanisms in members of both domains leads to the assumption that the separation of the hypothetical ancestor, the progenote, into the two prokaryotic lines of descent (Fig. 16.1) occurred within a rather 'short' period (within the first 700 million years) after the formation of the planet earth. It can however not be excluded that the carbon isotope discrimination was caused by the progenote(s) and that biological CO_2 fixation occurred

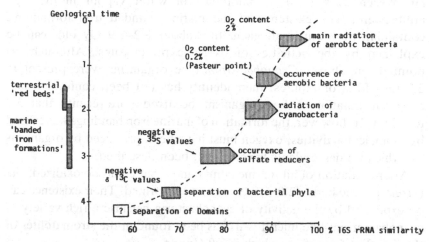

Fig. 16.2. Correlation between the occurrence of certain bacterial groups during evolution as derived from the palaeochemical record and 16S rRNA sequence similarities between the most unrelated members of these bacterial groups.

at a very early time but that the carbon isotopes were not trapped in sedimentary rocks.

A second valuable palaeochemical analysis is the determination of the ratio at which ^{32}S is enriched compared with the heavier ^{34}S during the biological reduction of sulphate (Monster *et al.*, 1979). No discrimination was detected in the 3.8 Gy old Isua sedimentary rocks. The earliest significant change in the ratio was found in rocks that are 3.2 to 2.8 Gy old. Sulphate respiration, as done by *Desulfotomaculum* (Gram-positive) and *Desulfovibrio* and relatives (proteobacteria) may therefore be a more recent adaptation than photosynthesis and methanogenesis. This is in accord with results of 16S rRNA analysis that separates the sublines embracing these sulphate reducers at around 78% similarity (Fig. 16.2). The main radiation of the Gram-negative sulphate reducers occurs even later in evolution (around 82% similarity).

Likewise, the history of atmospheric oxygen contains valuable events that allow a crude calibration of more recent time periods with sequence similarities. As indicated by Broda (1978), there is little doubt that the free oxygen content was minute before the emergence of plant photosynthesis. The source of oxygen in the early Precambian is unknown, although it is unlikely to have come from the free molecular oxygen as a result of photolysis of water vapour in the high atmosphere. The existence of the marine 'banded iron formation', containing magnetite and haematite, about 3.2–1.9 Gy old, can be explained by the activities of oxygenic prokaryotes. Although, as pointed out above, *Cyanobacterium*-type organisms were present in 2.7 Gy old stromatolites, their identity has not been confirmed, and the main radiation of these organisms occurred more recently than 3.2 to 2.7 Gy. If, however, the formation of marine iron bandings was caused by biological activities, oxygen must have been produced by organisms for which no descendents have, as yet, been described.

After oxidation of all marine compounds that could be oxidized, the terrestrial 'red beds' (2.0–1.8 Gy ago) were formed. Their existence can be explained by the activity of cyanobacteria because a rich variety of cyanobacterium-like microfossils has been found in the stromatolites of the Gunflint formation, which is 2.0 Gy old.

It is interesting to notice a correlation between the increased oxygen level, 1.9 to 1.5 Gy ago, and the emergence of microaerophilic organisms from anaerobic phototrophic and heterotrophic ancestors in various sublines of the bacterial domain (*Lactobacillus, Streptococcus, Actinomyces*, spirochetes, certain proteobacteria). Only after the Pasteur point (0.2% O_2) was reached (0.2%, 1.4–1.2 Gy ago) was it advantageous for organisms to switch over from fermentation to aerobic respiration (Berkner & Marshall, 1967). The evolution of aerobic species from microaerophilic fermentative ancestors correlates with rRNA similarities of around 87% while the main radiation of strictly aerobic organisms occurred at similarity values above 92%. According to Böger (1975), this event could have happened at an oxygen partial pressure of 2% in the lower Silurian, about 440 million years ago.

Towards a temporal scale for the evolution of 16S rRNA

Although the number of calibration points in Fig. 16.2 is too small, and the correlation too imprecise, to obtain a statistically significant plot, it is obvious that the relationship between geological time and similarity values of ribosomal RNAs is not linear but more or less curvilinear. Most likely is a staircase-type discontinuous correlation in which convergent phases (increase in sequence divergence and hence in actual information) are interrupted by divergent phases (maintenance of the present information content) (Fig. 16.3). The broken line was

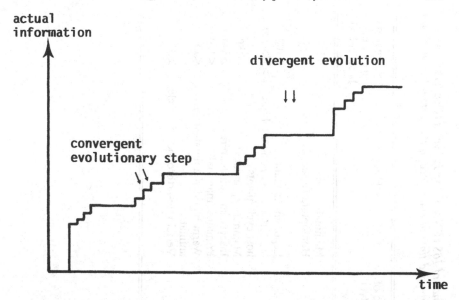

Fig. 16.3. Hypothetical course of evolution, in which periods of information increase (convergent phase) are followed by periods of information maintenance (divergent phase).

used for the determination of the average 16S rRNA substitution rate in the 40% or so of the nucleotide positions that tolerate mutations. Not unexpected is the finding that the substitution rate was greater during the first billion years after the origin of life ($1\% /6.2 \times 10^7$ years) than in the following two billion years (1.2×10^8 and 1.6×10^8). If the substitution rate is calculated for only the last 500 million years (during the main radiation of aerobic bacteria) the value is about $1\%/6 \times 10^7$ years. A similar value has recently been estimated ($1\%/5 \times 10^7$ years) from the relationships between several groups of recent bacteria and their eukaryotic hosts, for which the age is known (Ochman & Wilson, 1987). Assuming that bacteria and host co-evolved, this value is rather convincing for several bacterial groups (bifidobacteria, *E. coli, S. typhimurium*, mycobacteria), but deviations are sometimes significant (*Actinomyces, Rhizobium, Bradyrhizobium, Helicobacter, Rickettsiae*) (Table 16.1). Consequently, the Ochman and Wilson value cannot be considered to represent a general average substitution rate of the 16S rDNA, because evolution of prokaryotes began only 2 Gy ago ($40 \times 5 \times 10^7$ years), which is significantly different from the time of origin measured by palaeochemical data. If the Ochman and Wilson value is

Table 16.1. *Use of the temporal calibration value of 1% sequence divergence of 16S rRNA per 5×10⁷ years (Ochman & Wilson, 1987) to estimate the time of divergence between bacterial species. The estimated time of emergence of the current host is shown to allow conclusions about possible co-evolution between host and bacteria.*

Members of Taxa	% 16S rRNA Similarity	Estimated time of divergence (Myr)	Host	Estimated age of the hosts (Myr)
Actinomyces species	>85	700	Mammals	<150
Pea Aphid Symbionts-P/-S	88.1	595	Homoptera (insects)	<245
Pea Aphid Symbiont-S/*E. coli*	94.8	260		
Rhizobium/Bradyrhizobium	88–90	500–600	Leguminous plants	>100
Rochalimaea/Agrobacterium	91	450	Vertebrate and non-vertebrate host cycle/plants	>345/200
Helicobacter species	92.8–94.9	255–360	Mammals (primates)	<150
Bifidobacterium species	>93	350	Insects, mammals, birds	>345
Pathogenic mycobacteria	95–99.9	5–240	Primates, mammals	<150
*E. coli/S. typhimurium**	97.1	145	Warm blooded animals/man, animals	<150
Rickettsiae species (excl. *Rochalimaea*)	98.5–97.7	75–135	Vertebrate-arthropod host cycle	>345

Note: * Values taken from Ochman and Wilson (1987).

restricted to reflect the mutation rate of rRNA genes during the more recent epoch, it must indicate that the rate was significantly slower in more ancient times but accelerated during the last few hundred million years. Whether or not this is true, and if it is, whether the increased toleration of mutations correlates with the increased oxygen level in the atmosphere, remain open questions.

It may be the case that the Ochman and Wilson value is biased because host-associated organisms were selected for the determination of the mutation rate. Many obligate symbiotic and parasitic bacteria (as well as chloroplasts and mitochondria) have a significantly higher mutation rate than their free-living relatives which by and large evolve isochronically. The latter organisms will show a lower mutation rate (correlation with time is as yet not possible). In order to date back the origin of life to 4 Gy, one must assume a mutation rate of 1% fixed mutations (in the 40% of the 16S rRNA genes that tolerate mutations) per 10^8 years. All of these calculations, however, are highly speculative and prone to substantial errors: (i) the origin of life on earth is not known, (ii) as shown in Fig. 16.3, evolution is not a linear process, and (iii) it is unlikely that mutations were accumulated at the same rate (in parallel) in DNA of the various types of recent prokaryotes and their extinct ancestors. These problems are challenging and their solutions are far from being resolved. Nevertheless, within the last ten years, evolutionary biology achieved more than was accomplished over the preceding 90 years. The speed at which we obtain new insights into the natural relationships of organisms is breathtaking, and the topics of evolution and phylogeny have finally received their deserved place in biological science.

References

Berkner, L.V. & Marshall, L.C. (1967). *Adv. Geophys.*, **12**, 309–17.

Böger, P. (1975). *Naturwiss. Rundschau*, **28**, 429–35.

Broda, E. (1978). *The Evolution of the Bioenergetic Processes*, Pergamon Press, Oxford.

De Soete, G. (1983). *Psychomet.* **48**, 621–6.

Dobzhansky, T. (1973). *Am. Biol. Teacher*, **35**, 125–9.

Fitch, W.M. (1977). *Am. Nat.*, **111**, 223–57.

Fitch, W.M. (1981). *J. Mol. Evol.*, **18**, 30–7.

Jukes, T.H. & Cantor, C.R. (1969). In *Mammalian Protein Metabolism III*, H.N. Munro (Ed.) Academic Press, New York.

Lake, J.A. (1987). *Mol. Biol. Evol.*, **4**, 167.

Lake, J.A. (1988). *Nature*, **331**, 184–6.

Liesack, W., Söller, R., Stewart, T., Haas, H., Giovannoni, S. & Stackebrandt, E. (1992). *Syst. Appl. Microbiol.*, **15**, 357–62.

Margulis, L. (1970). *Science*, **161**, 1020–2.
Monster, J., Appel, P.W.U., Thode, H.G., Schidlowski, M., Carmichael, C.M. & Bridgwater, D. (1979). *Geochim. Cosmochim. Acta*, **43**, 405–13.
Ochman, H. & Wilson, A.C. (1987). *J. Mol. Evol.*, **26**, 74–86.
Oparin, A.I. (1924). Proiskhozhdenie zhizny. Moscow. Izd. Moskovshii Rabochii.
Schidlowski, M. (1988). *Nature*, **333**, 313–18.
Schleifer, K.H. & Ludwig, W. (1989). In *The Hierarchy of Life*, B. Fernholm, K. Bremer, and H. Jörnwall (eds.). Elsevier, Amsterdam, pp. 103–117.
Stackebrandt, E. (1991). In *The Prokaryotes*, 2nd. Ed. A. Balows, H.G. Trüper, M. Dworkin, W. Harder and K.H. Schleifer (Eds.) Springer, New York, pp. 19–47
Stackebrandt, E., Liesack, W. & Witt, D. (1992). *Gene*, **115**, 255–60.
Weisburg, W.G., Giovannoni, S.J. & Woese, C.R. (1988). *Syst. Appl. Microbiol.*, **11**, 128–34.
Woese, C.R. (1979). *J. Mol. Evol.*, **13**, 95–101.
Woese, C.R. (1987). *Microbiol. Rev.*, **51**, 221–71.
Woese, C.R. (1991). In *The Prokaryotes*, 2nd Ed. A. Balows, H.G. Trüper, M. Dworkin, W. Harder and K.H. Schleifer (Eds.) Springer, New York, pp. 3–18.
Woese, C.R., Achenbach, L., Rouviere, P. & Mandelco, L. (1991). *Syst. Appl. Microbiol.*, **14**, 364–71.
Woese, C.R. & Fox, G.E. (1977). *Proc. Natl. Acad. Sci. USA*, **74**, 5088–90.
Woese, C.R., Kandler, O. & Wheelis, M.L. (1990). *Proc. Natl. Acad. Sci. USA*, **87**, 4576–9.
Woese, C.R., Stackebrandt, E., Macke, T.J. & Fox, G.E. (1985). *Syst. Appl. Microbiol.*, **6**, 143–51.
Zuckerkandl, E. & Pauling, L. (1965). *J. Theor. Biol.*, **8**, 357–66.

17

Molecular systematics and seed plant phylogeny: a summary of a parsimony analysis of *rbc*L sequence data

KATHLEEN A. KRON AND MARK W. CHASE

Introduction

Plant systematics has experienced a methodological revolution in the last three decades. Parsimony analysis of character-state transformations and recognition of strictly monophyletic groups have given systematists the ability to explore the evolutionary relationships of plants with more rigour than in the past. Additionally, recent technological advances and developments in molecular biology have provided systematists with valuable new sources of data for phylogeny reconstruction. 'Molecular systematics' is a rapidly expanding field that includes the use of restriction endonuclease sites and nucleotide sequences, among other sources of information (Hillis & Moritz, 1990). Within the embryophytes and in some algal groups (e.g. Rhodophyta, Chlorophyta) chloroplast DNA is the molecule of choice for most plant systematists (Palmer *et al.*, 1988). The chloroplast genome is small, highly conserved, present in a large number of copies and usually inherited only from the female parent. All green plants possess a chloroplast genome that is structurally conserved across most major groups. Thus the chloroplast genome provides a source of data potentially comparable at high taxonomic levels.

Other sources of molecular data are the mitochondrial genome (mtDNA; Palmer, 1992) and ribosomal RNA and DNA (rRNA and rDNA; Hamby & Zimmer, 1992). The mitochondrial genome in plants is large, and its nucleotide sequence evolves slowly. However, it is subject to frequent structural rearrangements and insertions and deletions. These attributes render plant mtDNA of little systematic use in studies involving either restriction sites, which are dependent on a relatively stable genome, or sequences (Palmer, 1992).

The genes for rDNA code for ribosomal rRNAs, and are very conserved owing to structural features. Nucleotide sequences of these genes may be useful in reconstructing plant phylogeny at high taxonomic levels (Clegg & Durbin, 1990). At these levels, the non-transcribed spacer regions between the rRNA genes often exhibit size variation making alignment (i.e. determination of homologous base positions) questionable if not impossible. At lower taxonomic levels (i.e. population and species level), data from restriction site variation and nucleotide sequences in both the transcribed and non-transcribed spacer regions can be useful for systematics.

Both restriction endonuclease sites and sequence data from the chloroplast genome have been used to assess phylogenetic relationships. Restriction site characters usually are best suited for studying intergeneric or subgeneric relationships (Chase & Palmer, 1989, 1992; Wendel & Albert, 1992) and nucleotide sequences of protein-encoding or ribosomal genes for elucidating relationships at the family level and above (Chase *et al.*, 1993). However, these generalizations are based primarily upon available evidence. Relatively few restriction site studies have been published, considering the enormous diversity of green plants. The only chloroplast gene sequence to be investigated in any depth has been the *rbc*L gene, the large subunit of ribulose-1,5-bisphosphate carboxylase-oxygenase. This gene initially was suggested as appropriate for phylogenetic studies by Ritland and Clegg (1987) and Zurawski and Clegg (1987). Because *rbc*L codes for a protein that is essential to carbon fixation, it is conserved in all photosynthetic plants, apparently missing in some achlorophyllous taxa, thus making it an excellent starting point for molecular-based phylogenetic studies. Internal size variation is not a problem in *rbc*L, except at the 3' end of the gene; thus the nucleotide sequences can be aligned by sight. With the advent of thermal cycling and temperature resistant polymerases, a specific gene such as *rbc*L can be targeted and amplified, greatly speeding up the sequencing process.

To some workers, data obtained from DNA sequencing was seen as ideal, as the determination of character-states (A, T, G, C) is relatively straightforward, and, once the sequences are aligned, homology of characters is determined by the same base position from sequence to sequence. In addition, restriction sites and nucleotide sequences can provide a large number of characters when compared to morphological data sets (Donoghue & Sanderson, 1992). Whereas these are all attributes that render DNA data extremely useful for phylogenetic reconstruction, certain drawbacks still exist. Homoplasy, resulting from independent multiple origins (parallelisms) and reversals (losses), cannot be avoided by using molecular data. Sequence data with only four possible character states for any taxon at any position are subject to homoplasy. Indeed, the proportions of parallelisms and reversals in sequence-based phylogenies is nearly the same as that of morphology-based phylogenies (Donoghue & Sanderson, 1992).

In addition, because of the availability of comparable data molecular-based phylogenies using few, distantly related taxa (i.e. a few closely related taxa plus one or two distantly related taxa) have often suggested relationships that make little sense when compared to any other evidence. When divergence times are large, insufficient taxon sampling can result in extremely asymmetric character-state distributions and the inferred relationships can often be misleading. In practice, this can mean that distantly related taxa may end up as 'related' in a tree because character states (i.e. nucleotide changes) that have arisen independently are interpreted as inherited from a common ancestor. This 'branch attraction' problem is discussed in detail in Albert, Chase & Mishler (1993).

Methods

The analysis of *rbc*L nucleotide sequences summarized in this paper contains 475 taxa, most with 1428 characters (i.e. nucleotides), and is described and discussed in detail elsewhere (Chase *et al.*, 1993). Since these data have come from numerous collaborators who have contributed published and unpublished sequences, the actual methods varied from lab to lab and can be only generally characterized. References with detailed protocols for a variety of molecular methods can be found in Hillis and Moritz (1990).

In general, the *rbc*L gene was amplified using temperature cycling methods and primers that were near the ends of, or flanking, the gene. The resulting product was either cloned and then sequenced, or sequenced directly from the amplified product. Data were entered as nucleotide sequences with the cut-off at the 1428 base position (the consensus stop codon position) to ensure homology of position. Sequences were entered in a NEXUS file and analysed with parsimony using PAUP 3.0r (Phylogenetic Analysis Using Parsimony, by D. Swofford, Illinois Natural History Survey, Champaign, Illinois, USA). Characters (codon positions) and character states (transversions/transitions) were differentially weighted during the tree-building process via step matrices according to the criteria developed and discussed in detail by Albert and

Mishler (1992) and Albert *et al.* (1993). An explanation of the analysis shown here can be found in Chase *et al.* (1993). Because of the size of the matrix and the enormous amount of memory required to do the analysis, only the first 500 most parsimonious trees were saved. A summary (or consensus) tree was calculated using the combinable component method of Bremer (1990). An unrooted tree summarizing the major clades of this consensus tree is presented in Fig. 17.1. This tree has been depicted with the cycads as the basal group. However, arranging the tree with any of the other gymnosperms at the base does not affect the topology within the angiosperms. Detailed

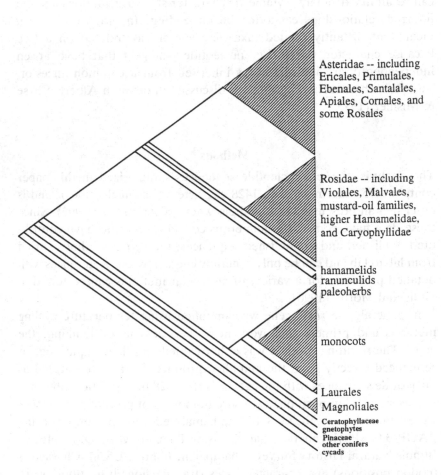

Fig. 17.1. Diagrammatic summary of the combinable component consensus tree of 500 most parsimonious trees resulting from an analysis of the *rbc*L nucleotide sequences of 475 seed plants (Chase *et al.*, 1993).

subtrees and pertinent discussions thereof can be found in Chase *et al.* (1993).

Although this is apparently the largest cladistic analysis performed to date, there are some important aspects of this tree that need to be emphasized. First, it must be regarded as a first approximation and useful for heuristic purposes only. It must not be considered the 'true' tree as that can never be known. However, because numerous representatives of the angiosperms (flowering plants) are included, it may provide clues to some evolutionary relationships at higher taxonomic levels. In addition, an analysis of this scope and character has brought to the fore a number of issues that have both practical and theoretical implications. These include the use of weighting criteria and problems of unequal branch lengths. The latter may be caused by differential rates of evolution of the gene, although in the case of *rbc*L this is probably insignificant because its sequence changes relatively slowly and varies within a narrow range (Doebley *et al.*, 1990; Wendel & Albert, 1992). Other problems may be caused by insufficient sampling of taxa, or extremely deep hierarchical relationships between some taxa but not others. These problems are not unique to molecular data, but this study has provided an opportunity to explore some aspects of these problems a bit more rigorously than in the past. The papers of Albert and Mishler (1992) and Albert *et al.* (1993) address the issues of weighting and branch lengths in detail.

Results

As mentioned above, this tree is unrooted with the cycads placed in the basal position after the analysis was completed. The apparent paraphyly of the conifers in this tree is due perhaps to poor sampling, as there is other evidence that the conifers are monophyletic, such as structural features and strict paternal transmission of the chloroplast genome (Raubeson & Jansen, 1992).

The general results (Fig. 17.1) indicate that the angiosperms are monophyletic (i.e. descended from a single common ancestor). This hypothesis is supported by cladistic analysis of morphological, embryological, and developmental evidence for fossil and living seed plants (Donoghue & Doyle, 1989; Doyle & Donoghue, 1986). *Ceratophyllum* is problematic because of its apparent morphological differences from all other flowering plants and yet basal-most position within the angiosperms.

The monocots appear to be a natural (monophyletic) group derived from what are traditionally considered the 'primitive dicots' (Cronquist, 1981). Although many of the taxa which occur in a relatively basal position in this *rbc*L-based tree are often grouped into one subclass (i.e. subclass Magnoliidae *sensu* Cronquist, 1981) they do not represent a natural group as presently recognized. Excluding *Ceratophyllum*, this tree indicates two major groups of flowering plants that correspond well to two basic types of pollen morphology found among the angiosperms, rather than the traditional monocot versus dicot split. Members of the terminal clade comprising Ranunculales, and above, possess triaperturate, or triaperturate-derived, pollen ('eudicots'). Members of the clade comprising the Magnoliales, Laurales, paleoherbs (e.g. Piperaceae, Saururaceae, Aristolochiaceae, Nymphaeaceae) and monocots possess uniaperturate or uniaperturate-derived pollen.

Within the eudicots, many taxa that are considered advanced based upon traditional interpretation of morphological data fall together in a terminal clade (Fig. 17.1) that roughly corresponds to the Asteridae of Cronquist (1981). Many of the taxa traditionally classified in the Rosidae (*sensu* Cronquist, 1981) also are indicated as a monophyletic group (Fig. 17.1). However, some traditionally recognized groups are not identifiable. Most notably, no support was found for the monophyly of the Dilleniidae (Cronquist, 1981) which include the heaths and their relatives (Ericales), the violets and passion-flowers (Violales), and the mustard-oil-producing taxa (Capparales, etc.). These groups are scattered throughout the *rbc*L-based tree. The position of the Ericales as basal to the most 'advanced' dicots is one of the most noticeable deviations from traditional classifications. However, this is supported by an independent cladistic analysis of morphological data (Hufford, 1992). The mustard-oil groups are more closely related to some members of the Rosidae than to other traditional dilleniids. Thus the Dilleniidae, as currently recognized, is not a natural group.

In the terminal clade (Fig. 17.1), the following groups (families) are indicated as closely related: Asteraceae (sunflower family) and Apiaceae (parsley family); Solanaceae (tomato family), Scrophulariaceae (foxglove family), Lamiaceae (mint family); and Ericaceae (heath family). Basal to these groups are Cornaceae (dogwoods) and members of the sandalwood family (Santalaceae) and its relatives.

The next major clade consists of members of the traditionally defined subclass Rosidae. This includes the Fabaceae (bean family) and Rosaceae (rose family) which are indicated as closely related.

The Malvaceae (hibiscus family), Brassicaceae (mustard family) and some of the wind-pollinated Hamamelidae are also members of this large 'rosid' group. In addition, the distinctive betalain-containing taxa and their relatives (Caryophyllidae) form a monophyletic group and are also members of this 'rosid' clade.

The three clades basal to the 'asterid' and 'rosid' clades include members of the wind-pollinated Hamamelidae, Proteaceae (protea family), and the Ranunculaceae (buttercup family), Papaveraceae (poppy family), and their relatives (Ranunculales).

The clade containing taxa with uniaperturate or uniaperturate-derived pollen taxa ('monosulcates') has the magnolias (Magnoliaceae), avocados (Lauraceae) and members of the pepper family (Piperaceae) and its relatives as basal to the monocots. Although these clades are shown as resolved on this consensus tree, in local analyses (Qiu *et al.*, 1993) the branches leading to the paleoherbs and the Laurales collapse at maximum parsimony. However, in these same taxon-subset experiments, the monocot clade remains intact in trees several steps less parsimonious.

Within the monocots, members of the grasses (Poaceae), sedges (Cyperaceae) and rushes (Juncaceae) are indicated as closely related. These are related to the bromeliads (Bromeliaceae). The bananas (Musaceae) and their relatives are a monophyletic group which, along with taxa such as *Zebrina* (Commelinaceae), form a sister group to the grass-bromeliad clade. The palms (Arecaceae) are basal to these clades. Many of the above taxa have been assigned traditionally to the subclass Commelinidae (Cronquist, 1981) and this group represents a monophyletic terminal clade within the monocots. The remaining taxa include most of what have at various times been included in the 'lilioid' monocots. This group has been in great taxonomic flux during the past few years due to different interpretations of morphological and chemical features (Cronquist, 1981; Thorne, 1983, 1992). No rigorous cladistic analyses have been done on these taxa at higher taxonomic levels (but see Dahlgren, Clifford & Yeo, 1985 for a 'Hennigian' approach to monocot relationships). The *rbc*L-based analysis indicates that the orchids (Orchidaceae) are more closely related to *Hypoxis* (Hypoxidaceae), *Iris* (Iridaceae) and other 'asparagoid' (*sensu* Dahlgren *et al.*, 1985) monocots than to the liliaceous monocots such as *Lilium* (Liliaceae), or *Alstroemeria* (Alstroemeriaceae). Within the monocots, the aroids (Araceae) and members of the Alismataceae and their relatives are basal.

Summary

Parsimony analysis of *rbc*L nucleotide sequences of 475 taxa of seed plants indicates that the angiosperms are a monophyletic (natural) group. Within the angiosperms two major groups are indicated: the 'eudicots', i.e. flowering plants possessing triaperturate or triaperturate-derived pollen, and the 'monosulcates', those possessing uniaperturate or uniaperturate-derived pollen. The monocots are a monophyletic group derived from within the monosulcate angiosperms. Notable deviations from traditional classifications of the angiosperms indicated in this analysis are: the placement of the Ericales as basal to the most advanced dicots, the split (i.e. polyphyly) of the wind-pollinated Hamamelidae, and the disintegration of the Dilleniidae. The relationships (or lack thereof) suggested in this *rbc*L-based tree for all three of these 'groups' have been supported, or at least suggested, by recent workers using morphological data. Hufford's (1992) analysis that supports the position of the Ericales near the higher asterids is the only available external cladistic support. But the heterogeneity (and subsequently inferred paraphyly or polyphyly) of the Hamamelidae and the Dilleniidae have been pointed out by several authors (e.g. Young & Watson, 1970).

Despite the problems of homoplasy, divergence times, and lack of adequate sampling of several important angiosperm groups, this analysis of *rbc*L nucleotide sequence data is a valuable beginning to our understanding of the contribution of molecular data to phylogeny reconstruction. Several of the major clades indicated in this analysis can be supported by consistent independent evidence, whether formally analyzed or not (e.g. uniaperturate vs. triaperturate pollen). However, the analysis presented here points to the need for more careful cladistic studies of morphological data as well as continued sampling of *rbc*L sequence data from poorly represented groups. Undoubtedly, molecular data from other genes will also contribute to our understanding of plant phylogeny, but all sources of data will need to be considered (and rigorously analysed, both separately and together) in order to obtain the best approximation of historical patterns in plant evolution.

Acknowledgements

The authors acknowledge the following support: National Science Foundation grants BSR-8821264 to KAK and BSR-8906496 to MWC. We thank V. A. Albert for helpful comments on the manuscript.

Additional texts

Phylogenetic Analyses of DNA Sequences, ed. by M. M. Miyamoto and J. Cracraft, (1991). Oxford University Press.

References

Albert, V.A., Chase, M.W. & Mishler, B.D. (1993). *Ann. Missouri Bot. Gard.*, **80** (in press).

Albert, V.A. & Mishler, B.D. (1992). *Cladistics*, **8**, 73–84.

Bremer, K. (1990). *Cladistics*, **6**, 369–72.

Chase, M.W. & Palmer, J.D. (1989). *Am. J. Bot.*, **76**, 1720–30.

Chase, M.W. & Palmer, J.D. (1992). In *Molecular Systematics in Plants*, P.S. Soltis, D.E. Soltis and J.J. Doyle (eds.), Chapman and Hall, New York, pp. 324–339.

Chase, M.W., Soltis, D.E., Olmstead, R.G., Morgan, D., Les, D.H., Mishler, B.D. Duvall, M.R., Price, R., Hills, H.G., Qiu, Y., Kron, K.A., Rettig, J.H., Conti, E., Palmer, J.D., Manhart, J.R., Sytsma, K.J., Michaels, H.J., Kress, W.J., Karol, K.G., Clark, W.D., Hedren, M., Gaut, B.S., Jansen, R.K., Kim, K., Wimpee, C.F., Smith, J.F., Furnier, G.R., Strauss, S.H., Xiang, Q., Plunkett, G.M., Soltis, P.S., Swensen, S.M., Williams, S.E., Gadek, P.A., Quinn, C.J., Eguiarte, L.E., Golenberg, E., Learn, G.H., Jr., Graham, S., Barrett, S.C.H., Dayanandan, S. & Albert, V.A. (1993). *Ann. Missouri Bot. Gard.*, **80**, 528–80.

Clegg, M.T. & Durbin, M. (1990). *Aust. Syst. Bot.*, **3**, 1–8.

Cronquist, A. (1981). *An Integrated System of Classification of the Flowering Plants*. Colombia Univ. Press, New York

Dahlgren, R.T., Clifford, H.T. & Yeo, P.F. (1985). *The Families of the Monocotyledons: Structure, Evolution, and Taxonomy*. Springer-Verlag, New York.

Doebley, J., Durbin, M., Golenberg, E.M., Clegg, M.T. & Ma, D.P. (1990). *Evolution*, **44**, 1097–108.

Donoghue, M.J. & Doyle, J.A. (1989). In *The Hierarchy of Life: Molecules and Morphology in Phylogenetic Analysis*, B. Fernholm, K. Bremer, and H. Jornvall (eds.) Elsevier Science Publishers, Amsterdam, pp. 181–193.

Donoghue, M.J. & Sanderson, M.J. (1992). In *Molecular Systematics in Plants*, P.S. Soltis, D.E. Soltis and J.J. Doyle (eds.), Chapman and Hall, New York, pp. 340–368.

Doyle, J.A. & Donoghue, M.J. (1986). *Bot. Rev.*, **52**, 321–431.

Hamby, R.K. & Zimmer, E.A. (1992). In *Molecular Systematics in Plants*, P.S. Soltis, D.E. Soltis and Doyle, J.J. (eds.) Chapman and Hall, New York, pp. 50–91.

Hillis, D.M. & Moritz, C. (eds.), (1990). *Molecular Systematics*, Sinauer Associates, Inc., Sunderland Mass.

Hufford, L. (1992). *Ann. Missouri Bot. Gard.*, **79**, 218–48.

Palmer, J.D. (1992). In *Molecular Systematics of Plants*, P.S. Soltis, D.E. Soltis and J.J. Doyle (eds.), Chapman and Hall, New York, pp. 36–49.

Palmer, J.D., Jansen, J.K., Michaels, H.J., Chase, M.W. & Manhart, J.R. (1988). *Ann Missouri Bot. Gard.*, **75**, 1180–206.

Qiu, Y-L., Chase, M.W., Les, D.H., Hills, H.G. & Parks, C.R. (1993). *Ann. Missouri Bot. Gard.*, **80** (in press).

Raubeson, L.A. & Jansen, R.K. (1992). *Biochem. Syst. and Ecol.*, **20**, 17–24.
Ritland, K. & Clegg, M.T. (1987). *Am. Naturalist*, **130**, S74–S100.
Thorne, R.F. (1983). *Nord. J. Bot.*, **3**, 85–117.
Thorne, R.F. (1992). *Aliso*, **13**, 365–89.
Wendel, J.F. & Albert, V.A. (1992). *Syst. Bot.*, **17**, 115–43.
Young, D.J. & Watson, L. (1970). *Aust. J. Bot.*, **18**, 387–433.
Zurawski, G. & Clegg, M.T. (1987). *Ann. Rev. Pl. Phys.*, **38**, 391–418.

Part VI
Case studies of viral taxa; their systematics and evolution

18

Evolution of poxviruses and African swine fever virus

R. BLASCO

Introduction

Poxviruses are a family of viruses with large DNA genomes that include pathogens of vertebrates (subfamily *Chordopoxvirinae*) and of invertebrates (subfamily *Entomopoxvirinae*) (Table 18.1). The prototype poxvirus is vaccinia virus, a member of the genus Orthopoxvirus (Dales & Pogo, 1981; Buller & Palumbo, 1991; Moss, 1990, 1991). Vaccinia virus was successfully used for almost 200 years for smallpox vaccination. The origin and passage history of vaccinia virus are obscure; however, analysis of vaccine strains indicate that the virus is most closely related to, but is distinct from, variola virus, which causes smallpox, and also cowpox and monkeypox viruses.

Poxvirus virions are large (300–400 nm) oval particles that contain at least two lipoprotein envelopes. The virion is composed of a DNA molecule and various amounts of more than 100 protein species, a number of which possess enzymatic activity. The poxviruses replicate in the cytoplasm of the cell, where their enzymes involved in nucleotide metabolism, DNA replication and transcription are expressed. The coding of these enzymes makes the virus independent of the corresponding host cell enzymes and, consequently, the virus is able to carry out gene expression and DNA replication in enucleated cells.

African swine fever virus (ASFV) occupies an anomalous taxonomic position and has not been assigned to any major virus family (Viñuela, 1985, 1987; Wilkinson, 1989). Although it was originally classified as a member of the family *Iridoviridae*, its unique characteristics justified its separation from this group. ASFV naturally infects warthogs and bush pigs in southern and eastern Africa. It also infects the domestic pig, causing severe economic losses in the affected countries. In addition to the mammalian host, the virus is able to infect ticks of the *Ornithodoros*

Table 18.1. *Poxvirus family*

Subfamily	Genera	Species
Chordopoxvirinae	Orthopoxvirus	Buffalopox, camelpox cowpox, ectromelia, monkeypox, rabbitpox, racoonpox, taterapox vaccinia, variola, volepox
	Parapoxvirus	Chamois contagious ecthyma, Orf, pseudocowpox, stomatitis papulosa
	Avipoxvirus	Canarypox, fowlpox, juncopox, pigeonpox, quailpox, sparrowpox, starlingpox, turkeypox
	Capripoxvirus	Goatpox, sheeppox, lumpy skin disease
	Leporipoxvirus	Hare fibroma, myxoma, rabbit (Shope) fibroma, squirrel fibroma
	Suipoxvirus	Swinepox
	Molluscipoxvirus	Molluscum contagiosum
	Yatapoxvirus	Tanapox, Yaba
Entomopoxvirinae	A	Melontha melontha
	B	Amsacta moori
	C	Chironimus luridus

genus, which serve as vectors for transmission of the disease. ASF virions are of icosahedral symmetry and contain a lipoprotein envelope which is necessary for infectivity. As in the poxviruses, purified virions contain a large number of polypeptide species, and several proteins with enzymatic activity are incorporated into mature virions.

Genome structure and variability

Given the difference in the virion structure of poxviruses and ASFV, it is surprising that they are remarkably similar in genome structure. Both encapsidate a molecule of linear, double-stranded DNA with inverted terminal repetitions (ITRs) and hairpin loop structures at the ends. In several poxviruses and in ASFV, the ITRs are composed of small tandem repeated sequences and unique sequences. DNA replication takes place through concatemer intermediates in which individual genomes are in a head-to-head and tail to tail orientation (Moyer & Graves, 1981; González *et al.*, 1986).

Comparison of the genomes of different poxviruses reveals that gross mutations tend to be located close to the ends of the viral genomes, resulting in large differences in the sizes of the terminal regions. In part, the observed differences of the genomic termini result from variability of the ITR sequences, which can be explained by unequal crossover events between the repeats present within the ITRs (Baroudy & Moss, 1982). In addition, variation is readily detected in the unique sequences close to the ITRs. Large, spontaneous deletions and gene rearrangements have been found in those regions in a vaccinia virus stock. Frequently, sequences adjacent to the ITR from one end of the genome are deleted and sequences from the opposite end are transposed (Moyer, Graves & Rothe, 1980; Esposito *et al.*, 1981; Archard *et al.*, 1984; Pickup *et al.*, 1984). In ASFV, variation also occurs predominantly in two regions located close to the DNA ends, which include the ITRs and unique sequences (Blasco *et al.*, 1989a). Large deletions are found in the left variable region in different virus isolates, in a region containing several repeated sequences (Blasco *et al.*, 1989b; De la Vega, Viñuela & Blasco, 1990). By contrast, the central portion of the genome (about 100–120 kb) is more conserved, with most changes caused by sequence divergence and small deletions or additions of sequence.

Genetic content of the variable regions

The isolation of virus variants containing deletions in the termini reveals that the DNA segments involved are not essential. Therefore, in both poxviruses and ASFV, a large number of inessential genes lie in the terminal regions. Although some genes in the central portion of the genome are dispensable for virus replication, no large deletions have been detected outside of the variable terminal regions. Therefore, it seems that most genes essential for virus replication are located in the centre of the genome. Accordingly, the uneven distribution of variability along the viral genome may reflect, primarily, the ability of the viral genes to change without affecting virus replication. In addition, the variability observed in the terminal sequences may be favoured by the existence of repeated sequences, or by the mechanism of replication of the viral genome.

Although the distribution of genome variability seems to be similar in ASFV and the poxviruses, the genetic content of the variable regions has not shown any clear similarity. It seems likely that the placement of inessential genes near the genome ends may be advantageous for the virus. It is well documented that changes in the variable regions of poxviruses cause changes in the biological characteristics of the virus (Turner & Moyer, 1990). It is possible that changes in genes involved in virus–host interactions provide a means for the fine-tuning of the biological characteristics of the virus. In this respect, the location of those genes in the vicinity of the genome ends may enhance the opportunities of generating successful virus variants.

Gene families

An interesting finding is the presence of large multigene families in the variable regions of ASFV DNA (Almendral *et al.*, 1990; González *et al.*, 1990). The number of genes present in each multigene family in a particular virus isolate may change, and at least one of the multigene families is dispensable for virus replication (Agüero *et al.*, 1990; De la Vega *et al.*, 1990). Probably, the ASFV multigene families originated by gene duplication and subsequent divergence, just like the cellular multigene families. The similarity between members of a family may vary from less than 30% to more than 90%, indicating that gene duplication within the multigene families is a continuous process. Asymmetrical homologous recombination provides a mechanism for gene duplication and hence, for expansion of a multigene family. On the other hand, deletions, which can arise either by homologous or non-homologous recombination mechanisms, result in the net loss of members of the multigene families and, occasionally, in the generation of new gene chimaeras. Multigene families also seem to be a general trait of poxviruses. Gene families within the inverted terminal repeats of Shope fibroma virus have been described (Upton & McFadden, 1986; Upton, DeLange & McFadden, 1987). Gene families are also present in similar locations in the genome of vaccinia virus (Kotwal & Moss, 1988a; Howard, Chan & Smith, 1991; Smith & Chan, 1991; Smith, Chan & Howard, 1991) and fowlpox virus (Tomley *et al.*, 1988). Duplicated genes are often found in the regions close to the genome ends. Transposition of terminal sequences can result in duplication of sequences contiguous to the ITR, thereby creating an enlarged ITR. When gene duplications are followed by divergence of the resulting genes, a gene family is generated. Occasionally, mutations of one of the members of the family can lead to a pseudogene, in which the integrity of the original gene has been lost. The presence of pseudogenes has been reported for several poxviruses (Upton & McFadden, 1986; Macauley, Upton & McFadden, 1987; Gershon & Black, 1989; Smith *et al.*, 1991).

Evolution of poxviruses

The relatedness of different poxviruses was first assessed by serological and cross-protection studies. However, restriction endonuclease mapping has proved to be more sensitive for comparing closely related poxviruses (Wittek *et al.*, 1977; Mackett & Archard, 1979; Esposito & Knight, 1985), and for estimating genetic distances between different poxviruses (Gibbs & Fenner, 1984; Gershon & Black, 1988). Also, viruses have been compared by cross-hybridization of restriction fragments (Gassman, Wyler & Wittek, 1985; Drillien *et al.*, 1987; Gershon, Ansell & Black, 1989). However, comparison of DNA or protein sequences is undoubtedly the most sensitive method available for studying virus evolution. In the last few years, much sequence data has been obtained for poxviruses, including the complete sequence of a vaccinia virus strain (Goebel *et al.*, 1990). Sequence information from other poxviruses is still fragmentary, and a comprehensive comparison of poxvirus genomes using molecular sequences is not possible yet.

The comparison of sequences from different poxvirus genera shows differences that mirror what was found when comparing isolates of a single species. It confirms that the genomes differ more in the terminal regions than in the central part of the genome. As a general rule, most genes in the central part of the genome are shared by different chordopoxvirus genera, and therefore the sequences from those regions can be easily aligned. By contrast, terminal sequences often contain genus- or strain-specific genes, and show extensive gene re-assortments in different poxviruses. In addition, the central region in entomopox virus and fowlpox virus DNA show extensive rearrangements with respect to the orthopoxviruses (Hall & Moyer, 1991; Mockett *et al.*, 1992).

The alignment of sequences from different poxvirus genera shows that the evolution of the viruses of the family from their common ancestor has involved not only sequence divergence, but also sequence transposition, deletion and duplication. To generate these mutations, several mechanisms must have played a role, including homologous and non-homologous recombination. There is considerable evidence for the occurrence of intra-genomic gene rearrangements and deletions. Recombination between different poxviruses belonging to the same genus has also been documented (Bedson & Dumbell, 1964a,b). In one instance, a recombination between Shope fibroma virus and myxoma virus resulted in the generation of a virus with some biological characteristics of each parent virus. The DNA of the recombinant virus consists of myxoma virus sequences and about 10 kb transposed from the terminus of Shope fibroma virus (Block *et al.*, 1985; Upton *et al.*, 1988).

Viral and host gene relationships

Most vaccinia virus genes are not similar to known genes from other organisms. However, a number of genes in vaccinia virus have been shown to be similar to cellular genes (Table 18.2). Those genes are distributed throughout the genome, rather than clustered, and are interspersed among genes with no known similarities to cellular genes. The degree of similarity of those vaccinia virus genes and their cellular counterparts varies greatly from case to case. For some genes, the differences are so large that their homology is not recognizable except for small conserved domains within the protein sequence. At the other end of the spectrum, some virus-encoded proteins are more than 75% identical to the cellular homologue.

Therefore, both the localization and the degree of similarity to the cellular genes favour the interpretation that those genes were transferred from the host to the viral genome as independent units. The divergence of those genes and their cellular homologues can be the result of different rates of evolution for different genes, but could also result from different times of divergence from the cellular lineage.

A striking finding is the presence of a viral gene that has domains derived from unrelated cellular genes (Koonin, Senkevich & Chernos, 1992). That gene was probably created by fusing portions of two cellular genes, possibly at the same time, or after, the incorporation of the genes into the viral genome.

The mechanism by which the poxviruses acquire genetic information from the host cell is a matter of speculation. Since the site of viral replication is the cytoplasm and the genes in the virus genome do not contain introns, it seems likely that mature mRNAs of cellular genes were somehow incorporated into the viral DNA.

Table 18.2. *Vaccinia virus genes with similarity to cellular genes*

Gene	Similar cellular gene	Reference
C12L, K2L, B13R	Serine protease inhibitor	Boursnell *et al.*, 1988 Kotwal & Moss, 1989 Smith, Howard & Chan, 1989
C11R	Epidermal growth factor	Blomquist *et al.*, 1984
C22L	Tumour necrosis factor receptor	Howard *et al.*, 1991
C3L, B5R	Complement control proteins	Kotwal & Moss, 1988*b* Goebel *et al.*, 1990
K1L, B18R	Ankyrin	Lux, John, Bennet 1990 Smith, Chan & Howard, 1991
K3L	Translation initiation factor 2-α	Goebel *et al.*, 1990
F2L	dUTPase	McGeoch, 1990
F4L, I4L	Ribonucleotide reductase	Tengelsen *et al.*, 1988 Slabaugh *et al.*, 1988 Schmitt & Stunnenberg, 1988
E4L	Transcription factor	Ahn *et al.*, 1990 Goebel *et al.*, 1990
E9L	DNA polymerase	Earl, Jones & Moss, 1986
G5.5R, J6R, A24R	RNA polymerase subunits	Broyles & Moss, 1986 Patel & Pickup, 1989 Amegadzie *et al.*, 1991 Amegadzie, Ahn & Moss, 1992
O2L	Glutaredoxin	Goebel *et al.*, 1990 Johnson *et al.*, 1991
J2R	Thymidine kinase	Weir & Moss, 1983 Hruby *et al.*, 1983
H6R	DNA topoisomerase	Shuman & Moss, 1987
D8L	Carbonic anhydrase	Niles *et al.*, 1986
A34R, A40R	Lectins	Smith, Chan & Howard, 1991
A42R	Profilin	Goebel *et al.*, 1990
A44L	3-β-hydroxysteroid dehydrogenase	Goebel *et al.*, 1990
A45R	Superoxide dismutase	Goebel *et al.*, 1990
A48R	Thymidylate kinase	Smith, De Carlos & Chan, 1989

Table 18.2. (cont.)

Gene	Similar cellular gene	Reference
A50R	DNA ligase	Smith, Chan & Kerr, 1989
A53R	Tumour necrosis factor receptor	Howard et al., 1991
A57R	Guanylate kinase	Smith, Chan & Howard, 1991
B1R, B12R	Protein kinase	Howard & Smith, 1989 Traktman, Anderson & Rempel, 1989
B16R, B19R	Interleukin-1 receptor	Smith & Chan, 1991

Gene phylogenies

When the viral and cellular genes are homologous and sequences from several species are known, it is possible to reconstruct their phylogeny. However, particularly for the viral members, those phylogenies would reflect the evolution of the genes rather than that of the organisms.

The best example to date of a host and poxvirus gene phylogeny reconstruction is for the thymidine kinase (TK) gene. Because of its usefulness as a selective marker, the TK gene has been sequenced from a number of poxviruses. There have been several attempts to trace the evolutionary history of the gene (Boyle et al., 1987; Blasco et al., 1990; Bockamp, Blasco & Viñuela, 1990; Koonin & Senkevich, 1992). A possible complication of these studies is that the rate of evolution of the viral and cellular enzymes may differ (Gentry et al., 1988). In spite of this limitation, however, dendrograms representing the degree of relatedness of different TK genes may help in delineating the evolutionary history of the gene. Fig. 18.1 shows such dendrograms, which include the recently published sequences from myxoma virus TK (Jackson & Bults, 1992) and an entomopoxvirus TK (Gruidl, Hall & Moyer, 1992). As noted in some studies, the trees obtained from the complete sequence (Fig. 18.1A) can be different from those obtained from the central, more conserved, region (Fig. 18.1B). Also, trees have been obtained by treating gaps as if they were the twenty first amino acid (Boyle et al., 1987). In spite of these differences, several conclusions can be drawn from the dendrograms. First, the groupings of the poxviruses fit well with their taxonomy determined in other ways (Table 18.1).

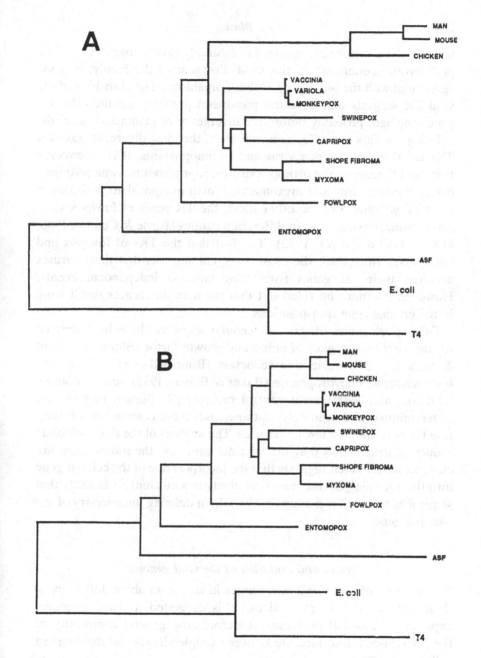

Fig. 18.1. Dendrograms showing the distances between different viral and cellular TK genes The dendrograms shown were derived from the complete TK sequences (A) or the central 'core' region of the TK sequences (B) (Boyle *et al.*, 1987). A progressive multiple sequence alignment (Feng & Doolittle, 1987) was used to compute the philogenetic distances, and a distance matrix algoritm (Fitch & Margoliash, 1967) was used to generate the dendrograms.

Secondly, the vertebrate enzymes consistently cluster together, and the prokaryotic sequences are the most divergent of the family, in good agreement with the phylogeny of those organisms. The branching of the viral TK suggests that ancestral poxviruses probably acquired the TK gene long ago, probably before the divergence of mammals and birds.

Using vaccinia virus TK as a reference, the most divergent poxvirus TKs are those of fowlpox virus and entomopoxvirus. It is noteworthy that the TK genes of the orthopoxviruses, capripoxvirus, swinepoxvirus, Shope fibroma virus and myxoma virus map in equivalent positions in the viral genome. On the other hand, the TK genes of fowlpox virus and entomopoxvirus map to different locations (Boyle & Coupar, 1986; Gruidl, Hall & Moyer, 1992). The fact that the TKs of fowlpox and entomopox viruses are the most divergent suggests that those viruses acquired their TK genes from their hosts in independent events. However, it cannot be ruled out that the map differences result from intra-genomic gene transpositions.

Other phylogenies of viral and cellular sequences have been derived for the viral homologues of epidermal growth factor (Blomquist, Hunt & Barker, 1984), nucleotide reductase (Boursnell *et al.*, 1991), 3-β-hydroxysteroid dehydrogenase (Baker & Blasco, 1992), serine protease inhibitors and complement control proteins (R. Blasco, unpublished observations). Unfortunately, in those cases there is sequence information for only one or a few poxviruses. The analysis of the ribonucleotide reductase gene shows that the vaccinia gene and the mouse gene are closely similar, which suggests that the incorporation of the cellular gene into the vaccinia genome was a relatively recent event. It is likely that sequences from other mammals will help in defining the ancestry of the vaccinia gene.

Origin and evolution of the viral genome

The genome of the poxviruses ranges in size from about 130 to more than 300 kb. For vaccinia virus, it is estimated to have a coding capacity of 200–300 proteins. Therefore, the genetic complexity of the poxviruses is intermediate between simple viruses and the simplest cellular organisms.

Two different scenarios can be envisioned to account for the origin of the poxvirus genome: in one, the primordial virus genome would have been much simpler, and would have evolved towards complexity, acquiring new genes and new functions. In the other, poxviruses would have originated from a cellular organism after adapting to intracellular replication. The subsequent evolution of this cellular organism to a viral lifestyle would imply the loss of many genes that would no longer be required for replication. Independently of which organism was the origin of poxviruses, it seems likely that the processes which generated the poxvirus ancestor took place in the remote past.

When more poxvirus sequences become available, it may be possible to determine the genes that were probably present in the common ancestor of the poxviruses. The phylogenetic study of those genes might be useful in tracing the ancestry of the different members of the poxvirus family. Genes that are essential for virus replication are good candidates for ancestral poxvirus genes, such as the individual subunits of the viral RNA polymerase. Interestingly, one of the subunits is equally related to those of archaebacteria and eukaryotes, suggesting a very ancient origin for the viral enzyme (Amegadzie *et al.*, 1991). Another example of distant relationships is provided by the vaccinia DNA polymerase gene, which is well conserved in different poxvirus genera (Binns *et al.*, 1987). Intriguingly, the vaccinia virus enzyme is more closely related to the bacteriophage T4 enzyme than to the eukaryotic DNA polymerases (Bernard *et al.*, 1987).

Other genes may have been acquired, or arisen *de novo* (Chapter 6) after the divergence of the different poxvirus genera, and consequently be only present in a particular lineage. Possible complications of this type of analysis are horizontal transfers between different poxviruses, and cellular genes independently acquired in different virus lineages.

Evolution of African swine fever virus

Several studies indicate that the basic genome organization of ASFV is similar to that found in poxviruses. As with poxviruses, variation tends to accumulate preferentially in the terminal regions. Smaller differences in the central region can be detected when comparing different virus isolates (Blasco *et al.*, 1989a; Sumption *et al.*, 1990). Variation in the terminal regions is caused by changes both within the terminal inverted repetitions and in unique sequences (Blasco *et al.*, 1989b; Dixon *et al.*, 1990). The most notable difference from the poxviruses is the prominent role that multigene families play in connection with genome variation. It is likely that the generation of a multigene family from a single copy gene dramatically increases the rate of evolution for that particular gene product. This would result from divergence of individual members of the gene family, and also from the generation of chimaeric genes by recombination between different members.

One feature shared by ASFV and the poxviruses is the structure of the termini of the DNA molecule, which are hairpin loop structures with unpaired nucleotides. Intriguingly, a similar structure is found at the ends of a linear plasmid found in the bacteria *Borrelia* (Hinnebusch & Barbour, 1991). This observation raises the possibility of common ancestry of the virus and the plasmid.

Several genes in ASFV resemble cellular and/or poxvirus genes. The sequences of the ASFV TK and ribonucleotide reductase are more divergent from the respective eukaryotic genes than are the poxvirus counterparts. Both the degree of divergence and the arrangement of genes in the viral genome suggest that they were acquired independently by poxviruses and ASFV. Therefore, it is likely that tracing of the phylogenetic relationship between ASFV and the poxviruses will have to rely on the identification of additional ancestral genes. However, whether those genes will exhibit sufficient similarity to allow a phylogenetic reconstruction is doubtful.

ASFV encodes a protein with similarity to type II topoisomerases, while vaccinia encodes a type I topoisomerase. As topoisomerase II sequences are highly conserved, it may be a reliable gene with which to establish distant evolutionary relationships. Analysis of the ASFV topoisomerase sequence suggests a very ancient origin for the viral gene (García-Beato *et al.*, 1992).

While poxviruses and ASFV share many basic common aspects of genetic organization, they show dramatic differences in capsid symmetry. This poses a very interesting question of whether both groups of viruses evolved from a common ancestor and, if so, how did they evolve to have such different capsids. It has recently been found that several ASFV virion proteins are generated by proteolytic processing of a polyprotein, similar to the strategy of certain plus-strand RNA viruses (Simon-Mateo, personal communication), opening the possibility that ASFV obtained its capsid genes from an RNA virus.

Future trends

The accumulation of sequence information will result, in the near future, in the assembling of the complete sequence of several poxviruses and ASFV. The sequencing of viruses belonging to different chordopoxvirus genera will likely allow the identification of a set of genes which was present in their common ancestor. Most likely, analysis of those genes will help in tracing the evolution of the different present-day vertebrate poxviruses. It is not yet possible to predict if the sequences from ASFV will allow a similar analysis.

Additionally, it may be possible to infer the series of genetic events leading to the radiation of the different poxviruses from the ancestral poxvirus. In any case, the analysis of large data sets will be necessary to precisely locate those events.

Acknowledgements

I want to thank S.N. Isaacs, P.D. Gershon, J. Dopazo, F. Fenner and A.J. Gibbs for critical reading of the manuscript, and R.W. Moyer, B-Y. Ahn, B.Y. Amegadzie and C. Simon-Mateo for making data available to me prior to publication. The bibliography for this chapter was searched in May 1992.

References

Agüero, M., Blasco, R., Wilkinson, P. & Viñuela, E. (1990). *Virology*, **176**, 195–204.

Ahn, B-Y., Gershon, P.D., Jones, E.V. & Moss, B. (1990). *Mol. Cell. Biol.*, **10**, 5433–41.

Almendral, J.M., Almazán, F., Blasco, R. & Viñuela, E. (1990). *J. Virol*, **64**, 2064–72.

Amegadzie, B.Y., Ahn, B-Y. & Moss, B. (1992). *J. Virol.*, **66**, 3003–10.

Amegadzie, B.Y., Holmes, M.H., Cole, N.B., Jones, E.V., Earl, P. & Moss, B. (1991). *Virology*, **180**, 88–98.

Archard, L.C., Mackett, M., Barnes, D.E. & Dumbell, K.R. (1984). *J. Gen. Virol.*, **65**, 875–86.

Baker, M. & Blasco, R. (1992). *FEBS Lett.*, **301**, 89–93.

Baroudy, B.M. & Moss, B. (1982). *Nucl. Acids Res.*, **10**, 5673–9.

Bedson, H.S. & Dumbell, K.R. (1964a). *J. Hyg.*, **62**, 141–6.

Bedson, H.S. & Dumbell, K.R. (1964b). *J. Hyg.*, **62**, 147–58.

Bernard, A., Zaballos, A., Salas, M. & Blanco, L. (1987). *EMBO J.*, **6**, 4219–25.

Binns, M.M., Stenzler, L., Tomley, F.M., Campbell, J. & Boursnell, E.G. (1987). *Nucl. Acids Res.*, **15**, 6563–73.

Blasco, R., Agüero, M., Almendral, J.M. & Viñuela, E. (1989a). *Virology*, **168**, 330–8.

Blasco, R., De la Vega, I., Almazán, F., Aguero, M. & Viñuela, E. (1989b). *Virology*, **173**, 251–7.

Blasco, R., López-Otín, C., Muñ, M., Bockamp, E-O., Simón-Mateo, C. & Viñuela, E. (1990). *Virology*, **178**, 301–4.

Block, W., Upton, C. & McFadden, G. (1985). *Virology*, **140**, 113–24.

Blomquist, M.C., Hunt, L.T. & Barker, W.C. (1984). *Proc. Natl. Acad. Sci. USA*, **81**, 7363–7.

Bockamp, E-O., Blasco, R. & Viñuela, E. (1990). *Gene*, **101**, 9–14.

Boursnell, M.E.G., Foulds, I.J., Campbell, J.I. & Binns, M.M. (1988). *J. Gen. Virol.*, **69**, 2995–3003.

Boursnell, M., Shaw, K., Yáñez, R.J., Viñuela, E. & Dixon, L.K. (1991). *Virology*, **184**, 411–16.

Boyle, D.B. & Coupar, B.E.H. (1986). *J. Gen. Virol.*, **67**, 1591–600.

Boyle, D.B., Coupar, B.E.H., Gibbs, A.J., Seigman, L.J. & Both, G.W. (1987). *Virology*, **156**, 355–65.

Broyles, S.S. & Moss, B. (1986). *Proc. Natl. Acad. Sci. USA*, **83**, 3141–5.

Buller, R.M.L. & Palumbo, G.J. (1991). *Microbiol. Rev.*, **55**, 80–122.

Dales, S. & Pogo, B.G. (1981). *Virol. Monogr.*, **18**, 1–109.

De la Vega, I., Viñuela, E. & Blasco, R. (1990). *Virology*, **179**, 234–46.

Dixon, L.K., Bristow, C., Wilkinson, P.G. & Sumption, K.J. (1990). *J. Mol. Biol.*, **216**, 677–688.

Drillien, R., Spehner, D., Villeval, D. & Lecocq, J. (1987). *Virology*, **160**, 203–9.

Earl, P.L., Jones, E.V. & Moss, B. (1986). *Proc. Natl. Acad. Sci. USA*, **83**, 3659–63.

Esposito, J.J., Cabradilla, C.D., Nakano, J.H. & Obijeski, J.F. (1981). *Virology*, **109**, 231–43.

Esposito, J.J. & Knight, J.C. (1985). *Virology*, **143**, 230–51.

Feng, D.F. & Doolittle, R.F. (1987). *J. Mol. Evol.*, **25**, 351–60.

Fitch, W.M. & Margoliash, E. (1967). *Science*, **155**, 279–84.

García-Beato, R., Freije, J.M.P., López-Otín, C., Blasco, R., Viñuela, E. & Salas, M.L. (1992). *Virology*, **188**, 938–47.

Gassmann, U., Wyler, R. & Wittek, R. (1985). *Arch. Virol.*, **83**, 17–31.

Gentry, G.A., Rana, S., Hutchinson, M. & Starr, P. (1988). *Intervirology*, **29**, 277–80.

Gershon, P.D., Ansell, D.M. & Black, D.N. (1989). *J. Virol.*, **63**, 4703–8.

Gershon, P.D. & Black, D.N. (1988). *Virology*, **164**, 341–9.

Gershon, P.D. & Black, D.N. (1989). *Virology*, **172**, 350–4.

Gibbs, A. & Fenner, F. (1984). *J. Virol. Methods*, **9**, 317–24.

Goebel, S.J., Johnson, G.P., Perkus, M.E., Davis, S.W., Winslow, J.P. & Paoletti, E. (1990). *Virology*, **179**, 247–66.

González, A., Calvo, V., Almazán, F., Almendral, J.M., Ramírez, J.C., De la Vega, I., Blasco, R. & Viñuela, E. (1990). *J. Virol.*, **64**, 2073–81.

González, A., Talavera, A., Almendral, J.M. & Viñuela, E. (1986). *Nucl. Acids Res.*, **14**, 6835–44.

Gruidl, M.E., Hall, R.L. & Moyer, R.W. (1992) *Virology*, **186**, 507–16.

Hall, R.L. & Moyer, R.W. (1991). *J. Virol.*, **65**, 6516–27.

Hinnebusch, J. & Barbour, A.G. (1991). *J. Bacteriol.*, **173**, 7233–9.

Howard, S.T., Chan, Y.C. & Smith, G.L. (1991). *Virology*, **180**, 633–47.

Howard, S.T. & Smith, G.L. (1989). *J. Gen. Virol.*, **70**, 3187–201.

Hruby, D.E., Maki, R.A., Miller, D.B. & Ball, L.A. (1983). *Proc. Natl. Acad. Sci. USA*, **80**, 3411–15.

Jackson, R.J. & Bults, H.G. (1992). *J. Gen. Virol.*, **73**, 323–8.

Johnson, G.P., Goebel, S.J., Pcrkus, M.E., Davis, S.W., Winslow, J.P. & Paoletti, E. (1991). *Virology*, **181**, 378–81.

Koonin, E.V. & Senkevich, T.G. (1992). *Virus Genes*, **6**, 187–96.

Koonin, E.V., Senkevich, T.G. & Chernos, V.I. (1992). *TIBS*, **17**, 213–14.

Kotwal, G.J. & Moss, B. (1988*a*). *Virology*, **167**, 524–37.

Kotwal, G.J. & Moss, B. (1988*b*). *Nature*, **335**, 176–8.

Kotwal, G.J. & Moss, B. (1989). *J. Virol.*, **63**, 600–6.

Lux, S.E., John, K.M. & Bennet, V. (1990). *Nature*, **344**, 36–42.

Macauley, C., Upton, C. & McFadden, G. (1987). *Virology*, **158**, 381–93.

McGeoch, D.J. (1990). *Nucl. Acids Res.*, **18**, 4105–10.
Mackett, M. & Archard, L.C. (1979). *J. Gen. Virol.*, **45**, 683–701.
Mockett, B., Binns, M.M. Boursnell, M.E.G. & Skinner, M.A. (1992). *J. Gen. Virol.*, **73**, 2661–8.
Moss, B. (1990). In *Virology*, 2nd edn. B.N. Fields, D.M. Knipe, R.M. Chanock, M.S. Hirsch, J.L. Melnick, T.P. Monath and B. Roizman (eds) Raven Press, New York, pp. 2079–2111.
Moss, B. (1991). *Science*, **252**, 1662–7.
Moyer, R.W. & Graves, R.L. (1981). *Cell*, **27**, 391–401.
Moyer, R.W., Graves, R.L. & Rothe, C.T. (1980). *Cell*, **22**, 545–53.
Niles, E.G., Condit, R.C., Caro, P., Davidson, K., Matusick, L. & Seto, J. (1986). *Virology*, **153**, 96–112.
Patel, D.D. & Pickup, D.J. (1989). *J. Virol.*, **63**, 1076–86.
Pickup, D.J., Bastia, D., Stone, H.O. & Joklik, W.K. (1982). *Proc. Natl. Acad. Sci. USA*, **79**, 7112–16.
Pickup, D.J., Ink, B.S., Parsons, B.L., Hu, W. & Joklik, W.K. (1984). *Proc. Natl. Acad. Sci. USA*, **81**, 6817–21.
Schmitt, J.F.C. & Stunnenberg, H.G. (1988). *J. Virol.*, **62**, 1889–97.
Shuman, S. & Moss, B. (1987). *Proc. Natl. Acad. Sci. USA*, **84**, 7478–82.
Slabaugh, M.B., Roseman, N.A., Davis, R.D. & Mathews, C.K. (1988). *J. Virol.*, **62**, 519–27.
Smith, G.L., De Carlos, A. & Chan, Y.S. (1989). *Nucl. Acids Res.*, **17**, 7581–90.
Smith, G.L., Chan, Y.S. & Kerr, S.M. (1989). *Nucl. Acids Res.*, **17**, 9051–62.
Smith, G.L., Howard, S.T. & Chan, Y.S. (1989). *J. Gen. Virol.*, **70**, 2333–43.
Smith, G.L. & Chan, Y.S. (1991). *J. Gen. Virol.*, **72**, 511–18.
Smith, G.L., Chan, Y.S. & Howard, S.T. (1991). *J. Gen. Virol.*, **72**, 1349–76.
Sumption, K.J., Hutchings, G.H., Wilkinson, P.J. & Dixon, L.K. (1990). *J. Gen. Virol.*, **71**, 2331–40.
Tengelsen, L.A., Slabaugh, M.A., Bibler, J.K. & Hruby, D.E. (1988). *Virology*, **164**, 121–31.
Tomley, F., Binns, M., Campbell, J. & Boursnell, M. (1988). *J. Gen. Virol.*, **69**, 1025–40.
Traktman, P., Anderson, M.K. & Rempel, R.E. (1989). *J. Biol. Chem.*, **264**, 21458–61.
Turner, P.C. & Moyer, R.W. (1990). *Curr. Top. Microbiol. Immunol*, **163**, 125–46.
Upton, C., DeLange, A.M. & McFadden, G. (1987). *Virology*, **160**, 20–30.
Upton, C., Macen, J.L., Maranchuk, R.A., DeLange, A.M. & McFadden, G. (1988). *Virology*, **166**, 229–39.
Upton, C. & McFadden, G. (1986). *Virology*, **152**, 308–21.
Viñuela, E. (1985). *Curr. Top. Microbiol. Immunol.*, **116**, 151–70.
Viñuela, E. (1987). In *African Swine Fever*. Y. Becker (ed.) Nijoff, Boston, pp. 31–49.
Weir, J.P. & Moss, B. (1983). *J. Virol.*, **46**, 530–7.
Wilkinson, P.J. (1989). In *Virus Infections of Porcines*. M.B. Pensaert (ed.) Elsevier science, Amsterdam, pp. 17–35.
Wittek, R., Menna, A., Schümperli, D., Stoffel, S., Müller, H.K. & Wyler, R. (1977). *J. Virol.*, **23**, 669–78.

19

Molecular systematics of the flaviviruses and their relatives

JAN BLOK AND ADRIAN J. GIBBS

Introduction

At least 70 flaviviruses, previously known as the Group B arboviruses, have been isolated. They share a group-specific antigen and can be further divided into subgroups or complexes depending on their serological cross-reactivity and type of arthropod vector. These viruses are transmitted by mosquitoes and ticks, although the vectors of some are, as yet, unknown. A number of them are human pathogens causing diseases such as yellow fever, dengue (including mild dengue fever, haemorrhagic fever and dengue shock syndrome), cases of central nervous system disease and fever due to infection with Japanese (JE), Murray Valley (MVE), and St Louis (SLE) encephalitis viruses or West Nile (WN) virus, and viruses causing tick-borne encephalitis (TBE). The prototype yellow fever (YF) virus gives the group its name, from the Latin *flavus* meaning yellow. The flaviviruses were grouped originally with the alphaviruses (previously known as the Group A arboviruses) in the family *Togaviridae*, which was defined as viruses whose virions contain single-stranded RNA of M_r 3×10^6 to 4×10^6, have isometric, probably icosahedral, nucleocapsids surrounded by a lipoprotein envelope containing host cell lipid and virus-specific polypeptides including one or more glycopeptides; the virions yield infectious RNA. In 1975 the ICTV approved the addition of two genera, rubiviruses and pestiviruses to the family *Togaviridae*.

However, closer analysis of the flaviviruses revealed that they differed in various fundamental ways from the alphaviruses. Flaviviruses produce small enveloped particles with icosahedral nucleocapsids surrounding the single-stranded *c.* 11 kb infectious RNA genome. The particles are smaller than those of the alphaviruses, and their replication strategy is also different from that of the alphaviruses which have a subgenomic 26S RNA species coding for the virion proteins, whereas the flaviviruses

270

do not. The complete nucleotide sequence of the YF virus revealed that the gene order was, from 5′ to 3′, the virion proteins, and then the nonvirion proteins, and this is opposite to that found for the alphaviruses. The cap structure at the 5′ end of flavivirus genomes is a type 1 cap rather than a type 0 cap which is found in the alphaviruses; a poly A tract is not present at the 3′ end of flavivirus genomes. Furthermore, they have no sequence homology except very distant motif similarities of at least two genes. In 1984, the ICTV therefore accepted the proposal that flaviviruses be placed into a separate family which is known as the *Flaviviridae* (Westaway *et al.*, 1985).

Based largely on cross-neutralization tests, there are eight serologically distinct clusters or antigenic subgroups within the flaviviruses; these are the tick-borne encephalitis, Rio Bravo, Japanese encephalitis, Tyuleniy, Ntaya, Uganda S, dengue and Modoc subgroups, and there are at least 17 viruses, including YF virus, which are ungrouped (Calisher *et al.*, 1989). The YF genome was the first from a flavivirus to be sequenced (Rice *et al.*, 1985), but now several more have been completely or partially sequenced (Table 19.1).

Genome and protein structures

All of the flaviviruses sequenced to date (Table 19.1) have a single-stranded RNA genome which is nearly 11 kb long (ranging from 10 468 nucleotides in TBE to 10t976 in JE). There is one long open reading frame (ORF) coding for a polyprotein which ranges in length from 10158 (DEN-4) to 10 302 (MVE) nucleotides. Flanking this coding region are two short untranslated regions at the 5′ and 3′ ends, which range in size from 80 (DEN-1) to 132 (TBE) nucleotides and 114 (TBE) to 585 (JE) nucleotides, respectively. The polyprotein is cleaved into the three virion proteins (C, capsid; M, membrane, which is synthesized as a precursor [prM]; and E, envelope) and seven non-virion proteins (NS1, NS2A, NS2B, NS3, NS4A, NS4B, NS5). No polyprotein is found in infected cells as it is processed both co-translationally and post-translationally by various proteolytic enzymes into 11 proteins. At least three different proteases cleave the polyprotein:

(i) a cellular signalase which is believed to cleave between C/prM/E, E/NS1 and NS4A/NS4B during translation of the polyprotein;
(ii) a serine-protease present at the N-terminal end of the viral NS3 protein which is believed to cleave between NS2A/NS2B/NS3/NS4A and NS4B/NS5, and;

Table 19.1. *Flaviviruses sequenced to date, their code names and references*

Antigenic subgroup	Virus strain	Code name	Partial or complete sequence	References
Tick-borne encephalitis	Neudorfl	TBEW	complete	Mandl et al., 1988; 1989
	Sofjin	TBEFE	complete	Pletnev et al., 1990
	Langat	LGT	coding region	Mandl et al., 1991; Iacono-Connors & Schmaljohn, 1992
	Louping ill		C-M-E-part NS1	Shiu et al., 1991
	Yelantsev		C-M-E	Mandl et al., 1991
Japanese encephalitis	JE-Nakayama	JE	C-M-E-NS1-NS2A-NS2B	McAda et al., 1987
	JE-OArS982		complete	Sumiyoshi et al., 1987
	JE-SA14*	JESA	complete	Nitayaphan, Grant & Trent, 1990
	JE-SA-14-14-2*		complete	Nitayaphan et al., 1990
	JE-Beijing		complete	Hashimoto et al., 1988
	Murray Valley	MVE	complete	Dalgarno et al., 1986 Lee et al., 1990
	St Louis	SLE	C-M-E-NS1-NS2A-NS2B	Trent et al., 1987
	West Nile	WN	complete	Castle et al., 1985, 1986 Wengler et al., 1985
	Kunjin	KUN	complete	Coia et al., 1988

Table 19.1. (cont.)

Antigenic subgroup	Virus strain	Code name	Partial or complete sequence	References
Dengue	D1-Nauru	D1-Nauru	C-M-E-NS1	Mason et al., 1987
	D1-Caribbean		C-M-E	Chu et al., 1989
	D1-Thailand		C-M-E	Chu et al., 1989
	D1-Philippines		C-M-E	Chu et al., 1989
	D1-Singapore	D1	complete	Fu et al., 1992
	D2-Jamaica	D2JAM	complete	Deubel et al., 1986, 1988
	D2-PR159S1	D2S1	complete	Hahn et al., 1988
	D2-NGC	D2NGC	complete	Irie et al., 1989
	D2-PUO218		C-M-E	Gruenberg et al., 1988
	D2-16681*		complete	Blok et al., 1992
	D2-16681PDK53*	D2PDK53	complete	Blok et al., 1992
	D3-H87	D3	complete	Osatomi & Sumiyoshi, 1990
	D4-814669	D4	complete	Mackow et al., 1987; Zhao et al., 1986
Ungrouped	YF 17D*	YF	complete	Rice et al., 1985
	YF Asibi*	ASIBI	complete	Hahn et al., 1987

Note: * viruses which are parent-vaccine pairs

Table 19.2. *Properties of the flavivirus proteins*

Gene	Protein	Size in amino acids	Function
C	Capsid	112–124	Nucleocapsid and RNA binding
M	Membrane	75–76	Virion spike component
E	Envelope	493–501	Major viral antigen, membrane fusion, virus assembly
NS1	Non-structural 1	352	Virus maturation?
NS2A	Non-structural 2A	218–231	Unknown
NS2B	Non-structural 2B	130–131	Unknown
NS3	Non-structural 3	615–623	Protease/helicase
NS4A	Non-structural 4A	149–150	Unknown
NS4B	Non-structural 4B	245–256	Unknown
NS5	Non-structural 5	900–905	RNA polymerase

(iii) a protease associated with the Golgi apparatus which cleaves prM
to its mature M protein.

The NS1/NS2A cleavage occurs post-translationally and even though the
site could be recognized by a signalase, its delayed processing suggests
another as yet unknown protease (Falgout, Chanock & Lai, 1989).
The cleavage points of the virion proteins have been determined by
sequencing the C- and N- terminal amino acids of these proteins from
various flaviviruses, including YF, WN, SLE, MVE, JE and DEN
(Chambers *et al.*, 1990). The cleavage points of all of the non-virion
proteins, NS1 to NS5, have been determined for Kunjin (KUN) virus
(Speight *et al.*, 1988; Speight & Westaway, 1989).

The flavivirus sequences can therefore be aligned and likely cleavage
sites determined using KUN virus as a standard. The encoded proteins,
their number of amino acid residues, as well as their functions are
shown in Table 19.2. It can be seen that most of the proteins differ
slightly in length, although these differences are small, and the only
one that seems not to is NS1. Its precise function is not clear but it
may be involved in virus maturation, and monoclonal antibodies to NS1
protect mice from lethal inoculum doses of the virus (Schlesinger *et al.*,
1986; Schlesinger, Brandriss & Walsh, 1987). This latter property has
made NS1 a target for vaccine development. There are 12 Cys residues
present in homologous positions in all flaviviruses except DEN-4; but
the potential N-glycosylation sites are variable and can be divided into

three types: (i) two sites for DEN, JE, YF, (ii) three sites for MVE, SLE, WN, KUN and (iii) another pattern of three sites for members of the tickborne encephalitis antigenic subgroup. One of the glycosylation sites is present in all flaviviruses.

The major virion antigen, E, is also a glycoprotein in some of the flaviviruses. It has been implicated in activities such as receptor binding, membrane fusion and virion assembly, and is the major target for neutralizing antibodies. There are 12 Cys residues which are present in all the flaviviruses sequenced to date, and the disulphide bonds linking some of the Cys residues have been determined for WN virus (Nowak & Wengler, 1987). The number of N-glycosylation sites varies from none in WN and KUN to three in DEN-3 and DEN-4, although one of the latter three sites is in the proposed membrane associated region.

The NS3 and NS5 proteins, which function together to replicate the viral genome, are closely similar in their primary sequences in different flaviviruses. However, the other non-virion proteins (NS2A and B, and NS4A and B), whose functions are unknown, have much more variable amino acid sequences although their hydrophobicity profiles are very similar (Fig. 19.5). This makes precise alignment of these proteins for further sequence analyses difficult. Similarly the C protein, which is found in virions, contains many basic amino acid residues and probably binds to the viral RNA genome, is very variable.

Evolution of flaviviruses

Relationships within the flaviviruses

Blok *et al.* (1992) aligned the eleven separate proteins (C, prM(-M), M, E, NS1, NS2A, NS2B, NS3, NS4A, NS4B, NS5) of the published flaviviruses by the progressive alignment program (Feng & Doolittle, 1987, 1990), and calculated all pairwise FJD distances (Feng, Johnson & Doolittle, 1985). Comparisons of the FJD distances show that the proteins differ in the amount of change that has occurred; compared with the rate of change of NS5, NS3 changes 1.25 times faster, NS1 1.38 times, E 1.58, prM/M 1.98, NS4A 1.99, NS4B 2.34, NS2B 2.58, C 3.02 and NS2A 3.64 times. As expected, these rates indicate that the proteins involved in viral replication (NS3 and NS5) are more conserved than those proteins whose function perhaps is merely to be hydrophobic (NS2 and NS4) or basic (C) in nature.

Neighbour-joining tree analyses of the FJD distances of the 11

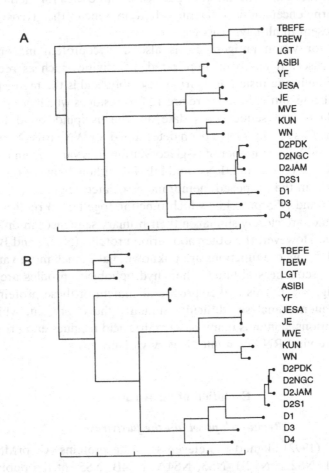

Fig. 19.1. Dendrograms illustrating the relationships of the amino acid sequences for the NS5 protein from 17 flaviviruses (Table 19.1). The sequences were aligned by the progressive-alignment method and their relationships determined. Tree A was constructed by the neighbour-joining method (Saitou & Nei, 1987) using FJD distance measures (Feng, Johnson & Doolittle, 1985) (scale units = 50), Tree B is one of two equally parsimonious trees produced using the PAUP 3.0 program (Swofford, 1991); the patristic distances of this tree correlated best with those of the NJ tree (scale units= 100 changes). C is a DIPLOMO comparison of the patristic distances in the NJ and parsimony trees in Fig 19.1A and 19.1B (see Chapter 36).

separate flavivirus proteins showed that the relatedness of the viruses was similar for each of the proteins (Blok *et al.*, 1992). This indicates that flaviviruses have probably evolved by divergent mutational change like many other virus groups, rather than by recombination as has been

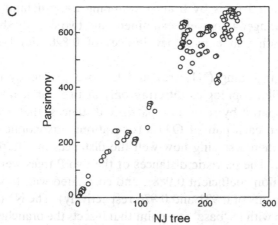

Fig. 19.1 c.

shown to be the case with the retroviruses (McClure *et al.*, 1988; Chapter 27) and the luteo-like viruses (Chapter 24). As the NS5 protein has the least rate of change of all flavivirus proteins analysed, it was used to obtain a detailed classification of 17 flavivirus sequences including the DEN-1, MVE, another JE, and two additional members of the tickborne encephalitis antigenic subgroup which were published since those analysed by Blok *et al.* (1992). Table 19.1 lists the viruses analysed, and their code names. Only one each of the JE and DEN-2 parent/vaccine pairs of viruses, namely JE-14–2 parent virus and D2 16681-PDK53 vaccine virus, was used in the analyses because the NS5 proteins of each pair differed only by 2 and 3 amino acids, respectively. Also, more than one JE and DEN-2 virus strains were used in the analyses, in order to allow correct placement of these viruses in the dendrogram. Both the parent yellow fever strain, ASIBI, and the vaccine 17D strain, YF, were included in the analyses as these were the only YF genomes sequenced and their NS5 proteins differed by 6 amino acids.

Two quite different methods were used to calculate and display the relationships of these NS5 protein sequences. The first was a distance method using the progressive alignment program of Feng and Doolittle (1987, 1990) to align the sequences and calculate their FJD distances. These were then displayed by the neighbour-joining method of Saitou and Nei (1987) as implemented by Studier and Keppler (1988) (Fig. 19.1A). The second method was a cladistic parsimony method. The alignments (each of 914 positions) produced during the first procedure were kindly analysed by J. Trueman (Australian National University) using the PAUP 3.0 program (Swofford, 1991) to make an heuristic search for unrooted trees by branch swapping using default settings. Over 4300 rearrangements were examined, and two equally short trees were found; both had consistency indices of 0.820 and homoplasy of 0.180.

The neighbour-joining (NJ) tree and the two parsimony (P) trees had closely similar topologies differing only at two minor nodes. The trees were compared by recording patristic distances, that is the path lengths, between each pair of OTUs (operational taxonomic units) in each tree, and then assessing how well the distances in different trees were correlated. The patristic distances of the two P trees were closely similar (correlation coefficient 0.988), and correlated with those of the NJ tree (coefficients of 0.931 and 0.926 respectively). The NJ tree (Fig. 19.1A) is drawn with its 'base' at a point that bisects the branches linking the most distant OTUs, and an earlier study (Blok *et al.*, 1992), using other viruses as outgroups, had shown that this is probably the natural root of this tree. To make comparison easy, the P tree (Fig. 19.1B) is drawn rooted at the midpoint of the same link, even though the branch lengths on either side of this point are very different. Fig. 19.1C is a DIPLOMO graph (Chapter 36) showing the relationships of the patristic distances of the NJ tree and the best P tree. It confirms that the trees have closely similar topologies, and that most of their differences are in the lengths of the basal branches.

Thus the oldest branching point, the origin of the flaviviruses, is probably on the node joining viruses belonging to the tick-borne encephalitis antigenic subgroup and the mosquito-borne flaviviruses. This was also indicated by a classification of the E genes of 8 tick-borne and 2 mosquito-borne flaviviruses (Fig. 19.2).

None of the flavivirus taxonomies indicates whether tick transmission is more ancient than mosquito transmission; the root seems to be between the vector groups. Viruses, such as hepatitis C and the pestiviruses, whose NS5-like genes are clearly related to those of the flaviviruses (see below) spread via a contaminated environment. This could suggest that arthropod transmission of the flaviviruses is an acquired trait, although the loss of arthropod transmission by the pestiviruses is equally possible, as a similar loss probably occurred when the rubiviruses split from the alphaviruses (Chapter 33).

The flavivirus trees discussed above are consistent; they cluster the replicate isolates of individual viruses (e.g. numerous dengue-2 isolates

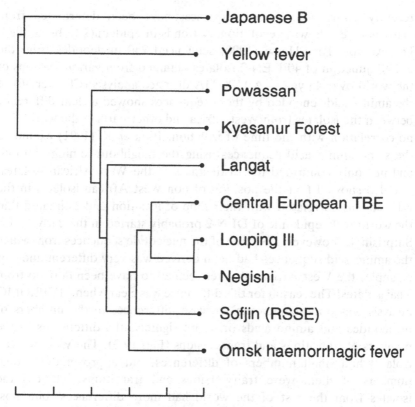

Fig. 19.2. Neighbour-joining tree calculated from the pairwise nucleotide differences of the E genes of 10 flaviviruses, 8 of them tick-borne (scale units =5%). Data kindly provided by E.A. Gould and K. Venu

from around the world) and the topology of the mosquito-borne virus clusters reflect their serological groupings (Calisher *et al.*, 1989). Thus these trees probably depict flavivirus phylogeny accurately.

Close relationships within flavivirus subgroups

Variation within flavivirus subgroups (e.g. dengue) has been shown using a variety of methods such as RNA oligonucleotide fingerprinting (Trent *et al.*, 1989), antigenic analysis (Monath *et al.*, 1986), RNA–DNA hybridization (Kerschner *et al.*, 1986), restriction enzyme mapping (Walker *et al.*, 1988) and sequencing regions of the genome (Rico-Hesse, 1990). All of these methods have been useful tools in determining the relationships of the virus isolates involved in outbreaks, but, more recently, short homologous sequences have been determined from many isolates allowing evolutionary trends in epidemics to be studied. For example Rico-Hesse (1990) sequenced 240 nucleotides from the E/NS1 junction of 40 DEN-2 isolates obtained from various regions of the world over 45 years. A UPGMA distance analysis (Chapter 36) of the amino acids encoded by these sequences showed a clear difference between the isolates from West Africa and other parts of the world, but no correlation with the time of isolation. Blok *et al.* (1991) examined the same amino acid sequences using the neighbour-joining method, and not only confirmed the distinctness of the West African isolates, but also showed that the position of non-West African isolates in the NJ tree was correlated with their time of isolation and estimated that the world-wide epidemic of DEN-2 probably started in the early 1940s. Surprisingly, however, a NJ tree of the nucleotide sequences from which the amino acid sequences had been derived was very different and, for example, the West African isolates seemed to have been derived from Thai isolates! The reason for this difference was clear when a DIPLOMO analysis was done comparing pairwise differences in the numbers of nucleotides and amino acids or, more significantly, differences in the numbers of transitions and transversions (Fig. 19.3). The West African isolates had small numbers of differences, but approximately equal numbers of them were transversions and transitions, whereas the isolates from the rest of the world had more differences, but most were 'silent' transitions. This indicates that the isolates from the world wide epidemic had been under different, probably less severe, selective constraints than those obtained from mosquitoes in West Africa, where the virus is probably endemic. When the nucleotide differences were corrected for bias, by converting transversion differences into transition equivalents, and used for analysis, a NJ tree (Fig. 19.4) was obtained that was almost identical to that calculated from amino acid differences (see Blok *et al.*, 1991), despite most of the nucleotide changes being silent!

Distant relationships

Recently, there have been several reports noting that flaviviruses resemble bovine viral diarrhoea (BVD) and hog cholera pestiviruses, and hepatitis virus C (Hep C) in the type of virion they produce and in the size, sequence and organization of their genomes. Also, some conserved motifs of their proteins are distantly related, but even more distantly related to those of como-, nepo-, picorna- and potyviruses (Bazan &

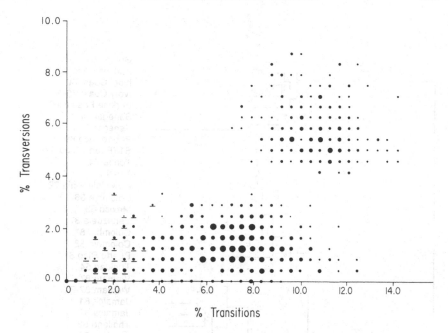

Fig. 19.3. DIPLOMO plot of the transversion and transition differences of a 240 nucleotide portion of the genomes of 40 DEN-2 isolates. The area of each spot corresponds to the number of comparisons with those co-ordinates. The underlined spots in the lower left side of the diagram are all from comparisons between West African isolates, and the broad swath of spots across the lower part represent comparisons between isolates from other parts of the world (the equation for the regression line calculated for these spots is: transitions% = 0.164 + 4.120× transversions%). The cluster of spots in the upper right corner of the diagram all represent comparisons between isolates from West Africa and those from the rest of the world.

Flatterick, 1989; Bruenn, 1991; Choo *et al.*, 1991; Collett, Moennig & Horzinek, 1989; Han *et al.*, 1991; Miller & Purcell, 1990). The pestiviruses have a single-stranded RNA genome *c.* 12.5 kb long. The genome of BDV lacks a poly A tract and has one long ORF coding for a polyprotein which consists of virion proteins at its N-terminus with non-virion proteins at its C-terminal end, as is found in the flaviviruses. The Hep C viral genome is similar but is *c.* 9 kb long and has a poly A tract. The pestiviruses as well as Hep C virus are now placed in the family *Flaviviridae* because of their similarity to the flaviviruses.

Sequences of proteins involved in virus replication (e.g., NS5 and NS3) further confirm these relationships. Viral RNA-dependent RNA

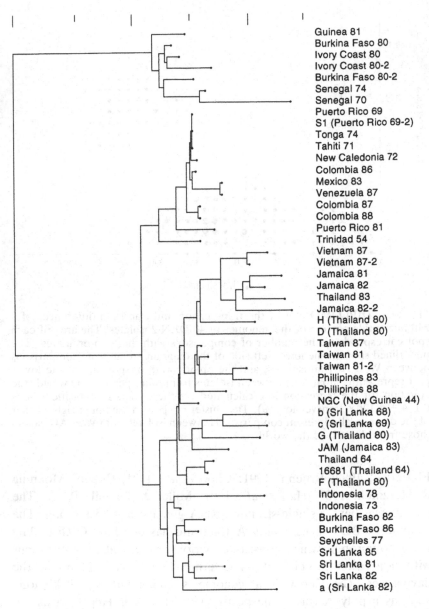

Fig. 19.4. Neighbour-joining tree calculated from the nucleotide differences of a portion of the E/NS1 genes of 40 DEN-2 flaviviruses (Rico-Hesse, 1990); for each pairwise comparison, the nucleotide difference was calculated as the sum of the transitions and 4.12 times the number of transversions (see text and Fig. 19.3: scale units = 5%).

polymerases, including the flavivirus NS5 protein, contain the characteristic sequence -GDD- or a close derivative of it and a classification
of this -GDD- region of 63 viral genomes representing 13 families
including the flaviviruses, alphaviruses, pestiviruses and Hep C virus
using several methods (Bruenn, 1991; Blok *et al.*, 1992) revealed that,
in most classifications, the flaviviruses and pestiviruses were sister
groups, and clustered next with Hep C virus before linking to the
other groups.

Other proteins showed more distant links. For example, NS3, the
protease/helicase protein of flaviviruses is clearly, though distantly,
related to those of the pestiviruses, Hep C and the cylindrical inclusion
proteins of the plum pox (Teycheney *et al.*, 1989), potato virus Y
(Robaglia *et al.*, 1989), tobacco etch (Allison, Johnston & Dougerty,
1986) and tobacco vein mottling (Domier *et al.*, 1986) potyviruses.
Similarly the E protein has a clear sequence similarity with the spike
glycoprotein precursors of vesicular stomatitis rhabdovirus, strains
Orsay and Glasgow (Gallione & Rose, 1985; Vandepol & Holland,
1986); however, a search of the Swissprot protein sequence database
(Release 16) with protein NS1 failed to find any related sequences
greater than 100 amino acids in length, or ones that were smaller but
consistently related.

Distant relationships with other viruses are only shown by the proteins
most conserved within the flaviviruses. What about the other more
variable proteins? These differ so much that it is unlikely that sequence
comparisons will show any distant relationships anyway. However, the
order and type of proteins encoded by the genomes indicate likely
similarities. For example, the hydrophobicity profiles of the amino acid
sequence encoded by the main flavivirus ORF are similar even though
the sequence homologies of the genomes are as little as 10% in some
regions. Fig. 19.5 shows the hydrophobicity profiles of the polyproteins
of two flaviviruses (YF and TBEFE), the BVD pestivirus, and the Hep
C virus. The eleven proteins of the flaviviruses are indicated and the
corresponding protein regions of BVD and Hep C (Choo *et al.*, 1991)
show that there are hydrophobic regions equivalent to the YF NS2 and
NS4 proteins in the other two viruses, although the NS4-like region
in BVD is less hydrophobic. These hydrophobic regions flanking the
protease/helicase (NS3) protein are not, however, found in potyviral
polyproteins.

There is little sequence similarity between these NS2 and NS4 regions
of flaviviruses (<20% similarity at the amino acid level), and they show

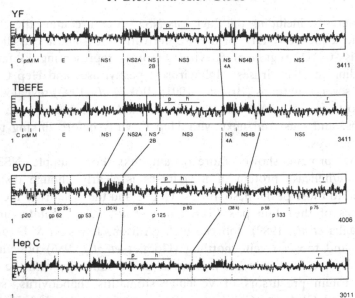

Fig. 19.5. Hydrophobicity profiles for two flaviviruses, YF and TBEFE, bovine viral diarrhoea pestivirus (BVD) and the hepatitis C (Hep C) virus were calculated by the Staden programs using the values of Kyte and Doolittle (1982). The eleven flavivirus proteins (C, capsid; M membrane [prM/M]; E, envelope; and non-structural proteins, NS1, NS2A, NS2B, NS3, NS4A, NS4B, NS5) are indicated for YF and TBEFE and the corresponding proteins for BVD and Hep C are marked as named by Choo *et al.* (1991); the protease, helicase and replicase domains are marked p, h and r, respectively. The regions corresponding to the NS2 and NS4 flavivirus regions are connected.

no significant similarity to any other proteins when tested against the Swissprot database. This led to the question: where did these genes arise? Analyses of several flaviviruses representing the various antigenic subgroups for potential open reading frames (ORFs) using the puRine-aNy-pYrimidine (RNY) method (Shepherd, 1981) revealed very

Fig. 19.6. The puRine-aNy-pYrimidine (RNY, Shepherd, 1981) profiles of two flaviviruses, YF and D2PDK (Table 19.1), and the bovine viral diarrhoea (BVD) pestivirus (Collett *et al.*, 1989) were calculated by the Staden programs. Each virus is represented by three boxes which correspond to the three reading frames (rf). The rf coding for the polyprotein is labelled 1 and the consecutive rfs are labelled 2 and 3, respectively. A single dot at the mid-height of a box indicates that this position has the highest score of the three rfs; points at consecutive positions in one rf thereby form a solid line. Stop codons are indicated by short vertical lines which bisect the mid-height of each box, and start codons (AUG) are indicated at the bottom of each box.

YF

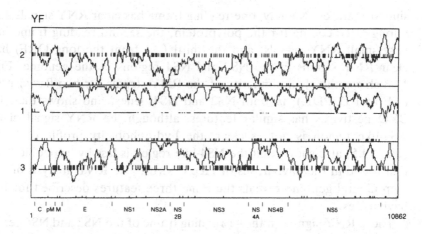

D2PDK

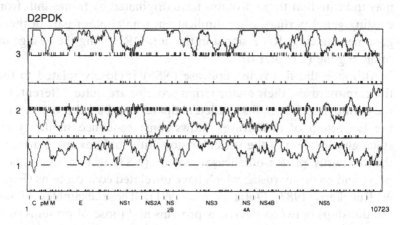

BVD

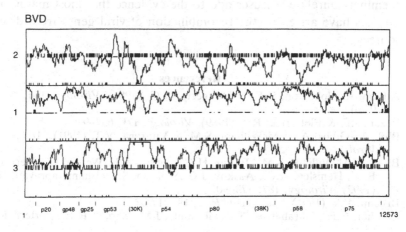

similar features. Namely, one reading frame has clear RNY signals and a long ORF coding for the polyprotein, the second reading frame has very small RNY signals, and the third (−1 from the long ORF) has clear RNY signals in the NS2 and NS4 regions of the genome. Fig. 19.6 shows the clear similarity of the RNY signals for two flaviviruses (YF and D2PDK), in both NS2 and NS4 genes, and shows that the BVD pestivirus has similar features, although the RNY signal of its NS4-like region is not as clear; the hydrophobicity profile of BVD (Fig. 19.5) also shows that the NS4-like region is less hydrophobic than that found in the flaviviruses or Hep C virus. The RNY analysis of the Hep C viral genome reveals the same three features described for the flaviviruses.

These RNY signals in the −1 reading frame of the NS2 and NS4 genes may indicate that these proteins have originated by frame-shift from an existing gene, perhaps after duplication from another part of the same genome, the clear RNY signal in the non-ORF being the vestige of the original gene (Chapter 6).

Although the flavivirus replicase (NS5) is closely related to that of the carmoviruses, their major virion proteins are quite different; that of the carmoviruses is an 8-stranded antiparallel β-barrel protein, whereas the E protein of flaviviruses shows some sequence similarity to the spike glycoprotein of the vesicular stomatitis rhabdovirus. Furthermore the flavivirus protease/helicase protein (NS3) is related to those of poty- and picornaviruses, which have unrelated coat proteins (Rossman & Rueckert, 1987; Dolja et al., 1991). These examples of distant relationships between flavivirus proteins and those of proteins of other seemingly unrelated viruses add to the evidence that most major virus families have arisen by the recombination of viral genes from diverse sources.

References

Allison, R., Johnston, R.E. & Dougerty, W.G. (1986). *Virology*, **154**, 9–20.
Bazan, J.F. & Fletterick, R.J. (1989). *Virology*, **171**, 637–9.
Blok, J., Gibbs, A.J., McWilliam, S.M. & Vitarana, V.T. (1991). *Arch. Virol.*, **118**, 209–23.
Blok, J., McWilliam, S.M., Butler, H.C., Gibbs, A.J., Weiller, G., Herring, B.L., Hemsley, A.C., Aaskov, J.G., Yoksan, S. & Bhamarapravati, N. (1992). *Virology*, **187**, 573–90.
Bruenn, J.A. (1991) *Nucl. Acids Res.*, **19**, 217–26.
Calisher, C.H., Karabatsos, N., Dalrymple, J.M., Shope, R.E., Porterfield,

J.S., Westaway, E.G. & Brandt, W.E. (1989). *J. Gen. Virol.*, **70**, 37–43.

Castle, E., Leidner, U., Nowak, T., Wengler, G. & Wengler, G. (1986). *Virology*, **149**, 10–26.

Castle, E., Nowak, T., Leidner, U., Wengler, G. & Wengler, G. (1985). *Virology*, **145**, 227–36.

Chambers, T.J., Hahn, C.S., Galler, R. & Rice, C.M. (1990). *Annu. Rev. Microbiol.*, **44**, 649–88.

Choo, Q.-L., Richman, K.H., Han, J.H., Berger, K., Lee, C., Dong, C., Gallegos, C., Coit, D., Medina-Selby, A., Barr, P.J., Weiner, A.J., Brandley, D.W., Kuo, G. & Houghton, M. (1991). *Proc. Natl. Acad. Sci. USA*, **88**, 2451–5.

Chu, M.C., O'Rourke, E.J. & Trent, D.W. (1989). *J. Gen. Virol.*, **70**, 1701–12.

Coia, G., Parker, M.D., Speight, G., Byrne, M.E. & Westaway, E.G. (1988). *J. Gen. Virol.*, **69**, 1–21.

Collett, M.S., Moennig, V. & Horzinek, M.C. (1989). *J. Gen. Virol.*, **70**, 253–66.

Dalgarno, L., Trent, D.W., Strauss, J.H. & Rice, C.M. (1986). *J. Mol. Biol.*, **187**, 309–23.

Deubel, V., Kinney, R.M. & Trent, D.W. (1986). *Virology*, **155**, 365–77.

Deubel, V., Kinney, R.M. & Trent, D.W. (1988). *Virology*, **165**, 234–44.

Dolja, V.V., Boyko, V.P., Agranovsky, A.A. & Koonin, E.V. (1991). *Virology*, **184**, 79–86.

Domier, L.L., Franklin, K.N., Shahabuddin, M., Hellmann, G.M., Overmyer, J.H., Hiremath, S.T., Shaw, M.F.E., Lommonssoff, G.P., Shaw, J.G. & Rhoads, R.E. (1986). *Nucl. Acids Res.*, **14**, 5417–30.

Falgout, B., Chanock, R. & Lai, C.-J. (1989). *J. Virol.*, **63**, 1852–60.

Feng, D.-F. & Doolittle, R.F. (1987). *J. Mol. Evol.*, **25**, 351–60.

Feng, D.-F. & Doolittle, R.F. (1990). In *Methods in Enzymology*, R.F. Doolittle (ed.), Academic Press, San Diego, Vol. 183, pp. 375–87.

Feng, D.-F., Johnson, M.S. & Doolittle, R.F. (1985). *J. Mol. Evol.*, **21**, 112–25.

Fu, J., Tan, B.-H., Yap, E.-H., Chan Y.-C. & Tan, Y.H. (1992). *Virology*, **188**, 953–8.

Gallione, C.J. & Rose, J.K. (1985). *J. Virol.*, **54**, 374–82.

Gruenberg, A., Woo, W.S., Biedrzycka, A. & Wright, P.J. (1988). *J. Gen. Virol.*, **69**, 1391–8.

Hahn, C.S., Dalrymple, J.M., Strauss, J.H. & Rice, C.M. (1987). *Proc. Natl. Acad. Sci. USA*, **84**, 2019–23.

Hahn, Y.S., Galler, R., Hunkapiller, T., Dalrymple, J.M., Strauss, J.H. & Strauss, E.G. (1988). *Virology*, **162**, 167–80.

Han, J.H., Shyamala, V., Richman, K.H., Brauer, M.J., Irvine, B., Urdea, M., Tekamp-Olson, P., Kuo, G., Choo, Q.-L. & Houghton, M. (1991). *Proc. Natl. Acad. Sci. USA*, **88**, 1711–15.

Hashimoto, H., Nomoto, A., Watanabe, K., Mori, T., Takezawa, T., Aizawa, C., Takegami, T. & Hiramatsu, K. (1988). *Virus Genes*, **1**, 305–17.

Iacono-Connors, L.C. & Schmaljohn, C.S. (1992). *Virology*, **188**, 875–80.

Irie, K., Mohan, P.M., Sasaguri, Y., Putnak, R. & Padmanabhan, R. (1989). *Gene*, **75**, 197–211.

Kerschner, J.A., Vorndam, A.V., Monath, T.P. & Trent, D.W. (1986). *J. Gen. Virol.*, **67**, 2645–61.

Kyte, J. & Doolittle, R.F. (1982). *J. Mol. Biol.*, **157**, 105–32.

Lee, E., Fernon, C., Simpson, R., Weir, R.C., Rice, C.M. & Dalgarno, L. (1990). *Virus Genes*, **4**, 197–213.

McAda, P.C., Mason, P.W., Schmaljohn, C.S., Dalrymple, J.M., Mason, T.L. & Fournier, M.J. (1987). *Virology*, **158**, 348–60.

McClure, M.A., Johnson, M.S., Feng, D.-F. & Doolittle, R.F. (1988). *Proc. Natl. Acad. Sci. USA*, **85**, 2469–73.

Mackow, E., Makino, Y., Zhao, B., Zhang, Y.-M., Markoff, L., Buckler-White, A., Guiler, M., Chanock, R. & Lai, C.-J. (1987). *Virology*, **159**, 217–28.

Mandl, C.W., Heinz, F.X. & Kunz, C. (1988). *Virology*, **166**, 197–205.

Mandl, C.W., Heinz, F.X. Stockl, E. & Kunz, C. (1989). *Virology*, **173**, 291–301.

Mandl, C.W., Iacono-Connors, L., Wallner, G., Holzmann, H., Kunz, C. & Heinz, F.X. (1991). *Virology*, **185**, 891–5.

Mason, P.W., McAda, P.C., Mason, T.L. & Fournier, M.J. (1987). *Virology*, **161**, 262–7.

Miller, R.H. & Purcell, R.H. (1990). *Proc. Natl. Acad. Sci. USA*, **87**, 2057–61.

Monath, T.P., Wands, J.A., Hill, L.J., Brown, N.V., Marciniak, R.A., Wong, M.A., Gentry, M.K., Burke, D.S., Grant, J.A. & Trent, D.W. (1986). *Virology*, **154**, 313–24.

Nitayaphan, S., Grant, J.A. & Trent, D.W. (1990). *Virology*, **177**, 541–52.

Nowak, T. & Wengler, G. (1987). *Virology*, **156**, 127–37.

Osatomi, K. & Sumiyoshi, H. (1990). *Virology*, **176**, 643–7.

Pletnev, A.G., Yamshchikov, V.F. & Blinov, V.M. (1990). *Virology*, **174**, 250–63.

Rice, C.M., Lenches, E.M., Eddy, S.R., Shin, S.J., Sheets, R.L. & Strauss, J.H. (1985). *Science*, **229**, 726–33.

Rico-Hesse, R. (1990). *Virology*, **174**, 479–93.

Robaglia, C., Durand-Tardif, M., Tronchet, M., Boudazin, G., Astier-Manifacier, S. & Casse-Delbart, F. (1989). *J. Gen. Virol.*, **70**, 935–47.

Rossman, M.G. & Rueckert, R.R. (1987). *Microbiol. Sci.*, **4**, 206–14.

Saitou, N. & Nei, M. (1987). *Mol. Biol. Evol.*, **4**, 406–25.

Schlesinger, J.J., Brandriss, M.W., Cropp, C.B. & Monath, T.P. (1986). *J. Virol*, **60**, 1153–5.

Schlesinger, J.J., Brandriss, M.W. & Walsh, E.E. (1987). *J. Gen. Virol.*, **68**, 853–7.

Shepherd, J.C.W. (1981). *Proc. Natl. Acad. Sci. USA*, **78**, 1596–600.

Shiu, S.Y.W., Ayres, M.D. & Gould, E.A. (1991). *Virology*, **180**, 411–15.

Speight, G., Coia, G., Parker, M.D. & Westaway, E.G. (1988). *J. Gen. Virol.*, **69**, 23–34.

Speight, G. & Westaway, E.G. (1989). *Virology*, **170**, 299–301.

Studier, J.A. & Keppler, K.L. (1988). *Mol. Biol. Evol.*, **5**, 729–31.

Sumiyoshi, H., Mori, C., Fuke, I., Morita, K., Kuhara, S., Kondou, J., Kikuchi, Y., Nagamatu, H. & Igarashi, A. (1987). *Virology*, **161**, 497–510.

Swofford, D.L. (1991). PAUP: Phylogenetic Analysis Using Parsimony, Version 3.0s. Computer program distributed by the Illinois Natural History Survey, Champaign, Illinois.

Teycheney, P.Y., Tavert, G., Delbos, R., Ravelonandro, M. & Dunez, J. (1989). *Nucl. Acids Res.*, **17**, 10115–6.

Trent, D.W., Grant, J.A., Monath, T.P. Manske, C.L., Corina, M. & Fox, G.E. (1989). *Virology*, **172**, 523–35.

Trent, D.W., Kinney, R.M., Johnson, B.J.B., Vorndam, A.V., Grant, J.A., Deubel, V., Rice, C.M. & Hahn, C. (1987). *Virology*, **156**, 293–304.

Vandepol, S.B. & Holland, J.J. (1986). *J. Gen. Virol.*, **67**, 441–51.

Walker, P.J., Henchal, E.A., Blok, J., Repik, P.M., Henchal, L.S., Burke, D.S., Robbins, S.J. & Gorman, B.M. (1988). *J. Gen. Virol.*, **69**, 591–602.

Wengler, G., Castle, E., Leidner, U., Nowak, T. & Wengler, G. (1985). *Virology*, **147**, 264–74.

Westaway, E.G., Brinton, M.A., Gaidamovich, S.Y., Horzinek, M.C., Igarashi, A., Kaarianinen, L., Lvov, D.K., Porterfield, J.S., Russell, P.K. & Trent, D.W. (1985). *Intervirology*, **24**, 183–92.

Zhao, B., Mackow, E., Buckler-White, A., Markoff, L., Channock, R.M., Lai, C.-J. & Makino, Y. (1986). *Virology*, **155**, 77–88.

20

Herpesviridae

ANDREW J. DAVISON AND DUNCAN J. McGEOCH

Introduction

In this chapter, we describe the biological, genomic and genetic diversity of the herpesviruses and then discuss the mechanisms that have been used to generate their different genetic complements. We consider how recent work has expanded our view of the possible origins of the family, and finish by sketching out the limitations that studies of present-day herpesviruses impose on our understanding of herpesvirus evolution. We have limited citations of the literature to appropriate reviews and particularly pertinent research papers.

Herpesviruses are complex viruses of higher eukaryotes which have large linear double-stranded DNA genomes and replicate in the nuclei of host cells (Matthews, 1982; Minson, 1989). They have a characteristic appearance in the electron microscope: an icosahedral capsid of 162 capsomeres embedded in an amorphous protein layer (the tegument) which is surrounded by a protein-containing lipid envelope. Agents of this type have been isolated from a wide range of vertebrates, including bony fish, amphibians, reptiles, birds, marsupials and placental mammals. It is possible that herpesviruses also infect invertebrates such as oysters (Farley *et al.*, 1972), but this has yet to be confirmed. Known hosts in addition to humans are largely those subjected to husbandry, and the best studied are infected by several different herpesviruses. For example, humans are hosts to at least seven herpesviruses, and horses to at least five. The ubiquity of members of this large virus family, coupled with the very high degree of host specificity exhibited by individual members, indicates that the herpesviruses have a long history of evolution in close association with their hosts. This virus family, then, comprises a large (probably extremely large) collection of viruses whose relationships are in general characterized by a high degree of divergence.

Biological diversity

Pathogenesis

Herpesviruses cause a wide spectrum of diseases, ranging from inapparent to epidemic, and some are associated with neoplastic diseases. Transmission occurs directly from infected to susceptible individuals, and there is no evidence for vector-mediated transmission. Also, in nature, these viruses are very host-specific, most infecting only a single species. There are, however, some circumstances in which unusual species may become infected: for example, pseudorabies virus (PRV), which infects pigs, may on occasion cause fatal encephalitis in other animals, including cattle (Wittmann, 1989), and infection of humans with B virus of old world monkeys has a similar outcome (Whitley, 1990). Nevertheless, inter-species transmission of herpesviruses is rare in nature, and the biological habits and degree of divergence of present-day viruses indicate that it is not a major component in evolution of the family.

Latency

Latency is a central feature of herpesvirus biology (Kieff & Licbowitz, 1990; Rock, 1993). After primary infection, a herpesvirus establishes a latent ('hidden') infection for the lifetime of the host. Sites of latency depend on the virus, but include sensory ganglia (for herpes simplex virus (HSV) and varicella-zoster virus (VZV)) and circulating lymphocytes (for Epstein–Barr virus (EBV)). Upon an appropriate stimulus, latent virus may reactivate and cause secondary disease. Some herpesviruses cause symptoms only on primary infection (e.g. malignant lymphoma caused by Marek's disease virus (MDV) of chickens), some are usually apparent only on reactivation (e.g. cold sores caused by HSV-1), some cause distinct diseases at both stages (e.g. chickenpox and shingles caused by VZV on primary infection and reactivation, respectively), and some are not associated with disease (e.g. equine herpesvirus 2 (EHV-2)). Thus, herpesviruses are presumed to be subject to evolutionary pressures not only during the relatively short-lived phase of primary infection but also during latency and reactivation. The evolutionary aspects of latency are particularly interesting, as described below, since several genetic systems for promoting latency appear to have arisen independently; indeed, there is no known example of a herpesvirus that lacks a latent cycle.

Classification

Classification of the Herpesviridae got under way before DNA sequence data became available, and was based largely on biological features which led to the definition of three subfamilies: the *Alpha-*, *Beta-* and *Gammaherpesvirinae* (Roizman *et al.*, 1981, 1992). The *Alphaherpesvirinae* have a relatively short growth cycle and are able to grow in cultured cells from the tissues of a relatively wide range of hosts; many establish latency in nervous tissue. Examples are HSV-1, VZV, PRV and equine herpesvirus 1 (EHV-1). The *Betaherpesvirinae* have longer growth cycles and usually grow in cells cultured from a much narrower range of hosts. They have the distinguishing mark of being able to cause enlargement of cells to give cytomegalia. They include human cytomegalovirus (HCMV) and murine cytomegalovirus (MCMV). The *Gammaherpesvirinae* are lymphotropic, and include EBV and herpesvirus saimiri (HVS).

This classification has been largely supported by the subsequent analyses of genetic content that underpin evolutionary studies; the few exceptions are dealt with below. For the purposes of this article, the family is considered in terms of corresponding genetic subfamilies termed the α-, β- and γ-herpesviruses, with the latter two groups subdivided into the β_1-, β_2-, γ_1- and γ_2-herpesviruses. The current classification of the *Gammaherpesvirinae* into the genera *Lymphocryptovirus* and *Rhadinovirus* is reflected in the subdivision of the γ-herpesviruses, but a genetic case for classification of the *Alphaherpesvirinae* into the genera *Simplexvirus* and *Varicellovirus* is rather less easy to support.

A single host may be infected by several different herpesviruses: humans, as the most intensively studied hosts, are host to three α-herpesviruses (HSV-1, HSV-2 and VZV), three β-herpesviruses (HCMV, HHV-6 and HHV-7) and one γ-herpesvirus (EBV). Some other well-studied hosts appear at present to lack a representative of one or other subfamily: for example, mice are host to a β-herpesvirus (MCMV) and a γ-herpesviruses (murine herpesvirus 68; Efstathiou *et al.*, 1990) and horses to three α-herpesviruses (EHV-1, EHV-3 and EHV-4) and two γ-herpesviruses (EHV-2 and EHV-5; Agius, Nagesha & Studdert, 1992; E.A.R. Telford, M.J. Studdert, C.T. Agius, M.S. Watson, H.C. Aird and A.J. Davison, manuscript in preparation). This may reflect the current state of virus identification in these animals; it seems likely that more herpesviruses await discovery in these hosts and perhaps also in humans.

Genomic diversity

General genome features

The DNA genome is packaged into the herpesvirus capsid in liquid crystalline form, apparently lacking intrinsic association with proteins (Booy *et al.*, 1991). Genomic sizes range from about 125 to 240 kbp, the smaller being found with the α- and γ$_2$-herpesviruses and the larger with the β$_1$-herpesviruses. The genomic termini are not covalently linked to protein (as they are in adenoviruses) and do not form covalently linked hairpins (as they do in poxviruses). The two herpesviruses that have been investigated in detail (HSV-1: Mocarski & Roizman, 1982; VZV: Davison, 1984) have single unpaired residues at the 3' genomic termini which are complementary to each other. It is held that the termini ligate after infection, and that DNA replication initiates at specific origins to generate long head-to-tail concatemers by a rolling circle mechanism. Concatemers are then cleaved to unit-length genomes at specific sites determined by sequences near potential termini (Roizman & Sears, 1990). In some herpesviruses, such as EBV, the genome is maintained as a circular plasmid during latency (Kieff & Liebowitz, 1990).

Genome structures

There appears to have been a great propensity for herpesvirus genomes to develop substantial repeats at the termini or at internal locations. Thus, they exhibit a diverse range of structures differing in the arrangements of unique and repeated regions. Fig. 20.1 illustrates the six types of genome arrangement that have been characterized. For most of the viruses listed below, the genome structures of several isolates have been examined and are identical.

The simplest genome arrangement is that of group 1, which consists of a unique sequence flanked by a direct repeat. It was first described for channel catfish virus (CCV: Chousterman, Lacasa & Sheldrick, 1979), which fits into none of the subfamilies (see below), and has since been found among the β$_2$- (human herpesvirus 6 (HHV-6): Martin *et al.*, 1991) and γ-herpesviruses (EHV-2: Browning & Studdert, 1989). The DNA of MCMV (a β$_1$-herpesvirus) also has this arrangement, but the direct repeats are only 30 bp in size (Marks & Spector, 1988).

So far, genomes in groups 2 to 4 have been found only among the γ-herpesviruses. Group 2 genomes, instead of containing a single direct repeat at the genome termini as in group 1, have a shorter sequence that is repeated a variable number of times. Members of the γ₂-herpesviruses typified by HVS have this structure (Bornkamm *et al.*, 1976). The presence of a variable number of the terminal repeats in inverse orientation at an internal location results in the group 3 genome structure, which has been described for cottontail rabbit herpesvirus (Cebrian, Berthebt & Laithier, 1989). This type of genome exists in four equimolar isomers in virion DNA by virtue of inversion of the two unique regions, presumably by recombination between inverted repeats. In another variation, the presence of an internal set of direct repeats unrelated to the terminal set results in the group 4 genome

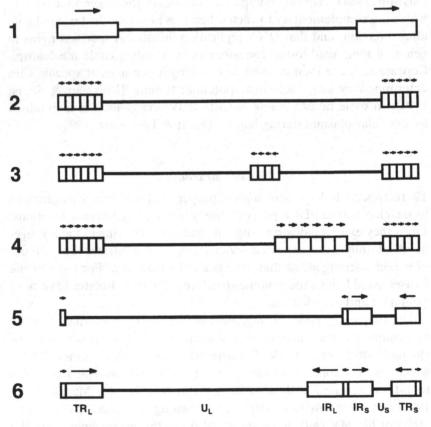

Fig. 20.1. Types of herpesvirus genome structure (not to scale). Unique and repeat regions are shown as horizontal lines and rectangles, respectively. The orientations of repeats are shown by arrows. Regions of the group 6 genome are indicated at the foot of the Figure.

structure, typified by the γ_1-herpesvirus EBV (Given & Kieff, 1979). The two unique regions do not invert in such a structure.

The group 5 structure has been identified to date only among the α-herpesviruses, such as PRV (Ben-Porat, Rixon & Blankenship, 1979), EHV-1 (Whalley, Robertson & Davison, 1981) and VZV (Dumas *et al.*, 1981). Two unique regions (U_L and U_S) are each flanked by an inverted repeat (TR_L/IR_L and TR_S/IR_S). The sets of inverted repeats are not related to each other. Group 5 contains the only herpesvirus genomes that are not terminally redundant. The two orientations of U_S are present in equimolar amounts in virion DNA, but U_L is found completely (EHV-1) or predominantly (VZV, PRV) in a single orientation.

Group 6 genomes are found among the α-herpesviruses (HSV-1: Sheldrick & Berthelot, 1974; bovine herpesvirus 2: Buchman & Roizman, 1978) and β_1-herpesviruses (HCMV: Weststrate, Geelen & van der Noordaa, 1980). This structure is similar to that of group 5, except that TR_L/IR_L is much larger and the two orientations of U_L are present in equimolar amounts in virion DNA. Group 6 genomes are terminally redundant, possessing a short sequence which is repeated directly at the genome termini and in inverse orientation at the junction between IR_L and IR_S.

Genome structure is clearly not a reliable guide in the development of evolutionary schemes, as it seems that similar structures have arisen independently more than once. A firm basis is rather to be found in the DNA sequences and relative arrangements of genes embedded therein.

Genetic diversity

Gene arrangement and expression

Complete genome sequences are now known for six mammalian herpesviruses: three α-herpesviruses (VZV: Davison & Scott, 1986; HSV-1: McGeoch *et al.*, 1988; EHV-1: Telford *et al.*, 1992), one β_1-herpesvirus (HCMV: Chee *et al.*, 1990), one γ_1-herpesvirus (EBV: Baer *et al.*, 1984) and one γ_2-herpesvirus (HVS: Albrecht *et al.*, 1992). A seventh complete sequence, that of CCV, is also known and has key implications in herpesvirus evolution (Davison, 1992), and is discussed in detail below. An extensive amount of sequence data is also available for other herpesviruses, including bovine herpesvirus 1 and

PRV (α-herpesviruses) and HHV-6 (β$_2$-herpesvirus). These sequences, in conjunction with a large amount of information on gene function, provide a firm basis on which evolutionary relationships can be assessed independently of biological criteria.

Analysis of the DNA sequences has facilitated a fairly comprehensive description of genetic content, and this has been backed up by ancillary data on gene expression, particularly for the more intensively studied viruses such as HSV-1 (McGeoch & Schaffer, 1993). Thus, herpesvirus genomes contain between 70 and 200 genes arranged about equally between the two coding strands. The great majority of the DNA codes for protein, but overlap between coding regions in different reading frames is rare. Most genes are transcribed by cellular RNA polymerase II as single exons from their own promoters, but it is common for families of adjacent genes in the same orientation to share a polyadenylation site downstream from the most 3′ member (Wagner, 1985). Only four of the 73 distinct HSV-1 protein-coding genes are spliced: two in their protein-coding regions and two in their 5′-non-coding leader sequences. The family of HSV-1 latency-associated transcripts, which are thought not to encode protein, are also spliced (Rock, 1993). In contrast, splicing is believed to be more common in HCMV, but many details are yet to be determined (Chee *et al.*, 1990). The majority of splicing in EBV specifically involves genes expressed during latency (Kieff & Liebowitz, 1990). Most strikingly, a family of latent cycle genes dispersed over 100 kbp is expressed by differential long-range splicing of transcripts from a common promoter.

Gene functions

Gene functions have been assigned from experimental data and from comparisons with genes of known function from other organisms. Herpesviruses are so widely diverged that the DNA sequences even of viruses in the same subfamily often show little similarity. The primary level of conservation is in protein function, and this is most usefully reflected for analytical purposes at the level of primary amino acid sequence. Gene functions may be considered in five categories, as illustrated for HSV-1 and other α-herpesviruses (McGeoch & Schaffer, 1993).

Firstly, herpesviruses encode proteins responsible for control of the replicative cycle. These include proteins that exert their influence at the transcriptional level in order to achieve coordinated control of gene expression (Everett, 1987), and proteins that modulate processes in the infected cell to facilitate viral replication, for example by modifying target cellular or viral proteins or by shutting off host cell macromolecular synthesis. A second set of seven genes encodes essential components of the DNA replicative machinery, including DNA polymerase, an associated subunit which promotes processivity, a protein which recognizes the origins of viral DNA replication, a single-stranded DNA-binding protein and three constituents of a helicase-primase complex (Challberg, 1991). Thirdly, several enzymes are involved in metabolism of DNA or nucleotides, including thymidine kinase, uracil-DNA glycosylase, dUTPase, ribonucleotide reductase (two subunits), a deoxyribonuclease and thymidylate synthase (in VZV and the γ_2-herpesviruses) (Morrison, 1991). Fourthly, about half of the HSV-1 genetic complement encodes virion proteins, including components of the capsid, tegument and envelope (Rixon, 1993). Lastly, some proteins are critical for pathogenesis or latency, and their absence may have little effect on the growth of virus in cell culture but may profoundly reduce virus virulence *in vivo*. This category also includes proteins which modulate the immune response of the host; these appear to be more numerous in the β- and γ-herpesviruses. It should be noted that certain proteins may have more than one function. The virion transactivator of HSV-1 is a salient example, since in addition to its role in activating immediate early transcription via interaction with cellular components, it is also an essential structural component of the tegument (O'Hare, 1993).

Genetic relationships

When the genetic contents of sequenced herpesviruses are compared, it is clear that members of the same subfamily share the great majority of genes in a very similar layout. For example, 64 of the 69 distinct VZV genes have counterparts, with varying degrees of relationship, at equivalent locations in the HSV-1 genome (McGeoch & Schaffer, 1993). Similarly, the majority of HVS genes have collinear counterparts in the EBV genome (Albrecht *et al.*, 1992). When, however, members of different subfamilies are compared, the degree of divergence is much greater, each subfamily having a sizeable subset of 'unique' genes. None the less, a set of about 40 'core' genes is conserved in all subfamilies (Davison & Taylor, 1987; Chee *et al.*, 1990; Albrecht *et al.*, 1992). Core genes are not confined to a particular functional group, and include members of most of the groups discussed above. Only one is known to be transcribed as a spliced mRNA.

Genetic comparisons largely support the biological classification of the herpesviruses, but with a few exceptions. MDV and the related herpesvirus of turkey were originally classified on the basis of their lymphotropic behaviour as members of the *Gammaherpesvirinae* (Roizman *et al.*, 1981). It is now clear from sequence data that they are α-herpesviruses (Buckmaster *et al.*, 1988). HHV-6 would also fit into the *Gammaherpesvirinae* biologically, but genetically is a β_1-herpesvirus (Lawrence *et al.*, 1990). These exceptions merely indicate that lymphotropism is not a property solely of the *Gammaherpesvirinae*. CCV was originally classified as a member of the *Alphaherpesvirinae*, but, as discussed below, it is clear that this virus does not fit into any of the present subfamilies of the Herpesviridae.

Like core genes, unique genes also have various functions, and presumably contribute to survival of the virus in its particular ecological niche. It is striking, however, that most genes involved in control processes are specific to each subfamily, and it is likely that each subfamily (and, in some cases, subdivisions within it) possesses a set of separately acquired latency genes. Thus, counterparts to EBV (γ_1) latency genes are not present in the α- and β-herpesviruses, and are absent even from HVS (γ_2). This intriguing conclusion, coupled with the observation that latency appears to be a feature of all herpesviruses, indicates that herpesviruses have encountered a strong evolutionary pressure to acquire and maintain latency systems suitable to the particular niche they occupy.

Many herpesvirus genes are not absolutely required for growth in cell culture; about half of the genetic complement of HSV-1 falls into this category (Roizman & Sears, 1990; McGeoch & Schaffer, 1993). Deletion of some of these genes is severely deleterious to pathogenicity, but removal of others seems not greatly to affect virulence in animal models. Obviously, these 'non-essential' genes are required for the life cycle in the natural host and for transmission in the long term, and experimental results from cell culture or animal models serve to highlight the limitations of these systems for studying herpesvirus genetics.

Mechanisms of divergence

Herpesviruses appear to have explored most major pathways to generate their characteristic divergence, including nucleotide substitution (and, by extension, small insertions and deletions) to modify protein functions or generate genes *de novo*, and recombinatory processes facilitating duplication, capture and rearrangement of genes.

Nucleotide substitution

One of the striking characteristics of herpesvirus DNAs is their wide range of base compositions. G+C contents vary from 32–74%, that is, the full range exhibited by most genomes, viral or otherwise (Honess, 1984). Genome composition does not seem to bear any relationship to biological characteristics or classification; indeed, viruses that are in the same subfamily and share substantial sets of properties may have widely different G+C contents. For example, HSV-1 and VZV are neurotropic α-herpesviruses which infect humans and share very similar gene layouts, and yet their genomes contain 68% and 46% G+C, respectively. Canine herpesvirus 1 and PRV are also α-herpesviruses, but their genomes contain 32% and 74% G+C, respectively. Clearly, certain herpesviruses have experienced strong pressures to extreme G+C contents at some point during their evolution (and may still be experiencing it), while retaining closely similar gene functions. These pressures have not been identified, but hypotheses have been put forward based on biased incorporation of nucleotides by the viral DNA polymerase or differing sizes of nucleotide pools (Honess, 1984).

In addition to their widely differing mean base compositions, some herpesvirus genomes exhibit a large degree of internal heterogeneity. Unique regions often differ markedly from flanking repeat regions in G+C content. The HVS DNA molecule, as a representative of group 2 genomes, comprises a unique region of 35% G+C flanked by non-coding tandem direct repeats at the genome termini of 71% G+C (Bornkamm *et al.*, 1976; Albrecht *et al.*, 1992). The inverted repeat regions (particularly TR_S and IR_S) in α-herpesviruses genomes, which are in groups 5 and 6, are significantly higher in G+C residues than the unique regions, regardless of mean base composition (Fig. 20.2; Davison & Scott, 1986; McGeoch *et al.*, 1988; Telford *et al.*, 1992). The observation that higher G+C content is not confined to coding regions indicates that the evolutionary pressures have not been exerted through the translational machinery (for instance, through the availability of tRNAs), but have operated directly on the DNA sequence (McGeoch *et al.*, 1986). The group 6 genome of HCMV (a

β_1-herpesvirus) also shows a higher G+C content in TR$_S$/IR$_S$ (Chee *et al.*, 1990). The nature of forces that have operated more strongly on repeat regions is unknown, but it is possible that biased gene conversion and recombination have promoted fixation of G:C base pairs (McGeoch *et al.*, 1986).

The way in which extreme G+C compositions have been attained in protein-coding regions has involved, as might be expected, substitutions of those codon positions that least affect coding potential. McGeoch *et*

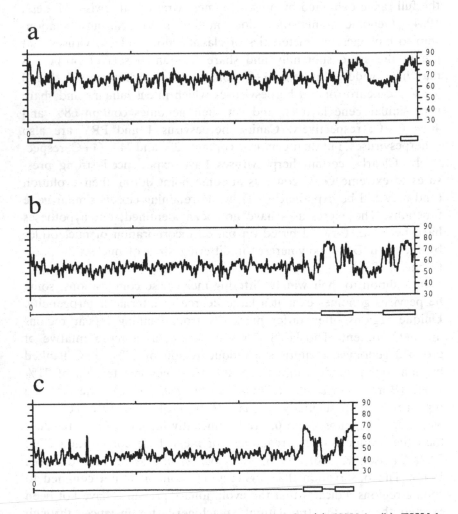

Fig. 20.2. Distribution of G+C residues along the (a) HSV-1, (b) EHV-1 and (c) VZV genomes. The percentage of G+C residues (vertical scale) was calculated for a window of 500 bp shifted by 100 bp intervals.

al. (1986) showed that a large protein-coding region in HSV-1 TR_S/IR_S exhibits preferential partition of G+C residues into the third and, to a lesser extent, first codon positions. Even so, the encoded protein is very rich in amino acids specified by G+C-rich codons (particularly alanine). This coding region contains 81.5% G+C, a composition which is likely to be at or near the maximum that can be attained while still retaining protein function.

Methylation of CG dinucleotides may affect base composition during evolution, as methylated C residues may be deaminated to U residues, leading to gradual replacement of CG by TG dinucleotides (Honess *et al.*, 1989). General CG depletion is a feature of γ-herpesvirus genomes, indicating that they are maintained as methylated plasmids during the latent cycle in lymphocytes. The genomes of viruses in the other subfamilies show no or only local CG depletion, presumably because methylation does not occur or is localized.

Gene duplication

Gene duplication followed by divergence and eventual partitioning of function has occurred to varying extents in different herpesvirus lineages. Examples are rare among the smaller α- and γ-herpesviruses, and more abundant in the larger $β_1$-herpesviruses. The HCMV genome contains nine sets of related genes that have probably arisen by gene duplication (Chee *et al.*, 1990): three consist of pairs of genes and six of larger families. There is also evidence that gene duplication has occurred in the U_S region of α-herpesvirus genomes (McGeoch, 1990*a*). HSV-1 and HSV-2 U_S contains five tandemly oriented genes (*US4*, *US5*, *US6*, *US7* and *US8*) which are specific to α-herpesviruses and encode glycoproteins gG, gJ, gD, gI and gE, respectively. Common structural elements among these glycoproteins are not generally apparent, but the similar arrangement of four conserved cysteine residues in gG (in HSV-2), gD and gI strongly suggests that *US4*, *US6* and *US8* have arisen by gene duplication. It has recently been demonstrated that these residues form two disulphide bridges which presumably form a conserved element in the three glycoprotein families (Long *et al.*, 1992).

Studies of individual genes have revealed processes that may occur subsequent to gene duplication. There is a hint that gene duplication was followed by inversion of one of the copies in the development of the two divergently transcribed genes *1* and *2* at the left end of the

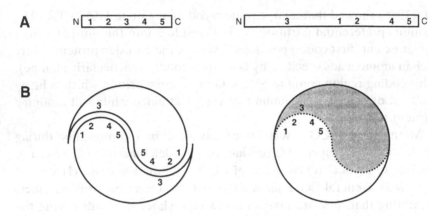

Fig. 20.3. Arrangement of motifs and model for dUTPase quaternary structure. (A) The linear arrangements of motifs in the E.coli and retroviral sequences is indicated on the left, and the herpesvirus arrangement on the right. (B) The left cartoon presents a model for the E. coli type of dUTPase. The active enzyme is shown as a dimer with two active sites, each composed of motifs 1, 2, 4 and 5 from one monomer and motif 3 from the other. The right cartoon represents a herpesvirus dUTPase monomer, with folding corresponding to the E. coli dimer; the N-terminal region is shaded. Reproduced from McGeoch (1990b) with permission.

EHV-1 genome (Telford et al., 1992). The VZV and HSV-1 lineages contain only one copy of potential counterparts of this gene (genes 2 and UL56, respectively) in opposite orientations with respect to each other. The presence of two large related genes in opposite orientations at opposite ends of the HVS genome suggests that similar processes may have occurred in this virus. McGeoch (1990b) suggested that gene duplication was followed by fusion in the development of the deoxyuridine triphosphatase (dUTPase) gene in mammalian herpesviruses, since the encoded proteins of α- and γ-herpesviruses are about twice the size of dUTPases in other organisms and have four conserved motifs in the C-terminal half and one in the N-terminal half (Fig. 20.3). This arrangement suggests that gene duplication and fusion were followed by differential loss of active site motifs in the two halves of the protein, so that the present-day pseudo-dimeric protein contains a single active site in comparison with the proposed two active sites of other dUTPase dimers. The model should be updated, since crystallographic analysis of the Escherichia coli enzyme shows that it is actually a homotrimer (Cedergren-Zeppezauer et al., 1992), but the principle remains valid.

Gene capture

The observation that all herpesviruses have several genes with clear cellular homologues indicates that gene capture via reverse transcription of cellular mRNAs or from genomic DNA has been employed on many occasions (in some cases more than once for a particular cellular gene) during evolution of the family. Some of these genes, being present in all herpesviruses and fundamental to their existence (e.g. that encoding DNA polymerase), are ancient components whose origins are hidden deep in the mists of time, and others appear to have been acquired more recently in certain herpesvirus lineages, where the mist is thinner. Most captured genes encode enzymes involved in nucleotide metabolism, such as DNA polymerase, thymidine kinase, ribonucleotide reductase, dUTPase, uracil-DNA glycosylase, thymidylate synthase (in VZV and the γ_2-herpesviruses), dihydrofolate reductase (in HVS), DNA helicases and protein kinases, or genes whose products probably function in immune modulation or pathogenesis, such as the HSV-1 *RL1* gene which encodes a protein similar to a cellular protein involved in differentiation (McGeoch & Barnett, 1991), the HVS genes whose products are homologous to complement control proteins (Albrecht *et al.*, 1992) and the HHV-6 gene which is related to an adeno-associated virus type 2 regulatory gene (Thomson, Efstathiou & Honess, 1991). The origins of genes encoding structural proteins, however, remain obscure as no cellular counterparts have been identified. Some may have been acquired from the host and have diverged drastically to fit their viral function, and some may have developed *de novo* in the viral genome.

Gene rearrangement

Evidence that genes have been rearranged during evolution of the α-herpesviruses comes from studies of the U_S region (Davison & McGeoch, 1986). Comparisons of HSV-1 and VZV indicate that TR_S/IR_S is capable of expanding or contracting during evolution, resulting in transfer of genes from or into U_S and even of genes from one end of U_S to the other. Thus, it is possible to relate these regions of the two genomes by a series of recombinational steps, as shown in Fig. 20.4. Data for other α-herpesviruses, including EHV-1, PRV and MDV are in full accord with this mechanism (for example, see Telford *et al.*, 1992).

Evidence is also available that the large-scale gene layout has been rearranged. A substantial part of PRV U_L is inverted with respect to other α-herpesviruses (Davison & Wilkie, 1983). The situation is more marked when viruses from different subfamilies are considered. Davison and Taylor (1987) noted that core genes of VZV and EBV are arranged in three rearranged blocks in U_L. Similarly, Chee *et al.* (1990) reported that core genes in EBV and HCMV are arranged in seven blocks. Thus, it appears that the most ancient part of present-day herpesvirus genomes

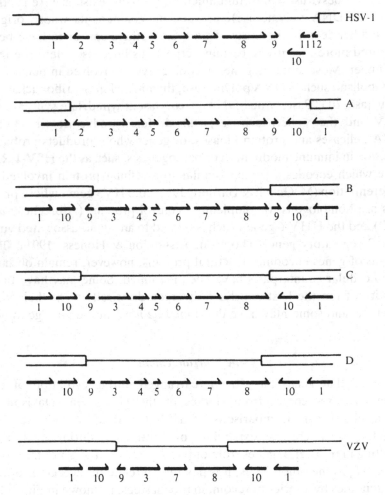

Fig. 20.4. Descent of the U_s regions of HSV-1 and VZV from that of an ancestral herpesvirus. The arrows indicate protein-coding regions numbered according to the HSV-1 *US* gene nomenclature. No commitment to direction in time has been made. Reproduced from Davison and McGeoch (1986) with permission.

is contained in the U_L region, and that large blocks of genes have been rearranged.

Fish herpesviruses

Although herpesviruses with mammalian or avian hosts represent a diverse evolutionary group, they are clearly related. This is taken as compelling evidence for their descent from a common ancestor. However, the recent analysis of the DNA sequence of CCV, a herpesvirus with a lower vertebrate host, poses an enigma (Davison, 1992). This agent has virions with a herpesvirus morphology, a group 1 genome similar in size to that of other herpesviruses, and certain aspects of gene arrangement (distribution of genes between the two DNA strands, degree of overlap and presence of 3'-coterminal gene families) like those of other sequenced herpesviruses. However, with the exception of a few captured genes, CCV bears no convincing genetic relationship to other herpesviruses. This leads to two possible conclusions. First, the two groups of viruses may have evolved by completely separate pathways. This is rather difficult to swallow, as CCV particles have the complex morphology of herpesvirions. Alternatively, mammalian/avian and fish herpesviruses may have evolved from a common source, but have diverged so far that a genetic relationship is no longer detectable. In this framework, however, the mammalian and avian viruses appear to be much more closely related than would be expected if they had co-evolved with their hosts. This could be resolved if interspecies transmission of α-herpesviruses has occurred, but would imply at least two separate events, since MDV and herpesvirus of turkeys are more closely related to HSV-1 than is infectious laryngotracheitis virus (Buckmaster *et al.*, 1988; Griffin, 1989).

Preliminary studies of another fish herpesvirus, salmonid herpesvirus 1 (SaHV-1), indicate a distant relationship to CCV but not to mammalian/avian herpesviruses (A.J. Davison, unpublished data). Thus, it is likely that CCV is a member of a group of relatively unstudied viruses comparable in genetic complexity and evolutionary breadth to the mammalian/avian herpesviruses.

Origins

As with all evolutionary studies, questions of distant origins are the most difficult to tackle, particularly in the absence of a viral fossil

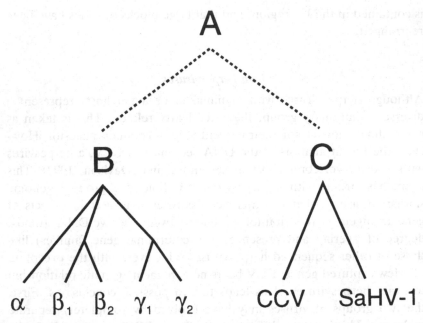

Fig. 20.5. A scheme for the evolution of present day herpesviruses from ancestors B and C, which may in turn have originated from ancestor A.

record; the more so since, as with other viruses, most significant events in herpesvirus evolution appear to have taken place a very long time ago. The possibility of setting schemes such as that shown in Fig. 20.5 within a reliable chronology is currently beyond our grasp. It is clear that mammalian/avian herpesviruses are related and have descended from an ancestral herpesvirus (B in Fig. 20.5) which itself was probably a recognizable herpesvirus with at least 40 genes. Fish herpesviruses have evolved from another progenitor (C in Fig. 20.5) of similar complexity. Nevertheless, if the mammalian/avian and fish herpesviruses share evolutionary origins, this places their ancestor (A in Fig. 20.5) much further in the past.

The available data are generally consistent with the hypothesis that radiation of the herpesviruses has followed that of their hosts. This notion might eventually help in assigning a chronology to herpesvirus evolution, even in indicating at what stage major radiations occurred, most notably the divergence of the α, β and γ lineages. It does not help us at present to rationalize the vast distance between the fish and mammalian/avian viruses or to speculate about the form of ancient ancestors. It is likely that additional studies of herpesviruses with amphibian, reptilian or (if any are found) invertebrate hosts will help us to decide whether all herpesviruses evolved from a common source. Nevertheless, it is difficult to see how we can even guess at, let alone determine, the evolutionary chronology and genetic complement of ancestor A and its (presumably simpler) antecedents by continued examination of present-day herpesviruses.

Our present state of knowledge, then, allows us only to speculate generally on the steps that may have occurred in early evolution of the herpesviruses. Important questions include the nature of the linkage between herpesvirus morphology and genetic complement, and the stage at which latency first entered the herpesvirus lifestyle. It is notable that the morphology of present-day herpesviruses is remarkably constant: virions contain at least ten viral proteins in the lipid envelope, perhaps 30 in the tegument and seven in the capsid. The documented sizes of mature or defective herpesvirus genomes indicate that at least 100 kbp of DNA are packaged into capsids (even if packaging signals are more frequent), and up to 240 kbp can be accommodated. We must assume that evolutionary ancestors preceding ancestor A were not as complex, that they had simpler structures and smaller genomes. Thus, we may envisage an early ancestor which contained fewer than ten genes encoding proteins involved in some aspects of control, DNA replication and capsid structure (as with present day smaller DNA viruses), but which relied on cellular proteins for other control and replication functions and lacked the tegument and envelope. At this stage, the genome may have been accommodated in a smaller capsid with a different morphology. Later stages include addition of genes encoding further control, replication and capsid proteins, acquisition of the herpesvirus capsid morphology, and development of the tegument and envelope and their associated proteins. Also, at some stage genes involved in latency were first added to the repertoire. Clearly, biological and genomic features of herpesviruses have been obtained in many stages, some more than once and some having profound influences on subsequent evolution. It is likely that if we knew details of what really occurred, even our most daring speculations would demonstrably lack imagination.

References

Agius, C.T., Nagesha, H.S. & Studdert, M.J. (1992). *Virology*, **191**, 176–86.
Albrecht, J-C., Nicholas, J., Biller, D., Cameron, K.R., Biesinger, B., Newman, C., Wittmann, S., Craxton, M.A., Coleman, H., Fleckenstein, B. & Honess, R.W. (1992). *J. Virol.*, **66**, 5047–58.

Baer, R., Bankier, A.T, Biggin, M.D., Deininger, P.L., Farrell, P.J., Gibson, T.J., Hatfull, G., Hudson, G.S., Satchwell, S.C., Séguin, C., Tuffnell, P.S. & Barrell, B.G. (1984). *Nature*, **310**, 207–11.

Ben-Porat, T., Rixon, F.J. & Blankenship, M.L. (1979). *Virology*, **95**, 285–94.

Booy, F.P., Newcomb, W.W., Trus, B.L., Brown, J.C., Baker, T.S. & Steven, A.C. (1991). *Cell*, **64**, 1007–15.

Bornkamm, G.W., Delius, H., Fleckenstein, B., Werner, F-J. & Mulder, C. (1976). *J. Virol.*, **19**, 154–61.

Browning, G.F. & Studdert, M.J. (1989). *Arch. Virol.*, **104**, 77–86.

Buchman, T.G. & Roizman, B. (1978). *J. Virol.*, **25**, 395–407.

Buckmaster, A.E., Scott, S.D., Sanderson, M.J.S., Boursnell, M.E.G., Ross, N.L.J. & Binns, M.M. (1988). *J. Gen. Virol.*, **69**, 2033–42.

Cebrian, J., Berthelot, N. & Laithier, M. (1989). *J. Virol.*, **63**, 523–31.

Cedergren-Zeppezauer, E.S., Larsson, G., Nyman, P.O., Dauter, Z. & Wilson, K.S. (1992). *Nature*, **355**, 740–3.

Challberg, M. (1991). *Seminars in Virology*, **2**, 247–56.

Chee, M.S., Bankier, A.T., Beck, S., Bohni, R., Brown, C.M., Cerny, R., Horsnell, T., Hutchison, C.A. III, Kouzarides, T., Martignetti, J.A., Preddie, E., Satchwell, S.C., Toplinson, P., Weston, K.M. & Barrell, B.G. (1990). *Curr. Top. Microbiol. Immunol.*, **154**, 125–69.

Chousterman, S., Lacasa, M. & Sheldrick, P. (1979). *J. Virol.*, **31**, 73–85.

Davison, A.J. (1984). *J. Gen. Virol.*, **65**, 1969–77.

Davison, A.J. (1992). *Virology*, **186**, 9–14.

Davison, A.J. & McGeoch, D.J. (1986). *J. Gen. Virol.*, **67**, 597–611.

Davison, A.J. & Scott, J.E. (1986). *J. Gen. Virol.*, **67**, 1759–816.

Davison, A.J. & Taylor, P. (1987). *J. Gen. Virol.*, **68**, 1067–79.

Davison, A.J. & Wilkie, N.M. (1983). *J. Gen. Virol.*, **64**, 1927–42.

Dumas, A.M., Geelen, J.L.M.C., Weststrate, M.W., Wertheim, P. & van der Noordaa (1981). *J. Virol.*, **39**, 390–400.

Efstathiou, S., Ho, Y.M. & Minson, A.C. (1990). *J. Gen. Virol.*, **71**, 1355–64.

Everett, R.D. (1987). *Anticancer Res.*, **7**, 589–604.

Farley, C.A., Banfield, W.G., Kasnic, G. & Foster, W.S. (1972). *Science*, **178**, 759–60.

Given, D. & Kieff, E. (1979). *J. Virol.*, **31**, 315–24.

Griffin, A.M. (1989). *J. Gen. Virol.*, **70**, 3085–9.

Honess, R.W. (1984). *J. Gen. Virol.*, **65**, 2077–107.

Honess, R.W., Gompels, U.A., Barrell, B.G., Craxton, M., Cameron, K.R., Staden, R., Chang, Y-N. & Hayward, G.S. (1989). *J. Gen. Virol.*, **70**, 837–55.

Kieff, E. & Liebowitz, D. (1990). In *Fields Virology*, 2nd edn., B.N. Fields and D.M. Knipe (eds.). Raven Press, New York, pp. 1889–1920.

Lawrence, G.L., Chee, M., Craxton, M.A., Gompels, U.A., Honess, R.W. & Barrell, B.G. (1990). *J. Virol.*, **64**, 287–299.

Long, D., Wilcox, W.C., Abrams, W.R., Cohen, G.H. & Eisenberg, R.J. (1992). *J. Virol.*, **66**, 6668–85.

McGeoch, D.J. (1990*a*). *J. Gen. Virol.*, **71**, 2361–7.

McGeoch, D.J. (1990*b*). *Nucl. Acids Res.*, **18**, 4105–10.

McGeoch, D.J. & Barnett, B.C. (1991). *Nature*, **353**, 609.

McGeoch, D.J., Barnett, B.C. & MacLean, C.A. (1993). *Seminars in Virology*, **4**, 125–34.

McGeoch, D.J., Dalrymple, M.A., Davison, A.J., Dolan, A., Frame, M.C., McNab, D., Perry, L.J., Scott, J.E. & Taylor, P. (1988). *J. Gen. Virol.*, **69**, 1531–74.

McGeoch, D.J., Dolan, A., Donald, S. & Bauer, D.H.K. (1986). *Nucl. Acids Res.*, **14**, 1727–45.

McGeoch, D.J. & Schaffer, P.A. (1993). In *Genetic Maps*, 6th edn., S. O'Brien (ed.). Cold Spring Harbor Press, New York, pp. 1.147–1.156.

Marks, J.R. and Spector, D.H. (1988). *Virology*, **162**, 98–107.

Martin, M.E.D., Thomson, B.J., Honess, R.W., Craxton, M.A., Gompels, U.A., Liu, M.Y., Littler, E., Arrand, J.R., Teo, I. & Jones, M.D. (1991). *J. Gen. Virol.*, **72**, 157–68.

Matthews, R.E.F. (1982). *Intervirology*, **17**, 1–199.

Minson, A.C. (1989). In *Andrewes' Viruses of Vertebrates*, 5th edn., J.S. Porterfield (ed.). Baillière Tindall, London, pp. 293–332.

Mocarski, E.S. & Roizman, B. (1982). *Cell*, **31**, 89–97.

Morrison, J.M. (1991). *Virus Induced Enzymes*. Wiley, Chichester.

O'Hare, P. (1993). *Seminars in Virology*, **4**, 145–55.

Rixon, F.J. (1993). *Seminars in Virology*, **4**, 135–44.

Rock, D.L. (1993). *Seminars in Virology*, **4**, 157–65.

Roizman, B., Carmichael, L.E., Deinhardt, F., de-The, G., Nahmias, A.J., Plowright, W., Rapp, F., Sheldrick, P., Takahashi, M. & Wolf, K. (1981). *Intervirology*, **16**, 201–17.

Roizman, B., Desrosiers, R.C., Fleckenstein, B., Lopez, C., Minson, A.C. & Studdert, M.J. (1992). *Arch. Virol.*, **123**, 425–49.

Roizman, B. & Sears, A.E. (1990). In *Fields Virology*, 2nd edn., B.N. Fields and D.M. Knipe (eds.). Raven Press, New York, pp. 1795–1841.

Sheldrick, P. & Berthelot, N. (1974). *Cold Spring Harbor Symp. Quant. Biol.*, **39**, 667–78.

Telford, E.A.R., Watson, M.S., McBride, K. & Davison, A.J. (1992). *Virology*, **189**, 304–16.

Thomson, B.J., Efstathiou, S. & Honess, R.W. (1991). *Nature*, **351**, 78–80.

Wagner, E.K. (1985). In *The Herpesviruses*, vol. 3, B. Roizman (ed.). Plenum Press, New York, pp. 45–104.

Weststrate, M.W., Geelen, J.L.M.C. & van der Noordaa, J. (1980). *J. Gen. Virol.*, **49**, 1–22.

Whalley, J.M., Robertson, G.R. & Davison, A.J. (1981). *J. Gen. Virol.*, **57**, 307–23.

Whitley, R.J. (1990). In *Fields Virology*, 2nd edn., B.N. Fields, D.M. Knipe (eds.). Raven Press, New York, pp. 2063–2075.

Wittmann, G. (1989). In *Herpesvirus Diseases of Cattle, Horses, and Pigs*, G. Wittmann (ed.). Kluwer Academic Publishers, Boston, pp. 163–175.

21

Aphthovirus evolution

J. DOPAZO, M.J. RODRIGO, A. RODRÍGUEZ,
J.C. SÁIZ AND F. SOBRINO

Introduction

The *Aphthovirus* genus, family *Picornaviridae*, is composed of seven distinct serotypes which cause foot-and-mouth disease (FMD), a severe disease of cloven-hoofed animals. FMD has serious economic consequences (Pereira, 1981) and is enzootic in most South American and African countries, as well as in regions of Asia and the Middle East. Foot-and-mouth disease virus (FMDV) usually causes a systemic acute infection with high morbidity and low mortality (Shanan, 1962). In ruminants, the virus may also produce an asymptomatic, persistent infection that involves limited viral amplification (Van Bekkum *et al.*, 1959). This type of infection has been proposed as an epidemiologically important reservoir of FMDV (Hedger & Condy, 1985).

The general structure and molecular features of FMDV are, in general, similar to those of other picornaviruses (Domingo *et al.*, 1990; Stanway, 1990). The capsid is composed of 4 proteins (VP1–4) and includes an RNA molecule of about 8500 nucleotides in length which encodes the structural proteins and at least 11 different, mature, non-structural polypeptides. The antigenic structure of FMDV includes continuous and discontinuous neutralizing epitopes located in exposed regions of the viral capsid, in which one or more of the capsid proteins, particularly VP1, are involved (Domingo *et al.*, 1990).

Aphthoviruses show considerable antigenic diversity; 7 serotypes, more than 65 subtypes and a multitude of variants have been identified mainly by *in vitro* cross-serum neutralization (Pereira, 1981) and, more recently, by the use of monoclonal antibodies (MAbs) (Domingo *et al.*, 1990). Immunization with viruses of one type does not confer protection against viruses of other serotypes, whereas cross-protection within serotypes is not always complete (Kitching *et al.*, 1989).

Results obtained during the last decade have strongly supported the

310

hypothesis that antigenic diversity is based on high genetic heterogeneity of FMDV populations. These data have contributed to an extension of the quasi-species concept to populations of RNA viruses (Chapter 13). Here we review what is known about the population structure of FMDV and the major evolutionary features of this pathogen.

The Aphthovirus genus

Aphthovirus is a well-defined taxonomic group within the *Picornaviridae*. After the first phylogenetic analysis of this family (Palmenberg, 1989) that modified its classical taxonomic grouping (Stanway, 1990; Rodrigo & Dopazo, 1994), aphthoviruses were the only picornavirus group that maintained its identity. A distinctive feature of FMDV RNA is that the G+C content of the third codon positions is higher than in the other picornaviruses, and more like the codon composition of mammalian genes (Aota & Ikemura, 1986; Marín *et al.*, 1988). Again, this appears to reinforce the taxonomic uniqueness of the aphthoviruses among the *Picornaviridae*.

Microevolution: High potential for virus variation

Genetic and antigenic heterogeneity of FMDV populations, as well as high rates of fixation of mutations, have been observed in populations derived from cloned viruses and then maintained as acute or persistent infections in cell culture. After about 30 independent serial passages of a plaque-purified virus, the average RNA nucleotide sequence of each population differed from the others in 14 to 57 nucleotide positions (Sobrino *et al.*, 1983). An analysis of individual clones showed that each infectious RNA deviated from the parental sequence by an average of two to eight mutations. The antigenic heterogeneity of FMDV populations was revealed by the high frequency of isolation (around 2×10^{-5} mutants per wild-type virus) of variant viruses resistant to MAbs estimated after minimal amplification of cloned viruses (Martínez *et al.*, 1991). Propagation of these heterogeneous populations can result in a rapid emergence of antigenic variants (Bolwell *et al.*, 1989; Díez, Mateu & Domingo, 1989; González *et al.*, 1991).

The analysis of viruses recovered from serial passages of BHK-21 cells persistently infected with FMDV demonstrated a gradual accumulation of nucleotide substitutions and phenotypic changes, as well as the rapid generation of heterogeneity (de la Torre *et al.*, 1985, 1988, 1989a, b; Díez et al., 1990). A co-evolution of both viruses, which showed increased virulence for the parental BHK-21 cells, and cells, which became progressively resistant to the initial virus, was documented in great detail (de la Torre et al., 1988, 1989a,b).

Similar results on genetic and phenotypic heterogeneity of FMDV were obtained after infection of swine with plaque-purified virus (Carrillo et al., 1990b). Likewise, an analysis of persistently infected cattle revealed the heterogeneity of the viral populations recovered over an 18-month period (Costa-Giomi et al., 1984; Gebauer et al., 1988). Nucleotide sequence comparison of the VP1 gene of sequential isolates from carrier animals allowed an estimate of a rate of fixation of mutations during such inapparent infection. It ranged from 0.9×10^{-2} to 7.4×10^{-2} substitutions per site and year (s/s and year) (Gebauer et al., 1988), similar to values reported for persistent infections of other viruses (Domingo & Holland, 1993). Thus, as for other RNA viruses there is good evidence that FMDV produces during its replication a high frequency of mutants which provide for a great potential for virus evolution and adaptability to different environmental conditions (Holland et al., 1982; Domingo et al., 1985, 1990, 1992; Domingo & Holland, 1988, 1993).

Evolution of FMDV in the field

Serotype comparison

The overall RNA sequence homology, estimated by nucleic acid competition-hybridization assays, ranged from 44 to 70% between RNA from viruses of serotypes A, O, or C, to greater than 70% within types A and O (Dietzshold et al., 1971; Robson, Harris & Brown, 1977). The considerable number of VP1 RNA sequences have been used to derive representative phylogenetic trees relating viruses of serotypes A, O, and C (Dopazo et al., 1988; Palmenberg, 1989; and Fig. 21.1), which showed a correlation between classical serological grouping (serotypes) and well-defined genetic groups. Assuming an overall constancy in the evolutionary rate along the evolutionary history of aphthoviruses (Martínez et al., 1992), an early lineage diversification was identified (Fig. 21.1), which gave rise to the three South African serotypes (SAT1, SAT2 and SAT3) as well as to a lineage which was the precursor of the present day A, C, O and Asia 1 serotypes. A further diversification which produced the different Euro-Asiatic serotypes (A, C, O and Asia

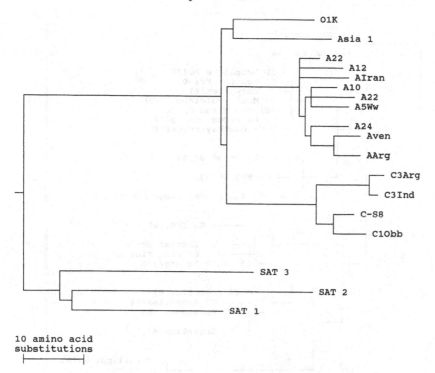

10 amino acid
substitutions

Fig. 21.1. Phylogenetic tree relating viruses from the seven aphthovirus serotypes, in which two main lineage diversification events, leading to serotype diversification, can be seen. The tree was constructed using the Neighbour-Joining method (Saitou & Nei, 1987), and VP1 amino acid sequences as data. The reliability of the tree was tested using the bootstrap procedure (Efron, 1982; Dopazo, 1994). The tree was rooted using sequences of other picornaviruses as *outgroups*.

1) circulating at present seems to have occurred more recently. These two major diversification events are likely to have taken place in a relatively short span of time (Fig. 21.1 and Martinez *et al.*, 1992).

Serotype diversification

A detailed view of the genetic diversification of FMDV type C over six decades has shown six major evolutionary lineages, and a complex network of sublines (Fig. 21.2 and Martínez *et al.*, 1992). A similar trend, with a high number of co-circulating lineages, has been found in comparing recent isolates of serotypes A and O from Asia and the Middle East

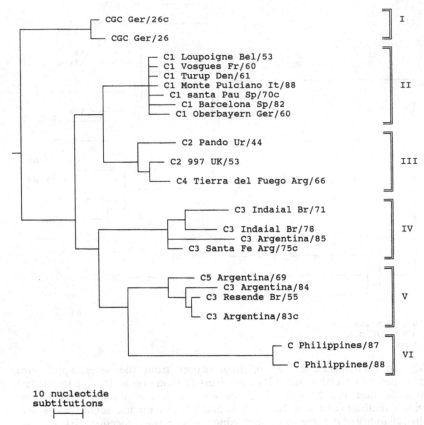

Fig. 21.2. Phylogenetic tree obtained for a representative sample of FMDV type C isolates over 60 years. The tree was constructed as indicated in Fig. 21.1, except that nucleotide sequences of the VP1 gene were used as data. The main evolutionary groups (I to VI) are indicated.

(Armstrong *et al.*, 1991; Saiz *et al.*, 1993). In contrast, European isolates are considerably more homogeneous (Fig. 21.2), probably because vaccines have reintroduced viruses (Beck & Strohmaier, 1987; Carrillo *et al.*, 1990*a*), although other reservoirs cannot be completely excluded (Domingo *et al.*, 1992). Occasionally, episodes of long-term genetic conservation (stasis) of C type South American viruses, for which a vaccine origin was not evident, have also been described (Piccone *et al.*, 1988).

When the viruses isolated in Europe after compulsory vaccination, started around 1965, are excluded, an analysis of the fixation of mutations with time gives a relatively constant evolutionary rate (1.43±0.16 ×

10^{-3} s/s and year) (Martínez *et al.*, 1992). Interestingly, while a linear accumulation of silent mutations with time was noticed, a fluctuation among a limited number of amino acid substitutions without net accumulation of amino acid replacements was apparent for the evolution of FMDV type C over 60 years (Martínez *et al.*, 1992).

Evolution during an epizootic

Due to the fast evolutionary rate exhibited by FMDV in experimental *in vivo* and *in vitro* models and the great heterogeneity of natural populations, a clear genetic diversification in the field is expected even during short time intervals. Indeed, sequence heterogeneity among individual cloned viruses recovered from a single animal has been documented (Domingo *et al.*, 1980; King *et al.*, 1981; Anderson *et al.*, 1985).

The analysis of C type viruses isolated in Spain from 1979 to 1982, by oligonucleotide fingerprinting of defined genomic segments, covering around 85% of the coding region, yielded rates of fixation of mutations of 4×10^{-4} to 4.5×10^{-2} s/s and year. The values varied depending on the time period and the genomic segment considered; the highest value corresponded to the RNA encoding capsid protein VP1 (Sobrino *et al.*, 1986). Even though no genomic segment remains genetically invariant, there appears to be a preferential localization of amino acid substitutions in VP1 (Martínez *et al.*, 1988; Palmenberg, 1989; Sobrino *et al.*, 1989). Several substitutions in capsid proteins are located on exposed surface domains and modify both the reactivity of the virus with neutralizing MAbs (Mateu *et al.*, 1988, 1990), and its ability to confer protection against challenge with related viruses (Martínez *et al.*, 1988). A phylogenetic tree based on VP1 gene sequences (Fig. 21.3) indicated that all the viruses isolated in Spain from 1979 until 1982 were closely related to a previous C type virus, C-S8, isolated in 1970. Two main evolutionary lineages were distinguished which corresponded to viruses isolated in different geographical areas. For both groups, a gradual accumulation of nucleotide substitutions of $5.8 \pm 0.9 \times 10^3$ s/s and year has been calculated (J. Dopazo, unpublished data).

Protein evolution

The most frequent mutational events observed during the evolution of aphthoviruses are point mutations. Insertions and deletions appear to

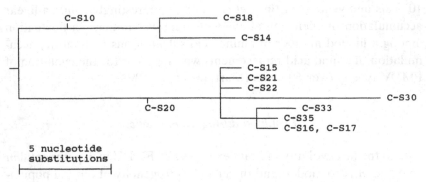

Fig. 21.3. FMDV type C evolution during an epidemic in Spain. The isolates were obtained from 1979 (C-S10) to the end of 1982 (C-S30). The tree was constructed as indicated in Fig. 21.1, except that nucleotide sequences of the VP1 gene were used as data. The tree was rooted using the sequence of C-S8 (not shown), a previous related isolate, which was being used as vaccine in 1979.

be fixed at low frequencies; they amount to less than 4% of the total number of mutations when VP1 RNA segments of viruses of serotype A, O, and C are compared, and they are restricted to the more variable regions encoding antigenic sites.

In vitro recombination of FMDV has been described and it occurs at high frequency among highly homologous strains in cell culture (King *et al.*, 1982). There has been only a recent description of a recombination event in the field (Krebs & Marquardt, 1992). The limited evidence of FMDV recombination *in vivo* may reflect the low probability of co-infection of animals with genetically (or phenotypically) distinguishable FMDV strains.

Most of the available evidence suggests that VP1 is the viral protein which contributes most to the antigenicity of FMDV (Domingo *et al.*, 1990), and therefore must constitute one of the main targets of the immune pressure of host animals. Accordingly, VP1 gene shows the highest rate of fixation of mutations (Sobrino *et al.*, 1986, 1989; Villaverde, Martínez-Salas & Domingo, 1988). Nevertheless, important constraints, probably structural requirements, may limit VP1 variation. Substitutions in VP1 occurred preferentially at a limited number of sites exposed on the virion surface (Dopazo *et al.*, 1988; Acharya *et al.*, 1989; Martínez *et al.*, 1992). Moreover, the evolution of serotype C during 60 years occurred without a net accumulation of amino acid substitutions with time, in spite of accumulation of synonymous nucleotide replacements (Martínez *et al.*, 1992). In addition, a decrease in the evolutionary rates is observed as the difference in the dates of isolation among the viruses analysed increases (compare the above-mentioned values for both the epizootic and the long-term evolution during 60 years). These observations suggest that, even in VP1, only a reduced number of sites are free to accept replacements.

Two mechanisms of diversification of a main antigenic determinant (site A) of FMDV type C VP1 have been distinguished. The most frequent mechanism involves a gradual accumulation of amino acid replacements. Alternatively, an abrupt antigenic change due to single, critical amino acid substitutions has also been documented both in the field and with several laboratory-derived mutants (Mateu *et al.*, 1990; Martínez *et al.*, 1991).

The molecular basis for the higher variation observed at a limited number of sites may rest on a combination of two factors: (1) the existence of a process of antigenic selection of variants (Carrillo *et al.*, 1989) and, (2) more relaxed constraints in the antigenic regions, which are located in exposed, flexible loops (Acharya *et al.*, 1989; Domingo *et al.*, 1992). The significance of the second factor is supported by the observation of generation and dominance of antigenic variants upon passage of virus in the absence of anti-FMDV antibodies (Díez *et al.*, 1989, 1990).

Concluding remarks

Several factors complicate the understanding of FMDV evolution over both short and long time periods. Among others, the molecular epidemiology of FMDV is affected by the following events: severe founder effects during transmission, unknown level of immunization of affected animals, large numbers of wild species affected, and frequent persistent infections in ruminants. In spite of such a complexity, it has been possible to suggest some consistent trends in the evolutionary history of the aphthoviruses. There is good evidence, at least for serotype C, of a relative overall constancy of evolutionary rates. Also, there is a sound correlation between antigenic and genetic divergence perturbed only by some critical point mutations and perhaps point deletions/insertions. The effect of the latter on antigenic specificity remains, however, largely unexplored. In addition, the available evidence supports the view that extensive vaccination of the main hosts may modify the pattern of evolution of FMDV. In fact, the incidence of the disease in Europe has been drastically reduced and most of the recorded outbreaks appear to be virus reintroductions from vaccines. These facts point to the convenience of developing synthetic vaccines, not involving live virus (Brown, 1989). The development of such vaccines faces, not only problems of efficient immunization, but also difficulties caused by the very variable nature of viral populations (Domingo, 1989). Therefore, a more extensive study of FMDV diversity in the field will not only throw light on the phylogenetic history of viruses, but also will open the possibility of a more rigorous selection of epitopes to be incorporated into synthetic vaccines, with the aim of minimizing potential breaks of immunity.

Acknowledgements

We wish to thank E. Domingo for his continuous support and for corrections of this manuscript. We are also indebted to M. A. Martínez, C. Carrillo, M. G. Mateu, A. Villaverde and A. Moya for their contributions to the study of FMDV evolution. Work supported by CICYT (Spain) and INIA (Spain).

References

Acharya, R., Fry, E., Stuart, D., Fox, G., Rowlands, D. & Brown, F. (1989). *Nature*, **337**, 709–17.

Anderson, E.C., Underwood, B.O., Brown, F. & Ngichabe, C.K. (1985). *Vet. Microbiol.*, **10**, 409–23.

Aota, S.-I & Ikemura, T. (1986). *Nucl. Acids Res.*, **14**, 6345–55.

Armstrong, R.M., Rendle, R.A.S.T., Ferris, N.P., Samuel, A.R. & Knowles, N.J. (1991). *Report of the European Commission for the Control of Foot-and-Mouth Disease, FAO.*, Rome, pp. 63–69.

Beck, E. & Strohmaier, K. (1987). *J. Virol.*, **61**, 1621–9.

Bolwell, C., Clarke, B.E., Parry, N.R., Ouldridge, E.J., Brown, F. & Rowlands, D.J. (1989). *J. Gen. Virol.*, **70**, 59–68.

Brown, F. (1989). In *Synthetic Peptides: Approaches to Biological Problems*, J.P. Tam and E.T. Kaiser (eds.). Alan R. Liss Inc., New York, pp. 127–142.

Carrillo, C., Dopazo, J., Moya, A., González, M., Martínez, M.A., Sáiz, J.C. & Sobrino, F. (1990a). *Virus Res.*, **15**, 54–6.

Carrillo, C., Plana, J., Mascarella, R., Bergadá, J. & Sobrino, F. (1990b). *Virology*, **174**, 890–2.

Carrillo, E.C., Rieder Rojas, E., Cavallaro, L., Schiappacassi, M. and Campos, R. (1989). *Virology*, **171**, 599–601.

Costa-Giomi, M.P., Bergmann, I.E., Scodeller, E.A., Augé de Mello, P., Gomes, I. & La Torre, J.L. (1984). *J. Virol.*, **51**, 799–805.

de la Torre, J.C., Dávila, M., Sobrino, F., Ortín, J. & Domingo, E. (1985). *Virology*, **145**, 24–35.

de la Torre, J.C., de a Luna, S., Díez, J. & Domingo, E. (1989a). *J. Virol.*, **61**, 2385–7.

de la Torre, J.C., Martínez-Salas, E. Díez, J. & Domingo, E. (1989b). *J. Virol.*, **63**, 59–63.

de la Torre, J.C., Martínez-Salas, E., Díez, J., Villaverde, A., Gebauer, F.,

Rocha, E., dávilia, E., Dávila, M. & Domingo, E. (1988). *J. Virol.*, **62**, 2050–8.

Dietzshold, B., Kaaden, O.R., Tokui, T. & Bohm, H.C. (1971). *J. Gen. Virol.*, **13**, 1–7.

Díez, J., Dávila, M., Escarmís, C., Mateu, M.G., Dominguez, J., Pérez, J.J., Giralt, E., Melero, J.A. & Domingo, E. (1990). *J. Virol*, **64**, 5519–28.

Díez, J., Mateu, M.G. & Domingo, E. (1989). *J. Gen. Virol.*, **70**, 3281–9.

Domingo, E. (1989). *Progr. Drug. Res.*, **33**, 93–133.

Domingo, E., Dávila, M. & Ortín, J. (1980). *Gene 11*:333–346.

Domingo, E., Escarmís, C., Martínez, M.A., Martínez-Salas, E., & Mateu, M.G. (1992). *Curr. Top. Microbiol. Immunol.*, **176**, 33–40.

Domingo, E. & Holland, J.J. (1988). In *RNA Genetics, Vol 3. Variability of RNA Genomes*, E. Domingo, J.J. Holland and P. Ahlquist (eds.). CRC Press, Boca Ratón, Florida, pp. 3–36.

Domingo, E. & Holland, J.J. (1994). In *Evolutionary Biology of Viruses*, S.S. Morse (ed.). Raven Press, New York. pp. 161–84.

Domingo, E., Martínez-Salas, E., Sobrino, F., de la Torre, J.C., Portela, A., Ortín, J., López-Galíndez, C., Pérez Breña, P., Villanueva, N., Nájera, R., VandePol, S., Steinhauer, D., DePolo, N. & Holland, J.J. (1985). *Gene*, **40**, 1–8.

Domingo, E., Mateu, M.G., Martínez, M.A., Dopazo, J., Moya, A. & Sobrino, F. (1990). In *Applied Virology Research, Vol. 2. Virus Variability, Epidermiology and Control*, E. Kurstak, R.G. Marusyk, F.A. Murphy and M.H.V. Van Regenmortel (eds.). Plenum Press, New York, pp. 233–266.

Dopazo, J. (1994). *J. Mol. Evol.*, **38**, 300–4.

Dopazo, J., Sobrino, F., Palma, E.L., Domingo, E. & Moya, A. (1988). *Proc. Natl. Acad. Sci. USA*, **85**, 6811–15.

Efron, B. (1982). *The Jackknife, the Bootstrap and Other Resampling Plans*. Society for Industrial and Applied Mathematics. Philadelphia.

Gebauer, F., de la Torre, J.C., Gomes, I., Mateu, M.G., Barahona, H., Tiraboschi, B., Bergman, I., Augéde Mello, P. & Domingo, E. (1988). *J. Virol.*, **62**, 2041–9.

González, M.J., Sáiz, J.C., Laor, O. & Moore, D.M. (1991). *J. Virol*, **65**, 3949–53.

Hedger, R.S. & Condy, J.B. (1985). *Vet. Rec.*, **117**, 205.

Holland, J.J., Spindler, K., Horodyski, F., Grabau, E., Nichol, S. & VandePol, S. (1982). *Science*, **215**, 1577–85.

King, A.M.Q., McCahon, D., Slade, W.R. & Newman, J.W.I. (1982). *Cell*, **29**, 921–8.

King, A.M.Q., Underwood, B.O., McCahon, D., Newman, J.W.I. & Brown, F. (1981). *Nature*, **293**, 479–80.

Kitching, R.P., Knowles, N.J., Samuel, A.R. & Donaldson, A.I. (1989). *Trop. Animal Health Production*, **21**, 153–66.

Krebs, O. & Marquardt, O. (1992). *J. Gen. Virol.*, **73**, 613–19.

Marín, A., Bertranpetit, J., Oliver, J.C. & Medina, J.R. (1988). *Nucl. Acids Res.*, **17**, 6181–9.

Martínez, M.A., Carrillo, C., González-Candelas, F., Moya, A., Domingo, E. & Sobrino, F. (1991). *J. Virol.*, **65**, 3954–7.

Martínez, M.A., Carrillo, C., Plana, J., Mascarella, R., Bergadá, J., Palma, E.L., Domingo, E. & Sobrino, F. (1988). *Gene*, **62**, 75–84.

Martínez, M.A., Dopazo, J., Hernández, J., Mateu, M.G., Sobrino, F., Domingo, E. & Knowles, N.J. (1992). *J. Virol.*, **66**, 3557–65.

Mateu, M.G., da Silva, J.L., Rocha, E., de Brum, D.L., Alonso, A., Enjuanes, L., Domingo, E. & Barahona, H. (1988). *Virology*, **167**, 113–24.

Mateu, M.G., Martínez, M.A., Capucci, L., Andreu, D., Giralt, E., Sobrino, F., Brocchi, E. & Domingo, E. (1990). *J. Gen. Virol.*, **71**, 629–37.

Palmenberg, A.C. (1989). In *Molecular Aspects of Picornaviral Infection and Detection*, B.L. Semler and E. Ehrenfeld (eds.). American Society for Microbiology, Washington, pp. 211–241.

Pereira, H.G. (1981). In *Virus Diseases of Food Animals, Vol. 2.*, E.P. J. Gibbs (ed.). Academic Press, London, pp. 333–363.

Piccone, M.E., Kaplan, G., Giavedoni, L., Domingo, E. & Palma, E.L. (1988). *J. Virol.*, **62**, 1469–73.

Robson, K.J.H., Harris, T.J.R. & Brown, F. (1977). *J. Gen. Virol.*, **37**, 271–6.

Rodrigo, M.J. & Dopazo, J. (1994). *J. Mol. Evol.*, in press.

Saitou, N. & Nei, M. (1987). *Mol. Biol. Evol.*, **4**, 406–25.

Sáiz, J.C., Sobrino, F. & Dopazo, J. (1993). *J. Gen. Virol.* **74**, 2281–5.

Shanan, M.S. (1962). *Ann. NY Acad. Sci.*, **101**, 445–54.

Sobrino, F., Dávila, M., Ortín, J. & Domingo, E. (1983). *Virology*, **128**, 310–8.

Sobrino, F., Martínez, M.A., Carrillo, C. & Beck, E. (1989). *Virus Res.*, **14**, 273–80.

Sobrino, F., Palma, E.L., Beck, E., Dávila, M., de la Torre, J.C., Negro, P., Villanueva, N., Ortín, J. & Domingo, E. (1986). *Gene*, **50**, 149–59.

Stanway, G. (1990). *J. Gen. Virol.*, **71**, 2483–501.

Van Bekkum, J.G., Frenkel, H.S., Frederiks, H.H.J. & Frenkel, S. (1959). *Tijdschr. Diergeneeskd.*, **84**, 1159–64.

Villaverde, A., Martínez-Salas, E. & Domingo, E. (1988). *Gene*, **23**, 185–94.

22

Evolution of the Bunyaviridae

RICHARD M. ELLIOTT

Introduction

The family Bunyaviridae is one of the largest groupings of viruses. More than 300 viruses, arranged in numerous serogroups, are contained in the family, and this number is taken as evidence of their evolutionary success and potential. Some of these viruses are associated with serious diseases of humans, such as haemorrhagic fever with renal syndrome, Crimean–Congo haemorrhagic fever, and Rift Valley fever. Hence the family warrants continual epidemiological surveillance. Characteristics that unite the Bunyaviridae include the following: tripartite, single-stranded RNA genome of negative- or ambi-sense polarity; enveloped virus particle comprising four structural proteins – two glycoproteins termed G1 and G2, a nucleocapsid protein, N and an RNA polymerase, the L protein; cytoplasmic site of viral replication with intracellular maturation at the Golgi; viral mRNA transcription primed by capped oligonucleotides cleaved from the 5' ends of host-cell mRNAs. Considering the large number of viruses that fulfil these criteria, it is perhaps not unexpected that considerable diversity exists at the molecular level in terms of genome coding and replication strategies, and at the biological level in terms of vector and host interactions. These factors will be briefly reviewed in this Chapter.

Biology of Bunyaviridae

The family Bunyaviridae is currently (Calisher, 1991) divided into five genera: *Bunyavirus, Hantavirus, Nairovirus, Phlebovirus* and *Tospovirus*. Such subdivision was originally made by serological comparisons but the availability of nucleotide sequence data has enabled more precise classification. (In this chapter 'Bunyaviridae' is used to refer to familial traits and the terms bunyavirus, hantavirus, etc, to generic traits.) Most of the viruses in the family can be classed as arboviruses, i.e. they are transmitted by arthropod vectors to vertebrate hosts. The exceptions are the hantaviruses and tospoviruses (see below). Typically, the maintenance cycles of these viruses in nature involve only one or a few arthropod species which are competent to transmit the virus, and only a restricted set of vertebrate species develop adequate titre viraemia to infect the blood-feeding vector. Gross generalizations can be made regarding the arthropods used as vectors by viruses in different genera: bunyaviruses are usually transmitted by mosquitoes or biting midges, nairoviruses by ticks, and phleboviruses by sandflies or mosquitoes or, for the Uukuniemi virus group, ticks. Hantaviruses are not transmitted by arthropod vectors, but are maintained in nature as persistent infections of rodents. Tospoviruses are plant-infecting members of the Bunyaviridae; over 350 plant species can be infected and the viruses are transmitted by arthropods, thrips. The family Bunyaviridae has been reviewed recently by a number of authors and the reader is referred to the following for more detailed information: Bishop (1990a), Elliott (1990), Schmaljohn and Patterson (1990), Bouloy (1991) and Kolakofsky (1991).

The ability to replicate in cells of disparate phylogeny is an important consideration for the evolutionary potential of arboviruses. Usually the viruses cause inapparent lifelong persistent infections of the vector, in contrast to the often lytic effect on vertebrate cells. In addition, it has been shown that some members of the Bunyaviridae can be transmitted venereally or transovarially between vectors. The capability of a particular arthropod to transmit a particular virus, vector competence, is determined by both arthropod and viral specific factors. The molecular determinants of vector-virus interaction remain poorly defined but it has been demonstrated that the gene products of the mRNA segment are major determinants of vector infectivity (Gonzalez-Scarano et al., 1988). Persistent infection of the vector allows the accumulation of point mutations in the viral genome, and subsequent transmission of the virus could lead to the emergence of a variant with altered pathogenicity or tropism. If the vector fed on separate vertebrates viraemic with different viruses the possibility of genetic interactions in the vector could be enhanced. Similarly, a change in the behaviour of the vector could lead to new vertebrate species being infected and subsequent adaptation of the virus to its novel environment. Clearly, the vector can exert an extreme influence on evolution of the virus, and of course ongoing evolution of the vector itself could also have an effect.

Genome-coding strategies

Throughout the family a generalization can be made regarding the coding strategy of the genome: the L RNA encodes the L protein, the M RNA encodes G1 and G2, and the S RNA encodes the N protein. Some viruses also encode non-structural proteins, and much variety in their coding strategy has evolved.

Complete genome sequences are available for representatives of all genera except *Nairovirus* where an L RNA sequence is lacking; using these data the coding strategies are summarized diagrammatically in Fig. 22.1. It can be seen that all L segments use a straightforward negative-sense strategy to encode a single gene product (L protein) in the complementary-sense RNA. The L protein is about 250 kD for the animal-infecting viruses, though is appreciably larger in the plant-infecting tospovirus. Of note is that the nairovirus L RNA segment is estimated to be about 11–13 kb, significantly larger than the L segments of viruses in the other genera, which may be indicative of additional coding information (Elliott, 1990).

The M and S segments show extreme diversity in their coding strategies. In the complementary-sense RNA (i.e. complementary to the genomic RNA) the bunya-, hanta- and phleboviruses encode, a precursor polyprotein of G1 and G2 (and perhaps a non-structural protein, NSm) which is probably co-translationally cleaved. The position of NSm in the precursor varies, sandwiched between G2 and G1 in bunyaviruses, at the N-terminus in some phleboviruses (but not in the uukuviruses), and absent in hantaviruses. The nairovirus M segment also encodes a glycoprotein precursor and regions corresponding to mature G1 and G2 have been identified. However, processing of the nairovirus precursor appears more complicated than in the above three genera and occurs over a period of several hours, i.e. it is post-translational (Marriott, El-Ghorr & Nuttall, 1992). Further diversity is displayed by the tospoviruses, which use an ambisense coding strategy for the M segment (Kormelink *et al.*, 1992; Law, Speck & Moyer, 1992). The structural glycoproteins are encoded as a precursor in the viral complementary-sense RNA but an equivalent of NSm is encoded in the same sense as the 5' end of the genome RNA. These proteins are translated from two subgenomic mRNAs; the NSm message must be transcribed from the complement of the genome RNA. Ambisense coding strategies have also been described for the S RNAs of phleboviruses and tospoviruses (see below), for both RNA segments

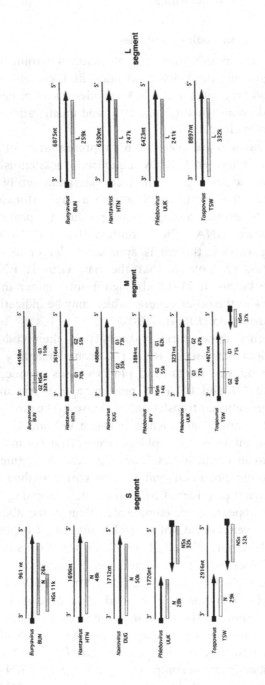

Fig. 22.1. Coding strategies of Bunyaviridae genome segments. Genomic RNAs are represented by thin lines (the number of nucleotides is given above) and mRNAs are shown as arrows (■ indicates host-derived primer sequence at 5′ end, ▲ indicates 3′ end). Gene products, with their size in kD, are represented by stippled boxes. Two examples of phlebovirus M segments are given which differ with respect to the presence or absence of NSm. Virus abbreviations: BUN, Bunyamwera; HTN, Hantaan; UUK, Uukuniemi; TSW, tomato spotted wilt; RVF, Rift Valley fever; DUG, Dugbe.

which comprise the arenavirus genome (Bishop, 1990*b*), and for two of the genome RNA segments of the plant-infecting tenuiviruses (Huiet, Tsai & Falk, 1992).

Variation is seen with the proteins encoded by the S segments. Nairo- and hantaviruses encode a single product, the *c*. 50 kD N protein, in the complementary-sense RNA. Bunyaviruses, phleboviruses and tospoviruses encode two proteins, N (25–30 kD) and a non-structural protein, NSs. For phlebo- and tospoviruses N is translated from a subgenomic complementary sense mRNA, whereas NSs is translated from a subgenomic genome sense RNA, i.e. in an ambisense arrangement. Both bunyavirus N and NSs proteins are translated from the same mRNA, the result of alternative initiation of translation at different AUG codons. It is of note that the animal-infecting Bunyaviridae which have a large N protein do not encode NSs, whereas those with a smaller N protein also produce a non-structural protein. It is tempting to speculate that the function of NSs (as yet unknown) may be contained within the larger N protein.

Mechanisms of evolution

RNA viruses with segmented genomes have two major means of evolving, through the acquisition of genome changes through point mutations, insertions, deletions, inversions, etc, or through genome segment re-assortment during the course of a mixed infection. Viral RNA-dependent RNA polymerases have a high error rate (10^{-2} to 10^{-4}), though there are no direct measurements of a Bunyaviridae polymerase, which gives rise to much heterogeneity in the population: quasi-species (Domingo & Holland, 1988; Chapter 13). *In vivo* the mutation rate is influenced by many factors, including immune pressure, lytic versus persistence infection, vectors, etc, and manifests differently in different genes.

Genetic drift

Although no large-scale systematic surveys of genome variation among Bunyaviridae have been undertaken, the accumulated evidence points to genetic drift as the major driving force in the evolution of these viruses. Several authors have reviewed instances of genetic drift in nature (Beaty, Trent & Roehrig, 1988; Beaty & Calisher, 1991; Kingsford, 1991) and a few examples are recorded here. Analysis of some 30 isolates of La

Crosse bunyavirus from different geographical locations at the same or different times, and from the same location also at different or the same times by RNase T1 oligonucleotide fingerprinting showed that every isolate had distinguishable L, M and S genome segments (El Said et al., 1979; Klimas et al., 1981). Many antigenic variants of Cache Valley and Tensaw bunyaviruses have been isolated in the United States (Calisher et al., 1988a,b) as have variants of bunyaviruses in the Gamboa serogroup (Calisher et al., 1988c). Persistent infection of the vector host is an excellent opportunity to promote genomic changes, as exemplified by the appearance of temperature-sensitive and plaque morphology mutants in the population of virus shed from mosquito cells persistently infected with Bunyamwera virus (Newton, Short & Dalgarno, 1981; Elliott & Wilkie, 1986).

The stability of viral genomes has also been demonstrated. No changes in the genome of Toscana phlebovirus were observed over the course of successive transovarial transmissions in its natural vector under laboratory conditions (Bilsel, Tesh & Nichol, 1988). Propagation of La Crosse and snowshoe hare bunyaviruses by lytic passage in different laboratories also did not result in detectable changes in their oligonucleotide fingerprints (Beaty & Calisher, 1991).

Genome segment re-assortment: genetic shift

The capability of the Bunyaviridae to exchange genome segments greatly enhances their evolutionary potential. Most data on re-assortment have come from laboratory experiments but there is some evidence for re-assortment in nature. A classic example is provided by analysis of six group C bunyaviruses isolated in the Utinga Forest in Brazil (Casals & Whitman, 1961; Shope & Causey, 1962). By haemaggutination-inhibition and neutralization tests, which assay M segment gene-products, the viruses segregate into three groups. However, by complement fixation, which assays the S segment gene products, the viruses fall into three alternate groups. These results are indicative of re-assortment occurring between these viruses; a detailed discussion of the natural history of these viruses is given by Shope (1985). Genotypic studies have shown that Shark River and Pahayokee bunyaviruses (Patois serogroup) have almost identical L and S segment oligonucleotide fingerprints but clearly distinguishable M segment fingerprints, suggesting that these viruses are natural re-assortants (Ushijima, Clerx-van Haater & Bishop, 1981).

In vitro experiments have indicated that re-assortment is confined to closely related viruses (i.e. between viruses in the same serogroup) suggesting that serogroups constitute gene pools that are evolving divergently (Iroegbu & Pringle, 1981; Bishop, 1990*a*; Pringle, 1991). Even within a serogroup, some viruses are genetically incompatible, though the reason for this cannot be explained by geographical or ecological isolation of the viruses. Maguari bunyavirus (from South America) and Batai bunyavirus (from Europe and Asia) are two members of the Bunyamwera serogroup that can exchange genome segments, whereas Main Drain and Northway bunyaviruses, which occur in the same area of North America, are genetically isolated (Elliott *et al.*, 1984). There is also evidence that segment exchange is non-random. In heterologous crosses of members of the Bunyamwera serogroup the L and S segments appeared to co-segregate (Pringle *et al.*, 1984), and, for re-assortants of La Crosse and snowshoe hare bunyaviruses (California serogroup), Urquidi and Bishop (1992) observed a preference for homologous L–M and M–S segment associations. The basis for the apparent genetic linkages is unknown but may limit the extent of evolution of Bunyaviridae by means of re-assortment.

Frequent re-assortment of bunyaviruses occurs in mosquitoes following oral or intrathoracic infection (Beaty *et al.*, 1981, 1985; Beaty & Bishop, 1988; Chandler *et al.*, 1990). Re-assortment has been demonstrated after simultaneous or interrupted feeding, but the time window for the latter is restricted to about two to three days between infections. Mosquitoes become resistant to superinfection with a second, closely related, virus after this time (Beaty *et al.*, 1985). Re-assortant viruses were found in the ovaries of dually infected mosquitoes, and the newly generated re-assortants could be transmitted transovarially to mosquito progeny and orally to mice (Chandler *et al.*, 1990). Obviously genome segment re-assortment is a major means by which viruses having altered host range, pathogenicity, virulence, tropisms, etc, can evolve though the extent to which this occurs for the Bunyaviridae in nature remains to be assessed.

Sequence comparisons

Nucleotide sequences

The terminal sequences of the three viral genome segments are conserved and the 3′ and 5′ ends are complementary; consensus sequences specific for each genus can be derived (Fig. 22.2). The complementarity presumably accounts for the circular and/or panhandle forms of RNA

Bunyaviridae

Bunyavirus	UCAUCACAU...
Hantavirus	AUCAUCAUCUG...
Nairovirus	AGAGUUUCU...
Phlebovirus	UGUGUUUCUG...
Tospovirus	UCUCGUUAG...

Orthomyxoviridae

Influenza A	UCGUUUUCGUCC...
Influenza B	UCGUCUUCGCUU...
Influenza C	UCGUUUUCGUCC...
Thogoto virus RNA 3	UCUCUUUAGUUU...

(*Tospovirus*)	UCUCGUUAG...

Arenaviridae

Tacaribe virus S RNA	GCGUGGCUCCUAGGA...
L RNA	GCGUGUCACCUAGGA...

Tenuivirus

Rice stripe virus	UGUGUUUCAG...

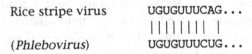

(*Phlebovirus*)	UGUGUUUCUG...

Fig. 22.2. 3′ terminal consensus sequences of the genome RNAs of negative or ambisense segmented viruses. The similarities between the terminal sequences of tospoviruses and Thogoto virus, and tenuiviruses and phleboviruses are indicated. (Modified from de Haan *et al.*, 1991.)

observed by electron microscopy (Hewlett, Pettersson & Baltimore, 1977; Pardigon *et al.*, 1982). The viral nucleocapsids are also circular, and it is thought that the complementary sequences may be involved in packaging of the genome by the N protein (Raju & Kolakofsky, 1987, 1989). Furthermore, the conserved termini may provide signals for recognition by the virus-coded polymerase. Interestingly, some similarities between the Bunyaviridae consensus sequences and those of other segmented genome viruses have been noted; these are also displayed in Fig. 22.2. De Haan *et al.* (1991) observed homology between the terminal sequences of tomato spotted wilt virus segments and RNA segment 3 of Thogoto virus (Orthomyxoviridae); and the plant-infecting tenuiviruses have similar terminal sequences to the phleboviruses (Kakutani *et al.*, 1990). These observations suggest distant ancestral relationships of these viruses (Kakutani *et al.*, 1990; de Haan *et al.*, 1991).

The 3' terminal nucleotide of hantavirus and nairovirus genome RNAs is a purine, which implies that synthesis of the complement of the genome would be initiated with a pyrimidine. However, this would be unusual for a viral polymerase (Banerjee, 1980). For Tacaribe arenavirus, which also has a 3' terminal purine in the genome RNA, it has been shown that genomic complements contain an apparently nontemplated G residue at their 5' ends, apparently added by the polymerase via a slippage mechanism (Garcin & Kolakofsky, 1992). If hantaviruses and nairoviruses also initiate synthesis of the genomic complements similarly this may suggest the evolution of a distinct function within the L proteins of these three groups of viruses.

Amino acid sequences

Comparison of the amino acid sequences of isofunctional proteins, in this case the structural polypeptides, encoded by viruses in different genera is not particularly instructive and little homology is found. In view of the fact that classification was originally based on serological properties this is not surprising, and hence it is satisfying that the subdivision of the family correlates with the phylogeny of these viruses. Attempts to identify common motifs in isofunctional proteins have also met with little success, an exception being the L protein.

L proteins

Seven L protein sequences are available and global comparison shows these to be highly diverged. The two phlebovirus (Rift Valley fever and Uukuniemi viruses) L proteins show 38% identity (Elliott *et al.*, 1992), and the three hantavirus (Hantaan, Seoul and Puumala viruses) L proteins show 69% amino acid identity overall (Antic, Lim & Kang, 1992). Sequences are also available for Bunyamwera bunyavirus and tomato spotted wilt tospovirus L proteins. Poch *et al.* (1989) described four conserved motifs, dubbed the polymerase module, in the RNA-dependent RNA polymerases, and this provides evidence for a common ancestral polymerase. An alignment of these putative polymerase domains in the Bunyaviridae L proteins is shown in Fig. 22.3 (Jin & Elliott, 1992). In motif C the triplet SDD should be noted;

```
                                           A
UUK  990  atsdDaakWnqchhvtkf....alMlchFtdplfhgfIirgcsmfmkkrimidqslididshttletsdaylqkihrgyhgslddqprwisrggafvqt
RVF  987  atsdDarkWnqghfvtkf....alMlceFtspkwwpllirgcsmftkkrmmnlnylkildghreldirddfvmdlfkayng..eaevpwafkgktylet
HTN  968  yvsaDatkWspgdnsakfrrftsmLhngLpnnklkncvidalkqvyktdfemsrklrnyidsmesldphikqfldffpd.......ghhge...v
SBO  968  yvsaDatkWspgdnsakfrrftaaLhngLpddrlkncvidalrhvktdfymsrklrhyidsmdtyephvrdflnffpd.......kvsas...i
NEV  968  yvsaDatkWspgdnsakfrrftqsLydgLrddklkncvdalrniyetdffisrklhryidnmgelsdevidflsffpn......glnynyvqi
BUN  1033 einaDmsKwsaqdvf.ykyfwliaMdpiLypaekt.rIiyfmcnymqklilipddlianildqkrpy..nddlilemtn.......gltntypv
TSW  1359 flsaDqstkwsasglttykyvlailLnpiLttgeaslmiecilmyvklkkvciptdiflnlrkaqqtfgenetaiglltk
Con       ---D--kW-----------L---L-------I--------

                                           C

etgmMQGiLhytSslLhtllqewlrtfsqrfirtrvsvdqrpdvlvdvlqssSDDsgmmisfpstdkgatgkyrylsalifkykvigkyl.giyssvKSt 1184
ttgmMQGiLhytSslLhtihqeyirslsfkifnlkvapemsksivcdmmggSDDssmlisfpaddekvltrckvaaaicfrmkkelgvyl.aiypsekst 1179
kgnwLQGnLnkcSslfgvam.....sllfkqwtnlfpel.dcffefahhsDDalfiyglepaddgtdwflfvsq..qiqaghlhwfsvntemwKSm 1141
rgnwLQGnLnkcSslfgvam.....slllfkeiwtlfpel.dcffefahhsDDalfiyglepaddgtdwflfvsq..qiqagklhwfnvntemwKSm 1141
kgnwLQGnLnkcSslfgaav.....sllfkrvwaklypel.ecffefahhsDDalfiyglepvddgtewfqyvtq..qiqagnfhwhavnqemwKSm 1141
krnwLQGnFnyiSSyvhsca.....mlvykdilkecmklldgdclinsmvhSDDnqtslaiiqnkvsdqiviqyaan...tfesvcltF.gcqanmkKTy 1207
smnwLQGnLnylSSvyhsca.....mkayhntl.ecyk..ncdfqtrwivhsDDnatsl..iasgevdkmltdfsss...slpem.lfR..sieahfKSf 1530
----LQG-L---SS----------------------------------SDD-------------------------------W------KS-

                                           B                                              D
```

Fig. 22.3. Alignment of the putative polymerase domains of Bunyaviridae L proteins. The alignment in this figure and those in Figs. 22.4–22.6 were made using the programs PILEUP and PRETTY in the UWGCG package (Devereux, Haeberli & Smithies, 1984). The regions corresponding to the four motifs (A to D) identified by Poch et al. (1989) in all RNA polymerases are overlined and 'Con' shows the residues conserved in the Bunyaviridae L proteins. The dots indicate residues mutated in the Bunyamwera virus L protein, and + or − whether the mutated protein retained or lost polymerase activity. Virus abbreviations as in Fig. 22.1 and 8EO, Seoul hantavirus; PUU Puumala hantavirus. (Modified from Jin & Elliott, 1992.)

this is characteristic of the polymerases of all negative and ambisense segmented genome viruses so far sequenced (orthomyxoviruses and arenaviruses), which suggests a distinct evolutionary lineage for these segmented genome viruses. Experimental evidence for the importance of the conserved amino acids in these motifs is provided by mutational analysis of the Bunyamwera virus L protein (Jin & Elliott, 1992).

On the basis of the alignment shown in Fig. 22.3, particularly with regard to the placement of gaps, the phleboviruses are more distant than the hanta-, bunya- and tospoviruses. Phleboviruses and tospoviruses share the property of ambisense coding arrangements of some genome segments, hence it was surprising that the tospovirus L protein seemed closer to that of the non-ambisense bunyavirus. When the L proteins of arenaviruses (arenaviruses have two ambisense RNAs) are included in the comparison no correlation emerges between ambisense strategy and conservation of polymerase (Tordo *et al.*, 1992).

Glycoproteins

The glycoproteins have been implicated in many of the important biological properties of the viruses, including virulence, tissue tropism, neutralization, haemagglutination and fusion, and are presumably involved in interaction with the host cell receptor(s). Bunyaviridae glycoproteins are rich in cysteine residues (>5%) and comparisons of glycoproteins within a genus show a high conservation of these residues, e.g. between the G1 and G2 proteins of Bunyamwera and California serogroup viruses (*Bunyavirus* genus) there is about 45% amino acid identity overall but greater than 95% of cysteines are conserved (Elliott, 1990). Similar observations can be made with the hantavirus glycoproteins, strongly suggesting that within a genus the glycoproteins adopt similar conformations (Antic *et al.*, 1992).

The only region of similarity so far detected between glycoproteins of viruses in different genera is a small region in the G1 proteins of bunyaviruses and tospoviruses (Kormelink *et al.*, 1992; Fig. 22.4). Whether this motif has a functional significance is unknown, though it seems more likely that this region would have a vector-associated function rather than a role associated with the plant or vertebrate host. However, this together with the data from the L proteins of these genera, suggest that bunyaviruses and tospoviruses diverged relatively later from the other viruses in the family.

```
INS   YtTaPiqsthtdffstCTGkCsdcrkeqpitgYqdFcitpTSyWGCEEvwCLAIneGatCGfCrni
TSW   YtTaPiqsthtdfystCTGnCdtcrknqaltgFqdFcvtpTSyWGCEEawCFAIneGatCGfCrnv
SSH   YtTgPtsgintkhdelCTGpCpa..kinhqtgWltFakerTSsWGCEEfgCLAIsdGcvFGsCqdi
LAC   YsTgPtsgintkhdelCTGpCpa..ninhqvgWltFarerTSsWGCEEfgCLAVsdGcvFGsCqdi
BUN   YdTgPtisintkhdehCTGqCps..nieheanWltFsqerTSrWGCEEfgCLAVntGcvFGsCqdv
GER   YdTgPtininskhdelCTGqCpk..kipadpnWltFsqerTSrWGCEEfgCLAIntGcvYGsCqdv
Con   Y-T-P-----------CTG-C-----------W--F----TS-WGCEE--CLAI--G--FG-C---
```

Fig. 22.4. Alignment of a short region in G1 conserved between bunyaviruses and tospoviruses. Virus abbreviations as in Fig. 22.1 and INS, Impatiens necrotic spot tospovirus; SSH, snowshoe hare bunyavirus; LAC, La Crosse bunyavirus; GER, Germiston bunyavirus. The residues compared are INS 652–717; TSW 672–737; SSH 1025–1088; LAC 1025–1088; BUN 1017–1080; GER 1021–1084.

N protein

The N proteins of viruses in the Bunyaviridae fall into two size classes: 25–30 kD for bunya-, phlebo- and tospoviruses, and about 50 kD for hanta- and nairoviruses. Most sequences are available in the *Bunyavirus* genus where 13 sequences covering viruses in three serogroups have been determined; the N proteins are either 233 or 235 amino acids in length (Elliott, 1989; Dunn, Pritlove & Elliott, 1994). Within a serogroup there is >62% identity and between serogroups >40%. However, identical residues are not distributed throughout the proteins but are clustered, particularly in the region between residues 90 and 105, showing almost total identity (Fig. 22.5). Comparison of the five available phlebovirus N proteins (from Rift Valley fever, Punta Toro, Toscana, sandfly fever Sicilian and Uukuniemi viruses) also shows an overall 40% similarity and again certain regions show total conservation, e.g. YQGFD, RGNK, KMSD and I/LTLSRV/I (Giorgi *et al.*, 1991). However, none of these strictly conserved motifs resembles those in the *Bunyavirus* genus, and comparison of all the 'small' N proteins indicates that, between the genera, the N protein sequences are highly diverged.

Three nairovirus (Dugbe, Hazara and Crimean–Congo haemorrhagic fever viruses) and five hantavirus (Hantaan, Seoul, Puumala (Hällnäs), Puumala (Sotkamo), and Prospect Hill viruses) N protein sequences are available (Antic *et al.*, 1992; Marriott & Nuttall, 1992). The nairoviruses show about 40% identity overall and the hantaviruses about 60%. A comparison of all these large N proteins (Fig. 22.6) indicates some conservation of particularly bulky or charged residues, and overall more similarity is seen than when comparing the small N proteins (data not shown). Thus the larger N proteins may be a distinct lineage.

```
        80                                   115
LAC     gnrnnpIgnndLTIHRLSGYLARWVldqynenddes
SSH     gnrnnpInsddLTIHRLSGYLARWVleqykenedes
LUM     gnrnnpIdnndLTIHRLSGYLARWVleqfkenedaa
CV      gnrnspVpddgLTLHRLSGYLARYLleki.lkvsdp
NOR     gnrnspVpddgLTLHRLSGFLARYLleki.lkvsep
MAG     gnrnspVpddgLTLHRLSGFLARYLleki.lkvsdp
BAT     gnrnspVpddgLTLHRLSGFLARYLleki.lkvsdp
BUN     gnrnnpVpddgLTLHRLSGFLARYLlekm.lkvsep
MD      gnrnspVpddgLTLHRLSGFLARYLleki.lkvsep
GER     gnrnnaVpdygLTFHRISGYLARYLlgky.laetep
GRO     gnrnspVlddsFTLHRVSGYLARYLlery.ltvsap
KRI     enrnmsVpddgLTLHRVSGYLARYLldrv.ysagep
AINO    qytanpVpdtaLTLHRLSGYLAKWVadqcktnqikl
Consensus ------V----LTLHRLSGYLARYL-----------
```

Fig. 22.5. Highly conserved domain in the N protein of 13 bunyaviruses representing three serogroups. California serogroup: LAC, SSH and LUM (Lumbo) viruses; Bunyamwera serogroup: CV (Cache Valley), NOR (Northway), MAG (Maguari), BAT (Batai), BUN, MD (Main Drain), GER, GRO (Guaroa), KRI (Kairi) viruses; Simbu serogroup, AINO virus.

Non-structural proteins

The functions of the Bunyaviridae encoded non-structural proteins are unknown and indeed not all viruses encode them. There is also considerable diversity in the manner in which NS proteins are encoded (Fig. 22.1). The NS proteins are the most variable proteins within a genus e.g. bunyavirus NSs proteins show about 25% amino acid identity overall (Elliott, 1989), and throughout the family the NS proteins show wide variation in size. No sequence homology can be detected between NS proteins of different genera and, until the roles of these proteins are delineated, it is unclear whether the proteins in different genera designated NSs, for example, are isofunctional. However, analysis of the non-structural protein sequences suggests that within a genus NSs and NSm proteins follow a common evolutionary pathway.

Evolutionary pathway of hantaviruses

Most complete genome sequences are available for hantaviruses and comparison of the structural protein sequences has revealed two distinct evolutionary pathways which are influenced most strongly by the rodent host (Antic *et al.*, 1991, 1992). One includes the field mouse (*Apodemus*

```
1                                                                                              100
HAN   ........matMeelqreinaheqglviarqkvrdaekqayekdpdelnkrtltdregvavsiqekIdElkrqlAdriatgKnlgkeqdptgvepgDhlk
SRV   ........matMeelqreisaheqglviarqkvdaedkqiarqiqksigkIdElkrqlAdriqqgRtsggdrdptgvepgDhlk
PUUH  .........msdttdiqeeltrtheqglviarqkikdaeravepddvnkstlqarqgtvsaledkLaDyKrlmAdavsrkKmdtkptdptgepdDhlk
PUUS  .........msdttdiqeeltrtheqglviarqkikdaeravepddvnkstlqsrisavstledkLaEfKrqlAdvisrqKmdekpvdptIeldDhlk
PH    .........msgiretqeeitrheqglviarqlkeaertvevdppdvnkstlqsrisavstledkLaEfKrqlAdvisrqKmdekpvdptIeldDhlk
CCHF  menkievnnkdeknkwfeefkkgnglvdtfnpysfcesvpnlerfvfqmasatddaqkdslyasaLvEatKfCApiyecaWvsstgivkkqLewFEkn.
HAZ   menkivastkeePntwykgfaaekhlnnkytessfcaeipqldtykymelastnerdalyssaLiEatRfCApimecaWasctgtvkrgLewFDknk
DUG   menqikannkkeFdewfkpfseeklqlrsnltnsaslcdrvpdialaemkmalatddkekdsvfsnaLvEatRfCApiyecaWtcstgvvqkslswZDknk
Con   --------------------------------------------------------------------L-E--R--A----K----D--

101                                                                                            200
HAN   ersmLsygnvlDlnhiddeptgqtadwlsilvyltsFvvpIlkalymlttrgqttkDnkgtrlrFkddssfedvngiRkpkhLyvslpnaqsmkae
SRV   ersaLsygntlDlnsiddeptgqtadwltiivyltsFvvpIlkalymlttrgrtskDnkgmrlrFkddssyedvngiRkpkhLyvsmpnaqsmkae
PUUH  erssLrygnvlDvnaidieepsgqtadwytigvvvigFtipIlkalymlstrgrtvkEnKgtrlrFkddtsfedingiRRpkhLyvsmptaqstmkae
PUUS  erssLrygnvlDvnaidieepsgqtadwygigvvvigFtipIlkalymlstrgrtvkEnKgtrlrFkddtsfedingiRRpkhLyvsmptaqstmkae
PH    ersaLgygnvlDvnsideeepsgqtadwlkigsvlieFalpIlkalymlstrgtvkEnKgtrlrFkddssyedvngiRRpkhLyvsmptaqstmkae
CCHF  agtiK...swDesyielkvevpkieqlanyqqaalkwrkdigFrvnantaalshkvlaEyKvpge1vmsvkemlsdmiRRnlllnrggdenp.rgpvs
HAZ   dsdtvk...vwDanyvkLrtetppeeallayqkaalnwrkdVpsigeytslikkavaEyvvpgtvlnnksmsdmirRRnriinggsddapkrgPvg
DUG   ..dfik...lwDakymdikkgipepeqlvsyvqqaqkWrkdVgyeinqftrslthpvvaEyKvppge1ll.msrmlsdmirRRnvlLngddgenagkkvlis
Con   ----L----I-----------------------F---F---L--E--X----F----L-------RR---L-

201                                                                                            300
HAN   eitpgvRytavcGlYpaqikarqmispvmsvIGfalakdwsdrieqwliepcklipdtaavsligpat....nrdylrqrvaLgnmetkeskairqh
SRV   eitpgRlRtavcGlYpaqikatranvspvmsvvGfalakdwtsrieewigpcktmaesliagsisgnpv...nrdylrqrgqaLagmepkefqalrqh
PUUH  eltpgRfRttvcGlFptqiqvrnimspvmgvlGfsfvkdwpekiresfixpdwperikrvfymrtqdVLdknhvadidklidy
PUUS  eltpgRfRttvcGlFptqiqvrnimspvmgvlGfsfvvdwperiremekecpfikpevkpgtpqeelemlkrnkyfmqrqdvLdknhvadidklidy
PH    eltpgRRttvcGlFpaqimarniispvmgvlGfafvewadkvkaflDqkcpfikaeprpgpageaeflssiravimrqavLdethlpdidalvel
CCHF  rehveweRefvkdkYlmafnppwgdinksgrsGialvatglak...laetegkgvfdeakktvealngidkhkdevdkasadmitnlkhkakaqel
HAZ   rehldwcRefasdKfinafnppwgeinkagksGypllatglak....lveleqkdvmdkakasiaglegwvkenkdqvdqkaedLlkgvresyktalal
DUG   rehvswgRelagdKfgvvfnppwgdinkcgksGiplaatamvk....vaeldgskkledirqalillkkwvednkdaledgkgneLvtmtkhlakhvel
Con   ----R--R-------G------G--------L---------------------------------------L---

301                                                                                            400
HAN   aeaagc..smlediEspsslwvPagapdrcpPtclfragiaelgaFFsliqdmRntlmasKtvgtseeklRkksfyqsyLrrtqsmgiqLgqrivL.
SRV   skdagc..tivehlEspsslwvPagapdrcpPtclfvggmeigaFFsliqdmRntlmasKtvgtadeklRkksfyqsyLrrtgsmgiqLdqriivM.
PUUH  aasgdp..tspddiEspnapwvPacapdrcpPtclyvagmeigaFFsliqdmRntimasKtvgteeelkksfyqsyLrrtqsmgiqLdqriiLL.
PUUS  aasgdp..tspdniDspnapwvPacapdrcpPtclyvagmeigaFFaliqdmRntimasKtvgteeelkksfyqsyLrrtqsmgiqLdqriiLM.
PH    aasgdp..tlpdsLEnphaawvPacapdrcpPtclyiagmeigaFFaliqdmRntimasKtvgtaeeklkksfyqsyLrrtqsmgiqLdqriiLM..
CCHF  yknsslaraggaqIDtfsssYywlykaqvtPeTfptvsqf....LPeLqxpRgtkkmkkaLlstpmkwgKk...lyeiFaddsfqgnriymhpaVLta
HAZ   akisnafraggaqIDtvfssYywpwkaqvtPvTfpsvsqf....LPeLknpKgqkmKaLintplKwgkr..llelFadnftenriymhpcVLts
DUG   skksnalraggaqIDtfsaYywawsagvkpeTtftsqf......LPeWqqsaRgqkkmikaLtstpLRwqKg...liniPaddfIgnrLymhpaVLtp
Con   ----W-F------P-T--V----FF-I---R------K-V----K--K-----L-----L----VL-

401                                                                                            500
HAN   ....fmvaWGkeavDnfhlgddmdpelRtLaqslidvkvkeisnqepkL*
SRV   ....fmvaWGkeavDnfhlgddmdpeLRsLaqilidqkvkeisnqepmkL
PUUH  ...ymleWGremvDhfhlgddmdpelRgtaqsliidqkvkeisnqepki.
PUUS  ...fmleWGkemvDhfhlgddmdpeIRgLaqaliidqkvkeisnqeplki
PH    ...ymleWGnevvnhfhlgddmdpeIRqLaqaliidqkvkeisnqeplki.
CCHF  grisemgveFGtipVanpddaaqsghtksilnirttetnnpcaktlvkLfeiqktgfniqdmdivasehllhqslvgkqspfqnaynvkgnatsanii
HAZ   grmselgisFGavpVspspddaaqsghtkavlnyktktevgnpccaclissLfeiqkagydiesmdivasehllhqslvgkrspfqnaylikgnatninii
DUG   grmsemgacFGvipVaspedailsgshsknlinfkiddsvqnpcastivqL.iqnpeiwl.................
Con   -------WG---V---------------R-L-
```

Fig. 22.6. Alignment of the N protein sequences of hantaviruses and nairoviruses. 'Con' represents amino acid residues conserved in all available sequences. Virus abbreviations as in Figs. 22.1 and 22.3, and CCHF, Crimean–Congo haemorrhagic fever; HAZ, Hazara virus; PUUH, Puumala Hällnäs; PUUS, Puumala Sotkamo; PH, Prospect Hill.

spp) and rat (*Rattus* spp) transmitted viruses, such as Hantaan and Seoul viruses, and the other vole (*Clethrionomys* spp, *Microtus* spp) transmitted viruses, such as Puumala and Prospect Hill viruses. Geographical factors do not appear to have been a major influence, because Puumala-like and Hantaan-like viruses have been isolated from the same region in Europe but from different rodent hosts. Antic *et al.* (1992) suggest that the divergence of these pathways is not recent, citing as evidence the high homology between the M segments of Hantaan-like viruses isolated from *Apodemus* and humans over a seven year period (Schmaljohn *et al.*, 1988).

Closing remarks

From the available sequence data the overall impression is that the current classification of the Bunyaviridae, placing viruses into genera which are quite different from each other is correct. Attempts to compare analogous structural proteins globally have not, in my view, provided convincing alignments with which to draw phylogenetic trees for the family. Such exercises merely reinforce that the viruses in the present genera are highly diverged. As a published example, comparison of the central third of the L proteins, which contains the polymerase domain, indicates the 'extreme dispersion' (Tordo *et al.*, 1992) of the Bunyaviridae polymerases. A clearer picture of the evolutionary relationships within this family may come from determination of functional homologies between proteins from viruses in different genera. For example we do not know whether the few conserved residues between hantavirus and nairovirus N proteins (Fig. 22.6) are significant, but research in this area is only just beginning.

References

Antic, D., Kang, C.Y., Spik, K., Schmaljohn, C., Vapalahti, O. & Vaheri, A. (1992). *Virus Res.*, **24**, 25–46.
Antic, D., Lim, B-U. & Kang, C.Y. (1991). *Virus Res.*, **19**, 47–58.
Banerjee, A.K. (1980). *Microbiol. Rev.*, **44**, 175–295.
Beaty, B.J. & Bishop, D.H.L. (1988). *Virus Res.*, **10**, 289–302.
Beaty, B.J. & Calisher, C.H. (1991). *Curr. Topics Microbiol. Immunol.*, **169**, 27–78.
Beaty, B.J., Rozhon, E.J., Gensemer, P. & Bishop, D.H.L. (1981). *Virology*, **111**, 662–5.
Beaty, B.J., Sundin, D.R., Chandler, L.J. & Bishop, D.H.L. (1985). *Science*, **230**, 548–50.
Beaty, B.J., Trent, D.W. & Roehrig, J.T. (1988). In *The Arboviruses: Epidemiology and Ecology, vol* I, T.P. Monath (ed.). CRC Press, Boca Raton, pp.59–85.

336 R.M. Elliott

Bilsel, P.A., Tesh, R.B. & Nichol, S.T. (1988). *Virus Res.*, **11**, 87–94.
Bishop, D.H.L. (1990a). In *Virology*, B.N. Fields and D.M. Knipe (eds). Raven Press, New York, pp. 1155–1173.
Bishop, D.H.L. (1990b). In *Virology*, B.N. Fields and D.M. Knipe (eds). Raven Press, New York, pp. 1231–1243.
Bouloy, M. (1991). *Adv. Virus Res.*, **40**, 235–75.
Calisher, C.H. (1991). *Arch. Virol. Suppl.*, **2**, 273–83.
Calisher, C.H., Lazuick, J.S., Lieb, S., Monath, T.P. & Castro, K.G. (1988b). *Am. J. Trop. Med. Hyg.*, **39**, 117–22.
Calisher, C.H., Lazuick, J.S. & Sudia, W.D. (1988c). *Am. J. Trop. Med. Hyg.*, **39**, 406–8.
Calisher, C.H., Sabattini, M.S., Monath, T.P. & Wolff, K.L. (1988a). *Am. J. Trop. Med. Hyg.*, **39**, 202–5.
Casals, J & Whitman, L. (1961). *Am. J. Trop. Med. Hyg.*, **10**, 250–8.
Chandler, L.J. Beaty, B.J., Baldridge, G.D., Bishop, D.H.L. & Hewlett, M.J. (1990). *J. Gen. Virol.*, **71**, 1045–50.
de Haan, P., Kormelink, R., Resende, R. de O., van Poelwijk, F., Peters, D. & Goldbach, R. (1991). *J. Gen. Virol.*, **71**, 2207–16.
Devereux, J., Haeberli, P. & Smithies, O. (1984). *Nucl. Acids Res.*, **12**, 387–95.
Domingo, E. & Holland, J.J. (1988). In *RNA Genetics, Vol. III.*, E. Domingo, J.J. Holland and P. Ahlquist, (eds.) CRC Press, Boca Raton, pp. 3–36.
Dunn, E.F., Pritlove, D.C., & Elliott, R.M. (1994). *J. Gen. Virol.*, **75**, 597–608.
El Said, L.H., Vorndam, V., Gentsch, J.R., Clewly, J.P., Calisher, C.H., Klimas, R.A., Thompson, W.H., Grayson, M., Trent D.W. & Bishop, D.H.L. (1979). *Am. J. Trop. Med. Hyg.*, **28**, 364–86.
Elliott, R.M. (1989). *J. Gen. Virol.*, **70**, 1281–5.
Elliott, R.M. (1990). *J. Gen. Virol.*, **71**, 501–22.
Elliott, R.M., Dunn, E., Simons, J.F. & Pettersson, R.F. (1992). *J. Gen. Virol.*, **73**, 1745–52.
Elliott, R.M., Lees, J.F., Watret, G.E., Clark, W. & Pringle, C.R. (1984). In *Mechanisms of Viral Pathogenesis: From Gene to Pathogen*, A. Kohn and P. Fuchs (eds). Martinus Nijhoff, Boston, pp. 61–76.
Elliott, R.M. & Wilkie, M.L. (1986). *Virology*, **150**, 21–32.
Garcin, D. & Kolakofsky, D. (1992). *J. Virol.*, **66**, 1370–76.
Giorgi, C., Accardi, L., Nicoletti, L. Gro, M.C., Takehara, K., Hilditch, C., Morikawa, S. & Bishop, D.H.L. (1991). *Virology*, **180**, 738–53.
Gonzalez-Scarano, F., Beaty, B.J., Sundin, D.R., Janssen, R., Endres, M.J. & Nathanson, N. (1988). *Microb. Path.*, **4**, 1–7.
Hewlett, M.J., Pettersson, R.F. & Baltimore, D. (1977). *J. Virol.*, **21**, 1085–93.
Huiet, L., Tsai, J.H. & Falk, B.W. (1992). *J. Gen. Virol.*, **73**, 1603–7.
Iroegbu, C.U. & Pringle, C.R. (1981). *J. Virol.*, **37**, 383–94.
Jin, H. & Elliott, R.M. (1992). *J. Gen. Virol.*, **73**, 2235–44.
Kakutani, T., Hayano, Y., Hayashi, T. & Minobe, Y. (1990). *J. Gen. Virol.*, **71**, 1427–32.
Kingsford, L. (1991). *Curr. Topics Microbiol. Immunol.*, **169**, 181–216.
Klimas, R.A., Thompson, W.A., Calisher, C.H., Clark, G.G., Grimstad, P.R. & Bishop, D.H.L. (1981). *Am. J. Epidemiol.*, **114**, 112–31.
Kolakofsky, D. (ed.) (1991). *'Bunyaviridae'*, *Curr. Topics Microbiol. Immunol.*, **169**.

Kormelink, R., de Haan, P., Meurs, C., Peters, D. & Goldbach, R. (1992). *J. Gen. Virol.*, **73**, 2795–804.

Law, M.D., Speck, J. & Moyer, J.W. (1992). *Virology*, **188**, 732–41.

Marriott, A.C., El-Ghorr, A.A. & Nuttall, P.A. (1992). *Virology*, **190**, 606–15.

Marriott, A.C. & Nuttall, P.A. (1992). *Virology*, **189**, 795–9.

Newton, S.E., Short, N.J. & Dalgarno, L. (1981). *J. Virol.*, **38**, 1015–24.

Pardigon, N., Vialet, P., Girard, M. & Bouloy, M. (1982). *Virology*, **122**, 191–7.

Poch, O., Sauvaget, I., Delarue, M. & Tordo, N. (1989). *EMBO J.*, **8**, 3867–74.

Pringle, C.R. (1991). *Curr. Topics Microbiol. Immunol.*, **169**, 1–25.

Pringle, C.R., Lees, J.F., Clark, W. & Elliott, R.M. (1984). *Virology*, **135**, 244–56.

Raju, R. & Kolakofsky, D. (1987). *J. Virol.*, **61**, 667–72.

Raju, R. & Kolakofsky, D. (1989). *J. Virol.*, **63**, 122–8.

Schmaljohn, C.S., Arikawa, J., Hasty, S.E., Rasmusssen, L., Lee, H.W., Lee, P.W. & Dalrymple, J.M. (1988). *J. Gen. Virol.*, **69**, 1949–55.

Schmaljohn, C.S. & Patterson, J.L. (1990). In *Virology*, B.N. Fields & D.M. Knipe (eds). Raven Press, New York, pp. 1175–1194.

Shope, R.E. (1985). In *Virology*, B.N. Fields (ed). Raven Press, New York, pp. 1055–1082.

Shope, R.E. & Causey, O.R. (1962). *Am. J. Trop. Med. Hyg.*, **11**, 283–90.

Tordo, N., de Haan, P., Goldbach, R. & Poch, O. (1992). *Semin. Virol.*, **3**, 341–57.

Urquidi, V. & Bishop, D.H.L. (1992). *J. Gen. Virol.*, **73**, 2255–65.

Ushijima, H., Clerx-van Haaster, C.M. & Bishop, D.H.L. (1981). *Virology*, **110**, 318–32.

23

Evolution of the tobamoviruses

AURORA FRAILE, MIGUEL A. ARANDA
AND FERNANDO GARCÍA-ARENAL

Introduction

The tobamovirus group of plant viruses has 12 recognized members (Francki *et al.*, 1991) sharing the following characteristics: infective particles are very stable, rigid rods, of about 300 × 18 nm, with a sedimentation coefficient of 190 S; each particle is built of about 2000 protein subunits of a single molecular species (M_r 17–18 × 10³), helically (pitch 2.3 nm) arranged together with a monopartite, single-stranded, genomic RNA (M_r 2 × 10⁶, or ~ 6.5 kb) of messenger polarity.

The genomic RNA (gRNA) encodes at least four proteins: gRNA is the mRNA for a 126 kD protein (126 K), whose open reading frame (ORF) initiates 60–71 nt from the capped 5′end, and for a readthrough protein of 183 kD (183 K). Both the 126 K and the 183 K are required for viral RNA replication, and found in them are consensus sequences for methyltransferases and helicases in the 126 K and for RNA-polymerases, in the 54 kD (54 K) readthrough portion of the 183 K. A third ORF encodes a 30 kD protein (MP), required for the cell-to-cell movement of the virus in the plant; the fourth ORF encodes the coat protein (CP). To the 3′ side of the CP ORF, there is a non-coding region (3′ncr) 179–414 nt long, that may adopt a tRNA-like configuration and can be amino-acylated, for most tobamoviruses, with histidine. Both MP, CP (and perhaps the 54 K readthrough part of the 183 K), are translated from 3′-coterminal subgenomic RNAs produced during replication. Whereas all tobamoviruses encapsidate the mRNA for the MP, only some members of the group encapsidate the mRNA for the CP, and this depends on the position in the gRNA of the sequence that acts as origin of assembly (OAS) of the particle.

Under natural conditions, most tobamoviruses have restricted host ranges, even though the experimental host ranges of most are very large. They are readily transmitted by mechanical inoculation. In nature transmission is by contact between plants, from contaminated soil or, for some virus : host combinations, through the seed, and no specific tobamovirus vectors are known.

The type member of the group, tobacco mosaic virus (TMV), is one of the most extensively studied plant viruses, and much progress has been made in the last decade understanding the molecular biology of its replication and its interactions with host plants. Also, there is much information on the molecular structure and biology of several other tobamoviruses. This wealth of knowledge has given rise to recent reviews dealing with different aspects of the biology of the tobamoviruses (Van Regenmortel & Fraenkel-Conrat, 1986; Culver, Lindbeck & Dawson, 1991; Dawson, 1992).

In this chapter we will not discuss the relationships between tobamoviruses and other viral groups, as this subject has been recently reviewed by Zimmern (1988), Matthews (1991), see also Chapter 4. We will, instead, discuss the long-term evolution of the tobamoviruses as well as the mechanisms by which genetic variability is generated and maintained in populations of specific tobamoviruses, trying to encompass both aspects in a general view of the evolution of these plant viruses.

Evolutionary relationships among the tobamoviruses

Relationships among the tobamoviruses may be assessed from a wide range of data, such as the nucleotide sequences of their nucleic acids, amino acid sequences, peptide mapping and amino acid composition of proteins, serological relationships, biological properties, etc (Gibbs, 1986). In this section we will focus on relationships established by sequence comparisons.

Table 23.1 lists those tobamoviruses for which the sequence of the whole gRNA (TMV, ToMV, TMGMV, PMMV and CGMMV), or parts of it (ORFs for MP and CP, and 3'ncr for SHMV and ORSV), or of the coat protein (RMV) has been reported. For TMV, ToMV, TMGMV, ORSV and CGMMV more than one strain has been totally or partially sequenced.

Table 23.2 shows estimates for the average number of amino acid substitutions per site (Nei, 1987) between any two tobamoviruses for the 126 K, 54 K, MP and CP proteins. Comparisons among strains of a virus give values that are below 5% of the smallest difference between two viruses (for TMV vs ToMV), and thus, individual viruses were

Table 23.1. *Tobamoviruses for which nucleotide or amino acid sequences have been reported, and sequences analysed in this work*

Virus	Strain	Natural hosts	Sequence*
Tobacco mosaic, TMV	vulgare	Solanaceae	gRNA (n) (1)
Tomato mosaic, ToMV	L	Solanaceae	gRNA (n) (2)
Pepper mild mottle, PMMV	S	Solanaceae	gRNA (n) (3)
Tobacco mild green mosaic, TMGMV	U2	Solanaceae	gRNA (n) (4)
Odontoglossum ringspot, ORSV	Japan	Orchidaceae	ORF for CP + 3'ncr (n) (5)
			ORF for MP (n) (6)
Ribgrass mosaic, RMV	HR	Plantaginceae	CP(a) (7)
Sunnhemp mosaic, SHMV	Cc	Leguminosae	ORF for CP + 3'ncr (n) (8)
			ORF for MP (n) (9)
Cucumber green mottle mosaic, CGMMV	SH	Cucurbitaceae	gRNA (n) (10)

Note: * (a) means amino acid sequence determined from the protein, (n) nucleotide sequence. Number in parenthesis gives reference.

Source: (1) Goelet et al. (1982); (2) Ohno et al. (1984); (3) Alonso et al. (1991); (4) Solís & García-Arenal (1990); (5) Isomura et al. (1991); (6) Isomura et al. (1990); (7) Wittmann, Hindennach & Witmann–Liebold (1969); (8) Meshi et al. (1982); (10) Ugaki et al. (1991).

Table 23.2. *Estimates of the average number of amino acid substitutions per site, d for different encoded proteins of the tobamoviruses[a], and estimates of the nucleotide substitution per site for the 3' ncr[b]*

(*a*) 126K

	ToMV	TMGMV	PMMV	CGMMV
TMV	0.0929	0.4246	0.3064	0.7967
ToMV		0.4108	0.2871	0.7824
TMGMV			0.4480	0.7899
PMMV				0.8007

(*b*) 54K

	ToMV	TMGMV	PMMV	CGMMV
TMV	0.0622	0.2999	0.2211	0.5236
ToMV		0.2992	0.1961	0.5477
TMGMV			0.2985	0.5677
PMMV				0.5327

(*c*) MP

	ToMV	TMGMV	PMMV	ORSV	SHMV	CGMMV
TMV	0.2384	0.5307	0.3729	0.4397	1.1865	0.9848
ToMV		0.4979	0.4054	0.4761	1.0303	0.9822
TMGMV			0.4214	0.5017	1.0026	0.9822
PMMV				0.3539	1.0671	0.9636
ORSV					0.9899	0.9899
SHMV						0.9004

(*d*) CP

	ToMV	TMGMV	PMMV	ORSV	SHMV	CGMMV	RMV
TMV	0.1785	0.3415	0.3558	0.3136	0.6424	0.7820	0.7459
ToMV		0.3264	0.3289	0.2792	0.8077	0.9618	0.7657
TMGMV			0.3835	0.3249	0.7722	0.9618	0.6868
PMMV				0.3585	0.8092	0.8625	0.7668
ORSV					0.8518	0.8625	0.7744
SHMV						0.7910	0.8692
CGMMV							0.9556

Table cont. overleaf.

Table 23.2. (cont.)

(e) 3'ncr

	ToMV	TMGMV	PMMV	ORSV	SHMV	CGMMV
TMV	0.3199	1.0578	0.6785	0.8080	1.3837	0.8084
ToMV		1.0692	0.6833	0.8317	1.3817	0.8996
TMGMV			1.0333	1.1382	1.7285	0.9948
PMMV				0.8863	1.7461	0.9192
ORSV					1.6126	0.7687
SHMV						1.7319

Note: a $d = -\log_e n_i/n$, where n = average number of amino acids in the protein, n_i = number of identical amino acids in the two proteins (Nei, 1987).
b Nucleotide substitutions per site calculated by Kimura's two-parameter method (Kimura, 1980).

represented by an arbitrarily chosen strain. The numbers of amino acid substitutions per site differ significantly for the four considered proteins: the 54 K is the most conserved protein for the tobamoviruses, the MP is the most divergent one, and there were no significant differences between the CP and the 126K in a Wilcoxon two-sample test (Sokal & Rohlf, 1981). When the corresponding RNA sequences were compared, nucleotide substitution values paralleled what was found for amino acid substitution values. On the other hand, values for synonymous nucleotide substitutions did not differ significantly among the four ORFs. We may consider synonymous mutations as not being subject to selection, although that may not be completely so (see below) and, thus, these data indicate that divergence between the tobamoviruses has occurred similarly throughout the whole genome. Differences in the degree of conservation of the four encoded proteins would be due to negative (purifying) selection related to functional constraints differing for each of them.

When the amino acid substitution values in Table 23.2 are analysed three levels of divergence are found for all four proteins: divergence is least between TMV and ToMV, the two most similar viruses; a second level of divergence, with amino acid substitution numbers 1.7–3.6 times higher depending on the protein than for TMV/ToMV is found among the viruses infecting Solanaceae (TMV, ToMV, PMMV and TMGMV) and ORSV; the third level of divergence is found among RMV, SHMV and CGMMV, and between any of these and the rest of the tobamoviruses: amino acid substitution numbers for these comparisons are about 2.5 times higher (for all proteins) than among tobamoviruses infecting Solanaceae. The same pattern is found for nucleotide substitutions in the 5' non-coding regions but not in the 3'ncr (Table 23.2). In the 3'ncr TMV and ToMV are still the two most similar viruses, but CGMMV is more similar to ToMV, TMV, PMMV and ORSV than TMGMV; SHMV differs widely from the rest, probably reflecting the fact that its 3'end accepts valine instead of histidine, and is able to adopt a structure similar to the 3'end of the valine-accepting turnip yellow mosaic virus (TYMV) (Meshi *et al.*, 1981). It is worth noting that ORSV, with natural hosts in the Orchidaceae, groups with the tobamoviruses infecting the Solanaceae, where this virus has experimental hosts. Assuming similar rates of amino acid substitution for each of the proteins, we may conclude that the divergence between SHMV, CGMMV and RMV, and between these three and the rest of the analysed viruses is 2–3 times older than the divergence between TMV, ToMV, TMGMV, PMMV and ORSV, and about 5 times older than the divergence of TMV and ToMV.

Sequence data, thus, show that the tobamoviruses have diverged from an ancestral type, that the whole genome has diverged as an unity, as shown by synonymous nucleotide substitution values in the different ORFs, and that most probably divergence occurred by sequential accumulation of mutations. There is no evidence for recombination having played a part in the evolution of the tobamoviruses, except for the 3'ncr of SHMV which could have been acquired from a tymo-like virus. This conclusion is further supported by the fact that the same phylogenctic tree is obtained by parsimony analysis for all four regions of the genome (Fig. 23.1A). The significance of this tree was analysed for MP and CP, and the probability of the tree being wrong is very small ($p \leq 0.02$ for CP, $p \leq 0.0001$ for MP), assuming rate-constancy and equal weights for all informative sites (Li & Gouy, 1990). The use of tobacco rattle virus (TRV) as an outgroup permits us to define two main evolutionary lines in the tobamoviruses, comprising respectively, those viruses that encapsidate the CP mRNA (Subgroup II, SHMV and CGMMV) and those that do not (Subgroup I, the rest). Viruses in Subgroup I have a pseudo OAS in the MP cistron, that in TMV is specifically recognized by the CP, and can adopt a structure similar to *bona fide* OAS (Zimmern, 1977). SHMV, in subgroup II, has also a pseudo OAS in the MP cistron, but it does not fold into the appropriate structure (Meshi, Ohno & Okada, 1982). It is possible that a duplication of the OAS occurred in the monopartite genome of

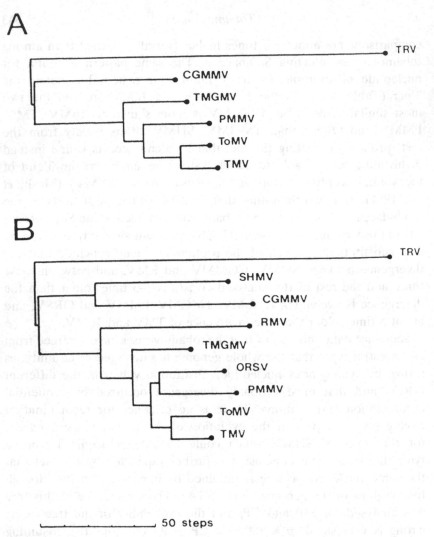

Fig. 23.1. Phylogenetic unrooted trees (Wagner parsimony) derived from protein sequences. (A) Tree obtained for all the viral-encoded proteins of the five completely sequenced viruses, that is shown here for CP. (B) Tree obtained for all the tobamoviruses with known CP sequences. TRV was used as an outgroup.

an ancestral tobamovirus or, alternatively, that the common ancestor of tobamo- and tobraviruses had a bipartite genome with two OAS, one in the ORF for MP (in RNA1) and a second one in the ORF for CP (in RNA2). Incidentally, we should mention that a mutant TMV with 2 OAS assembles efficiently (Lehto, Grantham & Dawson, 1990).

Based on the assumption of co-evolution between a virus and its host plants, Holmes (1951) has proposed a South American origin for TMV, as more species of host plants from the Solanaceae, especially in the genus *Nicotiana*, from South America, than from elsewhere have resistance genes for TMV. We are not aware of similar studies for other tobamoviruses.

The tree in Fig. 23.1A is congruent with a more complex tree based on the amino acid sequences of CP (Fig. 23.1B), which is itself congruent with published trees based on CP sequence or amino acid composition (Gibbs, 1986), or on peptide analysis of the 126 K (Fraile & García-Arenal, 1990). In all these trees, tobamoviruses having different natural hosts group apart, suggesting that a main driving force in their evolution has been related to their interaction with the host plant. This may be further supported by the MP being the most divergent of the tobamovirus-encoded proteins: although all four, 126 K, 183 K, MP and CP, may determine host range and symptom induction, the MP may have a particular interaction with host components, as it is tightly bound to the cell wall fraction, is associated with plasmodesmata and modifies the plasmodesmata to allow the virus to move from cell to cell (Dawson, 1992). In addition, for MP, large numbers of non-synonymous nucleotide substitutions are found among tobamoviruses with different host ranges. This is so not only for dispensable regions of the protein, where a relaxation of purifying selection could be expected (Li & Gojobori, 1983), but also for conserved domain II (Saito *et al.*, 1988), a region needed for the activity of MP (Berna *et al.*, 1991; Citovsky *et al.*, 1992), and this could be evidence for advantageous mutations.

The foregoing hypotheses on the evolution of the tobamoviruses are further supported by the study of the genetic variability and evolution of individual species.

Genetic variation and evolution of individual tobamoviruses

Studies of the mechanisms that generate and maintain genetic variability in populations of tobamovirus species have been of two different complementary kinds. Analysis of the variation, under controlled conditions, of viruses carrying specific genetic markers has allowed the identification and, to some degree, quantification of the mechanisms of variation. In addition, the comparison of field isolates representing natural populations of the tobamoviruses allows us to understand how variation is maintained and how it leads to the evolution of the viruses.

A primary source of variability in RNA viruses, and indeed in all replicating entities, is substitution mutation (and point insertion and deletion) caused by misincorporation of bases during nucleic acid replication. For viral RNA-dependent RNA-polymerase (RNA replicases) error rates have been shown to be much higher than for their hosts' DNA polymerases (10^{-3}–10^{-5} versus 10^{-8}–10^{-10} per nucleotide incorporated). These high error rates have been associated with a lack of proof-reading activity, and correspond to those predicted by molecular models of substitution mutation (Topal & Fresco, 1976) (see also Chapter 8). Donson *et al.* (1991) and Kearney, Donson and Dawson (1991) have quantified the error rates of TMV replication, using bacterial genes inserted into the TMV genome, assuming that such genes would be analogous to pseudo-genes and would not be subjected to purifying selection and thus, mutations in them would reflect misincorporation rates. The results show misincorporation rates in the range 10^{-3}–10^{-5}, as for other RNA viral genes.

The variability generated by mutation is usually greatly decreased by selection. It has been shown that mutants that have undetectable differences from wild-type (wt) TMV in efficiency of replication and spread within the plant were quickly outcompeted when co-infected with wt TMV (Lehto & Dawson, 1990*a*). The host plant must play an important part in selecting variants, as shown by different variants being selected in different host species (Aldaoud, Dawson & Jones, 1989). This can also be concluded from classical experiments that involved passaging in different hosts (Bawden, 1956; MacNeill & Boxall, 1974).

Homologous recombination seems to occur frequently in TMV, as shown by the elimination of long repeated sequences that had been inserted into mutant strains. The progeny of such mutants were wild type as the insertions were eliminated exactly (Dawson *et al.*, 1989; Lehto & Dawson, 1990*a*; Beck & Dawson, 1990). This was not always the case (Lehto & Dawson, 1990*b*), especially with heterologous inserts (Lehto *et al.*, 1990; Donson *et al.*, 1991). Heterologous recombination appears to occur much less frequently, except perhaps in situations of low selective constraints, as in defective deletion mutants, that are being maintained by co-infection with wt TMV (Raffo & Dawson, 1991). Thus, the work of Dawson and colleagues shows that recombination may be an important factor in TMV evolution, but also that recombination occurs less frequently the less related are the RNA variants.

In agreement with experiments showing the quick elimination through selection of mutants from TMV populations are reports of the great genetic stability of tobamoviruses. This was already indicated by various kinds of data ranging from the long term stability of the protection conferred against TMV and ToMV by some (but not all) of the resistance genes introduced in commercial cultivars (Harrison, 1981) to the similarity of sequence of stocks of TMV *vulgare* or U1 maintained in different laboratories (Dawson, 1992). This genetic stability is further shown by data from tobamovirus isolates: both field or single lesion-passaged isolates of TMGMV are heterogeneous populations of RNA species with a sequence distribution corresponding to a quasi-species (see Chapter 13), but one molecular species, the 'master copy', is the commonest in each population (Rodríguez-Cerezo & García-Arenal, 1989). When compared with similar data on the sequence variability of isolates of bacterial or animal viruses, it is clear that the heterogeneity of TMGMV isolates is considerably smaller, suggesting that negative selection is more actively checking genetic divergence generated by mutation.

The genetic stability of tobamoviruses is further shown by the comparison of isolates from natural (field) populations. Work on PMMV isolates from pepper (*Capsicum anuum* L.) during consecutive epidemics, and with TMGMV infecting wild *Nicotiana glauca* Grah. plants in Spain (Chapter 7) gives a similar view for both viruses: populations are formed of a large number of closely related variants, indicating a high rate of mutation, but genetic diversity in the populations does not increase with time (PMMV) or with distance between sites of isolation (TMGMV) (Rodríguez-Cerezo *et al.*, 1989, 1991). Comparison of TMGMV isolates from *N. glauca* plants from Australia, California and Spain by the pattern of mismatches with probes complementary to different parts of the genome of a reference strain shows that isolates from different parts of the world are outstandingly similar, and no conserved and variable regions could be defined in the genome (García-Arenal *et al.*, 1990). The analysis of RNA sequences of two regions of the 183K ORF for a set of TMGMV isolates from Australia, Spain and Greece (our unpublished results) show that negative selection (i.e. selection against change), probably resulting from functional constraints, operates to reduce variability. These constraints are related to the maintenance of the encoded protein sequence, but the analysis of synonymous mutations revealed selection that must be related to other kinds of constraints. Other factors contributing to TMGMV's genetic stability, notably periodic positive selection (i.e. selection resulting in change) have been proposed (Moya, Rodríguez-Cerezo & García-Arenal, 1993).

Thus, it may be concluded that the great potential of tobamoviruses to vary, due to high mutation and recombination rates, is checked by both negative and positive selection, resulting in populations that are very stable both in time and space. Host plant : viral interactions must be important in these selective forces.

Conclusions

Reported data show that the tobamoviruses have diverged from an ancestral type by accumulating mutations, and no major recombinational events among members of the group (or their ancestors) seem to have occurred. Adaptation to their host plants could be a major driving force in their evolution. Data also show that, in spite of a high potential to vary, tobamoviruses are genetically very stable, and that this reduction of genetic heterogeneity may result from both negative and positive selection. Again, interaction with the host plant seems to play an important role in these selective forces. Thus, it is possible that tobamoviruses have evolved by an alternation of periods of fast evolution as they colonize a new niche (new host plants), and of periods of high genetic stability.

Acknowledgements

This work was in part supported by grant AGR90–0152, from CICYT, Spain. M.A.A. was in receipt of a fellowship of Formación de Personal Investigador, Ministerio de Educación y Ciencia, Spain. Phylogenetic analyses were performed using the PHYLIP 3.4 package of J. Felsenstein, University of Washington.

References

Aldaoud, R., Dawson, W.O. & Jones, G.E. (1989). *Intervirology*, **30**, 227–33.
Alonso, E., García-Luque I., de la Cruz, A., Wicke, B., Avila-Rincón M.J., Serra, M.T., Castresana, C. & Díaz-Ruiz, J.R. (1991). *J. Gen. Virol.*, **72**, 2875–84.
Bawden, F.C. (1956). *Nature*, **177**, 302–4.
Beck, D.L. & Dawson, W.O. (1990). *Virology*, **177**, 462–9.
Berna, A., Gafny, R., Wolf, S., Lucas, W.J., Holt, C.A. & Beachy, R.N. (1991). *Virology*, **182**, 682–9.
Citovsky, V., Wong, M.L., Shaw, A.L., Ventakataram-Prasad, B.V. & Zambrisky, P. (1992). *Plant Cell*, **4**, 397–411.
Culver, J.N., Lindbeck, A.G.C. & Dawson, W.O. (1991). *Ann. Rev. Phytopathol.*, **29**, 193–217.
Dawson, W.O. (1992). *Virology*, **186**, 359–67.
Dawson, W.O., Lewandowski, D.J., Hilf, M.E., Bubrick, P., Raffo, A.J., Shaw, J.J., Grantham, G.L. & Desjardins, P.R. (1989). *Virology*, **172**, 285–92.

Donson, J., Kearney, C.M., Hilf, M.E. & Dawson, W.O. (1991). *Proc. Natl. Acad. Sci. USA*, **88**, 7204–8.

Fraile, A. & García-Arenal, F. (1990). *J. Gen. Virol.*, **71**, 2223–8.

Francki, R.I.B., Fauquet, C.M., Knudson, D.L. & Brown, F. (1991). *Classification and Nomenclature of Viruses. Fifth Report of the International Committee on Taxonomy of Viruses.* Springer-Verlag, Wien, pp. 357–359.

García-Arenal, F., Rodríguez-Cerezo, E., Aranda, M.A. & Fraile, A. (1990). In *Abstracts of the VIIIth International Congress of Virology*, p.89.

Gibbs, A. (1986). In *The Plant Viruses.2. The Rod-Shaped Plant Viruses*, M.H.V. Van Regenmortel and H. Fraenkel-Conrat (eds.). Plenum Press, New York, pp. 168–180.

Goelet, P., Lomonossoff, G.P., Butler, P.J.G., Akam, M.E., Gait, M.J. & Karn, J. (1982). *Proc. Natl. Acad. Sci. USA*, **79**, 5818–22.

Harrison, B.D. (1981). *Ann. Appl. Biol.*, **99**, 195–209.

Holmes, F.O. (1951). *Phytopathology*, **41**, 341–9.

Isomura, Y., Matumoto, Y., Maruyama, A., Chatani, M., Inouye, N. & Ikegami, M. (1990). *Nucl. Acids Res.*, **18**, 7748.

Isomura, Y., Matumoto, Y., Maruyama, A., Chatani, M., Inouye, N. & Ikegami, M. (1991). *J. Gen. Virol.*, **72**, 2247–9.

Kearney, C.M., Donson, J. & Dawson, W.O. (1991) Workshop on *Coevolution of Viruses, their Hosts and Vectors*, Madrid, December 1991.

Kimura, M. (1980). *J. Mol. Evol.*, **16**, 111–120.

Lehto, K. & Dawson, W.O. (1990a). *Virology*, **174**, 169–76.

Lehto, K. & Dawson, W.O. (1990b). *Virology*, **175**, 30–40.

Lehto, K., Grantham, G.L. & Dawson, W.O. (1990). *Virology*, **174**, 145–57.

Li, W.H. & Gojobori, T. (1983). *Mol. Biol. Evol.*, **1**, 94–108.

Li, W.H. & Gouy, M. (1990). *Meth. Enzym.* **183**, 645–59.

MacNeil, B.H. & Boxal, M. (1974). *Canad. J. Bot.*, **52**, 1305–7.

Matthews, R.E.F. (1991). *Plant Virology*, Academic Press, New York, pp. 673–682.

Meshi, T., Ohno, T., Iba, H. & Okada, Y. (1981). *Molec. Gen. Genet.*, **184**, 20–5.

Meshi, T., Ohno, T. & Okada, Y. (1982). *Nucl. Acids Res.*, **10**, 6111–17.

Moya, A., Rodríguez-Cerezo, E. & García-Arenal, F. (1993). *Mol. Biol. Evol.*, **10**, 449–56.

Nei, M. (1987). *Molecular Evolutionary Genetics*, Columbia University Press, New York, pp. 39–44.

Ohno, T., Aoyagi, M., Yamanashi, Y., Saito, H., Ikawa, S., Meshi, T. & Okada, Y. (1984). *J. Biochem.*, **96**, 1915–23.

Raffo, A.J. & Dawson, W.O. (1991). *Virology*, **184**, 277–89.

Rodríguez-Cerezo, E. & García-Arenal, F. (1989). *Virology*, **170**, 418–23.

Rodríguez-Cerezo, E., Moya, A., Elena, S.F. & García-Arenal, F. (1991). *J. Mol. Evol.*, **32**, 328–32.

Rodríguez-Cerezo, E., Moya, A. & García-Arenal, F. (1989). *J. Virol.*, **63**, 2198–203.

Saito, T., Imai, Y., Meshi, T. & Okada, Y. (1988). *Virology*, **167**, 653–6.

Sokal, R.R. & Rohlf, F.J. (1981). *Biometry*, Freeman and Co., New York, pp. 432–437.

Solís, I. & García-Arenal, F. (1990). *Virology*, **177**, 553–8.

Topal, M.D. & Fresco, J.R. (1976). *Nature*, **263**, 285–9.
Ugaki, M., Tomiyama, M., Kakutani, T., Hidaka, S., Kiguchi, T., Nagata, R., Sato, T., Motoyoshi, F. & Nishiguchi, M. (1991). *J. Gen. Virol.*, **72**, 1487–95.
Van Regenmortel, M.H.V. & Fraenkel-Conrat, H. (1986). *The Plant Viruses.2. The Rod-Shaped Plant Viruses*, Plenum Press, New York.
Wittmann, H.G., Hindennach, I. & Wittmann-Liebold, B. (1969). *Z. Naturforsch.* **24b**, 877–85.
Zimmern, D. (1977). *Cell*, **11**, 463–82.
Zimmern, D. (1988). In *RNA Genetics, Vol. II. Retroviruses, Viroids and RNA Recombination*, CRC Press, Boca Raton, Florida, pp. 211–240.

24

The luteovirus supergroup: rampant recombination and persistent partnerships

MARK GIBBS

Introduction

The evolutionary history of viruses, inferred from comparisons of genomic sequences, appears unlike that of most other life forms. Novel viruses seem often to arise through the conjunction of genes from very different viral lineages. Viruses with RNA genomes as well as viruses with DNA genomes have evolved in this way, and either homologous or non-homologous recombination may be involved. Botstein (1980) named this process modular evolution, and its importance is clear from evidence that, on several occasions, it has led to the evolution of new viral genera (Gibbs, 1987). However, little is known about the frequency of these events nor is it known if they occur in some pattern, and the interpretation of phylogenies disturbed by modular evolution is in its infancy.

In this chapter I present perhaps the first evidence of a pattern to viral modular evolution, and the first evidence of a connection between interviral gene transfer and an ecological interaction between viruses.

This evidence comes from the luteovirus supergroup which consists of plant viruses with small messenger sense RNA genomes and icosahedral virions (Goldbach & Wellink, 1988; Martin et al., 1990; Dolja & Koonin, 1991). Six viral groups and some ungrouped viruses belong to the supergroup (Brunt, Crabtree & Gibbs, 1990; Francki et al., 1991). Many of the viruses are agriculturally important (Duffus, 1977; Matthews, 1991). The genomes of 17 of them have been completely sequenced. Analysis of this data shows the supergroup to consist of two main clusters that are in some ways only distantly related.

In the first part of this chapter I describe an analysis of gene transfer in the supergroup. I estimate that at least seven, but perhaps as many as ten, transfer events have occurred among the ancestors of the

351

viruses whose genomes have been fully sequenced. This frequency is comparable to that found among retroviruses (McClure *et al.*, 1988 and Chapter 27). At least three transfer events have occurred between the two main clusters. This may be the first evidence of a pattern to modular evolution. Almost all of these recombinational events divide the genomes into 5' and 3' modules. The 5' module includes the RNA-directed RNA polymerase gene, and the 3' module the virion protein gene. Within each module gene phylogenies are congruent, but between the modules they are not. Several models of recombination can be developed from these phylogenetic incongruities.

In the second part of the chapter, I describe ecological interactions that link the two main clusters and may explain the recurring gene transfer between the main clusters. These interactions involve frequent or obligatory co-replication and transmission of a virus from one cluster with the help of a virus from the other cluster. Two homologous associations of this kind are known. The taxonomy of the participating viruses indicate that these relationships may be a conserved feature of the supergroup with a very early origin, and correlates to some degree with the pattern of intercluster transfer. The dependent and helper viruses are transmitted by aphids in an unusual way necessitating co-replication that in turn may facilitate gene transfer. Alternative explanations are also discussed.

Coherence and diversity in the luteovirus supergroup

The definition of the supergroup is complicated by conflicting evidence about its phylogeny. The virion proteins of the sequenced viruses are clearly related (Martin *et al.*, 1990; Dolja & Koonin 1991); however, the RNA-directed RNA polymerases comprise two distinct clusters that may not be related, except in the most distant way. Recent disagreements about the relatedness of these polymerase clusters emphasizes the evolutionary distance between them (Candresse, Morch & Dunez, 1990; Bruen, 1991; Koonin, 1991).

Fig. 24.1A and B are classifications of the RNA-directed RNA polymerase and virion protein shell domain sequences, respectively. These were inferred by the neighbour-joining method of Saitou and Nei (1987) from estimates of the divergence of every pair of sequences within each set, after first progressively aligning each set (Feng & Doolittle, 1987; Higgins & Sharp, 1989). Each tree contains two main clusters, one containing PEMV and most of the luteoviruses, which I call the enamo cluster, and the other containing RCNMV, MCMV, the carmoviruses, necroviruses, and tombusviruses, which I call the carmo cluster.

The relatedness of the viruses is supported by several features apart from virion protein sequence homology. The sequenced viruses express common genes in similar ways and have similar genome plans (Fig. 24.2). Most of the characterized members have monopartite genomes, but some are bipartite. None has a genome segment greater than six kilobases in length. They have only four to six coding open reading frames (ORFs), and all express their virion proteins from a subgenomic RNA co-terminal with the 3' end of the genome. The structure of the virions of tomato bushy stunt tombusvirus (TBSV), southern bean mosaic sobemovirus (SBMV), turnip crinkle carmovirus (TCV) and carnation mottle carmovirus (CarMV) have been determined by X-ray crystallography, confirming some of the virion protein relationships (Abad-Zapatero *et al.*, 1980; Harrison, 1983; Hogle, Maeda & Harrison, 1986; Rossmann & Johnson, 1989; D.I. Stuart, University of Oxford, UK, personal communication, 1992).

Some important features are not shared across the supergroup. Six of the viruses encode a pair of proteins essential for cell-to-cell movement (Hacker *et al.*, 1992). The smaller segment of the bipartite virus red clover necrotic mottle dianthovirus (RCNMV) also encodes a protein essential for cell-to-cell movement (Osman & Buck, 1987), but this gene appears unrelated to the more widely shared pair of movement-genes. Viruses of the enamo cluster probably encode a 3C-like protease (Gorbalenya & Koonin, 1990). Viruses of this same subset have a protein linked to the 5' end of their genomes (Mang, Ghosh & Kaesberg, 1982; Reisman & de Zoeten, 1982; Murphy, D'Arcy & Clark, 1989). Some members of the carmo cluster have methylated capped 5' ends (Nutter *et al.*, 1989), but others do not (Lesnaw & Reichmann, 1970). None of the characterized members of the supergroup has polyadenylated or tRNA-like 3' ends.

Data about non-molecular biological features of the viruses is patchy (Koenig, 1988). The experimental host-range and means of transmission of each of the characterized viruses have been investigated. Most of the viruses infect dicots, and a few infect monocots. Some have a very narrow host range, such as carrot red leaf luteovirus (CRLV) which only infects certain Apiaceae (Waterhouse & Murant, 1981). Others have very broad host ranges, such as turnip crinkle carmovirus (TCV) that infects species from more than 20 families (for review see Hollings & Stone, 1972). Some of the viruses are transmitted by insects, either certain Coleoptera or Homoptera, and others are transmitted by certain fungi, and some appear to spread through soil without the aid of a vector.

Fig. 24.1. Classifications of RNA-directed RNA polymerase and virion protein shell domain amino acid sequences from the seventeen sequenced members of the supergroup (dendrograms A and B, respectively). Dotted lines linking the two dendrograms represent likely gene transfer events thus constructing a model of gene transfer within and between the two main clusters of the supergroup. The dendrograms were deduced by the neighbour-joining method from estimates of the evolutionary distance between each pair of sequences calculated after the progressive alignment of each complete set of sequences. The pattern of transfer events was assessed from the incongruency of lineages given by different genes. The links are drawn arbitrarily at the midpoints of lines joining the nodes where incongruities are evident, as it is not possible to assess at what point a transfer event occurred within an incongruent lineage.

The lineages leading to PLRV and BWYV are assumed to be congruent as this minimizes a rate anomaly between the polymerase and virion protein trees within this subcluster. The lineages leading to CarMV, TCV, MNSV and TNV-D are assumed to be congruent within the carmo cluster.

The means of transmission of each virus is indicated next to the virion protein tree. Each luteovirus is transmitted by one or two Homoptera (aphid) species. Viruses transmitted by Coleoptera (beetles) can be transmitted by several species. Some viruses are transmitted by mechanical means possibly involving wounding and some of these are also transmitted without the aid of a vector through the soil. Fungi of the genus *Olpidium* are pathogenic.

Abbreviations: BWYV, beet western yellows luteovirus; PLRV, potato leaf roll luteovirus; BYDV-RPV, barley yellow dwarf luteovirus type RPV; PEMV, pea enation mosaic enamovirus; SBMV, southern bean mosaic sobemovirus; CarMV, carnation mottle carmovirus; TCV, turnip crinkle carmovirus; MNSV, melon necrotic spot carmovirus; MCMV, maize chlorotic mottle machlomovirus; TNV-A, tobacco necrosis necrovirus strain A; TBSV, tomato bushy stunt tombusvirus; CyRSV, cymbidium ring spot tombusvirus; CNV, cucumber necrosis tombusvirus; TNV-D, tobacco necrosis necrovirus strain D; BYDV-PAV, barley yellow dwarf luteovirus type PAV; BYDV-MAV, barley yellow dwarf luteovirus type MAV; RCNMV, red clover necrotic mosaic dianthovirus.

References to sequences: BWYV, Veidt *et al.* (1988); PLRV, Mayo *et al.* (1989); BYDV-RPV, Vincent *et al.* (1991); PEMV, Demler & de Zoeten (1991); SBMV, Wu, Rinehart & Kaesberg (1987); CarMV, Guilley *et al.* (1985); TCV, Carrington *et al.* (1989); MNSV, Riviere & Rochon (1990); MCMV, Nutter *et al.* (1989); TNV-A, Meulewaeter, Seurinck & Van Emmel (1990); TBSV, Hearne *et al.* (1990); CyRSV, Grieco, Burgyan & Russo (1989); CNV, Rochon & Tremaine, (1989); TNV-D, Coutts *et al.* (1991); BYDV-PAV, Miller *et al.* (1988); BYDV-MAV, Ueng *et al.* (1992); RCNMV, Xiong & Lommel, (1989).

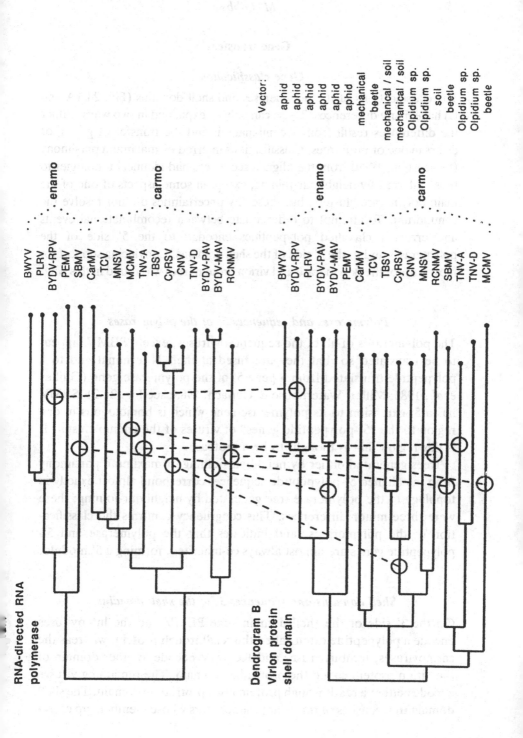

RNA-directed RNA polymerase

Dendrogram B
Virion protein shell domain

Vector:

BWYV	aphid
BYDV-RPV	aphid
PLRV	aphid
BYDV-PAV	aphid
BYDV-MAV	aphid
PEMV	aphid
CarMV	mechanical
TCV	beetle
TBSV	mechanical / soil
CyRSV	mechanical / soil
CNV	Olpidium sp.
MNSV	Olpidium sp.
RCNMV	soil
SBMV	beetle
TNV-A	Olpidium sp.
TNV-D	Olpidium sp.
MCMV	beetle

enamo

carmo

enamo

carmo

BWYV
PLRV
BYDV-RPV
PEMV
SBMV
CarMV
TCV
MNSV
MCMV
TNV-A
TBSV
CyRSV
CNV
TNV-D
BYDV-PAV
BYDV-MAV
RCNMV

Gene transfers

Gene classifications

The classifications of the polymerases and shell domains (Fig. 24.1A and B) have many differences. These can only be explained in two ways. Either the differences result from recombination, and the transfer of genes, or errors in one or both trees. Classifications inferred by maximum parsimony (Felsenstein, 1989) from the aligned sequences had identical topologies to those inferred by neighbour-joining, except in some aspects of one of the main polymerase clusters, but these few uncertainties did not resolve the conundrum. To attempt to differentiate between recombinational events and errors I classified polypeptides encoded to the 5' side of the polymerases and to the 3' side of the shell domains. Convincing support for most parts of the polymerase and virion protein trees was found in this way.

Polymerases and sequences 5' of the polymerases

The polymerases of all of the sequenced viruses, except SBMV, appear to be translated so that they are fused at their N-terminal end to a polypeptide translated from a gene 5' of the polymerase gene (Guilley *et al.*, 1985; Miller, Waterhouse & Gerlach, 1988; see Fig. 24.2). SBMV has a 5' extension to its polymerase gene which is homologous in one region to the "5'-polypeptide genes" of viruses of the enamo cluster. It is this region that is thought to be a protease.

Trees generated either by neighbour-joining or maximum parsimony from the aligned 5'-polypeptide sequences correspond almost exactly in topology to the polymerase tree generated by neighbour-joining; there were three minor differences. This congruency confirms the classification of the polymerases, and indicates that the polymerase and 5'-polypeptide genes are almost always co-inherited, forming a 5' module.

Shell domains and sequences 3' of the shell domains

On the 3' side of the shell domain gene PEMV and the luteoviruses encode a polypeptide often called the readthrough protein, whereas the carmoviruses, tombusviruses and RCNMV encode another domain of the virion protein called the protruding domain. The remaining viruses encode neither a readthrough protein nor a protruding domain. The shell domain tree consists of three large subclusters whose membership corre-

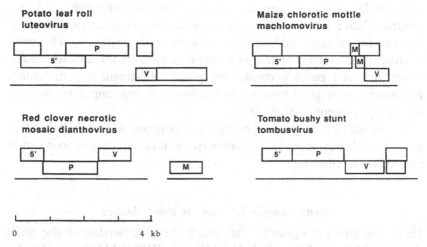

Fig. 24.2. Diagrams of genome plans of some members from the luteovirus supergroup. Boxes represent the major open reading frames (ORFs) and the lines beneath the boxes represent the genomes. ORFs labelled P encode RNA-directed RNA polymerases, labelled V encode virion proteins, labelled M encode movement proteins, and labelled 5' encode the 5'-polypeptides. References reporting genome sequences and plans are given in Fig. 24.1.

lates exactly with these features. Classifications of the readthrough proteins and protruding domains, by neighbour-joining and maximum parsimony, match exactly the classification of their partner shell domains, supporting these parts of the shell domain tree. This congruency confirms the shell domain tree and indicates that the shell domain gene is co-inherited with genes encoded on its 3' side forming a 3' module.

The clustering of shell domains from SBMV, tobacco necrosis necrovirus strain-A (TNV-A) and tobacco necrosis necrovirus strain-D (TNV-D) is supported beyond doubt by similarities between their sequences, but the position of the MCMV shell domain within the tree is less certain.

Groupings in the shell domain tree correlate weakly with the type of vector employed by the virus. Fig. 24.1 shows that viruses with sister shell domains are likely to have vectors of the same type.

Interpreting the differences between the trees

Gene transfer affects the co-evolution of sets of genes. Genes bound by co-inheritance are mostly involved in the same speciation events and have the same phylogeny. Some steps that affect the relationships of genes within a genome, such as duplication, are exceptions. When a genome obtains a new gene by gene transfer this is a speciation event for the transferred gene, adding a new lineage to its phylogeny. However, because it has exchanged genomes the phylogenies of the transferred and non-transferred genes become incongruous. Distinguishing congruent and incongruent gene lineages is the basis of detecting gene transfer through phylogenetic analysis.

Incongruities between the phylogeny of polymerase and shell domains are of two classes: those between the two main clusters and those within each of these clusters.

Gene transfer between the main clusters

There are two incongruities that affect the membership of the main clusters. The subcluster BYDV-PAV with BYDV-MAV is internally consistent, but it is rooted within the enamo cluster in the shell domain tree and within the carmo cluster in the polymerase tree. Conversely SBMV branches within the enamo cluster in the polymerase tree but within the carmo cluster in the shell domain tree.

A third incongruity between the clusters is found in the relative divergence of the polypeptides. As expected, within a cluster the polymerases diverge more slowly than the shell domains. However, the main clusters are far more distant from each other in the polymerase tree than in the shell domain tree. This incongruity is understated in Fig. 24.1 because the method used to estimate divergence does not account for multiple substitutions, and hence has underestimated the distance between the polymerase clusters.

The least complicated explanation is that each of these three incongruities between the main clusters indicates a recombinational transfer of a virion protein gene between the two clusters. Hence the lineage leading to BYDV-PAV and BYDV-MAV formed when a virion protein gene was transferred from an ancestor in the enamo cluster to an ancestor of the carmo cluster, and the lineage leading to SBMV formed when a virion protein gene was transferred in the opposite direction. Congruity between the virion protein tree and the readthrough protein tree indicates that these genes were transferred together as a module when the new BYDV lineage formed. The inconsistent divergence of the polymerase and shell domain clusters is best explained by a similar transfer early in the evolution of the supergroup, but the direction of this event cannot be deduced.

Gene transfer within the main clusters

Within each cluster incongruities can be interpreted in several ways, each assuming a different set of congruent lineages. Fortunately, the confusion is diminished because both main clusters contain internally consistent subclusters with congruent polymerase and shell domain phylogenies.

Within the enamo cluster, the pairing of PEMV and the line leading to BWYV, BYDV-RPV and PLRV is a feature of both trees. The subcluster BWYV with PLRV and BYDV-RPV is not internally consistent possibly indicating that, after the divergence leading to the extant taxa, gene transfer has occurred. Any two of these three lineages may be congruent. Within the carmo cluster two subclusters, CarMV with TCV, and TBSV with CyRSV and CNV, are congruent. Six larger sets of lineages in this cluster, each including more than three extant taxa, and one or both of the consistent subclusters, may also be congruent. One of these options implies three transfer events and each of the others imply four transfer events. Other smaller sets of lineages could be congruent but this is less likely as it would suggest even greater numbers of transfer events.

Hence, together the clusters add a further four or five transfer events to the three intercluster events. Fig. 24.1 presents one model of the transfer events mapped between the two trees. The lines linking the trees in the model represent likely gene transfer events. The pattern of transfer events is based on assumptions about the congruency of lineages. The polymerase and shell domain lineages linked by the lines are those claimed to be incongruent within the model. The parts of the trees claimed to be incongruent in the model do not coincide with the few uncertainties in the trees. One exception to this is the transfer event involving ancestors of BWYV, BYDV-RPV and PLRV. Although the different topologies of this cluster in the polymerase and shell domain trees were supported by the 5'-polypeptide and readthrough protein trees, the branch points that lead to these three taxa are so close that an error remains plausible.

Interviral associations

The frequency of interviral gene transfer in the evolutionary history of this supergroup appears greater than any previously reported among viruses. It is greater than that among plant viruses with a similar ecology, such as viruses of the alpha-like supergroup (Morozov, Dolja & Atebekov, 1989), far greater than that among viruses whose polymerase genes are closely related to those of the luteovirus supergroup, such as the flaviviruses (Blok *et al.*, 1992; see Chapter 19), and greater than any similar cluster of related viral groups (McClure *et al.*, 1988). Not only is the frequency of transfer extraordinary, but so is the pattern. Transfers have occurred between the two main clusters of the supergroup at least three times, although they are only distantly related, yet there is no evidence of transfer with other major groupings of viruses with which many share hosts.

However, viruses of the supergroup may have had unusual opportunities for transfer that could explain the pattern of recurring inter-cluster transfer.

Luteovirus associations

It has long been known that certain pairs of luteoviruses are often associated, and interact to allow dependent transmission; one virus is transmitted with the help of the other. This in itself is remarkable because luteoviruses are transmitted by aphids in a circulative non-propagative way (Harrison, 1987). Luteoviruses do not replicate in their aphid vectors. Virions ingested by the aphid pass through the gut wall into the haemolymph in the body cavity (Gildow, 1985). They are then absorbed from the haemolymph into the accessory salivary gland and excreted in the saliva and, once there, can infect another plant (Gildow & Rochow, 1980). This process is specific; each luteovirus is efficiently transmitted by only one or two aphid species (Waterhouse *et al.*, 1988).

Dependent transmission involving luteoviruses is achieved by transcapsidation. When the viruses replicate together in the same plant, genomes of the dependent luteovirus are encapsidated in capsids encoded by the helper luteovirus. This was detected when antiserum raised against virions of the helper virus blocked transmission of both viruses (Rochow, 1970), and was confirmed by immunohybridization (Creamer & Falk, 1990). Transcapsidation alters the vector specificity of the dependent virus (Rochow, 1965) indicating that the capsids provide the information that allows passage through the vector aphid.

There are at least five luteoviruses that interact in this way (Rochow, 1982). Confusingly, all five are called barely yellow dwarf luteovirus (BYDV). Each BYDV is transmitted by a different aphid species and they are distinguished accordingly (Rochow, 1970; Waterhouse *et al.*, 1988), for example, the BYDV transmitted by Rhopalosiphum padi (L.) is known as BYDV-RPV. The sequences of several BYDV genomes have been determined showing that they fall into two clusters. BYDV-RPV has a polymerase from the enamo cluster (Vincent, Lister & Larkins, 1991), whereas BYDV-PAV and BYDV-MAV have polymerases from the carmo cluster (Miller *et al.*, 1988; Ueng *et al.*, 1992). Serological tests have shown that all BYDVs fall into two sets that coincide with the polymerase phylogeny (Rochow, 1970; Waterhouse *et al.*, 1988).

Five combinations of BYDVs from the opposing clusters have been found to transcapsidate (Rochow, 1982; Wen & Lister 1991). Infections by some of these combinations are frequently found in the field (Rochow & Muller, 1974; Rochow 1979; P.M. Waterhouse unpublished results). Field plants naturally co-infected by BYDV-RPV with either BYDV-PAV or MAV have been found by Rochow (1979) and Creamer and Falk (1990) and virions produced by transcapsidation between these viruses have been detected in these plants by immunohybridization (Creamer & Falk, 1990). Transmission experiments indicate that either type of BYDV can act as helper (Creamer & Falk, 1990).

Umbravirus homology

The unusual interaction described above might mistakenly be thought to have evolved recently and to be a unique feature of the BYDVs. However, a similar interaction also occurs between another pair of viruses from the two polymerase clusters, and the phylogeny of these and the BYDVs indicate that dependent transmission by transcapsidation may be a conserved feature of the luteovirus supergroup and may have a very early origin.

One member of the pair is a luteovirus and the other a member of a newly recognized viral group, the umbraviruses. Each of the six known umbraviruses only occurs co-infecting with a certain luteovirus on which it depends for transmission (Falk & Duffus, 1981). The phylogeny of umbraviruses was previously unknown. I have cloned and sequenced the majority of the genome of carrot mottle umbravirus (CMotV) including its RNA-directed RNA polymerase gene and found this to belong to the carmo cluster. Carrot red leaf luteovirus (CRLV), CMotV's helper, is serologically related to BWYV, PLRV and BYDV-RPV (Waterhouse & Murant, 1981), and Dr R.R. Martin (Agriculture Canada, Vancouver,

Canada) has sequenced a large part of the CRLV polymerase gene confirming that this virus belongs to the enamo subcluster containing BWYV, PLRV and BYDV-RPV (R.R. Martin, personal communication 1992).

The relationship between umbra and luteoviruses seems obligatory for the umbravirus of each pair. When the viruses are separated the luteovirus helper is aphid transmissible, as before, but the umbravirus cannot be transmitted alone in this way (Falk & Duffus, 1981). Umbraviruses are manually transmissible but inocula are very unstable so this type of transmission is unlikely to occur in the field (Murant *et al.*, 1969). They seem not to produce their own capsids (Falk, Morris & Duffus, 1979). Like the dependent transmission of BYDVs, the basis of the interaction is transcapsidation. When the two viruses replicate together in a plant, umbravirus genomes are packaged in luteovirus capsids (Waterhouse & Murant, 1983).

Thus, like the interaction between the BYDVs, the interaction between CMotV and its helper luteovirus links the two main clusters and it is probable that other umbravirus–luteovirus pairs will do so too. The two types of interaction are clearly homologous, both involving transcapsidation of genomes of viruses from the carmo polymerase cluster in the capsids of viruses from the enamo polymerase cluster. There is a remote possibility that this relationship has evolved twice, possibly by gene transfer, but I have found no other close sequence similarities between CMotV sequences and luteovirus sequences apart from the relatedness of the CMotV polymerase gene to those of BYDV-PAV, BYDV-PAV and BYDV-MAV. Thus the relatedness of CRLV to BYDV-RPV and CMotV to BYDV-PAV and BYDV-MAV indicates that these partnerships have co-evolved in a parallel manner.

Fig. 24.3 is a classification of the polymerases and virion protein shell domains, including the phylogenies of CMotV and CRLV, and describes the interviral associations relative to the polymerase and virion protein phylogenies. The polymerase subcluster containing BYDV-PAV, BYDV-MAV and RCNMV is the most deeply rooted subcluster in the carmo cluster, and the next most deeply rooted taxon is CMotV. Hence dependent transmission through transcapsidation can be traced back to the base of the carmo cluster. In other words, an ancestor of the carmoviruses, dianthoviruses, tombusviruses, necroviruses, BYDV-PAV and BYDV-MAV, and CMotV may have been dependently transmitted by transcapsidation. Similarly the interaction can be traced to the lineage rooting BWYV, BYDV-RPV, CRLV and PLRV,

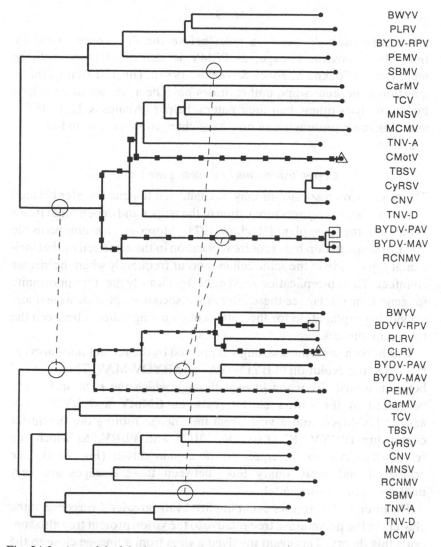

BWYV
PLRV
BYDV-RPV
PEMV
SBMV
CarMV
TCV
MNSV
MCMV
TNV-A
CMotV
TBSV
CyRSV
CNV
TNV-D
BYDV-PAV
BYDV-MAV
RCNMV

BWYV
BDYV-RPV
PLRV
CLRV
BYDV-PAV
BYDV-MAV
PEMV
CarMV
TCV
TBSV
CyRSV
CNV
MNSV
RCNMV
SBMV
TNV-A
TNV-D
MCMV

Fig. 24.3. A model of the interviral associations and the likely intercluster transfer events between the main clusters using the dendrograms of the polymerase and virion protein shell domain sequences as a framework. This model differs from Fig. 24.1 in that carrot mottle umbravirus (CMotV) and carrot red leaf luteovirus (CRLV) are included. The classification of the CRLV shell domain is inferred from a classification of the CRLV RNA-directed RNA polymerase. Interviral associations are formed between CMotV and CRLV, identified with triangles, and BYDV-PAV, BYDV-MAV and BYDV-RPV, identified by squares. These associations can be traced through the lineages marked by large blocks, and probably through the lineages marked by small blocks. Intercluster transfer events are represented by dotted lines. The intercluster event at the base of the trees is not shown in Fig. 24.1 because this event has not led to a topological incongruity between the trees but to an incongruity in the phylogenetic distance between the enamo and carmo clusters.

and maybe even deeper to a point before the divergence of PEMV from the luteovirus lineage, as PEMV is also often found with an umbravirus (Cockbain, Jones & Woods, 1986). This last step might be questioned because some umbraviruses have been shown to be helped by luteoviruses other than their natural helper (Adams & Hull, 1972), and thus some umbraviruses may have changed helpers in nature.

Co-replication may facilitate gene transfer

These interviral associations may be connected to gene transfer because transcapsidation requires replication of the helper and dependent viruses together in the same plant (Rochow, 1973). Moreover, it seems plausible that transcapsidation may require replication in the same cell or that incidental replication in the same cell may occur frequently when the viruses co-infect. This co-replication must offer significantly greater opportunity for gene transfer. Hence these interviral associations provide a clear and convincing explanation for the pattern of recurring transfer between the two main clusters.

This hypothesis is most strongly supported by the recombination event leading to the evolution of BYDV-PAV and BYDV-MAV. This occurred between a virus from the lineage through which the relationship can be traced in the carmo cluster, between CMotV and BYDV-PAV and BYDV-MAV, and a virus from the lineage rooting the subcluster comprising BWYV, BYDV-RPV, CRLV and PLRV, to which the relationship can also be traced in the enamo cluster (Fig. 24.3). The ecological and evolutionary links between these lineages are very unlikely to be co-incidental.

The intercluster transfer event implied by the greater divergence of the clusters in the polymerase tree relative to the virion protein tree also supports this theory. This event involved a virus from a lineage close to the base of the carmo cluster, a point to which the relationship can be traced.

The feature of interviral association may well have been conserved in lineages of the carmo cluster apart from those leading to BYDV-PAV, BYDV-MAV and CMotV. This could account for the transfer event leading to SBMV.

Viral helicases

An alternative or contributory reason for the frequency of transfer, both between and within the clusters, is that it is caused by some aspect of the replication of these viruses. This is supported by the frequent transfer among the lineages within each cluster. In fact, the recombination frequency within each cluster may even be greater than I have described, for although each of the sequenced viruses has only four to six genes, viruses of five of the six viral groups have at least one gene for which no homologous gene can be found in the other viral groups. These unique genes may have arisen through recombination, possibly with host nucleic acids, or have arisen *de novo* (Keese & Gibbs, 1992). The possibility of recombination with host nucleic acids is supported by the finding that some PLRV genomes in an isolate from Scotland appear to have a part of a chloroplastic gene at their 5' end (Mayo & Jolly, 1991). The natural generation of defective interfering RNAs by some members of the carmo cluster (Hillman, Carrington & Morris, 1987; Burgyan, Rubino & Russo, 1991) also supports this notion. Unlike among animal viruses, defective interfering RNAs are not commonly generated by plant viruses.

One possible reason for the frequency of recombination is that viruses of the luteovirus supergroup seem to lack the sequence motifs character-istic of nucleotide binding proteins from the helicase family found in the replication genes of most other messenger sense RNA viruses (Habili & Symons, 1989; Gorbalenya & Koonin, 1989). Some of the cellular homologues of these proteins are involved in DNA recombination and others are probably involved in RNA duplex unwinding (Gorbalenya & Koonin, 1989). Hence the helicases of other messenger sense RNA viruses may act as a barrier to non-homologous recombination by dissociating inadvertently paired RNA strands or in some other way as a side-effect of regulating homologous recombination.

This speculation gains circumstantial support from several sources. First, flaviviruses, that make up a similarly diverse cluster, and whose polymerases seem to be closely related to those of the luteovirus supergroup, have the helicase motifs but although seventeen members of the cluster have been sequenced no recombination has been found (Blok *et al.*, 1992; Chapter 19). Secondly, retroviruses seem to lack the helicase motifs (Gorbalenya & Koonin, 1989), and in a study similar to this, McClure *et al.* (1988) found that retroviruses have frequently recombined to exchange envelope protein genes. Thirdly, the recombinational events recorded in the genomes of this clus-ter seem only to involve other viruses of the supergroup; although they share hosts with viruses from other major groupings they seem not to have recombined with them, perhaps because such events have been suppressed by the helicases that most other viruses possess.

The replication of some messenger sense RNA viruses is known to be very prone to intratypic recombination, arguing against this suggestion (for review see Allison, Thompson & Alquist, 1990).

Problems for biotechnology

There is now much evidence that genetic recombination has had a major influence on the evolution of many, if not most, types of viruses, including those with RNA genomes. There has been surprisingly little comment on the implications this may have for the use of viral genes in biotechnology. For example, many laboratories are now attempting to protect crop plants against damaging viruses by genetically transforming plants with DNA encoding viral genes, particularly virion protein genes. Using this approach, laboratories in Australia, Europe and North America have made significant progress towards protecting crop plants from disease caused by certain luteoviruses. However, it may be possible for plants carrying such genes to become sources of genes for other superinfecting viruses, especially viruses from the luteovirus supergroup, and perhaps to produce particularly virulent hybrids. Successful recombination of this sort is rare; thus viruses from the luteovirus supergroup, given their unusual capacity for successful recombination, offer themselves as ideal subjects for experiments aimed at assessing this risk.

Acknowledgements

I thank J.I. Cooper and Peter Waterhouse for their vital support, George Weiller for assistance with preliminary classifications, and Adrian Gibbs for useful discussions, comments on several drafts and his great enthusiasm.

References

Abad-Zapatero, C., Abdel-Meguid, S.S., Johnson, J.E., Leslie, A.G.W., Rayment, I., Rossmann, M.G., Suck, D. & Tsukihara, T. (1980). *Nature*, **286**, 33–9.

Adams, A.N. & Hull, R. (1972). *Ann. Appl. Biol.*, **71**, 135–40.

Allison, R., Thompson, C. & Ahlquist, P. (1990). *Proc. Natl. Acad. Sci. USA*, **87**, 1820–4.

Blok, J., McWilliam, S.M., Butler, H.C., Gibbs, A.J., Weiller, G., Herring, B.L., Hemsley, A.C., Aaskov, J.G., Yoksan, S. & Bhamarapravati, N. (1992). *Virology*, **187**, 573–90.

Botstein, D. (1980). *Ann. NY Acad. Sci.*, **354**, 484–91.

Bruen, J.A. (1991). *Nucl. Acids Res.*, **19**, 217–25.
Brunt, A.A., Crabtree, K. & Gibbs, A.J. (1990). *Viruses of Tropical Crops.* C.A.B. International, UK.
Burgyan, J., Rubino, R. & Russo, M. (1991). *J. Gen. Virol.*, **72**, 505–9.
Candresse, T., Morch, M.D. & Dunez, J. (1990). *Res. Virol.*, **141**, 315–29.
Carrington, J.C., Heaton, L.A., Zuidema, D., Hillman, B.I. & Morris, T.J. (1989). *Virology*, **170**, 219–26.
Cockbain, A.J., Jones, P. & Woods, R.D. (1986). *Ann. Appl. Biol.*, **108**, 59–69.
Coutts, R.H.A., Rigden, J.E., Slabas, A.R., Lommonossoff, G.P. & P.J. Wise. (1991). *J. Gen. Virol.*, **72**, 1521–9.
Creamer, R. & Falk, B.W. (1990). *J. Gen. Virol.*, **71**, 211–17.
Demler, S.A. & de Zoeten, G.A. (1991). *J. Gen. Virol.*, **72**, 1819–34.
Dolja, V.V. & Koonin, E.V. (1991). *J. Gen. Virol.*, **72**, 1481–6.
Duffus, J.E. (1977). In *Aphids as Virus Vectors.* K.F. Harris, and K. Maramorosch (eds.), Academic Press, New York, pp. 361–383.
Falk, B.W. & Duffus, J.E. (1981). In *Plant Diseases and Vectors.* K. Maramorosch and K.F. Harris (eds.), Academic Press, New York, pp. 162–179.
Falk, B.W., Morris, T.J. & Duffus, J.E. (1979). *Virology*, **96**, 239–48.
Feng, D.F. & Doolittle, R.F. (1987). *J. Mol. Evol.*, **25**, 351–60.
Felsenstein, J. (1989). *Cladistics*, **5**, 164–6.
Francki, R.I.B., Fauquet, C.M., Knudson, D.L. & Brown, F. (eds.) (1991). *Archives of Virology*, Suppl. 2.
Gibbs, A. (1987). *J. Cell Sci. Suppl.*, **7**, 319–37.
Gildow, F.E. (1985). *Phytopathology*, **75**, 292–7.
Gildow, F.E. & Rochow, W.F. (1980). *Virology*, **104**, 97–108.
Goldbach, R. & Wellink, J. (1988). *Intervirology*, **29**, 260–7.
Gorbalenya, A.E. & Koonin, E.V. (1989). *Nucl. Acids Res.*, **17**, 8413–40.
Gorbalenya, A.E. & Koonin, E.V. (1990). *VIII International Congress of Virology Abstracts.*
Grieco, F., Burgyan, J. & Russo, M. (1989). *Nucl. Acids Res.*, **17**, 6383.
Guilley, H., Carrington, J.C., Balazs, E., Jonard, G., Richards, K. & Morris, T.J. (1985). *Nucl. Acids Res.*, **13**, 6663–77.
Habili, N. & Symons, R.H. (1989). *Nucl. Acids Res.*, **17**, 9543–55.
Hacker, D.L., Petty, I.T.D., Wei, N. & Morris, T.J. (1992). *Virology*, **186**, 1–8.
Harrison, S.C. (1983). *Adv. Virus Res.*, **28**, 175–240.
Harrison, B.D. (1987). In *Molecular Basis of Virus Disease*, W.C. Russell and J.W. Almond (eds.). Cambridge University Press, Cambridge.
Hearne, P.Q., Knorr, D.A., Hillman, B.I. & Morris, T.J. (1990). *Virology*, **177**, 141–51.
Higgins, D.G., & Sharp, P.M. (1989). *CABIOS*, **5**, 151–3.
Hillman, B.I., Carrington, J.C. & Morris, T.J. (1987). *Cell*, **51**, 427–33.
Hogle, J.M., Maeda, A. & Harrison, S.C. (1986). *J. Mol. Biol.*, **191**, 625–38.
Hollings, M. & Stone, O.M. (1972). C.M.I./A.A.B. *Descriptions of Plant Viruses.* No. 110.
Keese, P.K. & Gibbs, A. (1992). *Proc. Natl. Acad. Sci. USA*, **89**, 9489–93.
Koenig, R. (Ed.) (1988). *The Plant Viruses. Vol. 3.* Plenum Press, New York.
Koonin, E.V. (1991). *J. Gen. Virol.*, **72**, 2197–206.

Lesnaw, J.A. & Reichmann, M.E. (1970). *Proc. Natl. Acad. Sci. USA*, **66**, 140–5.

Mang, K., Ghosh, A. & Kaesberg, P. (1982). *Virology*, **116**, 264–74.

Martin, R.R., Keese, P.K., Young, M.J., Waterhouse, P.M. & Gerlach, W.L. (1990). *Ann. Rev. Phytopath.*, **8**, 341–63.

Matthews, R.E.F. (1991). *Plant Virology* 3rd edn. Academic Press. New York.

Mayo, M.A. & Jolly, C.A. (1991). *J. Gen. Virol.*, **72**, 2591–5.

Mayo, M.A., Robinson, D.J., Jolly, C.A. & Hyman, L. (1989). *J. Gen. Virol.*, **70**, 1037–51.

McClure, M.A., Johnson, M.S., Feng, D.F. & Doolittle, R.F. (1988). *Proc. Natl. Acad. Sci. USA*, **85**, 2469–73.

Meulewaeter, F., Seurinck, J.E.F. & Van Emmel, J. (1990). *Virology*, **177**, 699–709.

Miller, W.A., Waterhouse, P.M. & Gerlach, W.L. (1988). *Nucl. Acids Res.*, **16**, 6097–111.

Morozov, S.Y., Dolja, V.V. & Atebekov, J.G. (1989). *J. Mol. Evol.*, **29**, 52–62.

Murphy, J.F., D'Arcy, C.J. & Clark Jr., J.M. (1989). *J. Gen. Virol.*, **70**, 2253–6.

Murant, A.F., Goold, R.A., Roberts, I.M. & Cathro, J. (1969). *J. Gen. Virol.*, **4**, 329–41.

Nutter, R.C., Scheets, K., Panganiban, L.C. & Lommel, S.A. (1989). *Nucl. Acids Res.*, **17**, 3163–77.

Osman, T.A.M. & Buck, K.W. (1987). *J. Gen. Virol.*, **68**, 289–96.

Reisman, D. & de Zoeten, G.A. (1982). *J. Gen. Virol.*, **62**, 187–90.

Riviere, C.J. & Rochon, D.M. (1990). *J. Gen. Virol.*, **71**, 1887–96.

Rochon, D.M. & Tremaine, J.H. (1989). *Virology*, **169**, 251–9.

Rochow, W.F. (1965). *Phytopathology*, **55**, 1284–5.

Rochow, W.F. (1970). *Science*, **167**, 875–8.

Rochow, W.F. (1973). *Phytopathology*, **63**, 1317–22.

Rochow, W.F. (1979). *Phytopathology*, **69**, 655–60.

Rochow, W.F. (1982). *Phytopathology*, **72**, 302–5.

Rochow, W.F. & Muller, I. (1974). *Pl. Dis. Rept.*, **58**, 472–5.

Rossmann, M.G. & Johnson, J.E. (1989). *Ann. Rev. Biochem.*, **58**, 533–73.

Saitou, N. & Nei, M. (1987). *Mol. Biol. Evol.*, **4**, 406–25.

Ueng, P.P., Vincent, J.R., Kawata, E.E., Lei, C.-H., Lister, R.M. & Larkins, B.A. (1992). *J. Gen. Virol.*, **73**, 487–92.

Veidt, I., Lot, H., Leiser, M., Scheidecker, D., Guilley, H., Richards, K. & Jonard, G. (1988). *Nucl. Acids Res.*, **16**, 9917–32.

Vincent, J.R., Lister, R.M. & Larkins, B.A. (1991). *J. Gen. Virol.*, **72**, 2347–55.

Waterhouse, P.M., Gildow, F.E. & Johnson, G.R. (1988). *AAB Descriptions of Plant Viruses*. No. 339.

Waterhouse, P.M. & Murant, A.F. (1981). *Ann. Appl. Biol.*, **97**, 191–204.

Waterhouse, P.M. & Murant, A.F. (1983). *Ann. Appl. Biol.*, **103**, 455–64.

Wen, F. & Lister, R.M. (1991). *J. Gen. Virol.*, **72**, 2217–23.

Wu, S., Rinehart, C.A. & Kaesberg, P. (1987). *Virology*, **161**, 73–80.

Xiong, Z. & Lommel, S.A. (1989). *Virology*, **171**, 543–54.

25

The evolution of the Reoviridae

WOLFGANG K. JOKLIK AND MICHAEL R. RONER

Introduction

Although the sequences of the various genome segments of members of the eight genera that constitute the *Reoviridae* family have diverged completely, they still possess several common functional motifs, and the structural proteins that they encode have retained the ability to form structurally similar virus particles. The evolution of reoviruses has proceeded via genetic drift driven by positive and negative selection (acquisition of new hosts and ability of proteins to accommodate amino acid substitutions, respectively), which included not only point mutations but also partial genetic duplications, and proceeded in its early stages, at least under some conditions, via transitions rather than transversions. Evolution also proceeded via genetic shift which, for these viruses (rigid icosahedral particles) was successful only when the newly introduced genome segments resulted in the formation of virus particles no less stable than those generated by the parental homologous genome segment set.

Members of the *Reoviridae* family possess unique genomes that consist of 10, 11 or 12 segments of dsRNA that vary in size from 600 to 4000 bp. They form eight genera the genome segments of whose members have diverged to the point of complete randomness, do not re-assort (that is, cannot be incorporated into each others' genomes), and encode proteins which although retaining key functions, possess no common epitopes. The question of the evolution of reoviruses can be examined both within these eight genera, and among them.

The orthoreoviruses

Orthoreoviruses (or reoviruses for short) have been isolated from many species of animals, including both vertebrates (mammals, birds, reptiles and fish) and invertebrates (insects and molluscs). The only reoviruses that have been characterized to any extent are the mammalian and avian ones. The latter share weak group-specific antigenicity with the former, but only a very few of their genome segments have been sequenced and they will not be considered further here. Mammalian reoviruses have often been isolated and serotyped but, because they do not cause overt disease, only very few isolates have been characterized to any extent. It is known that their genome segments exhibit significant electrophoretic migration rate polymorphism (Hrdy, Rosen & Fields, 1979), but this has not been followed up by sequencing; it is likely, but not definitively established, that in most cases this polymorphism is caused by point mutations rather than by deletions/duplications/insertions. The only three reovirus strains whose genome segments have been extensively sequenced are the prototype strains of the three mammalian *Orthoreovirus* serotypes: serotype 1 (ST1) strain Lang, serotype 2 (ST2) strain D5/Jones, and serotype 3 (ST3) strain Dearing. For the latter all ten genome segments have been sequenced and for both the ST1 and ST2 prototype strains genome segments L1, M2, S1, S2, S3 and S4 have been sequenced. Together, these sequences provide information sufficient to provide a rather detailed picture of the evolutionary relationships of the three mammalian orthoreovirus serotypes to be assembled.

All reovirus genome segments encode one major protein; one (S1) also encodes a minor protein in the +1 reading frame (Ernst & Shatkin, 1985; Jacobs *et al.*, 1985; Sarkar *et al.*, 1985). Six of the proteins are components of the core ($\lambda 1$, $\lambda 2$, $\lambda 3$, $\sigma 2$, $\mu 1$ and $\mu 2$), three are components of the outer capsid shell ($\sigma 1$, $\sigma 3$, $\mu 1C$), and two are non-structural proteins that are associated with viral RNA and appear to function in the assembly of progeny genomes (μNS and σNS) (Antczak & Joklik, 1992).

Table 25.1 provides data concerning the relatedness of the ten reovirus genome segments and proteins. The following points are relevant.

1. Although reoviruses exhibit very efficient genome segment re-assortment in multiply infected cells (Cross & Fields, 1976), nine of the ten ST1 and ST3 genome segments are more closely related to each other than to the ST2 genome segments (Wiener & Joklik, 1989). The only exception is the S1 genome segment, for which the ST1 and ST2 genome segments are more closely related to each other than to the ST3 genome segment, the most likely explanation for which is that, at some time in the past, the ST2 and ST3 genome segments

Table 25.1. *The genetic relatedness of six reovirus genome segments and of the proteins that they encode*

Genome segment	RNA										Protein		Amino acid identity (%)	
	Extent of divergence (%)[a]													
	ST1:ST2			ST1:ST3			ST2:ST3			Protein	ST1:ST2	ST1:ST3	ST2:ST3
	1	2	3	1	2	3	1	2	3				
L1	17	5	77	2	1	13	17	5	77	λ3	92	98	92
L2										λ2			
L3										λ1			
M1										μ2			
M2	11	2	82	2	4	53	11	2	83	μ1	97	98	97
M3										μNS			
S1	55	40	79	68	68	96	85	68	91	σ1	48	25	25
S2	14	<1	73	6	0	53	14	<1	73	σ2	94	99	94
S3	25	8	79	5	2	48	25	8	74	σNS	86	97	86
S4	12	5	73	4	1	22	12	6	76	σ3	91	97	90

Note: [a] Percentage of mismatches multiplied by 1.333. This provides a measure of extent of divergence toward randomness.

re-assorted (Cashdollar *et al.*, 1985). With this unique exception, the genomes of the three prototype strains are homologous sets of ten genome segments that co-evolved. The reason for the stability of these sets is most probably that heterologous capsids are likely to be less stable than homologous capsids (Chen & Ramig, 1992). The emergence of the re-assortants with heterologous ST2 and ST3 S1 genome segments would then be ascribed to the fact that at some stage of evolution ST2 capsids with ST3 σ1 protein (and the reverse) gained a selective advantage, either because they were more stable than homologous capsids, or because they were selected when a new host species was acquired.

2. The ten genome segments differ greatly in the extent to which they have diverged toward randomness. The genome segment that has diverged by far the most extensively is the S1 genome segment which encodes protein σ1, the protein that in trimeric form is associated with the projections/spikes, is the cell attachment protein (Lee, Hayes & Joklik, 1981; Strong *et al.*, 1991) and possesses the epitopes that elicit formation of serotype-specific neutralizing antibodies (Weiner & Fields, 1977). Examination of the sequences of the three S1 proteins reveals that 79 of about 470 amino acid residues (17%) are still shared by all three proteins (Duncan *et al.*, 1990). Interestingly, these residues are clustered into five, about 30 residue-long rather well-conserved regions in which between 32% and 46% of residues are shared by all three proteins. Presumably the function that requires the highest extent of sequence conservation is the cell attachment function, because all three serotypes appear to use the same cellular receptor (Lee *et al.*, 1981). By contrast, the association of σ1 with the projections on reovirus particles is a function of the N-terminal approximately 150 amino acid long regions of the three σ1 proteins which exhibit essentially no amino acid identity, but are capable of assuming very similar coiled-coil structures because, in all of them, every seventh amino acid is I, L or V (Bassel-Duby *et al.*, 1985). Thus the most likely reason for the very extensive evolutionary divergence of the S1 genome segments is the fact that of the three known σ1 functions: interaction with receptors, serotype- specific antigenicity, and attachment to the virus particles, only the first requires sequence conservation.

The other genome segments are far less tolerant of sequence change. The most conserved of the six genome segments is genome segment M2 which encodes protein μ1C, one of the two components of the capsomers that make up the outer capsid shell. The three M2 alleles exhibit 3% or less mismatching (Jayasuriya, Nibert & Fields, 1988; Wiener & Joklik, 1988).

3. A great deal of information concerning evolution can be derived from consideration of the relative extent to which first, second and third base codon positions have diverged (Table 25.1) (Wiener & Joklik, 1989). Clearly ST2 diverged first from an ST1–ST3 progenitor, and ST1 and ST3 separated much later. Further, the extent of divergence figures of the three base codon positions reveals two distinct patterns. The first is the S1 pattern: divergence has been very extensive in all codon positions, even for the most closely related ST pairs. For the less closely related ST pairs, even the second base codon position has diverged 68% toward randomness, and the third base codon position 96%. For the other five genome segments, the extent of divergence figures for the first and second base codons is much lower, and those for the third base codon positions reveal a remarkable feature: for the less closely related ST pairs they are high and very similar (from 73% to 83%), whereas for the most closely related ST1 : ST3 pair they are much lower and vary widely (from 53% to 13%, and, for the first 94 codons of the L3 genome segment, 6% (Wiener & Joklik, 1989)). What is the significance of this extraordinary pattern of third base codon divergence, assuming that the reason is not differences in mutation rates? The closely similar third base codon divergence figures for the ST1 : ST2 and ST2 : ST3 L1, M2, S2, S3 and S4 genome segment pairs probably reflect the fact that these genome segments have evolved together in similar environments for similar lengths of time, and the considerably higher figures for the S1 genome segment may indicate that S1 is a more rapidly evolving gene (it is known that silent mutations accumulate at different rates in different genes (Britten, 1986; Bulmer, Wolfe & Sharp, 1991)). As for the widely differing ST1 : ST3 figures, the most likely explanation is that the divergence started at different times for each genome segment. It is conceivable that the precursor in each was a re-assortant in which an ST1 or ST3 genome segment had become associated in a 'stable' manner with the heterologous genome segment set; that is, a re-assortant arose at a time when the heterologous protein exercised its function at least as efficiently as the homologous protein. Presumably the L3 genome segments in the Lang and Dearing strains are descendents of a re-assortant that emerged rather recently, and the other genome segments are derived from re-assortants that arose progressively longer ago. Confirmation of this hypothesis will have to await sequencing other reovirus isolates from geographically separated

locations; more specifically, the descendants of the virus strains that were *not* re-assortants at the time when those discussed above arose should still exist. Their identification would give credence to the model proposed.

The incidence of transitions and transversions

Another interesting feature of the nucleotide substitution patterns of the various genome segments among the three serotype pairs is the incidence of transitions and transversions. For genome segment S1 the proportion of transitions for the most closely related ST pair, 1 : 2, is slightly above (43%) the expected value, and for the other two ST pairs it is practically normal (37%) (Duncan *et al.*, 1990). The situation is quite different for all the other five genome segments: here the proportion of transitions for the less related ST pairs is about 52%, very significantly above normal, whereas for the most closely related ST1 : ST3 pair, the proportion of transitions is in all cases more than 80%. In all cases all four types of transitions are represented to roughly equal extent.

An excess accumulation of transitions has also been noted during the course of persistent measles virus infections of the central nervous system when defective SSPE virus is generated, but this situation only involves one gene, the M gene, and one type of transition, U to C (Cattaneo *et al.*, 1988)). A tightly clustered U to C accumulation has also been observed in poliovirus (de la Torre *et al.*, 1992). A mechanism capable of accounting for such biased hypermutation based on the RNA-modifying activity of a cellular dsRNA unwinding activity which converts about 50% of A residues to I residues has been proposed; if such modified strands served as templates for replication, transitions would result (Bass & Weintraub, 1988; Bass *et al.*, 1989).

Whether this mechanism could account for the reovirus ST1 : 3 transitions is not clear. Here, the transitions are not clustered but are uniformly distributed throughout the genome segments; all four types of transitions are represented to approximately equal extent and the proportion of bases mutated is low (about 10%). Also, the unwinding/RNA modifying activity which most probably serves some cellular housekeeping function, is likely to be inhibited in infected cells (Morrissey & Kirkegaard, 1991). Further, here the genome segment that has diverged most extensively, the S1 genome segment, is not affected, but all others are, the effect being much more pronounced for the most closely related ST pair 1 : 3. It is known that polymerase errors result mostly in transitions (Kuge, Kawamura & Nomoto, 1989), and that many chemical mutagenic processes like deamination, and alkylation, and mutagens like nitrosoguanidine and bisulfite, also cause primarily transitions. However, the nature of the mechanism(s) that induce(s) transversions after the initial accumulation of transitions is not known.

The rotaviruses

Rotaviruses replicate almost exclusively in the differentiated enterocytes of the small intestine; they are the single most important cause of infectious gastroenteritis in infants and children, as well as in numerous animal species. Because they are responsible for a high degree of morbidity and mortality, their epidemiology is under intensive investigation. Innumerable strains of both human and animal rotaviruses have been isolated and their antigenic relationships characterized. Although the genome of only one, SA11, has been sequenced completely (Mitchell & Both, 1990), at least 100 genome segments of at least 50 rotavirus strains have been sequenced, and this has provided some measure of insight into their evolution. In contrast to the situation with orthoreoviruses, where a rather detailed comparison of the nature of three serotype prototype strains could be carried out, the primary objective of research on rotaviruses has been to classify isolates by serologic means; as a result, the genes encoding the proteins that are the principal antigens have been studied most extensively.

Rotaviruses are classified into seven groups, A to G, based on the immunologic structure of the major core capsid protein VP6, which accounts for about one-half the total virus particle protein. Only three of these groups, A, B and C, infect humans, of which group A rotaviruses (GARs) are by far the most common in children under five years of age. Most of the discussion that follows will deal with GARs because they have been studied far more extensively than GBRs and GCRs (see below).

The three genes on which most attention has been focussed are 6, 4 and 9, which encode VP6, VP4 and VP7, the major components of the inner (VP6) and of the outer (VP4 and VP7) capsid shells (Estes & Cohen, 1989). These are the three major rotavirus antigens. VP7 is the major neutralizing antibody eliciting antigen. Eleven different forms of GAR VP7 (G1 to G11) are now recognized; hence GARs can be grouped into eleven serotypes. The basis of this specificity is provided by several variable regions in VP7; one of these is region A, which elicits both ST- and group-specific antibodies, while another, region C, elicits the formation of only ST-specific antibodies (Estes & Cohen, 1989). The other outer capsid shell component, VP4, also elicits the formation of neutralizing antibodies, and here also there is a region of high sequence variation that contains five epitopes (P1, P2, etc.); five others are located in other parts of the molecule. Finally, VP6 also elicits the formation of antibodies; these are group-specific (for groups A–F, see above), and not neutralizing. The VP6 in GARs exists in two forms, SG1 and SG2, which give rise to sub-group specific antibodies (Greenberg et al., 1983). Variants of VP6 that react with both SG1 and SG2 antibodies, and others that react with neither, have also been identified (Gorziglia *et al.*, 1988).

Interestingly, the various serotypes exhibit a certain amount of host specificity: ST1, 2, 8 and 9 have only been found in humans, ST5, 6, 7, 10 and 11 only in animals, and only ST3 and 4 strains are found both in humans and in animals. Similarly the various types of VP4 exhibit host specificity: P4, 8 and 9 are only found in human, P1, 2, 3, 5 and 7 only in animals, and only P6 is found in both human and animal GARs. The host species-specific sequences have not yet been identified (see Nishikawa *et al.*, 1989).

Extent of divergence of the various genome segments

The GAR proteins are relatively closely related; the proportion of mismatches in the sequences of the VP1, VP2, NS2 and, interestingly, the VP6 proteins, of the various STs is less than 10%; for the sequences of the NS34 and NS26 proteins the figure is 10% to 20%; for those of VP4 and VP7 it is 20% to 30%; the most extensively diverged protein appears to be NS53 (about 60% mismatches). For strains of the same serotype, amino acid mismatches rarely exceed 5 percent.

The fact that many VP6-, VP4- and VP7-encoding genome segments have been sequenced has made it possible to construct phylogenetic trees for them. A phylogenetic tree for several ST3 VP7s and the VP7s of ST1, 2, 4, 5 and 6 is shown in Fig. 25.1.

Until the complete genomes of more rotavirus strains have been sequenced, it is impossible to determine whether any of them represent homologous genome segment sets, as do the three reovirus

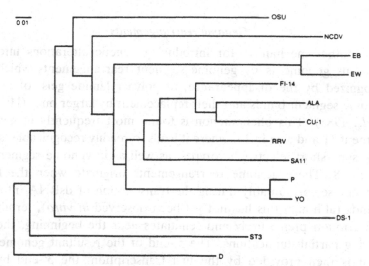

Fig. 25.1. Phylogenetic relationships among the VP7s of 14 rotavirus strains. OSU, serotype 5 (porcine); NCDV, serotype 6 (bovine); DS-1, serotype 2 (human); ST3, serotype 4 (human); D, serotype 1 (human). The rest are serotype 3 rotavirus strains: EB and EW, murine; FI-14, equine; ALA and CU-1, lapine; RRV and SA11, simian; P and YO, human. (From Nishikawa *et al.*, 1989).

serotype prototypes (see above). One clear example of successive point mutations is a porcine rotavirus strain with both G3 and G5 ST specificity (Nagesha *et al.*, 1990); another is the fact that the G11 ST VP7 protein shares striking similarity in the C region (see above) with the G3 ST VP7 protein, which is explained most readily by sequential stepwise mutations from a common ancestor (Hum, Dyall-Smith & Holmes, 1989). However, there is also evidence that rotaviruses evolve not only by genetic drift, but also by genetic shift (Green *et al.*, 1987), because several examples of naturally occurring re-assortants have been identified (Ward *et al.*, 1990; Blackhall *et al.*, 1992).

Several attempts have been made to determine how rapidly mutations accumulate in GARs. In one such study (Flores *et al.*, 1988) it was found that the mutation rate of rotaviruses in nature is likely to be lower than that of ssRNA-containing viruses such as poliovirus and influenza virus; only five mutations, all of them silent, accumulated in 10 000 bp of the VP7 genome segment of rotaviruses isolated from asymptomatically infected children over a four-year period, and none in 3000 nucleotides of the VP4 genome segment.

Genome rearrangements

Yet another mechanism for introducing genetic alterations into the rotavirus genome is by genome segment rearrangements which are recognized by the disappearance, in polyacrylamide gels, of normal genome segment bands and their replacement by larger ones (Hundley *et al.*, 1985). This phenomenon is found most frequently in genome segment 11 and also in 10, where it leads to readily recognizable 'short' and 'supershort' electropherotypes, as well as in genome segments 5, 6, and 8. These genome rearrangements originate when the RNA polymerase, presumably during the transcription of dsRNA into plus strands (although this has not yet been observed *in vitro*), terminates transcription prematurely and reinitiates near the beginning, thereby causing partial duplications. The 5'-end of the resultant genome segment is then provided by the first transcription, the 3'-end by the second. The first transcription can apparently terminate, and the second initiate, anywhere along the original template; however, in the products that are most usually encountered, the first transcription terminates in the 3'-untranslated region and the second initiates downstream of the initiation codon, so that only the normal protein is formed (Gonzalez *et al.*, 1989; Scott, Tarlow & McCrae, 1989; Ballard, McCrae & Desselberger, 1992). However, the formation of truncated proteins (when presumably new termination codons are created) has also been reported, and genome segments lacking termination codons would presumably not be detected. Remarkably, although re-initiation always occurs on the same genome segment, there is evidence that heterologous duplication occurs, that is, when the second transcription is that of the cognate genome segment of another rotavirus strain (Matsui *et al.*, 1990); the insertion of completely foreign AT-rich sequences has also been observed (Nuttall *et al.*, 1989). As many as 1800 base pairs can be added by this mechanism, the advantage of which, at least in one case that has been studied (Mattion *et al.*, 1990), is that the virus strain with the partially duplicated genome segment outgrows its normal counterpart. The significance of this type of genome segment, and viral, evolution remains to be defined, but could clearly be significant.

Group B and C rotaviruses (GBRs and GCRs)

GBRs and GCRs were originally recognized as 'atypical' rotaviruses. GBRs include viruses that are associated with annual epidemics of severe diarrhea primarily in adults in China; GCRs were first isolated from piglets, but also cause extensive disease in children. Many fewer strains of GBRs and GCRs than GARs have been studied and many fewer of their genome segments sequenced; however, it is clear that they all derived from a common ancestor. GAR proteins show 30% to 50% amino acid homology with their GCR counterparts, and both show only about 25% amino acid homology with their GBR counterparts. Thus GBRs and the GAR–GCR progenitor diverged first, and GARs and GCRs separated later.

The orbiviruses

There are some 14 serogroups of orbiviruses distributed throughout the world, classified on the basis of possession of group-specific antigens and ability to re-assort (Nuttall & Moss., 1989); and each serogroup is further subdivided into serotypes. The most intensively investigated serogroup is the bluetongue serogroup, primarily because many of its members cause an economically important disease in sheep, and most of the discussion that follows will be concerned with it.

Bluetongue virus particles possess ten genome segments that encode seven structural and three non-structural proteins. The particles comprise a core, the shell of which is made up of the major core components VP3 and VP7, both of which elicit the formation of group-specific non-neutralizing antibodies, and which contains three other minor proteins, namely VP1 (the RNA polymerase), VP4 (the guanylyltransferase), and VP6 (function unknown), and an outer capsid shell that consists of VP5 on the inner and VP2 on the outer surface. VP2 possesses epitopes (at least four, as determined with monoclonal antibodies) that elicit the formation of serotype-specific neutralizing antibodies; so far, 24 serotypes of Bluetongue virus have been recognized. There is evidence that VP5 also contributes to serospecificity, but exactly how is not known.

Many of the genome segments of the various bluetongue virus serotypes, five of which (2, 10, 11, 13 and 17) exist in the United States, eight in Australia (1, 3, 9, 15, 16, 20, 21 and 33) and 20 in Africa (1–16, 18, 19, 22 and 24), have been sequenced. With two exceptions, the proteins that they encode are very closely related (amino acid homologies greater than 90%). The two exceptions are the VP2 proteins which are related to the extent of 40% to 70% depending on the serotype pair, and the VP5 proteins, which are related to the extent of 75% to 95%. Not surprisingly, proteins of the same serotype derived from different parts of the world are related more closely.

The relatedness of the various serogroups

Cognate proteins of the various serogroups are generally less than 50% related. For example, the VP3 proteins (core components) of bluetongue viruses and members of the epizootic haemorrhagic disease and Warrego serogroups are 20% and 35% related, respectively; the NS2 proteins of bluetongue viruses and members of the African horse sickness and epizootic haemorrhagic disease serogroups are 33% and 46% related; the VP6 proteins of bluetongue viruses and members of the Palyam and African horse sickness serogroups are 52% and 50% related, respectively. Further, the genome segments of members of these serogroups are themselves related to a similar extent. However, not enough genome segments of viruses of these serogroups have been sequenced for phylogenetic trees to be constructed.

The role of genetic drift and genetic shift in orbivirus evolution

The rate of genetic drift has been assessed by sequencing the VP2 genome segment of the Australian bluetongue 1 virus both before and after 20 serial passages extending over a period of ten years and selecting for attenuated variants. This procedure caused ten nucleotide changes, six of which were silent and two generated conservative amino acid changes (Gould & Eaton, 1990).

Study of the role of genetic shift in bluetongue virus evolution has led to interesting results. First, whereas with certain parents re-assortment of genome segments was random, as is the case for orthoreoviruses, for other pairs of parents this is not the case. For example, characterization of re-assortants following mixed infection with two Kemerovo serogroup members, Broadhaven and Wexford, revealed that re-assortment of genome segments 2 and 10 was non-random, and that there was possible 'linkage' of segments 3 and 9 (Moss, Ayres & Nuttall, 1987). Further, analysis of re-assortants as a function of time between infection with the two parents, bluetongue 10 virus and bluetongue 17 virus, revealed that re-assortment was an early event and that superinfection exclusion was established within four hours (Ramig *et al.*, 1989). However, this effect, which was observed in cultured Vero cells, and which would clearly severely limit the incorporation of new genome segments into orbivirus genomes, was not observed in intact mosquitoes, for here re-assortant frequency when mosquitoes were fed the two parents simultaneously and separately at times 1, 3 and 5 days apart, was 48%, 67%, 71% and 17% (El Hussein *et al.*, 1989). This suggests that mosquitoes allow bluetongue viruses to evolve by genome segment re-assortment. Interestingly, in the former experiment, using Vero cells, evidence was again found for non-random re-assortment; in particular, segment 8 was strongly biased toward the bluetongue 17 parent, even when it was the minority parent, and segment 10 was biased toward the bluetongue 10 parent. This non-random segregation of these genome segments, both of which encode non-structural proteins, may reflect enhanced viability of viruses containing the bluetongue 10 NS3 and the bluetongue 17 NS2 proteins.

The phytoreoviruses, fijiviruses, cypoviruses, aquareoviruses and coltiviruses

The *Phytoreovirus* genus comprises three members: wound tumor virus (WTV), rice dwarf (RDV), and rice gall dwarf virus (RGDV). Most of the genome segments of the first two of these viruses have been sequenced (Anzola *et al.*, 1987, 1989*a*; Anzola, Xu & Nuss, 1989*b*; Omura *et al.*, 1989; Suzuki *et al.*, 1990*a,b*). Not only do their proteins share significant homology (ranging from about 50% for the outer shell component P5 and the capsid protein P8 to 20% to 30% for several non-structural proteins), but they also share a unique sequence feature, namely an inverted repeat that is up to 9 bp long and is immediately adjacent to their 5'-terminal conserved hexanucleotide and their 3'-terminal conserved tetranucleotide (Anzola *et al.*, 1987). Interestingly, the inverted repeat at the 3'-end can influence the conformational and functional properties associated with the 5'-end (Xu *et al.*, 1989). Further, this feature is retained in fijiviruses which, however, possess at least one novel feature of gene expression strategy (as would be expected of viruses assigned to different genera): their genome segment S6 possesses two long non-overlapping open reading frames (Marazachi, Boccardo & Nuss, 1991).

Finally, there is no doubt that cypoviruses, aquareoviruses and coltiviruses also derive from the common *Reoviridae* progenitor, but too few of their genome segments have been sequenced to provide sufficient information for profitable speculation concerning their evolutionary history as compared to that of the other Reoviridae.

Observations concerning the evolutionary relationships of orthoreoviruses, rotaviruses and orbiviruses

Although there is little doubt that all members of the *Reoviridae* family evolved from a common ancestor, the sequences of their genome segments have by now diverged to randomness. It has been suggested that there is some degree of homology between certain genome segments, but this is only found with the introduction of an impermissibly large number of gaps. Certain recognizable functional motifs have, however, persisted; for example, the VP1 proteins of rotaviruses and orbiviruses and protein λ3 of orthoreoviruses all possess the well-conserved RNA polymerase motif GDD (Roy *et al.*, 1988; Bruenn, 1991; Morozov, 1989); the VP4 protein of BTV and the λ2 protein of orthoreoviruses both possess guanylyltransferase activity; the NS2 proteins of orbiviruses, the NS34 protein of rotaviruses and protein σNS of orthoreoviruses all possess a motif thought to specify ability to bind ssRNA (van Staden *et al.*, 1991). Additional searches are likely to reveal more such motifs.

References

Antczak, J.B. & Joklik, W.K. (1992). *Virology*, **187**, 760–76.

Anzola, J.V., Dall, D.J., Xu, Z. & Nuss, D.L. (1989*a*). *Virology*, **171**, 222–8.

Anzola, J.V., Xu, Z., Asamizu, T. & Nuss, D.L. (1987). *Proc. Natl. Acad. Sci. USA*, **84**, 8301–5.

Anzola, J.V., Xu, Z. & Nuss, D.L. (1989*b*). *Nucl. Acids Res.*, **17**, 3300.

Ballard, A., McCrae, M.A. & Desselberger, U. (1992). *J. Gen. Virol.*, **73**, 633–8.

Bass, B.L. & Weintraub, H. (1988). *Cell*, **55**, 1089–98.

Bass, B.L., Weintraub, H., Cattaneo, R. & Billeter, M.A. (1989). *Cell*, **56**, 331.

Bassel-Duby, R., Jayasuriya, A., Chatterjee, S., Sonenberg, N., Maizel, J.V. & Fields, B.N. (1985). *Nature*, **315**, 421–3.

Blackhall, J., Bellinzoni, R., Mattion, N., Estes, M.K., La Torre, J.L. & Magnusson, G. (1992). *Virology*, **189**, 833–7.

Britten, R. (1986). *Science*, **231**, 1393–8.

Bruenn, J.A. (1991). *Nucl. Acids Res.*, **19**, 217–26.

Bulmer, M., Wolfe, K.H. & Sharp, P.M. (1991). *Proc. Natl. Acad. Sci. USA*, **88**, 5974–8.

Cashdollar, L.W., Chmelo, R.A., Wiener, J.R. & Joklik, W.K. (1985). *Proc. Natl. Acad. Sci. USA*, **82**, 24–8.

Cattaneo, R., Schmidt, A., Eschle, D., Baczko, K., ter Meulen, V. & Billeter, M.A. (1988). *Cell*, **55**, 255–65.

Chen, D. & Ramig, R.F. (1992). *Virology*, **186**, 228–37.

Cross, R.K. & Fields, B.N. (1976). *Virology*, **74**, 345–62.

de la Torre, J.C., Giachetti, C., Semler, B.L. & Holland, J.J. (1992). *Proc. Natl. Acad. Sci. USA*, **89**, 2531–5.

Duncan, R., Horne, D., Cashdollar, L.W., Joklik, W.K. & Lee, P.W.K. (1990). *Virology*, **174**, 399–409.

El Hussein, A., Ramig, R.F., Holbrook, F.R. & Beaty, B.J. (1989). *J. Gen. Virol.*, **70**, 3353–62.

Ernst, H. & Shatkin, A.J. (1985). *Proc. Natl. Acad. Sci. USA*, **82**, 48–52.

Estes, M.K. & Cohen, J. (1989). *Microbiol. Rev.*, **53**, 410–49.

Flores, J., Sears, J., Green, K.Y., Perez-Schael, I., Morants, A., Daoud, G., Gorziglia, M., Hoshino, Y., Chanock, R.M. & Kapikian, A.Z. (1988). *J. Virol.*, **62**, 4778–81.

Gonzalez, S.A., Mattion, N.M., Bellinzoni, R. & Burrone, R. (1989). *J. Gen. Virol.*, **70**, 1329–36.

Gorziglia, M., Hoshino, Y., Nishikawa, K., Maloy, W.L., Jones, R.W., Kapikian, A.Z. & Chanock, R.M. (1988). *J. Gen. Virol.*, **69**, 1659–69.

Gould, A.R. & Eaton, E.T. (1990). *Virus Res.*, **17**, 161–72.

Green, K.Y., Midthun, K., Gorziglia, M., Hoshino, Y., Kapikian, A.Z., Chanock, R.M. & Flores, J. (1987). *Virology*, **161**, 153–9.

Greenberg, H.B., Flores, J., Kalica, A.R., Wyatt, R.T. & Jones, R. (1983). *J. Gen. Virol.*, **64**, 313–20.

Hrdy, D.D., Rosen, L. & Fields, B.N. (1979). *J. Virol.*, **31**, 104–11.

Hum, C.P., Dyall-Smith, M.L. & Holmes, I.H. (1989). *Virology*, **170**, 55–61.

Hundley, F., Biryahwaho, B., Gow, M. & Desselberger, U. (1985). *Virology*, **143**, 88–103.

Jacobs, B.L., Atwater, J.A., Munemitsu, J.M. & Samuel, C.E. (1985). *Virology*, **147**, 9–18.

Jayasuriya, A.K., Nibert, M.L. & Fields, B.N. (1988). *Virology*, **163**, 591–602.

Kuge, S., Kawamura, N. & Nomoto, A. (1989). *J. Mol. Biol.*, **207**, 175–82.

Lee, P.W.K., Hayes, E.C. & Joklik, W.K. (1981). *Virology*, **108**, 156–63.

Marazachi, C., Boccardo, G. & Nuss, D.L. (1991). *Virology*, **180**, 156–63.

Matsui, S.M., Mackow, E.R., Matsuno, S., Paul, P.S. & Greenberg, H.B. (1990). *J.Virol* **64**, 120–4.

Mattion, N.M., Bellinzoni, R.C., Blackhall, J.O., Estes, M.K., Gonzalez, S., La Torre, J.L. & Scodeller, E.A. (1990). *J. Gen. Virol.*, **71**, 355–62.

Mitchell, D.B. & Both, G.W. (1990). *Virology*, **177**, 324–31.

Morozov, S.U. (1989). *Nucl. Acids Res.*, **17**, 5394.

Morrissey, L.M. & Kirkegaard, K. (1991). *Mol. Cell Biol.*, **11**, 3719–25.

Moss, S.R., Ayres, C.M. & Nuttall, P.A. (1987). *Virology*, **157**, 137–44.

Nagesha, H., Huang, J., Hum, C.P. & Holmes, I.H. (1990). *Virology*, **174**, 319–22.

Nishikawa, K., Hoshino, Y., Taniguchi, K., Green, K.Y., Greenberg, H.B., Kapikian, A.Z., Chanock, R.M. & Gorziglia, M. (1989). *Virology*, **171**, 503–15.

Nuttall, S.D., Hum, C.P., Holmes, I.H. & Dyall-Smith, M.L. (1989). *Virology*, **171**, 453–7.

Nuttall, P.A. & Moss, S.R. (1989). *Virology*, **171**, 156–61.

Omura, T., Ishikawa, K., Hirano, H., Ugaki, M., Minobe, Y., Tsuchizaki, T. & Kato, H. (1989). *J. Gen. Virol.*, **70**, 2759–64.

Ramig, R.F., Garrison, C., Chen, D. & Bell-Robinson, D. (1989). *J. Gen. Virol.*, **70**, 2595–603.

Roy, P., Fukusho, A., Ritter, G.D. & Lyon, D. (1988). *Nucl. Acids Res.*, **16**, 11759–67.

Sarkar, G., Pelletier, J., Bassel-Duby, K., Jayasuriya, A., Fields, B.N. & Sonenberg, N. (1985). *J. Virol.*, **54**, 720–5.

Scott, D.E., Tarlow, O. & McCrae, M.A. (1989). *Virus Res.*, **14**, 119–28.

Strong, J.E., Leone, G., Duncan, R., Sharmer, R.K. & Lee, P.W.K. (1991). *Virology*, **184**, 23–32.

Suzuki, N., Watanabe, Y., Kusano, T. & Kitagawa, Y. (1990*a*). *Virology*, **179**, 446–454.

Suzuki, N., Watanabe, Y., Kusano, T. & Kitagawa, Y. (1990*b*). *Virology*, **179**, 455–9.

van Staden, V., Theron, J., Greyling, B.J., Huismans, H. & Nel, L.H. (1991). *Virology*, **185**, 500–4.

Ward, R.L., Nakagomi, O., Knowlton, D.R., McNeal, M.M., Nakagomi, T., Clemens, J.D., Sack, D.A. & Schiff, G.M. (1990). *J. Virol.*, **64**, 3219–25.

Weiner, H.L. & Fields, B.N. (1977). *J. Exp. Med.*, **146**, 1305–10.

Wiener, J.R. & Joklik, W.K. (1988). *Virology*, **163**, 603–13.

Wiener, J.R. & Joklik, W.K. (1989). *Virology*, **169**, 194–203.

Xu, Z., Anzola, J.V., Nalin, C.M. & Nuss, D.L. (1989). *Virology*, **170**, 511–22.

26

Genetic variation and evolution of satellite viruses and satellite RNAs

G. KURATH AND C. ROBAGLIA

Introduction

Satellite viruses and satellite RNAs are sub-viral microbes which depend on another virus, referred to as the 'helper virus', for their replication within a host cell. The basis of satellite dependence is not completely understood, but it is presumed that satellite genomes are replicated by the RNA-dependent RNA polymerase encoded by the helper virus. Along with viroids, satellites are the smallest and simplest biological entities known, with single-stranded RNA genomes ranging from 194 to 1376 nt. Satellites are uniquely prevalent in plant hosts, and have been found associated with 28 different plant viruses from diverse virus groups (Roossinck, Sleat & Palukaitis, 1992). To date, no examples of satellites of animal viruses are known. The only satellite-like entity associated with an animal virus, hepatitis delta virus, is not a true satellite because it is capable of autonomous replication, and appears to need its helper virus only for encapsidation and transmission (Taylor, 1990).

The classical definition of a satellite involves three main features (Murant & Mayo, 1982). These are that satellites are not capable of replicating in the absence of the helper virus, they are not required for the life cycle of the helper virus, and that they share little or no sequence similarity with the helper virus genome. Thus, satellites differ from other subviral entities such as viroids, which are capable of independent replication in a host cell, and defective interfering (DI) RNAs, which are comprised of helper virus genetic sequences. Satellites often interfere with the replication of their helper viruses, and the symptoms caused by helper virus/satellite co-infections can differ dramatically from those of the helper virus alone. In the last decade, fascinating biological and molecular characteristics of many satellite systems have been elucidated. These have shown that satellites

385

can occur in complex mixtures and can include intermediate subviral forms which do not fit perfectly into the classical definition described above. For general information readers are referred to several excellent reviews on plant viral satellites (Murant & Mayo, 1982; Francki, 1985; Fritsch & Mayo, 1989; Roossinck *et al.*, 1992). This chapter will only deal with information that has direct relevance to the molecular evolution and origins of satellites.

Sequence relationships among satellites

The characterization of different isolates of many satellites has often shown the existence of strains and variants with different biological properties. In general, different isolates show a large amount of sequence similarity, and sequence comparisons provide clear evidence that single base substitutions, insertions, deletions, and recombination events have all been active mechanisms of genetic change during satellite evolution. The following are general descriptions of several satellite groups for which sequence information with evolutionary implications is available.

Satellite viruses are satellites that encode their own coat proteins and form 17 nm icosahedral virions serologically distinct from those of their helper viruses. The genomes of the four satellite viruses known have all been sequenced and are 825–1250 nt plus-sense RNA molecules which do not share significant sequence similarities with each other or with their helper virus genomes. The only exception is the satellite virus (STMV) of tobacco mild green mosaic virus (TMGMV, previously TMV-U5), which has two 50 nt regions within its 3' terminal 150 nt that are nearly identical to the corresponding 3' sequences of its helper virus (Mirkov *et al.*, 1989). Although only the type strain STMV sequence has been reported, RNase protection analyses of the genomic RNA of STMV field isolates show that this virus is genetically very diverse (Kurath *et al.*, 1993). The only satellite virus for which the genomes of two strains have been sequenced is satellite tobacco necrosis virus (STNV). The sequences are nearly identical in the 30 nt 5' untranslated leader region, but the remaining 1200 nt show only approximately 60% sequence similarity (Danthinne *et al.*, 1991). Despite this divergence, both 3' non-coding regions are predicted to fold into similar secondary structures, suggesting conservation of information and/or function in these structures.

Satellite RNAs differ from satellite viruses in that they are encapsidated in the coat proteins of their helper viruses, and thus depend on their helpers for transmission as well as replication. The most diverse and widely studied group of satellite RNAs is that associated with strains of cucumber mosaic cucumovirus (CMV). Many CMV satellite RNAs have been characterized, and they can modify the symptoms of CMV in various ways including amelioration, chlorosis, or lethal necrosis. These satellites are generally 330–340 nt long and share 85–99% sequence similarity including ten and six conserved residues at the 5′ and 3′ termini respectively (Francki, 1985). They contain no conserved open reading frames and no protein coding activity has been demonstrated *in vivo*, suggesting that their genomes are expressed in novel ways that would be subject to different selection pressures from those operating on coding sequences. All CMV satellite RNA sequences can be folded into stable secondary structures with 3′ terminal tRNA-like structures similar to those at the 3′ end of CMV genomic RNAs (Gordon & Symons, 1983).

A phylogenetic comparison of 25 CMV satellite RNA sequences showed that they clustered into three main groups which did not correlate with their geographic origins (Fraile & García-Arenal, 1991). The sequence differences between CMV satellite RNAs suggest that small substitutions, deletions, and insertions have occurred during the evolution of these satellites. Two unusual Japanese CMV satellite RNAs, Y-sat (368 nt) and OY2-sat (386 nt), show evidence of multiple insertions in the central regions of the genome (Hidaka *et al.*, 1984, 1988). A recent study of 20 CMV satellite RNAs isolated from field epidemics in Spain showed that they diverged along a single evolutionary line through accumulation of mutations. There seemed to be a rapid rate of evolution, with up to 4% nucleotide difference between isolates from one tomato field (Aranda, Fraile & García-Arenal, 1990).

Turnip crinkle carmovirus (TCV) is able to support the replication of several dependent RNAs, sometimes in complex mixtures. These include satellite RNAs which have no significant sequence similarity with the helper virus genome, DI RNAs derived from helper virus sequences, and chimaeric RNAs composed of satellite RNA sequences at the 5′ end linked to helper virus sequences at the 3′ end (Simon & Howell, 1986; Li *et al.*, 1989). The chimaeric RNAs, found only with TCV to date, have been used to study RNA recombination in the TCV system and are described later in this chapter. The TCV-M strain supports a mixture of sat-RNA D (194 nt), sat-RNA F (230 nt), and the chimaeric sat-RNA C (356 nt), all of which contain seven bases at their 3′ ends which are identical to the 3′ end of the helper virus genome. Sat-RNA D may be considered as the progenitor of the other satellite RNAs, since sat-RNA

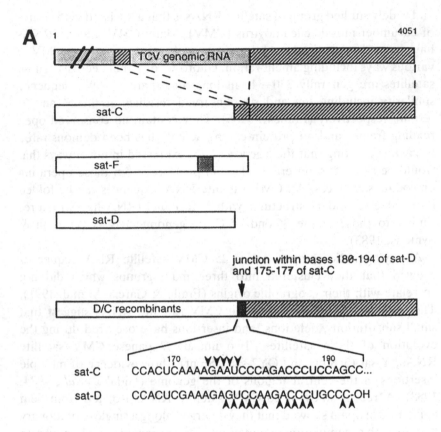

junction within bases 180-194 of sat-D
and 175-177 of sat-C

sat-C   CCACUCAAAAGAAUCCCAGACCCUCCAGCC...

sat-D   CCACUCGAAAGAGUCCAAGACCCUGCCC-OH

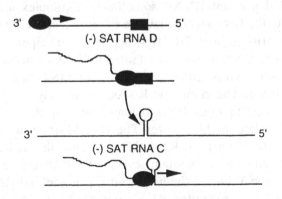

F is equivalent to sat-RNA D with a 36 base insertion, and sat-RNA C is a chimaeric molecule of sat-RNA D sequences at the 5' end linked to two segments of TCV genomic sequences at the 3' end (Simon & Howell, 1986) (Fig. 26.1A).

Nepoviruses are nematode transmitted plant viruses in which the two genome RNAs have a 3' poly(A) tail and a 5'-linked VPg protein. They support a diverse group of satellite RNAs that fall into two different types referred to as 'large' and 'small' nepovirus satellite RNAs. The large satellite RNAs are linear RNAs, approximately 1.1–1.4 kb in length, which have terminal structures, and thus presumably replication strategies, similar to those of their helper virus genomic RNAs. They encode a 38–48 kD protein which is expressed *in vivo* and appears to be necessary for replication (C. Fritsch, personal communication) although its function is unknown. Proteins from different large nepovirus satellite RNAs are unrelated in amino acid sequence but all are relatively basic with positive charges as large as histone-like proteins (Hemmer *et al.*, 1987). The best studies of the large nepovirus satellite RNAs are those of tomato black ring virus (TBRV) satellites. Comparison of the satellite sequences from five TBRV isolates showed that they fall into two groups which correlate with the serogrouping of the helper viruses (Hemmer *et al.*, 1987). This correlation suggests that the satellite RNAs had co-evolved with their respective helper viruses, rather than transferring between them more recently. A short UGAAAA sequence is found at the 5' ends of TBRV and grapevine fanleaf virus (GFLV) genomic RNAs and at the 5' ends of TBRV, GFLV, CYMV, and ArMV satellite RNAs (Fuchs *et al.*, 1989).

Fig. 26.1. Recombination in the turnip crinkle virus system. (A) Schematic representation of sequence similarities between TCV and its satellite RNAs. Hatched portions show similarities between TCV RNA and sat-C RNA, and unshaded portions show similarities in the 5' regions of the three satellite RNAs. Black arrows show the various crossover points in over 150 D/C recombinants. The underlined 'motif 1' indicates a conserved sequence found at the right sides of D/C junctions and the defective interfering RNA G junction, and at the 5' end of the TCV genome. This is suggested as a replicase recognition sequence that forms part of a hairpin structure and facilitates RNA recombination. (B) A model for formation of D/C recombinants. During the synthesis of plus strand sat-D RNA, the replicase (represented by a disc) dissociates from the minus strand template due to yet undetermined constraints (represented by a black box) and reinitiates at a recognition signal (e.g. motif 1, represented by a hairpin structure) on sat-C RNA without releasing the nascent plus strand. (Modified from Cascone *et al.*, 1990, and Cascone, Haydar & Simon, 1993).

The small satellite RNAs of nepoviruses are 300–450 nt long and are found in linear, circular, and multimeric forms, without the poly(A) and VPg terminal structures found in the helper virus genome. The multimeric molecules cleave autocatalytically and are thought to be produced by a rolling circle mechanism similar to that of viroids and the circular satellite RNAs of sobemoviruses (Symons, 1989). The best studied example of this type of satellite is the satellite of tobacco ringspot nepovirus (STobRV). A comparison of four STobRV sequences from different helper virus isolates showed that they differ significantly only between nucleotides 100–140 of the 359 nt sequence (Buzayan *et al.*, 1987). In this region, three nearly identical blocks of sequence appear in inverted order in different satellites, suggesting that there had been recombinational events in their evolution.

Three small satellite RNAs associated with different nepoviruses are STobRV, the satellite of arabis mosaic virus (ArMV), and the small satellite of chicory yellow mottle virus (CYMV). These satellites share the same self cleavage structural domains and have extensive sequence similarities only in these domains (Rubino *et al.*, 1990b; see later section).

Sequence heterogeneity within satellite populations

RNA viruses have been shown to exist as heterogeneous populations of variants all related to a consensus sequence (Domingo *et al.*, 1985; Chapter 13). This 'quasi-species' nature of RNA virus genome populations applies also to satellite populations. Sequence microheterogeneity in the form of minor sequence differences between individual cDNA clones from satellite populations has been reported for the sat-C of TCV (Simon *et al.*, 1988), three CMV satellite RNAs (Kurath & Palukaitis, 1989), and the satellite viruses STNV (Danthinne *et al.*, 1991) and STMV (Kurath *et al.*, 1992).

The most detailed study of the quasi-species nature of a satellite population reported to date involves genetic heterogeneity within the type strain of STMV (Kurath *et al.*, 1992). In this study 16 variant RNase protection patterns were found within 42 cDNA clones of the STMV type strain. Mapping and sequencing of the heterogeneity sites showed that they occurred throughout the 1059 nt STMV genome, with no apparent clustering or preference for non-coding sequences (Fig. 26.2). Nearly all of them were single base substitutions, of which 60% were changes of a G in the consensus sequence to an A

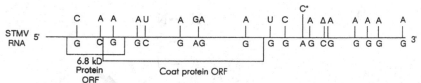

Fig. 26.2. Cumulative map of heterogeneity sites within a single population of satellite tobacco mosaic virus (STMV). The horizontal line represents the 1059 nt consensus sequence of the STMV type strain with the two open reading frames shown as boxes below the line (Mirkov *et al.*, 1989). Locations of sequence differences identified in variant cDNA clones from an STMV type strain population are shown as vertical lines with the consensus nucleotide at the bottom and the nucleotide in the variant sequence at the top. The symbol Δ indicates a single base deletion. Most sites represent sequence microheterogeneity found in only one clone, with the exception of the major heterogeneity site at position 751 (*), where approximately 20% of the population varied from the consensus sequence (Modified from Kurath, Rey & Dodds, 1992).

variant, suggesting a biased mechanism operating in the generation of heterogeneity in STMV. In addition to this sequence microheterogeneity a major heterogeneity site was identified where approximately 20% of the molecules in the population varied from the consensus sequence. A similar example of a major heterogeneity site has been described in the CMV D-sat RNA population (Kurath & Palukaitis, 1989). The significance of such sites, if any, is not known, but in both cases they have been shown to be sites of rapid genetic change as described later in this chapter.

Self-cleavage structures and satellite evolution

Several satellite RNAs produce circular intermediates during their replication cycles and are thought to replicate by a rolling circle mechanism involving autocatalytic cleavage of multimeric precursors. Most of these RNAs have conserved sequences flanking the self-cleavage sites which form conserved predicted secondary structures. The 'hammerhead' structural domain is found in the positive strands of the circular satellites of sobemoviruses and the small satellite RNAs of nepoviruses, in both strands of the lucerne transient streak sobemovirus satellite (Symons, 1989), and in the barley yellow dwarf luteovirus satellite RNA (Miller *et al.*, 1991). There is also a cellular analogue of the hammerhead domain in a self-cleaving RNA in newts (Epstein & Gall, 1987).

Ruffner, Stormo and Uhlenbeck (1990), after defining the minimal

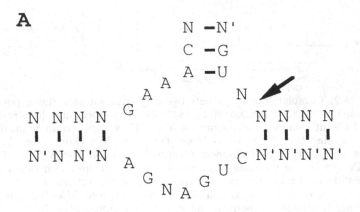

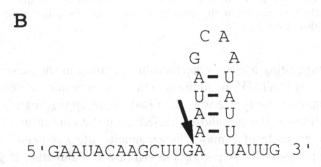

Fig. 26.3. Simplest known self-cleaving RNA structures. (A) Hammerhead domain found in 12 satellite RNAs. N represents any base and N' the corresponding Watson–Crick paired base. (B) RNA hairpin found at the 5' end of the Tetrahymena group I ribosomal intron, which self-cleaves in the presence of manganese. In both diagrams arrows show the sites of cleavage. (Redrawn from Dange, Van Hatta & Hecht, 1990).

sequence requirements for a self-cleaving RNA, state that their frequency in cellular RNAs can be one in each 11 000 nucleotides. Other results (Dange *et al.*, 1990) show that self-cleaving RNA domains can take very simple forms (Fig. 26.3). This suggests that they could be relatively easily acquired during satellite evolution by either infidelity or non-templated polymerization by the replicating enzyme, or by recombination with host RNAs containing self-cleaving structures. These apparently conserved catalytic structures could result from convergent evolution selected by the rolling circle mode of replication, or they could indicate a monophyletic origin as suggested by Elena *et al.* (1991) who made a phylogenetic study of viroids, four sobemovirus circular satellite

RNAs, two small nepovirus satellite RNAs, and the viroid-like domain of HDV.

RNA Recombination in turnip crinkle virus satellites

As described earlier, TCV supports multiple satellite RNAs and DI RNAs, as well as the chimaeric sat-RNA C which has sat-RNA D sequences at the 5' end linked to TCV sequences at the 3' end (Simon *et al.*, 1988). In a recent study, non-viable sat-RNA C transcripts with mutations generated *in vitro* in the 5' domain were shown to recombine with sat-RNA D in plants to generate chimaeric D/C molecules analogous to the natural sat-RNA C (Cascone *et al.*, 1990; Fig. 26.1). Analyses of numerous independent recombination junctions showed that unequal crossing over had occurred at specific regions of sequence similarity between the two satellite RNAs. Seven out of 20 recombinants also had 1–3 non-template encoded nucleotides in the junctions. Sequence comparisons of the 3' sides of these recombinant junctions, junction sites of TCV DI RNAs, and the 5' ends of TCV and TCV sat-RNAs, revealed conserved sequence motifs which could be recognition sites for the viral polymerase. This suggested that RNA recombination in the TCV system involves a replicase-driven copy choice mechanism in which the replicase and a nascent plus strand can dissociate from a template and re-associate at specific recognition sites to generate recombinants (Fig. 26.1B). Further characterization of the putative replicase recognition site has recently defined an approximately 60 base region in sat-C which includes the conserved recognition sequence motif and at least one stem-loop structure, both of which are necessary for recombination to occur (Cascone *et al.*, 1993). Evidence for recombination between sat-RNA D and the 3' end of the TCV helper genome has also been described (Zhang, Cascone & Simon, 1991). This work explains how TCV infections can generate DI RNAs and chimaeric sat-RNAs, and clearly shows the important role of RNA recombination in the evolution of these subviral entities.

Experimental evolution of satellites

Several studies have been reported in which satellite genomes generated from cDNA clones have been used to initiate infections in plants, and progeny have been analysed for evidence of genetic change. These can be called experimental evolution studies because they attempt to recreate and investigate, under controlled conditions, the processes of change involved in satellite evolution.

In one such study RNA transcribed from a cDNA clone of the CMV D-sat was combined with satellite-free CMV to inoculate serial passage lines in five host plant species (Kurath & Palukaitis, 1990). Genetic change at the major heterogeneity site found in the natural CMV D-sat population (Kurath & Palukaitis, 1989) was detected by RNase protection assays within 3–4 passages in four out of five hosts, suggesting that mutation and selection at that site occurred rapidly. A host-specific influence on this evolutionary process was also observed in that the new variant accumulated and formed a much higher proportion of the population in tobacco than it did in the other host species.

A similar study of the progeny of STMV clone transcripts in multiple serial passage lines in tobacco has recently been completed (G. Kurath & J.A. Dodds, unpublished data). Genetic changes were detected within 1–5 passages in many lines, and occurred at both the major heterogeneity site (Kurath *et al.*, 1992) and at other novel sites in a non-random manner. An example of the genetic change detected in one of these lines is shown in Fig. 26.4.

Many other examples of experimental evolution involve replication of mutant satellite genomes created *in vitro* by manipulation of infectious cDNA clones. Although various small substitution, deletion, and insertion mutations can be stably maintained in some satellites such as STNV (Van Emmelo, Ameloot & Fiers, 1987), TCV sat-RNA C (Simon *et al.*, 1988), and CMV Y-sat (Masuta & Takanami, 1989), other modifications are either lethal or lead to genetic change in the progeny. This genetic change can sometimes result in reversion to a non-mutant sequence. Examples of this include reversion when growing in plants of several insertion and deletion mutations of the TCV sat-RNA D (Collmer *et al.*, 1991), a point mutation of the CMV D-sat (Collmer & Kaper, 1988), and four linker insertion mutations of the CMV D-sat (G. Kurath, P.E. Aeschleman & P. Palukaitis, unpublished data). A tendency to revert to a wild-type sequence is not unexpected, as standard mechanisms of evolutionary change would create variants in the mutant progeny, and variants more similar to the wild-type sequence would presumably often have a selective advantage over the mutant sequence.

There are other cases where the replication of mutant satellite sequences has generated variants which provide information about mechanisms of genetic change. Recently, linker insertion mutants of STobRV were expressed as transgenic sequences in plants which were

PASSAGE NUMBER

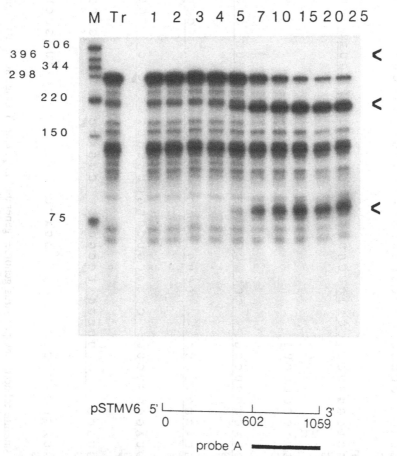

Fig. 26.4. Genetic change in a tobacco serial passage line inoculated with RNA transcribed from an STMV cDNA clone. RNase protection patterns of the original inoculum RNA (lane Tr, transcript of clone pSTMV5) and genomic RNAs of STMV populations from passages 1 to 25 (numbered above the lanes) show that genetic change has occurred, detected as the gradual appearance of two new cleavage fragments (two lower arrows at right). The change in this passage line maps to approximately base 850 of the 1059 base STMV genome, while other passage lines showed different changes or no change (Kurath & Dodds, unpublished observations). The hybridization probe for these assays was complementary to the 3' region of the STMV sequence of clone pSTMV6 (lower diagram). The intact probe migrated at the position indicated by the top arrow on the right, and it was cleaved in assays of pSTMV5 transcript RNA due to a sequence difference at base 751 (Kurath *et al.*, 1992). Lane M contains size markers as noted (in nt) at the left margin.

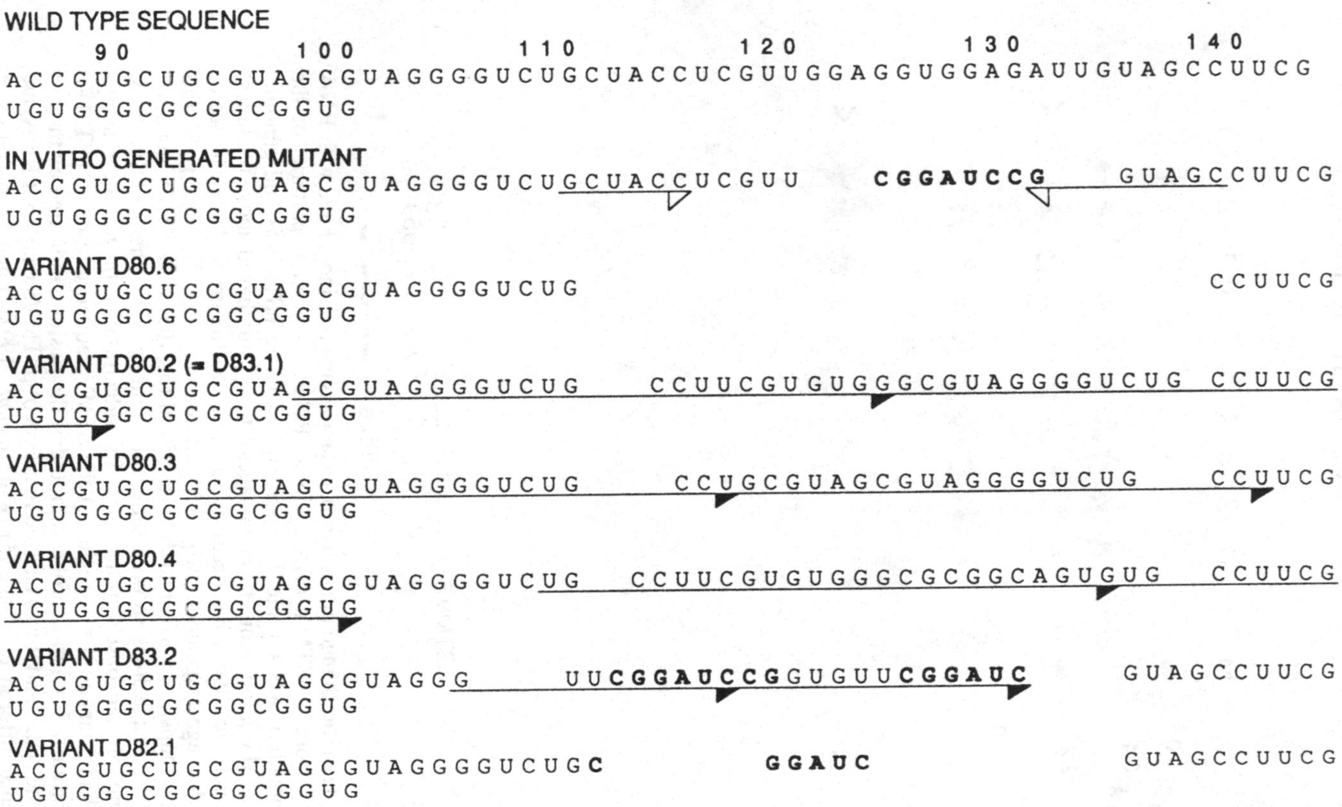

WILD TYPE SEQUENCE

```
        90          100         110         120         130         140
ACCGUGCCUAGCCGUAGGGGUCUGCUGCUACCUCGUUGGAGGUGGAGAUUGUAGCCUUCG
UGUGGGGCGCGGGUG
```

IN VITRO GENERATED MUTANT

```
ACCGUGCCUAGCCGUAGGGGUCUGCUGCUACCUCGUU  CGGAUCCG  GUAGCCUUCG
UGUGGGGCGCGGGUG
```

VARIANT D80.6

```
ACCGUGCCUGCGUAGCCGUAGGGGUCUG                         CCUUCG
UGUGGGGCGCGGGUG
```

VARIANT D80.2 (= D83.1)

```
ACCGUGCCUAGCCGUAGGGGUCUG  CCUUCGUGUGGGGCGUAGGGGUCUG  CCUUCG
UGUGGGGCGCGGGUG
```

VARIANT D80.3

```
ACCGUGCCUAGCCGUAGGGGUCUG       CCUGCGUAGCCGUAGGGGUCUG  CCUUCG
UGUGGGGCGCGGGUG
```

VARIANT D80.4

```
ACCGUGCCUAGCCGUAGGGGUCUG  CCUUCGUGUGGGGGCGCCGGCAGUGUG  CCUUCG
UGUGGGGCGCGGGUG
```

VARIANT D83.2

```
ACCGUGCCUGCGUAGCCGUAGGG  UUCGGAUCCGGUGUUCGGAUC  GUAGCCUUCG
UGUGGGGCGCGGGUG
```

VARIANT D82.1

```
ACCGUGCCUGCGUAGCCGUAGGGGUCUGC  GGAUC  GUAGCCUUCG
UGUGGGGCGCGGGUG
```

Fig. 26.5. Sequences of variant satellite tobacco ringspot virus genomes generated *in planta*. Transgenic plants expressing a mutated STobRV sequence (13 deleted bases replaced with an 8 base Bam HI linker) were infected with TobRV to support satellite RNA replication. Variant sequences of STobRV molecules recovered from individual plants D80, D82, and D83 are shown. Black arrows indicate sequence duplications, white arrows indicate inverted repeats, and bold typed letters highlight bases derived from the linker insertion (Redrawn from Robaglia et al., 1993).

subsequently infected with a TobRV helper virus to support satellite replication. Several mutants were found to undergo sequence changes after replicating in plants for a few days as shown in Fig. 26.5 (Robaglia *et al.*,1993). Some modifications were limited to a few bases surrounding the mutation site, and others involved various extents of deletion and/or duplication either around or at a distance from the linker insertion. New genetic material representing more than 10% of the genome in size was acquired by sequence duplication in some cases. Up to four different co-replicating molecules were isolated from a single plant, and the same original mutant could evolve into as many as seven different sequences in different plants. This study demonstrates the potential of the STobRV molecule for adaptation, since most of the sequences altered *in vitro* could evolve into efficient replicons. The mechanisms of this rapid evolution appear to include deletion, incorporation of non-templated bases, and intra- or inter-molecular recombination leading to sequence duplications.

Selection pressures on satellite evolution

The discernible selection pressures that act on satellite evolution include some shared with all plant viruses, such as the host plant species in which replication occurs. Other pressures are specific to the nature of satellites, including the need to interact with specific helper viruses and the maintenance of the highly ordered secondary structures of satellite genomes.

Indications that the host plant influences satellite evolution are abundant. There are several reports showing that satellites, in combination with a specific helper virus, can replicate with different efficiencies in different host plants (Roossinck *et al.*, 1992). An example is CMV satellite RNAs, which typically replicate to high concentrations in *Nicotiana* species, but replicate very poorly in squash. There are also two reports demonstrating the tendencies of different host species to select different variants from mixtures of necrogenic and non-necrogenic CMV satellite RNAs (García-Luque *et al.*, 1984; Kaper, Tousignant & Steen, 1988).

The first characterization of genetic changes associated with a host shift was a study in which the satellite virus STNV, normally propagated in tobacco, was serially passaged in mung bean and the progeny were analysed for sequence changes by RNase H cleavage patterns (Donis-Keller, Browning & Clark, 1981). Increasing passage number in mung bean was correlated with an increase in sequence heterogeneity of the populations and with specific, reproducible sequence changes that occurred in two independent lines. This suggests that selection of the same pre-existing variant from the original STNV population resulted from replication in a new host plant species. The influence of host species on selection of variants in CMV satellite RNA populations was shown in the experimental evolution study described earlier (see also Kurath & Palukaitis, 1990). Another clear example of this was reported recently, in which transfer of a CMV satellite RNA population with a U residue at position 102 from tomato to squash consistently resulted in selection and amplification of a satellite variant with a C at position 102 (Moriones, Fraile & García-Arenal, 1991). All these studies show specific genetic changes in a satellite population, and demonstrate how minor variants can evolve into major population components by replicating in different host plants.

A universal observation in satellite literature is that satellite genomic RNA sequences are predicted to fold into very ordered secondary structures with 49–73% of their nucleotides involved in base-pairing (Roossinck et al., 1992). An example of such a complex secondary structure is shown in Fig. 26.6. It is thought that these secondary structures provide for the exceptional stability of satellite genomes and also fulfil functional roles, especially for small satellite RNAs and the 3' regions of satellite virus genomes which do not appear to encode proteins. Thus, it is reasonable to suggest that satellite genomes would be selected to develop and maintain these secondary structures. This was seen in a study of the pattern of variation among 25 CMV satellite RNA sequences (Fraile & García-Arenal, 1991). In this analysis, the numbers of substitutions per site, deletions and insertions, and substitutions that would disrupt base-pairing, were significantly greater for unpaired versus base-paired positions in the secondary structure model of CMV satellite RNAs. Another example comes from a study of mutated STobRV satellite RNAs, where most altered molecules could generate replicating satellites, but a mutation which destroyed the plus-strand hammerhead catalytic domain was unable to replicate. These data indicate that the maintenance of secondary structure does indeed constrain the evolution of satellite RNAs.

Another major influence on satellite populations is likely to be the specific helper virus which supports the satellite, although relatively few studies of this phenomenon have been described. There are two reports in which different CMV strains supported different levels of a particular CMV satellite RNA in squash (Kaper & Tousignant, 1977; Roossinck

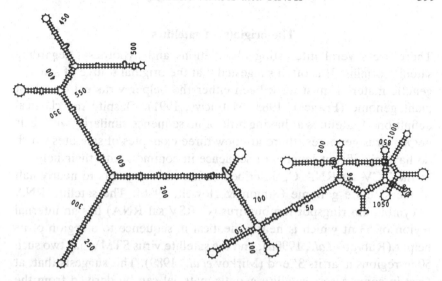

Fig. 26.6. Secondary structure model for satellite tobacco mosaic virus RNA. The computer-predicted secondary structure of the STMV RNA consensus sequence illustrates the highly ordered structure typical of satellite genomes in that 68% of the nucleotides are involved in base-pairing (Kurath, Rey & Dodds, 1993). For a review of satellite RNA secondary structures see Roossinck & Palukaitis, 1992.

& Palukaitis, 1991). In one of these reports, pseudorecombinants of the two helper virus strains were used to map the satellite RNA replication function to CMV RNA1 (Roossinck & Palukaitis, 1991). The first direct demonstration of genetic change in a satellite population resulting from helper virus influence involves STMV. This satellite virus is found naturally associated with tobacco mild green mosaic tobamovirus (TMGMV), but STMV can be experimentally supported by several different tobamoviruses (Valverde, Heick & Dodds, 1991). There is now evidence that replication of STMV with any of several helper viruses other than TMGMV is accompanied by a specific genetic change in the population. This is the deletion of a single G from five consecutive G residues 61–65 bases from the 5′ end of the STMV genome (Kurath, Rey & Dodds, 1993). The same change was repeatedly observed when RNA transcribed from STMV clones replicated with an alternative helper virus, showing that both mutation and selection at this 'helper adaptation domain' can occur within a single passage in plants. This clearly demonstrates that the helper virus exerts selection pressure on this satellite population.

The origin(s) of satellites

There are several interesting observations and hypotheses regarding satellite origins. It is often suggested that the original source of satellite genetic material must have been either the helper virus(es) or a host plant genome (Francki, 1985; Matthews, 1991). Despite the classical definition of satellites as having little or no sequence similarity with their helper virus genomes, there are now three examples of satellites which do have significant lengths of sequence in common with their helpers. In the TCV sat-RNA C, described earlier, this amounts to nearly half of the satellite genome (Simon & Howell, 1986). The satellite RNA of cymbidium ringspot tombusvirus (CyRSV sat-RNA) has an internal region of 53 nt which is nearly identical in sequence to a region of its helper (Rubino et al., 1990a), and the satellite virus STMV has two such 50 nt regions near its 3' end (Mirkov et al., 1989). This suggests that, at least in some cases, satellite genetic material can be derived from the helper virus. It may be that this is a common phenomenon, but in other cases the acquired sequence has diverged beyond recognition owing to the change in selection pressures on the sequence once it is part of a satellite genome.

There are no examples of host plant genomic sequences with extensive sequence similarity to any significant length of a satellite sequence. However, this could also be due to rapid divergence expected to occur under the radically different selection pressures of a satellite and plant genome. Transgenic plant studies have shown that both CMV satellite RNA and STobRV RNA populations can be derived from genomic DNA sequences when plants are infected with the appropriate helper viruses (Gerlach, Llewellyn & Haseloff, 1987; Harrison, Mayo & Baulcombe, 1987).

General hypotheses proposed by Francki (1985) suggested two different origins for different types of satellites. Sequence and structural similarities between certain small satellite RNAs, viroids, and cellular introns have been cited as evidence that these satellites may have evolved from introns or other small RNAs of the host plant (Collmer, Hadidi & Kaper, 1985; Francki, 1985; Matthews, 1991), or from free-living molecules of the pre-cellular world (Elena et al., 1991). This line of origin applies to satellite RNAs with no mRNA function, but which have viroid-like features associated with a rolling circle mechanism of replication. These include circular satellites of sobemoviruses and small satellite RNAs of nepoviruses and luteoviruses.

The second theoretical line of origin is descent from conventional viruses, which could have given rise to satellite viruses by loss of replication functions (Francki, 1985; Matthews, 1991). Further modifications and loss of coat protein expression could have generated large nepovirus satellite RNAs and other satellite RNAs, such as those of CMV, TCV, and tombusviruses, which have terminal structures and replication features similar to those of the helper virus. It is interesting to note that this group of satellites, which theoretically descended from independent viruses, includes the three noted earlier as having significant lengths of sequence nearly identical to sequences of their helper viruses. These similarities could be the result of recent RNA recombination, as is likely for the TCV satellite RNA, or they could represent regions of an entirely helper-derived satellite genome which did not diverge along with the rest of the genome because of functional constraints. The regions of near-identity between STMV and TMGMV are conserved among other tobamoviruses (Mirkov *et al.*, 1989), and the sCyRSV region in common with CyRSV is conserved among other tombusviruses (Rubino *et al.*, 1990a).

Conclusions

The study of sequence relationships between satellite strains and variants has provided clear evidence that nucleotide substitutions, insertions, deletions, and RNA recombination are mechanisms of genetic change which have facilitated satellite evolution. Characterization of progeny from experimental evolution studies has elucidated a molecular mechanism of recombination, and has added sequence duplication and incorporation of non-template-encoded nucleotides to the list of known mechanisms of change. Through serial passage experiments we know that genetic change in satellite populations can be observed within days or weeks. This provides an optimal experimental system for further investigations into the patterns and mechanisms of satellite evolution, with the ultimate goal of understanding, and even predicting, how satellite populations respond genetically to changing environmental and biological selection pressures.

References

Aranda, M.A., Fraile, A. & García-Arenal, F. (1990). *Abstracts of the VIIIth International Congress of Virology*, Berlin, p. 130.

Buzayan, J.M., McNinch, J.S., Schneider, I.R. & Bruening, G. (1987). *Virology*, **160**, 95–9.

Cascone, P.J., Carpenter, C.D., Li, X.H. & Simon, A.E. (1990). *EMBO J.*, **9**, 1709–15.
Cascone, P.J., Haydar, T.F. & Simon, A.E. (1993). *Science*, **260**, 801–5.
Collmer, C.W., Hadidi, A. & Kaper, J.M. (1985). *Proc. Natl. Acad. Sci. USA*, **82**, 3110–14.
Collmer, C.W. & Kaper, J.M. (1988). *Virology*, **163**, 293–8.
Collmer, C.W., Stenzler, L., Fay, N. & Howell, S.H. (1991). *Virology*, **183**, 251–9.
Dange, V., Van Hatta, R.B. & Hecht, S.M. (1990). *Science*, **248**, 585–7.
Danthinne, X., Seurinck, J., Van Montagu, M., Pliej, C.W.A. & Van Emmelo, J. (1991). *Virology*, **185**, 605–14.
Domingo, E., Martinez-Salas, E., Sobrino, F., de la Torre, J.C., Portela, A., Ortin, J., Lopez-Galindez, C., Perez-Brena, P., Villanueva, N., Najera, R., VandePol, S., Steinhauer, D., DePolo, N. & Holland, J.J. (1985). *Gene*, **40**, 1–8.
Donis-Keller, H., Browning, K.S. & Clark, J.M.Jr. (1981). *Virology*, **110**, 43–54.
Elena, S.F., Dopazo, J., Flores, R., Diener, T.O. & Moya, A. (1991). *Proc. Natl. Acad. Sci. USA*, **88**, 5631–4.
Epstein, L.M. & Gall, J.G. (1987). *Cell*, **48**, 535–43.
Fraile, A. & García-Arenal, F. (1991). *J. Mol. Biol.*, **221**, 1065–9.
Francki, R.I.B. (1985). *Ann. Rev. Microbiol.*, **39**, 151–74.
Fritsch, C. & Mayo, M.A. (1989). In *Plant Viruses, vol. I. Structure and Replication*, C.L. Mandahar (ed.). CRC Press, Inc., Boca Raton, Fla., pp. 289–321.
Fuchs, M., Pinck, M., Serghini, M.A., Ravelonandro, M., Walter, B. & Pinck, I. (1989). *J. Gen. Virol.*, **70**, 955–62.
García-Luque, I. Kaper, J.M., Diaz-Ruiz, J.R. & Rubio-Huertos, M. (1984). *J. Gen. Virol.*, **65**, 539–47.
Gerlach, W.L., Llewellyn, D. & Haseloff, J. (1987). *Nature*, **328**, 802–5.
Gordon, K.H.J. & Symons, R.H. (1983). *Nucl. Acids Res.*, **11**, 947–60.
Harrison, B.D., Mayo, M.A. & Baulcombe, D.C. (1987). *Nature*, **328**, 799–802.
Hemmer, O., Meyer, M., Greif, C. & Fritsch, C. (1987). *J. Gen. Virol.*, **68**, 1823–33.
Hidaka, S., Hanada, K., Ishikawa, K. & Miura, K. (1988). *Virology*, **164**, 326–33.
Hidaka, S.K., Ishikawa, K., Takanami, Y., Kubo, S. & Miura, K. (1984). *FEBS Lett.*, **174**, 38–42.
Kaper, J.M. & Tousignant, M.E. (1977). *Virology*, **80**, 186–95.
Kaper, J.M., Tousignant, M.E. & Steen, M.T. (1988). *Virology*, **163**, 284–92.
Kurath, G. & Palukaitis, P. (1989). *Virology*, **173**, 321–40.
Kurath, G. & Palukaitis, P. (1990). *Virology*, **176**, 8–15.
Kurath, G. Heick, J.A. & Dodds, J.A. (1993). *Virology*, **194**, 414–18.
Kurath, G., Rey, M.E.C. & Dodds, J.A. (1992). *Virology*, **189**, 233–44.
Kurath, G., Rey, M.E.C. & Dodds, J.A. (1993). *J. Gen. Virol.*, **74**, 1233–43.
Li, X.H., Heaton, L.A., Morris, T.J. & Simon, A.E. (1989). *Proc. Natl. Acad. Sci. USA*, **86**, 9173–7.
Masuta, C. & Takanami, Y. (1989). *The Plant Cell*, **1**, 1165–73.
Matthews, R.E.F. (1991). In *Plant Virology*, 3rd edn, Academic Press Inc., San Diego, California, pp. 650–682.

Miller, W.A., Hereus, T., Waterhouse, P.M. & Gerlach, W.L. (1991). *Virology*, **183**, 711–20.
Mirkov, T.E., Mathews, D.M., Du Plessis, D.H. & Dodds, J.A. (1989). *Virology*, **170**, 139–46.
Moriones, E., Fraile, A. & García-Arenal, F. (1991). *Virology*, **184**, 465–8.
Murant, A.F. & Mayo, M.A. (1982). *Ann. Rev. Phytopath.*, **20**, 49–70.
Robaglia, C., Bruening, G., Haseloff, J. & Gerlach, W.L.(1993). *EMBO J.*, **12**, 2969–76.
Roossinck, M.J. & Palukaitis, P. (1991). *Virology*, **181**, 371–3.
Roossinck, M.J., Sleat, D. & Palukaitis, P. (1992). *Microbiol. Rev.*, **56**, 265–79.
Rubino, L., Burgyan, J., Grieco, F. & Russo, M. (1990*a*). *J. Gen. Virol.*, **71**, 1655–60.
Rubino, L., Tousignant, M.E., Steger, G. & Kaper, J.M. (1990*b*). *J. Gen. Virol.*, **71**, 1897–903.
Ruffner, D.E., Stormo, G.D. & Uhlenbeck, O.C. (1990). *Biochemistry*, **29**, 10695–702.
Simon, A.E., Engel, H., Johnson, R.P. & Howell, S.H. (1988). *EMBO J.*, **7**, 2645–51.
Simon, A.E. & Howell, S.H. (1986). *EMBO J.*, **5**, 3423–8.
Symons, R.H. (1989). *Trends in Biol. Sci.*, **14**, 445–50.
Taylor, J. (1990). *Semin. Virol.*, **1**, 135–41.
Valverde, R.A., Heick, J.A. & Dodds, J.A. (1991). *Phytopathology*, **81**, 99–104.
Van Emmelo, J., Ameloot, P. & Fiers, W. (1987). *Virology*, **157**, 480–7.
Zhang, C., Cascone, C.J. & Simon, A.E. (1991). *Virology*, **184**, 791–4.

27

Molecular evolution of the retroid family

MARCELLA A. McCLURE

Introduction

Membership in the retroid family (Fuetterer & Hohn, 1987) is defined by the shared amino acid sequence identity to the conserved motifs in the reverse transcriptase (RT) segment of the RNA-directed DNA polymerase. The retroid family is composed of retroviruses, two different classes of DNA viruses, two distinct types of retrotransposons, retroposons, group II introns and plasmids of cellular organelles, an orphan group, the retrons (Temin, 1989) of bacteria (Inouye *et al.*, 1989; Lampson, Inouye & Inouye, 1989; Lim & Maas, 1989) and the telomere elongation protein (EST-1) of yeast (Table 27.1) (Lundblad & Blackburn, 1990).

The variability of gene content and sequence similarity in extant retroid elements (Table 27.1) can be used to infer a model of the evolutionary history of the retroid family (Fig. 27.1). The phylogenetic tree topology, based on RT similarities, indicates an initial bifurcation leading to lineages I and II. Independently derived phylogenies for each set of homologous proteins in the capsid/ribonuclear protein/protease/ reverse transcriptase/ribonuclease H/histidine-cysteine motif/integrase (CA/NC/PR/RT/RH/H-C/IN) element and capsid/reverse transcriptase/tether (CA/RT/T) element produce trees that are congruent with the RT tree, suggesting that the ancestral units evolved as linked genes (Doolittle *et al.*, 1989; McClure, 1992; McClure, unpublished data). The hypothetical ancestral unit of lineage I is further supported by the congruency of the phylogenetic tree for the H-C/IN segment, even though it is found in a position in the *copia*-like retrotransposons different from its position in all other retroid elements. Subsequent acquisition or deletion of gene units may have occurred as indicated along the various branches leading to the tree tips (Fig. 27.1). There

Table 27.1. *Regulatory features and gene complement of retroid elements*

Retroid elements	Example	LTRs	PBS	CA	NC	PR	RT	T	RH	H/C	IN	ENV	PolyA
Retroviruses	MoMLV	+	+	+	+	+	+	+	+	+	+	+	+
Caulimoviruses	CaMV	-	+	+[b]	+	+	+	-	+	-	-	-	+
Hepadnaviruses	HePB	+[a]	-	+[b]	-	-	+	-	+	-	-	-	+
Retrotransposons	17.6	+	+	+[b]	+[c]	+	+	-	+	+	+	-[d]	+
	Copia	+	+	+[b]	+[c]	+	+	+	+	+	+	-	+
Retroposons	LINEs	-	-	+	-	-	+	+	+	-	-	-	+
	Cin4	-	-	+	+[e]	-	+	+	-	-	-	-	+
	R2Bm	-	-	+	-	-	+	-	-	+	+	-	+
	I Factor	-	-	+	+[e]	-	+	+	+	-	-	-	+
	Ingi	-	-	-	-	-	+	+	+	+	+	-	+
Group II introns	INT-SC1	-	-	-	-	-	+	+	-	-	-	-	-
Group II plasmids	Mauriceville	-	-	-	-	-	+	+	+	-	-	-	-
Orphan class	Dirs-1	+[f]	-	-	-	-	+	-	-	-	-	-	-
Retrons	MX65	-	-	-	-	-	+	+[g]	-	-	-	-	+
Telomerase protein	EST-1	-	-	-	-	-	+	-	-	-	-	-	+

Note: [a] There is a region of the *HEPB* genome suggested to be related to retroviral LTRs (Miller & Robinson 1986). [b] Members of the retrotransposon, caulimovirus, hepadnavirus and copia-like retrotransposon lineages have capsid proteins but convincing similarity to retrovirus capsid sequences has not been demonstrated. [c] The presence of the NC gene varies among different members of the retrotransposon and copia-like retrotransposon lineages. [d] Some member activity and sequence similarity to the retrovirus RH is variable. All other abbreviations are as in Fig. 27.1.

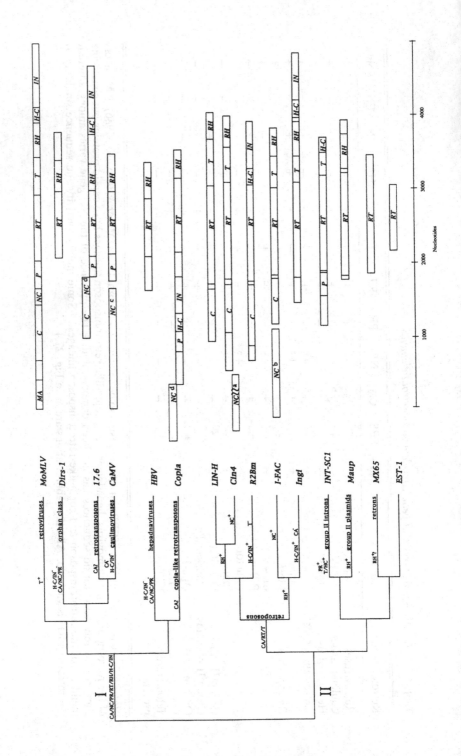

are several findings supporting the independent acquisition of genes at various evolutionary times: two cases of topological incongruency for the RH sequence analysis and the lack of significantly closer similarity between the H-C/IN sequences of the *ingi* and *R2Bm* transposable elements than to any other retroid H-C/IN sequence, as would be expected, from the sequence relationships of the *Ingi* and *R2Bm* RTs (McClure, 1992). The hepadnaviruses, *Dirs-1* element and the caulimoviruses each have less than a full complement of the ancestral unit genes. Both of these members of the retroid family have deleted some functions, or their genomic plan indicates that subsets of the ancestral unit can move independently. These observations suggest that the evolutionary history of the retroid family involves genes that have either been acquired or lost by xenologous recombination (i.e. the replacement of a homologous resident gene with a homologous foreign gene), or independent gene assortment.

Fig. 27.1. Schematic representation of the eukaryotic retroid family phylogeny determined by RT sequence similarity (left side) and summary maps of retrovirus gene complement found in representatives of the major lineages. The hypothetical common ancestors, representing the majority of genes found in the daughter retroid elements at the tree tips, are positioned at critical nodes. Acquisition or deletion of additional gene units is indicated on the descending limbs by a plus or minus sign next to the name of the gene. [a]The NC of *Cin4* is found in a non-overlapping ORF 933 residues long. [b]The NC of the *I-FAC* and its relatives varies in position and repeat number. [a,b,c]The sequence similarity of the *Cin4*, *I-FAC* and CaMV NC is limited to the CXXCXXXXHXXXXC motif of the retroviral NC. [d]The NC of the LTR-retrotransposons and the copia-like LTR retrotransposons is not found in all members of these lineages. Since there are no retroid homologues for retrovirus envelope gene, it is not shown. The retroid element key is as follows: *MoMLV* is Moloney murine leukaemia virus; *Dirs-1* is a transposable element from *D. discoideum*; 17.6 is a transposable element from *D. melanogaster*; *CaMV* is cauliflower mosaic virus; *HBV* is human hepatitis B virus (ayw strain); *Copia* is a transposable element from *D. melanogaster*; *LIN-H* is the human LINE; *Cin4* from *Z.mays*; *R2Bm* from *B. mori*; *I-FAC* from *D. melanogaster*; and *Ingi* from *T.brucei* are other transposable elements. *INT-SC1* is the first intron of cytochrome oxidase subunit 1 from *S. cerevisiae*; *Maup* is the Mauriceville plasmid-1c strain of *Neurospora crassa*; *MX65* is retron 65 of *Myxococcus xanthus*; and *EST-1* is the telomere elongation protein of *Tetrahymena*. The gene key is as follows: MA is the matrix protein, C is the capsid protein, NC is the ribonuclear protein, P is the protease, RT is the reverse transcriptase, T is a region connecting the RT and RH, RH is the ribonuclease H, H-C is the histidine and cysteine motif of the IN and the IN is the integrase.

Retroid family members

Retroviruses

The retrovirus genome generally encodes structural proteins: a matrix protein (MA), a capsid protein (CA), that forms an icosahedral-like shell, a ribonuclear protein (NC) that associates with the viral RNA and an envelope protein that inserts into the host cell membrane to provide the virus coat, and enzymatic proteins: an aspartic protease (P), which cleaves the polyprotein, an RNA-dependent DNA polymerase (RT) connected by a tether (T) to a segment with ribonuclease H activity (RH), and an integrase (IN), which exhibits cleavage specificity for LTR junctions. Both the NC and IN protein have different histidine/cysteine motifs that are potential zinc binding regions (Covey, 1986; Johnson *et al.*, 1986). Spliced mRNAs encode the envelope proteins as well as various regulatory proteins not found in all retroviruses.

DNA viruses

The caulimoviruses have circular, double-stranded DNA genomes that replicate through an RNA intermediate. The mRNA has eight open reading frames (ORFs), only one of which encodes proteins with similarity to retrovirus proteins (P, RT and RH), while the ORF immediately upstream has a small region that can be aligned to the cysteine/histidine motif of the NC (Covey, 1986). To date, the capsid proteins of the plant viruses have not been shown to be related to the analogous protein found in retroviruses. The caulimoviruses appear to share common ancestry with the retrotransposons (Fig. 27.1).

The hepadnaviruses are the causative agents of hepatitis and hepato-cellular carcinoma which is widespread in Asia and Africa. These viruses exhibit a relatively restricted host range; humans, small mammals and birds (Tiollais, Pourcel & Dejean, 1985). They have circular, partially double-stranded DNA genomes that replicate through a RNA inter-mediate. The genomes have four ORFs; only one encodes a protein with similarity to the RT/RH of retroviruses (Miller & Robinson, 1986; Doolittle *et al.*, 1989). One ORF encodes a capsid protein that is claimed to be related, on the one hand, to the retroviral capsid protein (Miller & Robinson, 1986) and on the other, that of the picornaviruses (Argos & Fuller, 1988). By conventional criteria these suggestions of relationship are marginal and will not be considered

Table 27.2. *Location of retroid elements*

Retroid element	Cellular location	Integration into host genome
Retroviruses	Nuclear	+
Caulimoviruses	Nuclear	−
Hepadnaviruses	Nuclear/cytoplasmic	+[a]
Retrotransposons	Nuclear	+[b]
Retroposons	Nuclear	+
Group II introns	Organelle	+
Group II plasmids	Organelle	−
Orphan class	Nuclear	+
Retrons	Bacterial genome	+
EST-1	Nuclear	+[c]

Note: [a] Integration is not necessary for replication. [b]Exceptions: Tst1 is found as a plastid intron, TED is integrated into a mutant strain of baculovirus. [c]EST-1 is a cellular gene.

further. The hepadnaviruses share common ancestry with the copia-like retrotransposons (Fig. 27.1).

Transposable elements

There are two distinct classes of transposable elements that encode various retrovirus protein homologues and are found primarily in the nuclear genome of a variety of organisms (Table 27.2). The class I retrotransposons comprise two separate groups in one of the two major lineages of the retroid family, while the retroposons or class II retrotransposons are found in the other major lineage (Fig. 27.1).

Retrotransposons

Retrotransposons are bounded by LTRs, have primer binding sites (PBS) and transpose via a RNA intermediate. These elements are not known to be infectious. Their gene content is very similar to that of retroviruses, however (Table 27.1). There are two distinct lineages of retrotransposons based on the order of gene complement and RT sequence relationships. The retrotransposons have the same gene order as the extant retroviruses (Fig. 27.1). This group includes several transposable elements: *17.6, 412, gypsy* and *297* of *Drosophila melanogaster* (Saigo et al., 1984; Inouye, Yuki & Saigo,

1986; Yuki, Ishimaru & Saigo, 1986), *Ty3* of *Saccharomyces cerevisiae* (Hansen, Chalker & Sandmeyer, 1988), *del* of *Lilium henryi* (Smyth, Kalitsis & Sentry, 1989), *dre* of *Dictyostelium discoideum* (Marschalek *et al.*, 1989), *surl* found in eight different sea urchin species (Springer, Davidson & Britten., 1991) and *ted* which is found integrated into a mutant strain of a baculovirus (Lerch & Friesen, 1990) (Table 27.2).

The other group, *copia*-like retrotransposons, has a gene order different from all other retroid family members: the integrase protein is found between the protease and polymerase genes (Fig. 27.1). This group is comprised of *copia* from *D. melanogaster*, *Ty1* and *Ty2* from *S. cerevisiae*, *Ta1* from *Arabidopsis thaliana* and *Tnt1* from *Nicotiana tabacum* (Boeke, 1989). *Ty1*, *Ty2* and *copia* make intracellular virus-like particles which are necessary for transposition. A novel member of this group, *Tst1*, is found on the complementary strand of the fifth intron of the potato starch phosphorylase plastid gene (Table 27.2) (Camirand *et al.*, 1990).

Various retrotransposons appear to have both sequence and regional specificities for integration. Some are found in AT-rich regions, inserted into copies of themselves or are found associated with tRNA genes (Sandmeyer, Hansen & Chalker, 1990).

Retroposons

The other group of nuclear transposable elements, that lacks LTRs, is referred to as the retroposons (Rogers, 1985). The retroposons can be further classified into those that integrate into the host genome in a non-sequence-specific manner and those that are target site specific. The first group consists of the long interspersed nuclear elements (LINEs) of higher eukaryotes, the *I,F,G* and *jockey* factors of *D. melanogaster* and *Tx1* of *Xenopus laevis* (Garrett, Knutzon & Carroll, 1989), while the second group includes the *Ingi* factor of *Trypanosoma brucei*, the spliced leader associated conserved sequences (SLACS) of RNA gene clusters of trypanosomes and trypanosomatids (Villanueva *et al.*, 1991), the *Cin 4* element of *Zea mays* and the *R2Bm* element of *Bombyx mori*. There is no correlation between the ability to integrate in a sequence-specific manner and the position of an element in the RT phylogeny (McClure, 1992). LINEs, *I* factor, *Ingi*, *Cin4* and *R2Bm*, representative of this group, are described in more detail (Table 27.1, Fig. 27.1).

The LINEs are found throughout the genomes of mammals, comprising at least 5% of the genomic DNA. They can be found inserted into genes associated with several human diseases; breast adenocarcinoma, haemophilia and teratocarcinoma (Fanning & Singer, 1987). The haemophilia A producing LINE is an active element with demonstrated RT activity that has occupied a chromosome 22 locus for at least 6 million years (Dombroski *et al.*, 1991; Mathias *et al.*, 1991). The *I* factor is responsible for hybrid dysgenesis of *D. melanogaster*. The *Ingi* factor of the parasitic protozoan, *T. brucei*, is found embedded in a 512 bp ribosomal mobile element (RIME) (Kimmel, Ole-Moiyoi & Young, 1987). *Cin4* of the corn plant, *Z. mays*, is found within the A-1 gene and is the first sequence with RT coding capacity of presumed non-viral origin found in plants (Schwarz-Sommer *et al.*, 1987). In the silk moth, *B. mori*, there are two related elements *R2Bm* and *R1Bm* (Burke, Calalang & Eickbush, 1987; Xiong & Eickbush, 1988), which insert into different positions of the 28S rRNA gene.

Group 2 introns and plasmids

The mitochondrial introns and plasmids are not found in all species of yeast, nor in all strains of a given species of yeast (Natvig, May & Taylor, 1984; Lang & Ahne, 1985). These types of elements have not been found in the mitochondria of any other organisms. An intron of a plastid gene from a green alga, however, shows similarity to the RT of the group II yeast mitochondrial introns (Kuck, 1989).

The group II plasmids, Mauriceville-Ic, Varkud and Labelle-1b, are closely related (Pande, Lemie & Nargang, 1989) and their presence in different species of *Neurospora* varies (Natvig, May & Taylor, 1984). The RT sequences of the group II introns and plasmids of yeast mitochondria share a common ancestor and are more closely related to the retrons, EST1 and the retroposons than to any other retroid element (Fig. 27.1).

Orphan class

The *DIRS-1* element of *Dictyostelium discoideum* also belongs to the caulimovirus and LTR-retrotransposon lineage (Fig. 27.1) (Doolittle *et al.*, 1989; Xiong & Eickbush, 1990). This element has only one retroviral-like gene, the polymerase, and this has both RT and RH regions.

Retrons

The RT sequences of bacteria are called retrons. Myxobacteria contain approximately 600 copies of a multicopy single-stranded DNA (msDNA) with an attached RNA that serves as a primer for the RT activity required in msDNA synthesis. An ORF upstream of the msDNA gene encodes the RT. In contrast only 6% of clinical and 13% of wild-type *E. coli* strains encode an analogous RT (Inouye & Inouye, 1991). Another RT has also been found in a P4-like cryptic prophage integrated into the selenocystyl tRNA gene (Sun, Inouye & Inouye, 1991). These data suggest the presence of various classes of retrons in prokaryotic organisms.

Telomere elongation protein

The telomerase of yeast is a ribonucleoprotein enzyme with an endogenous RNA that acts as the template for telomeric repeats and a protein, EST1, with sequence similarity to the retroid RT (Lundblad & Blackburn, 1990).

Conclusions

There has been much speculation on the origin of the retroid family and the possible role or roles of various members in eukaryotic genomes (Finnegan, 1983; Baltimore, 1985; Rogers, 1985, 1986). The rapidly increasing number of retroid elements that are known allows us to begin to answer several questions. Can 'ancient' elements be distinguished from 'modern' counterparts? The *I* factor is found in two forms in all *Drosophila* species. In all strains they are present at constant locations as inactive and incomplete elements. In the active and mobile form the *I* factor causes hybrid dysgenesis and differentiates between inducer and reactive strains (Simonelig *et al.*, 1988).

Can the absence or presence of retroid elements be used to determine the phylogeny of organisms? The presence of the *R1Bm* and *R2Bm*, two related but distinct types of 28S insertional elements, in two different classes of insects (flies and moths) suggests that these types of elements could have been part of an ancestral genome (Jakubczak, Xiong & Eickbush, 1990). The LINEs elements, found in all mammals, appear to have been present since the mammalian radiation (Fanning & Singer, 1987). Is any one type of retroid element restricted to a particular species or genomic 'state'? The retrotransposons are found in insects, yeast and plants, while the retroposons are found in mammals, insects, amphibians, plants and protozoans. Most retroid elements are not restricted to a specific genome state. They can be found as endogenous and exogenous retroviruses, as plant and animal DNA viruses, and as various host-bound transposable elements (one retrotransposon is found integrated into the genome of a baculovirus). In contrast, the intron retroid elements appear to be restricted to organelle genomes. They are not restricted, however, to a specific class of retroid element: the potato plastid intron belongs to the retrotransposons of lineage I, whereas the group II yeast mitochondrial and green alga plastid introns are from lineage II.

Are there multiple classes of retroid elements within the same genome? Obviously, endogenous retroviruses can be found in the same genomes as LINEs. Yeasts have group II mitochondrial introns and plasmids, retrotransposons and copia-like retrotransposons. *Drosophila* have both retrotransposons and copia-like retrotransposons. Have the more 'ancient' members of the retroid family become intimately associated with basic cellular processes? Obviously, the EST1 function to maintain telomere length is critical. The *I* factor is one of the two elements responsible for hybrid dysgenesis of *Drosophila*. The LINEs may play a role in differentiation and development, perhaps by way of gene inactivation/activation (Fanning & Singer, 1987). The first active human LINEs element has been isolated. Its insertion into the factor VIII gene produces haemophilia A (Dombroski *et al.*, 1991). As more evidence accumulates, more retroid elements that have critical genomic functions should be found. Such information will help answer the most basic question about the evolutionary history of the retroid family. At what point in evolutionary time did the initial RT-containing element come into existence? The great variety of organisms that host retroid elements indicates that it is unlikely that a single mode of transmission, either horizontally or vertically, has been responsible for the dispersal of these elements.

It can be imagined that at some point in evolutionary time a limited gene pool provided different combinations of genes, the most successful of which happened to have a retroviral-like gene order. The various retroid genomes could be descendants of this recombinant pool, which by chance, have survived in various evolutionary lineages. Asking which came first, the LTR-containing retroid elements, both extracellular (as viruses) and cellular-bound (as retrotransposons) or non-LTR bounded cellular elements may be a meaningless question. The different retroid lineages may merely reflect the viable, co-evolving states a given modular gene pool can occupy over evolutionary time. Some members of the retroid family may be original descendants of the RNA genetic pool that was instrumental in the conversion of RNA-based systems to DNA-based ones, while other members reflect the ongoing evolution of this remarkable variety of segmentally related genetic elements.

Acknowledgements

I would like to thank Lance Palmer and Sandy Perrine for technical support. Support was provided by NIH grant AI-28309.

References

Argos, P. & Fuller, S.D. (1988). *EMBO J.*, **7**, 819–24.
Baltimore, D. (1985). *Cell*, **40**, 481–2.
Boeke, J.D. (1989). In *Mobile DNA*, D.E. Berg and M.M. Howe (eds.) American Society of Microbiology, Washington, DC, pp. 336–368.
Burke, W.D., Calalang, C.C. & Eickbush, T.H. (1987). *Mol. Cell. Biol.*, **7**, 2221–30.
Camirand, A., St-Pierre, B., Marineau, C. & Brisson, N. (1990). *Mol. Gen. Genet.*, **224**, 33–9.
Covey, S.N. (1986). *Nucl. Acids Res.*, **14**, 623–33.
Dombroski, B.A., Mathias, S.L., Nanthakumar, E., Scott, A.F. & Kazazian, J.H.H. (1991). *Science*, **254**, 1805–8.
Doolittle, R.F., Feng, D.-F., Johnson, M.S. & McClure, M.A. (1989). *Quart. Rev. Biol.*, **64**, 1–30.
Fanning, T.G. & Singer, M.F. (1987). *Biochim. Biophys. Acta*, **910**, 203–12.
Finnegan, D.J. (1983). *Nature*, **302**, 105–6.
Fuetterer, J. & Hohn, T. (1987). *Trends Biochem. Sci.*, **12**, 92–5.
Garrett, J.E., Knutzon, D.S. & Carroll, D. (1989). *Mol. Cell. Biol.*, **9**, 3018–27.
Hansen, L.J., Chalker, D.L. & Sandmeyer, S.B. (1988). *Mol. Cell. Biol.*, **8**, 5245–56.
Inouye, S., Hsu, M.-Y., Eagle, S. & Inouye, M. (1989). *Cell*, **56**, 709–17.
Inouye, M. & Inouye, S. (1991). *TIBS*, **16**, 18–21.
Inouye, S., Yuki, S. & Saigo, K. (1986). *Eur. J. Biochem.*, **154**, 417–25.
Jakubczak, J.L., Xiong, Y. & Eickbush, T.H. (1990). *J. Mol. Biol.*, **212**, 37–52.
Johnson, M.S., McClure, M.A., Feng, D.-F., Gray, J. & Doolittle, R.F. (1986). *Proc. Natl. Acad. Sci., USA*, **83**, 7648–52.
Kimmel, B.E., Ole-Moiyoi, O.K. & Young, J.R. (1987). *Mol. Cell. Biol.*, **7**, 1465–75.
Kuck, U. (1989). *Mol. Gen. Genet.*, **218**, 257–65.
Lampson, B.C., Inouye, M. & Inouye, S. (1989). *Cell*, **56**, 701–7.
Lang, B.F. & Ahne, F. (1985). *J. Mol. Biol.*, **184**, 353–66.
Lerch, R.A. & Friesen, P.D. (1990). *J. Virol.*, **66**, 1590–601.
Lim, D. & Maas, W.K. (1989). *Cell*, **56**, 891–904.
Lundblad, V. & Blackburn, E.H. (1990). *Cell*, **60**, 529–30.

Marschalek, R., Brechner, T., Amon-Bohm, E. & Dingermann, T. (1989). *Science*, **244**, 1493–6.

Mathias, S.L., Scott, A.F., Kazazian, J.H.H., Boeke, J.D. & Gabriel, A. (1991). *Science*, **254**, 1808–10.

McClure, M.A. (1992). *Mol. Biol. Evol.*, **8**, 835–56.

Miller, R.H. & Robinson, W.S. (1986). *Proc. Natl. Acad Sci, USA*, **83**, 2531–5.

Natvig, D.O., May, G. & Taylor, J.W. (1984). *J. Bact.*, **159**, 288–93.

Pande, S., Lemie, E.G. & Nargang, F.E. (1989). *Nucl. Acids Res.*, **17**, 2023–42.

Rogers, J. (1985). *Int. Rev. Cytol.*, **93**, 187–279.

Rogers, J. (1986). *Nature*, **319**, 725.

Saigo, K., Kugimiya, W., Matsuo, Y., Inouye, S., Yoshioka, K. & Yuki, S. (1984). *Science*, **312**, 659–61.

Sandmeyer, S.B., Hansen, L.J. & Chalker, D.L. (1990). *Ann. Rev. Genet.*, **24**, 491–518.

Schwarz-Sommer, Z., Leclercq, L., Gobel, E. & Saedler, H. (1987). *EMBO J.*, **6**, 3873–80.

Simonelig, M., Bazin, C., Pelisson, A. & Bucheton, A. (1988). *Proc. Natl. Acad. Sci., USA*, **85**, 1141–5.

Smyth, D.R., Kalitsis, P. & Sentry, J.W. (1989). *Proc. Natl. Acad. Sci., USA*, **86**, 5015–19.

Springer, M.S., Davidson, E.H. & Britten, R.J. (1991). *Proc. Natl. Acad. Sci., USA*, **88**, 8401–4.

Sun, J., Inouye, M. & Inouye, S. (1991). *J. Bacteriol.*, **173**, 4171–81.

Temin, H.M. (1989). *Nature*, **339**, 252–5.

Tiollais, P., Pourcel, C. & Dejean, A. (1985). *Nature*, **317**, 489–95.

Villanueva, M.S., Williams, S.P., Beard, C.B., Richards, F.F. & Aksoy, S. (1991). *Mol. Cell. Biol.*, **11**, 6139–48.

Xiong, Y. & Eickbush, T.H. (1988). *Mol. Cell. Biol.*, **8**, 114–23.

Xiong, Y. & Eickbush, T.H. (1990). *EMBO J.*, **9**, 3353–62.

Yuki, S., Ishimaru, S., S., I. & Saigo, K. (1986). *Nucl. Acids Res.*, **14**, 3017–30.

28

Adaptation of members of the Orthomyxoviridae family to transmission by ticks

PATRICIA A. NUTTALL, MARY A. MORSE,
LINDA D. JONES AND AGUSTIN PORTELA

Introduction

The influenza viruses of the Orthomyxoviridae family totally depend on higher vertebrates as hosts, and are transmitted by a respiratory or a faecal–oral route. In contrast, two new members of the family, Thogoto (THO) and Dhori (DHO) viruses, here called the orthoacarivirus group, replicate in both vertebrate and tick cells, and are transmitted by tick bite. This chapter compares orthoacariviruses with orthomyxoviruses (influenza viruses), and speculates on their evolutionary origins.

Natural history of Thogoto and Dhori viruses

The first reported isolation of THO virus was from a pool of ticks removed from cattle in Thogoto forest near Nairobi, Kenya, in 1960 (Haig, Woodall & Danskin, 1965). To date, the virus has been isolated from ixodid tick species collected in several countries extending across central Africa, and in Egypt, Iran, Sicily, and Portugal (Davies, Jones & Nuttall, 1986). Dhori virus was first isolated in 1971 from ticks removed from camels in north-west India (Anderson & Casals, 1973). Subsequent isolations have been made from ticks in the former USSR, Portugal, and Egypt (Jones et al., 1989). Experimental studies confirmed that both THO and DHO viruses are true arboviruses (Davies et al., 1986; Jones et al., 1989) in that uninfected ticks become infected as they feed on THO or DHO virus-infected vertebrate hosts; the viruses replicate in the ticks, and are then transmitted by tick bite when the infected ticks take their next bloodmeal.

Both THO and DHO viruses have a wide vertebrate host range (Table 28.1). THO virus infections of sheep cause febrile illness and

Table 28.1. *Tick and vertebrate host associations*

Virus	Ticks	Vertebrates
Thogoto	*Amblyomma variegatum, Boophilus annulatus, B. decoloratus, Hyalomma a. anatolicum, H. truncatum, Rhipicephalus appendiculatus, R. bursa, R. evertsi, R. sanguineus*	Virus isolations: cattle, camels, humans Antibody: cattle, goats, donkey, buffalo, camel, rats, humans
Dhori	*Dermacentor marginatus, Hyalomma dromedarii, H.m. marginatum*	Virus isolations: none Antibody: cattle, sheep goats, horses, dog, buffalo, camel, rodents birds, humans

possibly abortions (Haig *et al.*, 1965, Davies, Soi & Wariru, 1984). In 1966, two human cases of THO virus infection were reported in Ibadan, Nigeria (Moore *et al.*, 1975). Virus was isolated from the cerebrospinal fluid of one patient who had bilateral optic neuritis, and from the blood of the second patient with a fatal meningitis that was complicated by sickle cell anaemia. DHO virus has not been isolated from vertebrates despite the wide range of species in which antibodies have been detected. However, accidental laboratory infections of five humans resulted in febrile illness, and two developed encephalitis (Butenko *et al.*, 1987).

Properties of Thogoto and Dhori viruses

Many of the biological, biophysical, and biochemical properties of THO and DHO viruses are similar to those of the influenza viruses (Table 28.2). The viral proteins are broadly comparable in size (M_r range 28 to 92 K) and function to those of the influenza viruses (Clerx & Meier-Ewert, 1984; Portela, Jones & Nuttall, 1992). No antigenic relationships have been shown between THO and DHO virions, and neither is related antigenically to those of the influenza viruses. Furthermore, haemagglutination by THO and DHO virions occurs at acidic pH values and not at neutral values as with influenza virions.

THO and DHO viruses possess a segmented genome of negative sense, single-stranded RNA, ranging in size from approximately 800

Table 28.2. *Comparison of virus properties*

Property	THO/DHO	Influenza A,B	Influenza C
Segmented, negative sense, single-stranded RNA genome	+	+	+
Lipoprotein membrane	+	+	+
Virion buds from plasma membrane at cell surface	+	+	+
Nuclear antigen	+	+	+
Inhibited by actinomycin D	+	+	+
mRNA polyadenylated	+	+	+
Haemadsorbes	+	+	+
Haemagglutinates	+[a]	+	+
Neuraminidase activity	−	+	−
Esterase activity	−	−	+
Fusion from within	+[b]	+	+

Note: [a] No haemagglutination at physiological pH.
[b] No data for DHO virus.

to 2250 nucleotides. Re-assortment of THO *ts* mutants has been demonstrated in dually infected ticks and vertebrates (Davies *et al.*, 1987; Jones *et al.*, 1987*a*). Sequences of the ends of the viral RNA segments are partially complementary and resemble those of influenza viruses (Clerx, Fuller & Bishop, 1983; Fuller, Freedman-Faulstich & Barnes, 1987; Staunton, Nuttall & Bishop, 1989; Lin *et al.*, 1991; Clay & Fuller, 1992). DHO virus is reported to have a genome comprising seven segments (Lin *et al.*, 1991) whereas THO virus has six segments (Staunton *et al.*, 1989); influenza A and B viruses have eight segments, and influenza C virus has seven segments. Pooling the genetic data for THO and DHO viruses reveals that three of the five genomic segments sequenced to date have small but significant amino acid identity with segments of the genome of influenza viruses (Table 28.3). However, the PA-equivalent protein of THO virus (597 amino acids) is considerably smaller than those of influenza viruses (approx. 720 amino acids) and, when gaps in the alignment comparison are taken into consideration, the level of identity (11–12%) is much less than that of other segments (A.C. Marriott, personal communication). Furthermore, although the probable viral matrix (M) protein of DHO virus is similar in size and charge to the M proteins of influenza viruses, it has no sequence similarity (Clay & Fuller, 1992). It is not yet known whether THO and DHO viruses exploit the

Table 28.3. *Comparison of amino acid identities*

THO/DHO genome segment	Amino acid identity with influenza A, B, C proteins		Equivalent influenza A viral protein	Comments
	THO	DHO		
1	*	*	*	No data
2	*	27–31%	PB1	Lin *et al.*, 1991
3	18–20%	*	PA	Staunton *et al.*, 1989
4	None	None	HA	Unrelated to influenza proteins, See Morse *et al.*, 1992; Freedman-Faulstich & Fuller, 1990
5	*	18–20%	NP	Fuller *et al.*, 1987
6	*	None	M	Clay & Fuller, 1992
7	–	*	*	No data

Note: * no data; – no segment.

coding strategies used by influenza viruses, e.g. splicing, translation from overlapping reading frames (Lamb & Horvath, 1991). A second overlapping reading frame exists in the DHO M protein gene sequence but neither subgenomic mRNA nor a corresponding gene product have been identified (Clay & Fuller, 1992).

Two striking differences have been observed between orthoacariviruses and influenza viruses. First, there is no evidence of a 'cap snatching' mechanism similar to that found in other segmented, negative sense RNA viruses (Staunton *et al.*, 1989; Portela, unpublished data). If such a mechanism is utilized, the sequence data, and the similarity in length of vcRNA and mRNA for THO segment 4, indicate that transcription initiation does not involve the 'stealing' of 10–12 nucleotides from cellular mRNAs, as found for influenza viruses. Second, the single glycoproteins of DHO and THO viruses, encoded by segments 4, show significant identity to each other but are unrelated to any of the influenza glycoproteins (Freedman-Faulstich & Fuller, 1990; Morse, Marriott & Nuttall, 1992) (Fig. 28.1). The significance of these observations is undetermined but evidence suggests that the glycoprotein is involved in their specific adaption to ticks.

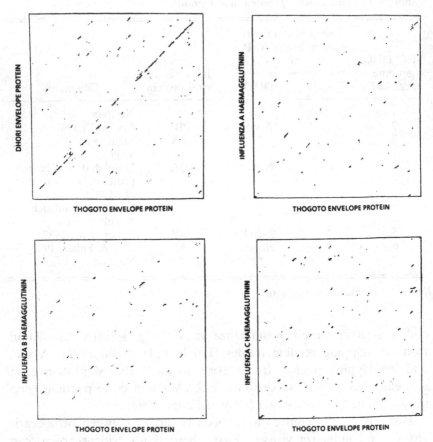

Fig. 28.1. Dot matrix comparison of amino acid sequences of THO glycoprotein with the DHO glycoprotein, and the haemagglutinin of influenza A, B and C viruses.

Adaptation of Thogoto and Dhori viruses to ticks

Infection

The glycoproteins of DHO and THO viruses share about 30% amino acid identity with the envelope glycoprotein (gp64 or gp67) of some baculoviruses (Morse *et al.*, 1992). Multiple alignment of the glycoprotein sequences of THO and DHO viruses, and the gp64/67 sequences of the nuclear polyhedrosis viruses of *Autographa californica* (ACNPV), *Orgyia pseudotsugata* (OPNPV) and *Galleria mellonella* (GMNPV) revealed homology between all five proteins (Fig. 28.2). Baculoviruses have large DNA genomes and have been isolated from arachnids and crustaceans although most members of the family infect insects of the order Lepidoptera (Blissard & Rohrmann, 1990; Murphy *et al.*, 1994). The baculovirus glycoprotein is found on the surface of budded virions and is believed to function in cell-to-cell spread, within the insect host, by mediating fusion with the membrances of endosomes (Volkman & Goldsmith, 1985). Endocytotic fusion is also involved in the early

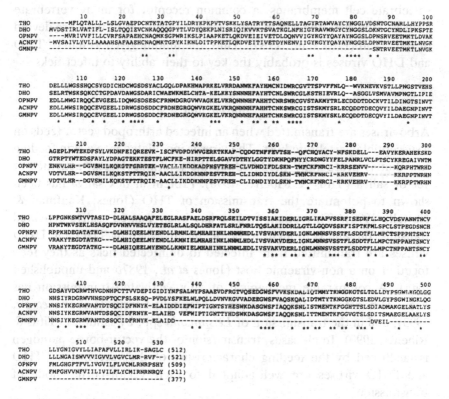

Fig. 28.2. Multiple alignment of amino acid sequences. Amino acids common to all five viral proteins are indicated by *. Data from: Morse *et al.*, 1992 (THO virus); Freedman-Faulstich & Fuller, 1990 (DHO virus); Blissard & Rohrmann, 1989 (OPNPV); Whitford *et al.*, 1989 (ACNPV); Blinov *et al.*, 1984 (GMNPV).

stages of cell infection by influenza viruses (Huang, Rott & Klenk, 1981). Evidence that THO virus induces cell fusion at acidic pH, and that fusion is inhibited by neutralizing monoclonal antibodies directed against the THO viral glycoprotein (Portela *et al.*, 1992), indicate that THO virus utilizes the endocytotic pathway to infect cells.

If THO and DHO viruses use a similar mechanism of cell infection to that of influenza viruses, why do they each use a glycoprotein that appears to have been acquired from a different ancestor? In the context of infection, the answer may lie in the nature of the cell receptor of the virus. Infection of cells by influenza viruses is initiated by binding of the viral haemagglutinin to sialyloligosaccharides on the surface of vertebrate cell membranes, a common receptor for many vertebrate viruses. However, the cell membrance glycoproteins of ticks are devoid of sialic acid (Pino *et al.*, 1989). Thus, the unique glycoprotein of THO and DHO viruses is probably the key to their ability to infect ticks.

Transmission

Arboviruses are transmitted when an infected arthropod vector feeds on a susceptible vertebrate host. The virus is inoculated into a skin feeding site that is profoundly modified by the pharmacological activities of vector saliva (Titus & Ribeiro, 1990). Protein in tick saliva has been shown to potentiate the transmission of THO (Jones, Kaufman & Nuttall, 1992). This phenomenon has been named 'saliva-activated transmission' and is thought to be the means whereby THO and DHO viruses are transmitted from infected to uninfected ticks as they feed together on a non-viraemic host (Jones *et al.*, 1987*b* and unpublished data). A comparable mechanism of saliva-activated transmission has been demonstrated for tick-borne encephalitis virus (Labuda *et al.*, 1993), and in the transmission of *Leishmania* spp. by sandflies (Titus & Ribeiro, 1990). In all cases, transmission of the vector-borne pathogen is facilitated by the feeding characteristics of the vector. Thus THO and DHO viruses are well-adapted to their vector-borne mode of transmission.

Classification and evolutionary origins of Thogoto and Dhori viruses

THO and DHO viruses are classified as species of the new genus *Orthoacarivirus* of the family Orthomyxoviridae (Murphy *et al.*, 1994).

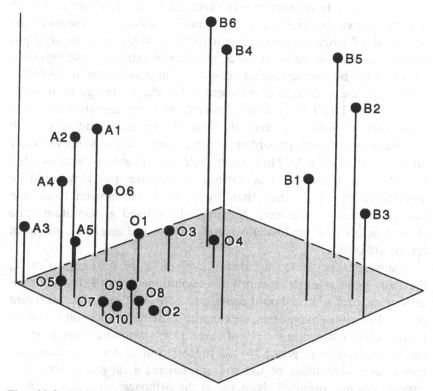

Fig. 28.3. Diagram representing an ordination of the gene sequences of several genes of orthoacariviruses, orthomyxoviruses and baculoviruses. The percentages of each of the 16 nearest neighbour nucleotide pairs was calculated for each gene. Then, using the NTSYS-pc 1.70 package of programs (Rohlf, 1992), these were used to calculate 'average taxonomic distances' for all pairs of genes, and the distances ordinated by multidimensional scaling (stress level 2). The gene sequences came from the Genbank database and were: Thogoto orthoacarivirus gp75, A1; segment 3, A3; Dhori orthoacarivirus gp75, A2; segment 2, A5; segment 5, A4; *Autographa californica* nuclear polyhedrosis baculovirus polyhedrin, B1; p10, B3; gp67, B5; *Orgyia pseudotsugata* nuclear polyhedrosis baculovirus polyhedrin, B2; p10, B4; gp67 B6; Influenza A/NT/60/68 (H3N2) haemagglutinin, 01; neuraminidase, 03; nucleoprotein, 05; polymerase 1, 07; polymerase 2, 09; polymerase 3, 010; Influenza A/PR/8/34 (H0N1) haemagglutinin, 02; neuraminidase, 04; matrix protein, 06; polymerase 1, 08.

They show similarities to influenza viruses in virion structure, morphogenesis and strategy of replication, but differ notably in their mode of transmission and properties of the viral glycoprotein.

Are the orthoacariviruses ancestors of the influenza viruses or descendents, or have both genera diverged from a common ancestor? If orthoacariviruses evolved first, the major evolutionary events in the origins of influenza viruses (i.e. 'punctuations' in the evolutionary equilibrium) must have included the acquisition of extra genomic segments encoding the neuraminidase and possibly non-structural proteins (NS1, NS2), and the exchange of segment 4 for the haemagglutinin gene. If THO and DHO viruses are descendents of an ancestral influenza virus, their evolution involved the loss of viral genes and acquisition of a baculovirus gene, possibly through recombination events in dually infected arthropod cells. However, baculoviruses have not been isolated from ticks and, indeed, it is difficult to conceive how they could be maintained by ticks other than by vertical transmission from one tick generation to the next. Possibly, the related glycoprotein gene of orthoacariviruses and baculoviruses was derived independently from arthropod cells.

Assuming that orthoacariviruses evolved from orthomyxoviruses, then any gene acquired recently (in evolutionary terms) from a non-orthomyxoviral source should contain features that reflects its different origin. To test this hypothesis, we examined the base ratios, dinucleotide frequencies, and codon usages of several orthoacarivirus, baculovirus, and orthomyxovirus genes. In all these features, the orthoacariviral genes resembled those of the orthomyxoviruses suggesting that they have a common origin. This included the orthoacarivirus glycoprotein gene even though in the analysis it was on the edge of the cluster that the orthomyxovirus genes formed (Fig. 28.3). Thus the sequence analyses show that the orthoacarivirus and baculovirus glycoprotein genes have a common ancestor, although none of the other genes seems to be related; however, the orthoacarivirus glycoprotein gene now has all the compositional features of an orthomyxoviral gene. At present, we can only speculate on 'which acquired what, from whom, and when'?

Acknowledgements
Many thanks to Adrian Gibbs for Fig. 28.3 and all that went with it.

References

Anderson, C.R. & Casals, J. (1973). *Ind J. Med. Res.*, **61**, 1416–20.

Blinov, V.M., Gutorov, V.V., Holodilov, N.G., Iljichev, A.A., Karginov, V.A., Mikrjukov, N.N., Mordivinov, V.A., Nikonov, I.V., Petrov, N.A., Urmanov, I.H. & Vasilenko, S.K. (1984). *FEBS Lett.*, **167**, 254–8.

Blissard, G.W. & Rohrmann, G.F. (1989). *J. Mol. Biol.*, **198**, 327–37.

Blissard, G.W. & Rohrmann, G.F. (1990). *Ann. Rev. Entomol.*, **35**, 127–55.

Butenko, A.M. *et al.*, (1987). *Vop. Virusol.*, **32**, 724–9.
Clay, W.C. & Fuller, F.J. (1992). *J. Gen. Virol.*, **73**, 2609–16.
Clerx, J.M.P., Fuller, F. & Bishop, D.H.L. (1983). *Virology*, **127**, 295–319.
Clerx, J.P.M. & Meier-Ewert, H. (1984). In *Segmented Negative Strand Viruses*, R.W. Compans and D.H.L. Bishop (eds.), Academic Press, New York, pp. 320–329.
Davies, C.R., Jones, L.D., Green, B.M. & Nuttall, P.A. (1987). *J. Gen. Virol.*, **68**, 2331– 8.
Davies, C.R., Jones, L.D. & Nuttall, P.A. (1986). *Am. J. Trop. Hyg.*, **35**, 1256–62.
Davies, F.G., Soi, R.K. & Wariru B.N. (1984). *Vet. Rec.*, **115**, 654.
Freedman-Faulstich, E.Z. & Fuller, F.J. (1990). *Virology*, **175**, 10–18.
Fuller, F.J., Freedman-Faulstich, E.Z. & Barnes, J.A. (1987). *Virology*, **160**, 81–7.
Haig, D.A., Woodall, J. & Danskin, D. (1965). *J. Gen. Microbiol.*, **38**, 389–94.
Huang, R.T.C., Rott, R. & Klenk, H-D. (1981). *Virology*, **110**, 243–7.
Jones, L.D., Davies, C.R., Green, B.M. & Nuttall, P.A. (1987*a*). *J. Gen. Virol.*, **68**, 1299–306.
Jones, L.D., Davies, C.R., Steele, G.M. & Nuttall, P.A. (1987*b*). *Science*, **237**, 775–7.
Jones, L.D., Davies, C.R., Steele, G.M. & Nuttall, P.A. (1989). *Med. Vet. Entomol.*, **3**, 195–202.
Jones, L.D., Kaufman, W.R. & Nuttall, P.A. (1992). *Experientia*, **48**, 779–82.
Labuda, M., Jones, L.D., Danielova, V. & Nuttall, P.A. (1993). *J. Med. Ent.*, **30**, 295–9.
Lamb, R.A. & Horvath, C.M. (1991). *TIG*, **7**, 261–6.
Lin, D.A., Roychoudhury, S., Palese, P., Clay, W.C. & Fuller, F. (1991). *Virology*, **182**, 1–7.
Moore, D.L., Caisey, O.R., Carey, D.E., Reddy, S., Cooke, A.R., Akinkugbe, F.M., David-West, T.S. & Kemp, G.E. (1975). *Ann. Trop. Med. Parasitol.* 69:49–64.
Morse, M.A., Marriott, A.C. & Nuttall, P.A. (1992). *Virology*, **186**, 640–6.
Murphy, F.A., Fauguet, C.M., Bishop, D.H.L., Ghabrial, S.A., Jarvis, A.W., Martelli, G.P., Mayo, M.P. & Summers, M.D. (1994). *Virus taxonomy: Sixth Report of the International Committee on Taxonomy of Viruses*. Sprunger Verlag, Wien, New York, pp. 587.
Pino, C., del, Treis Trindade, V.M., Guna, F. & Bernard, E.A. (1989). *Insect Biochem.*, **19**, 657–61.
Portela, A., Jones, L.D. & Nuttall, P.A. (1992). *J. Gen. Virol.*, **73**, 2823–30.
Rohlf, F.J. (1992). *Numerical Taxonomy and Multivariate Analysis System*, Version 1.70.
Staunton, D., Nuttall, P.A. & Bishop, D.H.L. (1989). *J. Gen. Virol.*, **70**, 2811–17.
Titus, R.G. & Ribeiro, J.M.C. (1990). *Parasitol. Today*, **6**, 157–60.
Volkman, L.E. & Goldsmith, P.A. (1985). *Virology*, **143**, 185–95.
Whitford, M., Stewart, S., Kuzio, J. & Faulkner, P. (1989). *J. Virol.*, **63**, 1393–9.

29

The Order *Mononegavirales*: evolutionary relationships and mechanisms of variation

C.R. PRINGLE

Introduction

Animal viruses with non-segmented negative stranded RNA genomes have several properties in common in addition to the sense and form of the nucleic acid sequestered in the virion (Table 29.1). The distinguishing features which account for the classification of these viruses into the three separate families *Filoviridae*, *Paramyxoviridae* and *Rhabdoviridae* are predominantly morphological and biological (Francki *et al.*, 1991) and are summarized in Table 29.2. Recent progress in the cloning, and nucleotide sequencing of the genomes of representative viruses from these three families, has revealed the anomaly that some paramyxoviruses appear to be more closely related to members of the family *Rhabdoviridae* than to other members of the family *Paramyxoviridae*. To accommodate these emerging inter-family relationships, the three families of viruses with non-segmented negative stranded genomes have been grouped together as the Order *Mononegavirales* (mono: undivided; nega: negative strand;

Table 29.1. *Unsegmented genome negative stranded RNA viruses*

- Negative sense RNA in the virion
- Virion-associated RNA polymerase mediates transcription and replication
- Genome transcribed into 6–10 separate mRNAs from a single promoter
- Replication occurs by synthesis of a complete positive-sense RNA anti-genome
- Nucleoprotein is the functional template for synthesis of replicative and mRNA
- Independently assembled nucleocapsids are enveloped at the cell surface at sites containing virus proteins
- Mainly cytoplasmic
- Occur in invertebrates, vertebrates and plants

Table 29.2. *The classification into three families of animal viruses with non-segmented negative stranded RNA genmoes.*

The family Rhabdoviridae

- Bullet-shaped or bacilliform; 45–100 nm diameter × 100–430 nm length
- Surface spikes composed of G protein alone; 5–10 nm length × 3 nm diameter
- Helical nucleocapsid; 50 nm diameter, unwinding to helical structure of 20 × 700 nm (VSV).
- Genome size; 11 161 kb for vesicular stomatitis virus and 11 932 kb for rabies virus
- 5–6 open reading frames coding 5–7 polypeptides. Proteins: N, NS (P or M1), M (or M2), G and L common to all; variable proteins include one of unknown function encoded in P, and a non-structural protein sc4 in the plant virus SYNV, and a non-virion protein in the fish virus IHNV.
- May be vertically transmitted in arthropods, but otherwise spread horizontally
- Infect invertebrates, vertebrates and plants

The family Paramyxoviridae

- Pleomorphic, some roughly spherical, 150 nm or more in diameter, filamentous forms common
- Envelope derived from cell membrane, incorporating 1 or 2 viral glycoproteins and 1 or 2 unglycosylated proteins
- Helical nucleocapsid 13–18 nm diameter, 5.5–7 nm pitch
- Genome size uniform; range 15 156 kb (NDV) to 15 892 kb (measles virus)
- 7–8 functional ORF (genes) encoding 10–12 polypeptides, of which 4–5 may be derived from 2–3 overlapping RF in the P gene
- Proteins: N (or NP), P, L, M, F, and an attachment protein (G, H or HN) common to all genera; variable proteins include the non-structural C, 1C (or NS1) and 1B (or NS2), a cysteine-rich V, a small integral membrane protein SH, and a second inner envelope protein M2 (or 22K)
- Horizontal transmission, mainly airborne
- Found only in veterbrates, no vectors

The family Filoviridae

- Filamentous forms with branching; sometimes U-shaped, 6-shaped or circular
- Uniform diameter of 80 nm and varying lengths upto 14 000nm. Infectious particle length is 790 nm for Marburg virus and 970 nm for Ebola virus
- Surface spikes of 10 nm length
- Helical nucleocapsid; 50 nm diameter, with an axial space of 20 nm diameter and helical periodicity of about 5 nm
- Genome; negative sense single stranded RNA of M_r 4.2×10^6
- At least 5 proteins; a large (polymerase) protein, a surface glycoprotein, 2 nucleocapsid-associated proteins, and at least one other protein of unknown function
- Biology enigmatic; only two antigenically unrelated viruses known; blood-borne infection of humans and monkeys

virales: viruses), the taxonomic designation Order being preferred to any modification of an existing taxon, such as super-family, which might imply direct evolutionary descent. At the same time, the family *Paramyxoviridae* was split into the two sub-families *Paramyxovirinae* and *Pneumovirinae* in recognition of the distinctiveness of the pneumoviruses (Pringle, 1991*a*). Recognition of distant taxonomic relationships in the case of these three families may be a consequence of the absence (or extreme rarity) of genetic recombination among non-segmented negative strand RNA viruses (for review see Pringle, 1991*b*). Consequently, disruption of lineages by introduction of extraneous genetic information has probably not occurred and distant evolutionary relationships can be discerned.

Genome structure and coding capacity

The genomes of all negative strand viruses with undivided genomes exhibit certain regular features. The 3'–5' orientation of the individual genes in relation to the 3'-promoter region is similar. The envelope protein genes are located centrally between two or more core protein genes on the 3'-terminal side and the L (polymerase) protein gene on the 5'-terminal side. The majority of genes possess a single open reading frame and use a single start site. No ambisense encoding of information has been recognized. The phosphoprotein (P) gene is exceptional and may encode several polypeptides (Lamb & Paterson, 1991). In the paramyxovirus Sendai virus, for example, at least six polypeptides are expressed from three overlapping reading frames (Kolakofsky, Vidal & Curran, 1991). The P open reading frame is the largest with a single start site, whereas the C open reading frame has several alternate start sites, and a form of RNA editing (non-templated insertion of nucleotides in transcripts) is involved in accessing the complete V protein reading frame (Lamb & Paterson, 1991). In some other paramyxoviruses (the genus *Rubulavirus*), on the other hand, it is the P reading frame which is accessed by RNA editing, and the V protein which is translated from an uninterrupted reading frame. Additional products may be derived by internal initiation of translation, and in the pneumoviruses and rhabdoviruses this may be the only mechanism for expansion of the coding potential of the P gene (Herman, 1987; Curran & Kolakofsky, 1988; Caravokyri, Zajac & Pringle, 1992).

The out-of-phase encoding of the C protein in relation to the P protein is a conserved feature in paramyxoviruses with overlapping P/C cistrons, and the third position of the C codons corresponds to the first position of the P codons (Galinski & Wechsler, 1991). Because of the redundancy of the genetic code, mutations occurring in the third position are less likely to result in an amino acid replacement than mutations in the first position. Consequently the two different proteins encoded in the same nucleotide sequence may diverge at different rates; and indeed it is observed that the C protein amino acid sequence of related viruses is more conserved than that of the P protein which is the least conserved of all paramyxovirus proteins.

There is an apparent progression in complexity of organization of the genome from the basic pattern of five transcriptional units of the vesiculoviruses through the six of the paramyxoviruses to the ten of the pneumoviruses (Fig. 29.1, 29.2, and 29.3). It is not possible, however, to deduce from such comparisons alone whether the direction of evolution is towards greater complexity or towards greater economy.

Comparison of the amino acid sequences of the major structural proteins of paramyxoviruses and rhabdoviruses has suggested that there has been differential evolution of the different transcriptional units, with strong selective pressure for maintenance of the mechanism of replication and transcription. The order of conservation of the structural proteins of paramyxoviruses is L > M > F > N > H(N) > P (Rima, 1989). The same descending order is observed in comparison of paramyxoviruses and rhabdoviruses with the L protein being the most conserved and the P protein the least (Spriggs & Collins, 1986; Tordo *et al.*, 1988). Indeed, the L protein is remarkably conserved throughout all non-segmented negative strand RNA viruses with six separate domains clearly recognizable, confirming the common evolutionary ancestry of these viruses (Tordo *et al.*, 1988; Poch *et al.*, 1990). At the domain level, clear structural homologies have been identified also between the nucleocapsid (N) proteins of viruses belonging to the families *Filoviridae*, *Paramyxoviridae* and *Rhabdoviridae* (Barr *et al.*, 1991; Sanchez *et al.*, 1992).

Mechanisms of variation and genome evolution

Comparisons of the genome organizations of those individual viruses for which complete gene sequences are available (see Figs 29.1, 29.2 and 29.3) suggest that the non-segmented negative stranded RNA viruses evolve predominantly by expansion (or contraction) of intergenic regions. The intergenic regions range from a minimal dinucleotide separation observed in some rhabdoviruses, to the non-uniform pattern observed in pneumoviruses and some paramyxoviruses. In addition, rearrangement of genes within the envelope protein region may occur rarely (Fig. 29.3).

The rhabdoviruses

In the family *Rhabdoviridae* (Fig. 29.1) both the animal vesiculovirus vesicular stomatitis Indiana virus (VSV-I) and the plant *Sonchus*

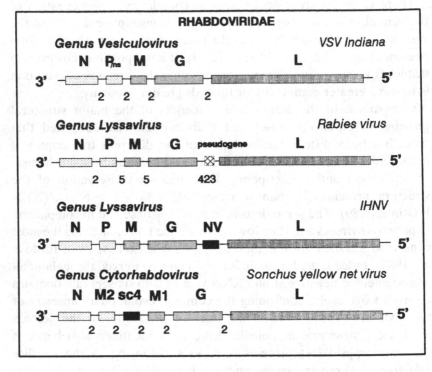

Fig. 29.1. Gene order maps: the family *Rhabdoviridae*. The figures below are the number of intergenic nucleotides. N: nucleoprotein gene; P: phosphoprotein gene; M: matrix protein gene; G: surface glycoprotein gene; L: polymerase gene. P/ns indicates a phosphoprotein gene including a non-structural protein gene accessed by internal initiation. NV and sc4 are presumptive non-structural protein genes.

Key: ▭ – core protein genes
▨ – matrix and envelope protein genes
▦ – polymerase gene
▧ – pseudogene
■ – new gene

Scale approximate.

yellow net virus exhibit the minimal configuration of two intergenic nucleotides. However, the plant virus has an additional gene (sc4) coding for a presumptive non-structural protein located between the M2 and M1 genes. The fish lyssavirus infectious haematopoietic necrosis virus also has a sixth gene (NV) coding for a non-virion protein, in this case located between the G (glycoprotein) and L (polymerase) genes. The genome of rabies virus may represent a transitional stage between these two states. In the virulent Pasteur strain of rabies virus the intergenic region between G and L extends to 423 nucleotides and includes an untranslated pseudogene (Tordo *et al.*, 1986, 1988). In a proportion of transcripts, however, the pseudogene is transcribed by read-through of the G protein gene termination signal (Tordo, cited in Conzelmann *et al.*, 1990). In the attenuated SAD B19 strain of rabies virus, the pseudogene is incorporated into the non-translated tail of the G protein gene as a result of deletion of three A residues in the latter, and as a consequence the size of the intergenic region is reduced to 24 nucleotides. Again, such comparisons do not indicate whether there is loss or gain of coding sequence by this mechanism. However, since all rabies virus strains so far examined transcribe the pseudogene lying between G and L, there appears to be strong selection towards reduction of intergenic distances, favouring the interpretation that the pseudogene represents a redundant gene rather than an emerging gene (Conzelmann *et al.*, 1990).

The paramyxoviruses

The viruses now classified in the sub-family *Paramyxovirinae* of the family *Paramyxoviridae* exhibit a similar uniformity of genome organisation with intergenic distances of generally 3 nucleotides, but varying between 1 and 7 nucleotides (Fig. 29.2). The major difference from the rhabdoviruses is the complex encoding of information in the P gene and the replacement of the G protein gene by the F and H (or HN) gene pair, which delegates the attachment and fusion functions of the surface glycoprotein to separate molecules. The genus paramyxoviruses proper can be sub-divided into two sub-groups (Stec, Hill & Collins, 1991) represented, on the one hand, by mumps virus, parainfluenza virus types 2 and 4, and Newcastle disease virus (now designated the genus *Rubulavirus*), and on the other by Sendai virus and parainfluenza virus types 1 and 3 (now designated the genus *Paramyxovirus*). The former is characterized by the presence of a small hydrophobic (SH) protein

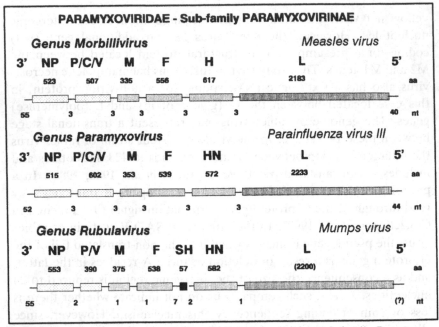

Fig. 29.2. Gene order maps: the family *Paramyxoviridae*, sub-family *Paramyxovirinae* As Fig.29.1. Figures above the genes are number of amino acids encoded. NP: nucleoprotein; P/C/V: phosphoprotein gene; F: fusion protein gene; SH: small hydrophobic protein gene; H and HN: attachment protein gene.

gene lying between F and HN and a different pattern of processing of information from the P gene. The members of the genus *Morbillivirus* (measles, canine distemper, phocine distemper, rinderpest and pestedes-petite ruminants viruses) are a group of antigenically related viruses. They lack an SH gene and in terms of genome structure resemble the second group of paramyxoviruses, except that the attachment protein lacks a neuraminidase function.

The filoviruses

Characterization of several members of the family *Filoviridae* is proceeding rapidly and the organization of the genomes of these viruses seems to be very similar to the patterns described above (Sanchez *et al.*, 1992). However, less is known about these viruses because of problems inherent in working with such lethal agents; they will not be considered further here.

The pneumoviruses

The second sub-family of the paramyxoviruses, the *Pneumovirinae*, is markedly different from the viruses described so far, although the absolute size of the genomes of all paramyxoviruses is highly conserved (Fig. 29.3). The pneumoviruses possess several additional genes (NS1 or 1C, NS2 or 1B, 1A or SH, and 22K or M2), non-uniform intergenic distances and in the case of respiratory syncytial (RS) virus an overlap of the 5'-proximal 22K and L genes (Fig. 29.3). There are three glycosylated membrane protein genes (F, SH, and G) and two unglycosylated membrane-associated protein genes (M, and 22K). The G protein is a heavily glycosylated protein with a high proportion of O-linkages that has no counterparts in other negative stranded RNA viruses (for review see Collins, 1991). The 3'–5' relative orientation of the core, envelope, and polymerase genes, however, is maintained, but there is variation in the order of the transcriptional units within the envelope

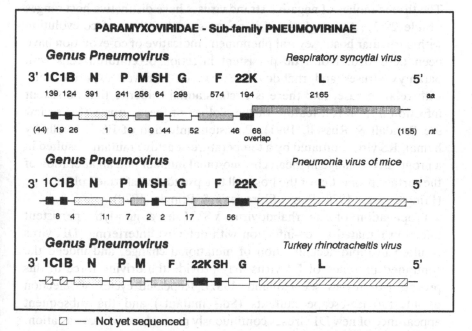

Fig.29.3. Gene order maps: the family *Paramyxoviridae*, sub-family *Pneumovirinae* As Fig. 29.1. 1C and 1B: non-structural protein genes; SH: small hydrophobic protein gene; 22K: non-glycosylated envelope protein gene. Figures above are number of amino acids, and figures below are the number of intergenic nucleotides.

protein gene region (Fig. 29.3). The order of the attachment (G) and fusion (F) protein genes in pneumoviruses is inverted with respect to the corresponding genes in other paramyxoviruses, and in addition the SH and G genes of the avian pneumovirus, turkey rhinotracheitis virus (TRTV), appear to have been transposed relative to their location in the other two pneumoviruses (Ling, Easton & Pringle, 1992). It is worthy of note that TRTV is an avian virus associated with a new epidemic disease, whereas the other two pneumoviruses have long associations with their mammalian hosts. As emphasized already, genetic recombination has not been demonstrated directly in any non-segmented genome RNA virus and the SH-G transposition in TRTV probably could not have been generated by a single recombinational event. It may be, however, that there are rare circumstances that allow recombination and that such events may be associated with transmission to a new host.

Co-evolution

The three families of negative strand viruses have distinctive host ranges (Table 29.2). Although there is no consistent pattern of co-evolution with particular hosts, several phenomena indicative of co-evolution have been described. Non-lytic persistent infection of cultured cells with paramyxoviruses and rhabdoviruses can be achieved experimentally with relative ease, and there is circumstantial evidence that persistent infection *in vivo* is a feature of the biology of these viruses (for review see Randall & Russell, 1991). Persistent infection of BS-C-1 cells by human RS virus, initiated by a temperature-sensitive mutant, resulted in a progressive change in mean chromosomal number and an alteration of the surface properties of the host cell to a pseudotransformed phenotype (Pringle *et al.*, 1978; Parry, Shirodaria & Pringle, 1979).

Propagation of the rhabdovirus VSV-I as a non-lytic persistent infection initiated by co-infection with defective interfering (DI) virus resulted in rapid accumulation of mutational change, and indeed the continued presence of DI virus may provide the driving force in this process (for review see Holland, 1987). Repeated cycles of selection of interference-escape mutants (Sdi^2 mutants) and the subsequent appearance of new DI viruses continuously generate sequence variation. Nucleotide sequencing of the terminal regions of both non-defective and DI virus revealed progressive accumulation of mutations with 5'-terminal mutations exceeding 3'-terminal mutations. VSV-I propagated as a persistent infection in mouse L cells in the absence of DI virus did not accumulate mutations preferentially in the 5'-terminal region, suggesting that the 5'-terminal mutations were selected by the DI virus. The generation of DI virus is strongly influenced by the host cell, being rare in some cells and frequent in others. As a consequence DI virus-driven evolution of a virus population is not independent of the genetic make-up of the tissue or cells harbouring the virus. Sequencing of non-defective and DI virus from persistent infections revealed that the same mutations were accumulating in the termini of each simultaneously, indicating that the non-defective and the defective virus were affected by similar selective pressures (O'Hara *et al.*, 1984). Biased hypermutation (U → C and A → G transitions) has been reported by Murphy, Dimock Yong Kang (1991) in the 3'-terminal region of parainfluenza virus type 3 genomic RNA from persistently infected cell cultures.

The neurological diseases, which are rare consequences of measles virus infection in humans, are the best examples of the consequences of persistent infection *in vivo* (for review see Billeter & Cattaneo, 1991). Virus rescued from the brains of individuals dying as a result of SSPE (sub-acute sclerosing panencephalitis) appeared to have undergone extensive mutation with, in one case, a cluster of A → G changes (biased hypermutation) in the H gene. M gene-specific biased hypermutation (with replacement of about 50% of all U residues by C residues) has been observed in the cloned genome of measles virus originating from the brain of a MIBE (measles inclusion body encephalitis) patient and at a lower frequency in several SSPE cases. Biased hypermutation is believed to be due to a double-stranded RNA unwindase activity present in some eukaryotic cells which converts adenosine to inosine, and this cellular activity could play a role in virus co-evolution. Rataud, Hirano and Wong (1992) have demonstrated that the matrix protein gene of a measles virus variant accumulated numerous U → C transitions during passage in IMR-32 human neuroblastoma cells, not found during passage of the same virus in Vero cells. IMR-32 cells contained higher levels of a nuclear RNA unwindase activity than Vero cells, which supports the hypothesis that this activity is responsible for the biased hypermutation observed in measles virus recovered from neural tissue.

Genetic evidence of the involvement of host factors in the functional activity of the viral RNA polymerase provides another opportunity for co-evolution of the negative strand RNA viruses and their hosts. In rhabdoviruses, mutations affecting the polymerase may have marked effects on host range. Both temperature-dependent and non-conditional host range mutants of vesicular stomatitis New Jersey virus and Chandipura virus have been isolated which exhibited different extents of restriction in certain cell lines and secondary cultures of embryonic cells from several types of animals and birds (Pringle, 1978). The link between viral replicative ability and the genetic constitution of the host provides a mechanism for co-evolution of virus and host. The conventional live virus vaccines developed to control canine distemper, mumps, measles and rabies viruses have all been derived by modification of pathogenic properties by prolonged sequential passage in a foreign cell type. However, although antigenic variation in these viruses is common and may be an important factor in evasion of the host immune response, progressive and cumulative antigenic variation of the type associated with influenza A virus has been difficult to demonstrate, despite the fact that the attachment (G) protein of human RS virus exhibits comparable variability to the HA of influenza A virus (Cane *et al.*, 1991). Recent analysis of the molecular epidemiology of antigenic subgroup A of human RS virus has revealed that several strains (lineages) of virus co-circulate during the course of a single epidemic and that the predominance of particular strains changes from one annual epidemic to the next. Furthermore the same strains are present world-wide, although varying in predominance at any one time (Cane *et al.*, unpublished data). It is likely that the other paramyxoviruses will prove to be similar in this respect.

Conclusion

Little is known about the natural routes of transmission of the non-segmented negative stranded RNA viruses, apart from those rhabdoviruses which are usually associated with a specific arthropod vector. Until there is greater understanding of the factors which influence the spread of these viruses, the study of co-evolution will remain predominantly at the level of the comparison of taxonomic relationships described above.

References

Barr, J., Chambers, P., Pringle, C.R. & Easton, A. (1991). *J. Gen. Virol.*, **72**, 677–85.

Billeter, M.A. & Cattaneo, R. (1991). In *The Paramyxoviruses*, D.W. Kingsbury (ed.), Plenum Press, New York, pp. 323–346.

Cane, P.A., Matthew, D.A. & Pringle, C.R. (1991). *J. Gen. Virol.*, **72**, 2545–9.

Caravokyri, C., Zajac, A.J. & Pringle, C.R. (1992). *J. Gen. Virol.*, **73**, 865–73.

Collins, P.L. (1991). In *The Paramyxoviruses*, D.W. Kingsbury (ed.), Plenum Press, New York, pp. 103–160.

Conzelmann, K-L., Cox, J.H., Schneider, L-G. & Thiel, H-J. (1990). *Virology*, **175**, 485–99.

Curran, J & Kolakofsky, D. (1988). *EMBO Journal*, **7**, 2869–74.

Francki, R.I.B., Fauquet, C.M., Knudsen, D.L. & Brown, F. (1991). *Classification and Nomenclature of Viruses: Fifth Report of the International Committee on Taxonomy of Viruses.* Archives of Virology, Suppl. 2, Springer-Verlag, Wien and New York.

Galinski, M.S. & Wechlser, S. (1991). In *The Paramyxoviruses*, D.W. Kingsbury (ed.), Plenum Press, New York, pp. 41–82.

Herman, R.C. (1987). *Biochemistry* **26**, 8346–8350.

Holland, J.J. (1987). In *The Rhabdoviruses* R.R. Wagner (ed.), Plenum Press, New York, pp. 297–360.

Kolakofsky, D., Vidal, S. & Curran, J. (1991). In *The Paramyxoviruses* D.W. Kingsbury (ed.), Plenum Press, New York, pp. 181–214.

Lamb, R.A. & Paterson, R.G. (1991). In *The Paramyxoviruses* D.W. Kingsbury (ed.), Plenum Press, New York, pp. 181–214.

Ling, R., Easton, A.J. & Pringle, C.R. (1992). *J. Gen. Virol.*, **73**, 1709–24.

Murphy, D.G., Dimock, K. & Yong Kang, C. (1991). *Virology*, **181**, 760–3.

O'Hara, P.J., Horodyski, F.M., Nicol. S.T. & Holland, J.J. (1984). *Journal of Virology*, **49**, 793–8.

Parry, J.E., Shirodaria, P.V. & Pringle, C.R. (1979). *J. Gen. Virol.*, **44**, 479–91.

Poch, O., Blumberg, B.M., Bougueleret, L. & Tordo, N. (1990). *J. Gen. Virol.*, **71**, 1153–62.

Pringle, C.R. (1978). *Cell*, **15**, 597–606.

Pringle, C.R. (1991*a*). *Archives of Virology*, **117**, 137–40.

Pringle, C.R. (1991*b*). In *The Paramyxoviruses*, D.W. Kingsbury (ed.), Plenum Press, New York, pp. 1–40.

Pringle, C.R., Shirodaria, P.V., Cash, P., Chiswell, D.J. & Malloy, P. (1978). *Journal of Virology*, **28**, 199–216.

Randall, R.E. & Russell, W.C. (1991). In *The Paramyxoviruses*, D.W. Kingsbury (ed.), Plenum Press, New York, pp. 299–322.

Rataud, S.M., Hirano, A. & Wong, T.C. (1992). *Journal of Virology*, **66**, 1769–73.

Rima, B.K. (1989). In *Genetics and Pathogenicity of Negative Strand Viruses*, D. Kolakofsky and B.W.J. Mahy (eds.), Elsevier, Amsterdam, pp. 254–263.

Sanchez, A., Kiley, M.P., Klenk, H-D. & Feldmann H. (1992). *J. Gen. Virol.*, **73**, 347–57.

Spriggs, M.K. & Collins P.L. (1986). *J. Gen. Virol.*, **67**, 2705–19.

Stec, D.S., Hill III, M.G. & Collins, P.L. (1991). *Virology*, **183**, 273–87.

Tordo, N., Poch, O., Ermine, A., Keith, G. & Rougeon, F. (1986). *Proc. Natl. Acad. Sci. USA*, **83**, 3914–18.

Tordo, N., Poch, O., Ermine, A., Keith, G. & Rougeon, F. (1988). *Virology*, **165**, 565–76.

30

The molecular evolution of the human immunodeficiency viruses

PAUL M. SHARP, CONAL J. BURGESS
AND BEATRICE H. HAHN

Introduction

Two major groups of immunodeficiency-associated retroviruses are presently known to infect man. These are human immunodeficiency virus type 1 (HIV-1) which is the recognized agent of AIDS in Central Africa, Europe, the United States, and most countries worldwide (Barre-Sinoussi *et al.*, 1983; Popovic *et al.*, 1984), and the more geographically restricted human immunodeficiency virus type 2 (HIV-2), which is most predominant in West Africa (Barin *et al.*, 1985; Clavel *et al.*, 1987). Although a causative relationship between HIV-1, HIV-2 and clinical AIDS is well established (Fauci, 1988), the pathogenic mechanisms by which these viruses induce immunosuppression and disease remain unknown. Ten million individuals are currently infected with HIV worldwide and this number is expected to rise to 40 million by the year 2000 (Mann, 1992). Given the magnitude of the AIDS pandemic, new insights into the natural history of HIV-1 and HIV-2 infection as well as a better understanding of the events responsible for their recent epidemic spread are critically needed for a rational drug design and the development of effective vaccines.

Molecular characterization of HIV-1 and HIV-2, including cloning and sequence analysis of several full-length genomes, has revealed features that distinguish human AIDS viruses from the traditional RNA tumour viruses. First, the HIV-1 and HIV-2 genomes are relatively complex, each including at least six genes (in addition to the common *gag, pol* and *env* genes) termed *vif, vpr, vpu* (HIV-1 only), *vpx* (HIV-2 only), *tat, rev* and *nef*, which collectively regulate viral transcription, translation, latency and other properties that contribute to viral pathogenesis (Cullen, 1991). Two of these (*tat* and *rev*) are absolutely essential for viral replication; all other accessory proteins are dispensable for viral growth *in vitro*, although they are likely to be responsible for the ability of HIV to establish and maintain a persistent infection in the presence of an (at least partially) effective immune response *in vivo*. Secondly, HIV-1 and HIV-2 undergo rapid genetic change within infected individuals over time (Hahn *et al.*, 1986; Balfe *et al.*, 1990; Gao *et al.*, 1992), primarily as a result of reverse transcriptase misincorporations. This variability allows the virus to escape host immune responses and adapt to any selection pressure with great flexibility, leading to the rapid emergence of drug resistant mutants in individuals undergoing chemotherapy and posing a major obstacle to the design of protective vaccines. As a consequence, HIV-1 and HIV-2 'isolates' consist of complex mixtures of genotypically distinguishable variants and thus represent a viral quasi-species rather than individual viruses (Saag *et al.*, 1988; Meyerhans *et al.*, 1989; Gao *et al.*, 1992). Biological properties of these viruses *in vitro* as well as *in vivo* must then be interpreted as the weighted sum of the biological properties of all genotypic variants representing them.

The AIDS viruses HIV-1 and HIV-2 belong to the sub-family *Lentivirinae* ('slow viruses') of the family *Retroviridae* (Chiu *et al.*, 1985; Doolittle *et al.*, 1989; Chapter 27). The first lentivirus isolated was Visna virus, from sheep (see Leigh Brown, 1990). Lentiviruses have subsequently been isolated from several orders of mammals, including other *Artiodactyla* (bovine immunodeficiency-like virus from cattle, and caprine arthritis encephalitis virus from goats), *Perissodactyla* (equine infectious anaemia virus from horses), *Carnivora* (feline immunodeficiency virus from cats) and *Primates*. This distribution is almost certainly coloured by bias of ascertainment, and it is possible that all orders of mammals harbour lentiviruses. In the following, we will review current knowledge of the evolution of the human and primate immunodeficiency viruses, and discuss areas of uncertainty which will require further investigation.

Molecular phylogenetic methodologies

Before discussing lentiviral phylogeny, a few comments concerning the evolutionary interpretation of molecular data may be of use to the general reader. Many authors have presented tables of sequence identity values for comparisons among different AIDS viruses. These values are often useful in indicating approximately to which group a virus belongs, but may sometimes be misleading because of evolutionary rate variations; in those cases a phylogenetic tree is more informative.

Although it is not appropriate here to go into the details of the methods for computing phylogenetic trees from molecular data, some points are particularly relevant to the analysis of AIDS viruses. The majority of trees of AIDS viruses have been produced by the maximum parsimony method (Hirsch *et al.*, 1989; Dietrich *et al.*, 1989; Talbott *et al.*, 1989; Huet *et al.*, 1990; Kirchhoff *et al.*, 1990; Allan *et al.*, 1991; Khan *et al.*, 1991; Myers, MacInnes & Korber, 1992). Others have used distance matrix based methods, (Li, Tanimura & Sharp, 1988; Yokoyama, Chung & Gojobori, 1988; Tsujimoto *et al.*, 1989; Garvey *et al.*, 1990; Gojobori *et al.*, 1990). The trees produced by these two types of methods can differ in important respects. In distance matrix methods, the differences (distance values) between sequences can be 'corrected' for hidden multiple hits (superimposed mutations), but in the maximum parsimony approach, there is no attempt to correct for such superimposed mutations. This can lead to an underestimation of the length of deep (ancestral) branches in the parsimony tree. The maximum parsimony method is also known to have a tendency to yield incorrect branching orders when evolutionary rates vary among lineages (Felsenstein, 1988). This is because highly divergent sequences may have sites which are similar merely due to chance convergence; this may also lead to inaccurate branch length estimation (see the discussion of HIV-2_{D205} below).

Different genes, and different sites within genes (specifically silent sites as opposed to sites where mutation causes an amino acid replacement) evolve at different rates. For example, of the three major genes in HIV-1, the *env* gene evolves faster than *gag* or *pol*, because the polypeptides encoded by the two latter genes are more constrained (Li *et al.*, 1988; Leigh Brown & Monaghan, 1988). As a consequence, the parts of the viral sequence which are appropriate for use in any particular phylogenetic analysis will depend on the level of divergence of the viruses compared. That is, for comparisons among closely related HIV-1 isolates it is appropriate to consider silent sites, but in comparisons among more distantly related lentiviruses these sites have become saturated with nucleotide substitutions and evolutionary trees must be based on either the non-silent sites or the predicted protein sequences; in distant comparisons it is also better to use the *gag* or *pol* proteins, because even at the protein level *env* sequences contain some highly divergent regions.

An important aspect of drawing a tree is the identification of the 'root', i.e., the deepest ancestral point of the tree. Unless an assumption of a constant molecular clock is made, this rooting can only be done by reference to an outgroup, i.e. a sequence which is known (or assumed) to branch outside the group being investigated.

Finally, since any phylogenetic tree is merely an estimate of the true relationships, it is also important to have some feeling for the reliability of the branching order depicted. Statistical methodologies have been adapted and developed for this purpose, such as the use of the bootstrap approach (Felsenstein, 1988). Here we emphasize two general points. First, the lengths of short branches are often not significantly greater than zero, so that the inferred order of branching among lineages separated only by short branches may not be accurate. Secondly, the more sites considered (i.e. for a given level of variability, the longer the sequence) the more reliable the branching order should be.

Primate lentiviruses

Lentiviruses have been found in a range of primate species (Desrosiers, 1990); the non-human viruses are collectively known as simian immuno-deficiency viruses (SIV) with a subscript to denote the species of origin (Table 30.1). Of these primate species, macaques, African green monkeys, mandrills and sooty mangabeys are all Old World monkeys (*Cercopithecidae*), whereas the chimpanzee is an ape closely related to man. The primate (including human) lentiviruses are all more closely related to each other than to lentiviruses of non-primate origin (Fig. 30.1). Note that the branches near the root of the tree (at the left)

Table 30.1. *Primate lentiviruses*

Virus	Host (Genus)	Natural	Pathogenic
HIV-1	Humans (*Homo*)	No	Yes
HIV-2	Humans (*Homo*)	No	Yes
SIV$_{agm}$	African green monkeys (*Cercopithecus*)	Yes	No
SIV$_{mac}$	Macaques (*Macaca*)	No	Yes
SIV$_{smm}$	Sooty mangabeys (*Cercocebus*)	Yes	No
SIV$_{mnd}$	Mandrills (*Papio*)	Yes	No
SIV$_{cpz}$	Chimpanzees (*Pan*)	Yes?	No?

Note: Only those species from which viral sequence data are available are included.

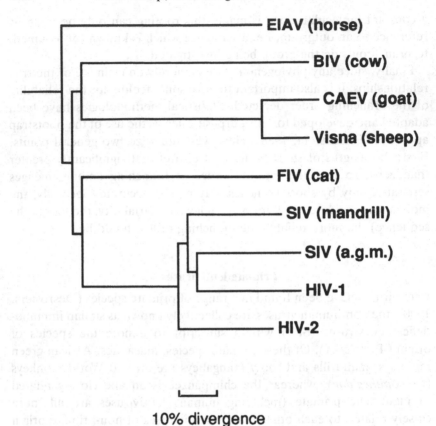

10% divergence

Fig. 30.1. Phylogenetic tree of lentiviruses, based on *pol* protein sequence comparisons. Species from which the viruses were isolated are indicated in brackets. The tree was derived by the neighbor-joining method (Saitou & Nei, 1987) applied to amino acid sequence divergence corrected for multiple hits (Kimura, 1983; p. 75), and was rooted by reference to Moloney murine leukaemia virus. Horizontal branch lengths are to scale (the bar = 10% amino acid replacement); vertical separation and order is for clarity only.

are all very short, indicating that the order of this branching cannot be determined with any certainty; thus there are five major, approximately equidistant, lineages; Myers *et al.* (1992) have recently discussed the question of whether such apparently sudden radiations within lentiviral phylogenies represent real evolutionary events, or whether they may be an artefact caused by current molecular evolutionary methods. Interestingly, BIV does not appear to be significantly more closely related to CAEV and Visna virus than to (for example) the primate viruses, even though cow and goat (and sheep) are quite closely related artiodactyls.

The primate viruses fall into four groups (Fig. 30.2): (1) HIV-1 and SIV_{cpz}, (2) HIV-2, SIV_{smm} and SIV_{mac}, (3) SIV_{agm} and (4) SIV_{mnd} (see also Gojobori *et al.*, 1990; Johnson, Hirsch & Myers, 1991; Myers *et al.*, 1992). Note that the branching order among these four lineages varies between the analyses in Fig. 30.1 and Fig. 30.2; this should not be too surprising, because the branches separating these four lineages (i.e. at the left in Fig. 30.2) are short. Thus, again, we can only conclude that these four groups are approximately equidistantly related. Note also that the HIV/SIV nomenclature is not useful in a phylogenetic or classificatory sense, since the HIVs and the SIVs do not form distinct groups.

Interestingly, much more sequence diversity is seen among the viruses from African green monkeys than within the lineages containing HIV-1 or HIV-2 (Fig. 30.2). There are four groups of African green monkeys, which are referred to variously as separate species, or as sub-species of *Cercopithecus aethiops*. Viral strain 677 was isolated (Fomsgaard *et al.*, 1991) from a grivet (which, in the separate species nomenclature, retains the name *C. aethiops*) while TYO, 3 and 155 were isolated from vervets (*C. pygerithrus*). Partial characterization of viruses from a third species (*C. sabaeus*) indicates that, though they cluster with other SIV_{agm} strains, they may be even more divergent than 677 (Allan *et al.*, 1991).

More recently, a virus has been isolated from Sykes' monkey (Emau *et al.*, 1991), another member of the genus *Cercopithecus*, but distinct from the African green monkey group. This virus is not particularly closely related to SIV_{agm}; rather it appears to be a member of a fifth distinct lineage (P.N. Fultz, personal communication). Note also that, although the mandrill and mangabey are thought to be quite closely related (Disotell, Honeycutt & Ruvolo, 1992), the viruses isolated from these species do not cluster together (Fig. 30.2).

Thus, the phylogenetic relationships among the lentiviruses are not always concordant with those among their hosts, immediately implying the occurrence of cross-species transmission. Where information is available, it appears that the primate lentiviruses are only pathogenic when in 'unnatural' hosts (Table 30.1), a fact that may be instructive in elucidating their origins and the direction of cross-species transmission.

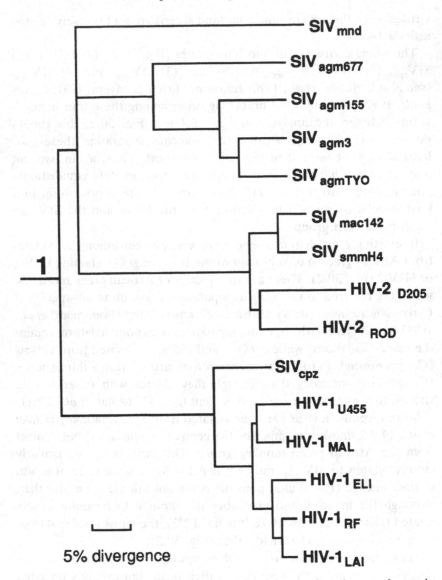

Fig. 30.2. Phylogenetic tree of primate lentiviruses, based on *pol* protein sequence comparisons. The tree was rooted by reference to FIV (see Fig. 30.1). Node (branching point) 1 is referred to in the text. The scale bar = 5% amino acid replacement; see Fig. 30.1 for methodological details.

Origin and evolution of HIV-2

One of the subgroups of primate lentiviruses includes strains isolated from man (HIV-2), sooty mangabeys (SIV$_{smm}$) and macaques (SIV$_{mac}$). SIV$_{smm}$ is found in wild-caught mangabeys (Marx *et al.*, 1991; P.N. Fultz, personal communication), but SIV$_{mac}$ is restricted to macaques in captivity. HIV-2 is most prevalent in West Africa, which is also the natural range of sooty mangabeys. SIV$_{smm}$ does not appear to be pathogenic, whereas HIV-2 and SIV$_{mac}$ are. Taken together, these observations strongly suggest that the sooty mangabey is a natural reservoir, and that HIV-2 and SIV$_{mac}$ have resulted from recent cross-species transmissions from mangabeys, to humans in West Africa, and to macaques in captivity.

Most of the earliest HIV-2 strains that were isolated (e.g. ROD and NIHZ) fall into one distinct group (Fig. 30.3), which we refer to as 'prototypic' HIV-2s. These viruses, which were all derived from symptomatic individuals and were propagated in cell culture, are closely related to each other. In contrast, a rather divergent isolate (D205) was described by Dietrich *et al.* (1989), and another strain (GH2), which has been partially characterized, appears to be quite closely related to D205 (Miura *et al.*, 1990). This led Miura *et al.* (1990) to suggest that HIV-2s fall into two distinct groups.

We have recently described three new HIV-2 isolates from West Africa; phylogenetic analyses of these viruses provide a new perspective on this group (Gao *et al.*, 1992). One virus (HIV-2$_{F0784}$) is more closely related to the SIVs than it is to any other HIV-2 (Fig. 30.3), and so appears to have arisen from a mangabey to human transfer quite separate from the transfer leading to the prototypic HIV-2 group. A second virus, HIV-2$_{2238}$, clusters with, though is quite divergent from, D205 (Fig. 30.3). The third new HIV-2 isolate, HIV-2$_{7312A}$, is seen to cluster with D205 when comparisons are made based on the *pol* gene (Fig. 30.3), but with HIV-2ST when comparisons are based on the *env* gene (Gao *et al.*, 1992). In each case the clustering is significant, and this indicates that 7312A is probably a recombinant virus. We note that recombination has been seen to occur at a high rate in retroviral replication (Hu & Temin, 1990). Since the phylogenetic positions for the *pol* and *env* sequences for 7312A are so different, the implication is that one individual must have been simultaneously co-infected with two quite divergent strains of HIV-2. This is particularly interesting because no such case of co-infection has yet been described.

In the light of these new HIV-2 variants, it is not easy, and may not be meaningful (at the present time) to identify subgroups among the HIV-2s other than the prototypic group, which does appear to be distinct (with a long branch to the left in Fig. 30.3). It seems clear that the prototypic HIV-2s and F0784 result from separate cross-species transfers. Given the phylogenetic separation of D205 and 2238 from

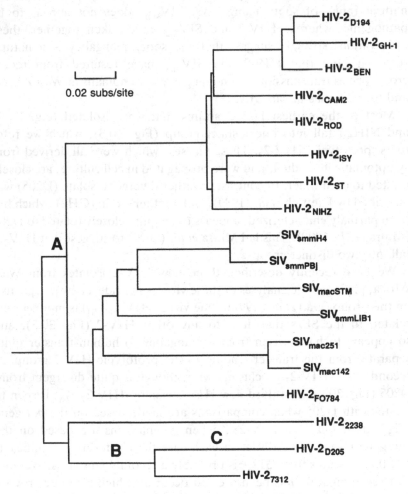

Fig. 30.3. Phylogenetic tree of the HIV-2/SIV$_{smm}$/SIV$_{mac}$ group, based on partial *pol* gene sequences. The tree was derived by the neighbor-joining method applied to DNA sequence divergence corrected for multiple hits (Kimura, 1983; p. 92), and was rooted by reference to members of each of the other primate lentivirus lineages (see Fig. 30.2). The scale bar = 2% nucleotide sequence divergence. Branches A, B and C are referred to in the text.

other HIV-2s, and from each other (Fig. 30.3), these viruses may also have arisen through distinct transmission events. Thus, the phylogenetic analyses seem to indicate multiple cross-species transmissions (including also probably two separate transfers from mangabey to macaque). It is important to realize that sooty mangabeys are abundant in West Africa where they are hunted as an important source of food, and are often kept as household pets: monkey scratches and bites, and concentrated exposure to monkey blood associated with food preparation, are common. Thus, there are likely to be many opportunities for mangabey to human transmission.

Another interesting feature of these phylogenetic analyses is the length of the branch separating D205 from the other viruses (branch A in Fig. 30.3). In the first description of D205, this branch was quite long (Dietrich *et al.*, 1989; Johnson *et al.*, 1991), implying that the ancestor of D205 had diverged at a very early stage from the virus ancestral to the prototypic HIV-2 and SIV_{smm}/SIV_{mac} lineages. However, in Fig. 30.3, branch A is seen to be quite short. This discrepancy seems to result from the different phylogenetic methods used: earlier analyses utilized the maximum parsimony method, whereas we have used the neighbor-joining method applied to a matrix of pairwise distances corrected for multiple hits. As discussed above, the latter approach may produce more reliable branch lengths – it is possible that mutations which have occurred in the more recent ancestry of D205 (on branches B or C) are convergent with mutations in the branch to the outgroup, and so were incorrectly attributed to branch A by the parsimony method.

Origin and evolution of HIV-1

Viruses in this group infect humans and chimpanzees. A large number of different HIV-1 strains have been isolated and characterized: a phylogenetic analysis of representative HIV-1 strains is shown in Fig. 30.4. All viruses of non-African (largely North American) origin fall into one distinct cluster. However, much more divergence is seen among HIV-1s isolated from central Africa. Because central African isolates are found on both sides of the initial divergence (at the root of the tree), this strongly indicates that the ancestral HIV-1 existed in central Africa. Although, as with HIV-2, any attempt to subdivide the HIV-1s may be quite arbitrary, there appear to be multiple lineages (defined by deep branches to the left of Fig. 30.4), of which the North American cluster is just one (Myers *et al.*, 1992).

Within the 'North American' cluster there are also viruses isolated from Europe (e.g. LAI and HAN) and Japan, which have probably spread from North America. Of particular interest, there is an isolate from Gabon (OYI), which clearly groups within the North American cluster (Fig. 30.4). There are two obvious alternative explanations; either this represents movement of a North American strain back to Africa, or perhaps OYI is closely related to the African virus which was ancestral to the North American cluster.

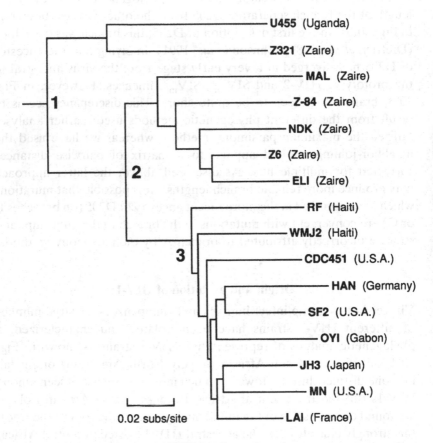

Fig. 30.4. Phylogenetic tree of HIV-1 isolates, based on *env* gene sequences. The tree was derived by the neighbor-joining method applied to DNA sequence divergence corrected for multiple hits (Li, Wu & Luo, 1985), and was rooted by reference to SIV$_{cpz}$ (see Fig. 30.2). The scale bar = 2% nucleotide sequence divergence. Nodes 1–3 are referred to in the text.

Among the HIV-1 strains of African origin, there is another probable example of recombination: MAL appears to be a mosaic virus (Li *et al.*, 1988). In the phylogenetic tree based on the *env* gene (Fig. 30.4), MAL clusters with several other isolates from Zaire (e.g. ELI and NDK), but in trees derived from the other major genes (*gag* and *pol*; not shown) MAL clusters more (though not very) closely with U455.

The rate of nucleotide substitution in HIV-1 has been estimated by making the simple assumption that strains isolated at later dates should have diverged more (with respect to an outgroup) than strains isolated at earlier dates (Li *et al.*, 1988). Several comparisons yielded rate estimates averaging about 10×10^{-3} nucleotide substitutions per silent site per year. This contrasts with a rate of about 5×10^{-9} nucleotide substitutions per silent site per year in a wide range of animal and plant genes (Wolfe, Sharp & Li, 1989). So, at least recently, HIV-1 has evolved about one million times faster than genes in the mammalian genome, and at a similar rate to influenza A viruses (Buonagurio *et al.*, 1986). By applying this rate estimate to the branches of the HIV-1 tree, we estimated that the African lineage including ELI diverged from the North American lineage (at node 2 in Fig. 30.4) around 1969, and that the various branches within the North American lineage diverged (at node 3) around 1975 (Li *et al.*, 1988); the root of the tree (node 1) would be estimated at about 1960. These estimates were derived without reference to (and are thus independent of) epidemiological data, with which they seem to agree quite well.

The phylogenetic tree of HIV-1s (Fig. 30.4) was rooted using SIV_{cpz} as an outgroup, since SIV_{cpz} is much more closely related to HIV-1 than to any other viruses (Fig. 30.2), but all HIV-1s which have been characterized in detail are more closely related to each other than to SIV_{cpz}. Given the pathogenicity of HIV-1 in humans, it is not likely that humans are the natural host of this group of viruses. Chimpanzees then represent a candidate species for the origin of HIV-1 (Huet *et al.*, 1990; Johnson *et al.*, 1991). However, as those authors recognize, this conclusion is not yet firm, since it is not clear how common SIV_{cpz} is in wild chimpanzees, nor whether the virus is pathogenic in that host. Nor is it clear whether chimpanzee populations in the recent past were large enough to act as a natural reservoir for such viruses. Two recently reported viruses, not shown in Fig. 30.4 because their sequences are not yet published, are of particular interest. A strain of HIV-1 (ANT70) has been isolated that appears to be more divergent from other HIV-1s than is SIV_{cpz} (De Leys *et al.*, 1990; Myers *et al.*, 1992), suggesting that as with

HIV-2 there have been multiple (at least two) cross-species transfers in this lineage. A second, highly divergent, strain of SIV_{cpz} has also been reported (Vanden Haesevelde *et al.*, 1992), perhaps increasing the likelihood that chimpanzees were the source of HIV-1. On the other hand, if they were not, this would lead to the conclusion that both humans and chimpanzees received this virus (each on at least two occasions) from another species of Old World monkey. Given the picture that is emerging for the origins of HIV-2 (discussed above), such a scenario cannot be ruled out. Several other Old World monkey species are known to be sero-positive for lentiviruses (Johnson *et al.*, 1991), but these viruses have not yet been characterized.

Time-scale of primate lentivirus origins

In molecular evolutionary studies, time-scales can often be placed on phylogenetic trees. This normally relies on having evidence about the divergence times of various species from the fossil record. Then one can extrapolate across the evolutionary tree because molecular evolution is generally observed to occur in a clock-like manner (Kimura, 1983). Can a time-scale be placed on the primate lentivirus phylogeny? The only well-characterized molecular clock for these viruses is that for HIV-1 (see above), but viruses could evolve a million times slower if they spend long periods of time integrated into the host genome.

Since some viruses (e.g. papova viruses, Soeda *et al.*, 1980) appear to have evolved in a host-dependent fashion, an alternative approach is to estimate the dates of viral divergence from those of the hosts. When the first full-length SIV_{agm} sequence was determined (Fukasawa *et al.*, 1988), it was suggested that SIV_{agm} and HIV-1 are sufficiently divergent that they might have evolved concurrently with their hosts, implying that the ancestral virus at the base of the primate lentivirus tree (node 1 in Fig. 30.2) existed in the common ancestor of Old World monkeys and man, i.e. about 25 million years ago (Mya). Almost contemporaneously, others extrapolated from the evolutionary divergence within HIV-1 to suggest that HIV-1 and HIV-2 might have had a common ancestor as recently as 40 years ago (Smith *et al.*, 1988). We pointed out that these two rather discrepant time estimates (25 Mya, and 40 years ago) both refer to effectively the same point in the evolutionary tree (Sharp & Li, 1988). In fact, the amount of divergence among HIV-1s, which are thought to date back to at least 1960 (see above) indicates that 40 years ago is far too recent a date for the divergence of HIV-1 and HIV-2. *If* the rate of evolution has been constant throughout the divergence of the HIV-1, HIV-2 and SIV$_{agm}$ lineages from their common ancestor, that virus would have existed about 150 years ago (Sharp & Li, 1988). Yokoyama *et al.* (1988) suggested about 300 years ago, on the assumption that lentiviruses evolve at a rate similar to oncoviruses.

However, it seems more likely that the viruses currently infecting man, i.e. HIV-1 and HIV-2, have resulted from recent transmissions from other species, in which case each of the major primate lentivirus lineages may have spent most of the time since their divergence in a simian host. Thus, it is possible that the SIVs (excluding SIV$_{mac}$, and possibly SIV$_{cpz}$) have been evolving in a largely host-dependent fashion, and that the ancestral primate lentivirus may have existed in ancestors of these monkeys, perhaps a few million years. That they have not diverged to a greater extent than is observed could be explained by a much slower rate of evolution within a naturally infected host species.

Conclusions

As more primate lentiviruses are isolated and characterized at the sequence level, phylogenetic analyses are leading to a picture of multiple cross-species transmission events. Thus, while we have argued that HIV-1 and HIV-2 have arisen in humans recently, there is no reason to believe that monkey to human transmissions have not occurred in the past. There is no current evidence for earlier transmissions, but this may simply be due to the lack of spread of such viruses: that HIV-1 and HIV-2 have spread, at a time coincidental with the development of biochemical techniques allowing their detection, most probably reflects recent massive increases in urbanization and travel. It is tempting to apply the molecular clock approach to estimate when ancestral viruses existed, and thus to estimate the times of origin of these viruses. This approach seems to be quite successful when dealing with closely related viruses, e.g. among the HIV-1s, but is probably not appropriate when comparing the various major lineages of primate lentiviruses because rates of evolution might vary enormously (Doolittle *et al.*, 1989). It should be anticipated that our view of the origins and evolution of AIDS viruses will be modified, and hopefully clarified, as more primate viruses are examined.

Note added in proof

This review was written in late 1992, and the literature cited reflects that. Since then, considerably more HIV-1, HIV-2 and SIV strains have been characterized, adding to the known diversity of these viruses (see, e.g. Sharp, P.M. *et al.* (1994). *AIDS*, **8**, S27–S42), without invalidating any of the conclusions reached above.

Acknowledgements

This is a publication from the Irish National Centre for Bioinformatics. This work was supported in part by grants from the National Institutes of Health, the US Army Medical Research Acquisition Activity, and the Birmingham Center for AIDS Research.

References

Allan, J.S., Short, M., Taylor, M.E., Su, S., Hirsch, V.M., Johnson, P.R., Shaw, G.M. & Hahn, B.H. (1991). *J. Virol.*, **65**, 2816–28.

Balfe, P., Simmonds, P., Ludlam, C.A., Bishop, J.O. & Leigh Brown, A.J. (1990.) *J. Virol.*, **64**, 6221–33.

Barin, F., M'Boup, S., Denis, F., Kanki, P.J., Allan, J.S., Lee, T.H. & Essex, M. (1985). *Lancet*, **ii**, 1387–90.

Barre-Sinoussi, F., Chermann, J.C., Rey, F., Nugeyre, M.T., Chamaret, S., Gruest, J., Dauguet, C., Axler-Blin, C., Vezinet-Brun, F., Rouzioux, C., Rozenbaum, W. & Montagnier, L. (1983). *Science*, **220**, 868–70.

Buonagurio, D.A., Nakada, S., Parvin, J.D., Krystal, M., Palese, P. & Fitch, W.M. (1986). *Science*, **232**, 980–2.

Chiu, I.-M., Yaniv, A., Dahlberg, J.E., Gazit, A., Skuntz, S.F., Tronick, S.R. & Aaronson, S.A. (1985). *Nature*, **317**, 366–8.

Clavel, F., Mansinho, K., Chamaret, S., Guetard, D., Favier, V., Nina, J., Santos-Ferreira, M.-O., Champalimaud, J.-L. & Montagnier, L. (1987). *N. Engl. J. Med.*, **316**, 1180–5.

Cullen, B.R. (1991). *J. Virol.*, **65**, 1053–6.

De Leys, R., Vanderborght, B., Vanden Haesevelde, M., Heyndrickx, L., van Geei, A., Wauters, C., Bernaerts, R., Saman, E., Nijs, P., Willems, B., Taelman, H., van der Groen, G., Piot, P., Tersmette, T., Huisman, J.G. & Van Heuverswyn, H. (1990). *J. Virol.*, **64**, 1207–16.

Desrosiers, R.C. (1990). *Annu. Rev. Immunol.*, **8**, 557–8.

Dietrich, U., Adamski, M., Kreutz, R., Seipp, A., Kuhnel, H. & Rbsamen-Waigmann, H. (1989). *Nature*, **342**, 948–50.

Disotell, T.R., Honeycutt, R.L. & Ruvolo, M. (1992). *Mol. Biol. Evol.*, **9**, 1–13.

Doolittle, R.F., Feng, D.-F., Johnson, M.S. & McClure, M.A. (1989). *Quart. Rev. Biol.*, **64**, 1–30.

Emau, P., McClure, H.M., Isahakia, M., Else, J.G. & Fultz, P.N. (1991). *J. Virol.*, **65**, 2135–40.

Fauci, A.S. (1988). *Science*, **239**, 617–22.

Felsenstein, J. (1988). *Ann. Rev. Genet.*, **22**, 521–565.

Fomsgaard, A., Hirsch, V.M., Allan, J.S. & Johnson, P.R. (1991). *J. Virol.*, **182**, 397–402.

Fukasawa, M., Miura, T., Hasegawa, A., Morikawa, S., Tsujimoto, H., Miki, K., Kitamura, T. & Hayami, M. (1988). *Nature*, 333, 457–61.

Gao, F., Yue, L., White, A.T., Pappas, P.G., Barchue, J., Hanson, A.P., Greene, B.M., Sharp, P.M., Shaw, G.M. & Hahn, B.H. (1992). *Nature*, 358, 495–9.

Garvey, K.J., Oberste, M.S., Elser, J.E., Braun, M.J. & Gonda, M.A. (1990.) *Virology*, 175, 391–409.

Gojobori, T., Moriyama, E.N., Ina, Y., Ikeo, K., Miura, T., Tsujimoto, H., Hayami, M. & Yokoyama, S. (1990). *Proc. Natl. Aacd. Sci. USA*, 87, 4108–11.

Hahn, B.H., Shaw, G.M., Taylor, M.E., Redfield, R.R., Markham, P.D., Salahuddin, S.Z., Wong-Staal, F., Gallo, R.C., Parks, E.S. & Parks, W.P. (1986). *Science*, 232, 1548–53.

Hirsch, V.M., Olmsted, R.A., Murphey-Corb, M., Purcell, R.H. & Johnson, P.R. (1989). *Nature*, 339, 389–92.

Hu, W.-S. & Temin, H.M. (1990). *Science*, 250, 1227–33.

Huet, T., Cheynier, R., Meyerhans, A., Roelants, G. & Wain-Hobson, S. (1990). *Nature*, 345, 356–9.

Johnson, P.R., Hirsch, V.M. & Myers, G. (1991). *AIDS Res. Rev.*, 1, 47–62.

Khan, A.S., Galvin, T.A., Lowenstine, L.J., Jennings, M.B., Gardner, M.B. & Buckler, C.E. (1991). *J. Virol.*, 65, 7061–5.

Kimura, M. (1983). *The Neutral Theory of Molecular Evolution*. Cambridge University Press, Cambridge, UK.

Kirchhoff, F., Jentsch, K.D., Stuke, A., Mous, J. & Hunsmann, G. (1990). *AIDS*, 4, 847–57.

Leigh Brown, A.J. (1990). *Trends Ecol. Evol.*, 5, 177–81.

Leigh Brown, A.J. & Monaghan, P. (1988). *AIDS Res. Human Retroviruses*, 4, 399–407.

Li, W.-H., Tanimura, M. & Sharp, P.M. (1988). *Mol. Biol. Evol.*, 5, 313–30.

Li, W.-H., Wu, C.-I. & Luo, C.-C. (1985). *Mol. Biol. Evol.*, 2, 150–74.

Mann, J.M. (1992). *J. Infect. Dis.*, 165, 245–50.

Marx, P.A., Li, Y., Lerche, N.W., Sutjipto, S., Gettie, A., Yee, J.A., Brotman, B.H., Prince, A.M., Hanson, A., Webster, R.G. & Desrosiers, R.C. (1991) *J. Virol.*, 65, 4480–5.

Meyerhans, A., Cheynier, R., Albert, J., Seth, M., Kwok, S., Sninsky, J., Morfeldt-Manson, L., Asjo, B. & Wain-Hobson, S. (1989). *Cell*, 58, 901–10.

Miura, T., Sakuragi, J.-i., Kawamura, M., Fukasawa, M., Moriyama, E.N., Gojobori, T., Ishikawa, K.-i., Mingle, J.A.A., Nettey, V.B.A., Akari, H., Enami, M., Tsujimoto, H. & Hayami, M. (1990). *AIDS*, 4, 1257–61.

Myers, G., MacInnes, K. & Korber, B. (1992) *AIDS Res. Human Retroviruses*, 8, 373–86.

Popovic, M., Sarngadharan, M.G., Read, E. & Gallo, R.C. (1984) *Science*, 224, 497–500.

Saag, M., Hahn, B.H., Gibbons, J., Li, Y., Parks, E.S., Parks, W.P. & Shaw, G.M. (1988) *Nature*, 334, 440–4.

Saitou, N. & Nei, M. (1987). *Mol. Biol. Evol.*, 4, 406–25.

Sharp, P.M. & Li, W.-H. (1988). *Nature*, 336, 315.

Smith, T.F., Srinivasan, A., Schochetman, G., Marcus, M. & Myers, G. (1988). *Nature*, 333, 573–5.

Soeda, E., Maruyama, T., Arrand, J.R. & Griffin, B.E. (1980). *Nature*, **285**, 165–7.

Talbott, R.L., Sparger, E.E., Lovelace, K.M., Fitch, W.M., Pedersen, N.C., Luciw, P.A. & Elder, J.H. (1989). *Proc. Natl. Acad. Sci. USA.*, **86**, 5743–7.

Tsujimoto, H., Hasegawa, A., Maki, N., Fukasawa, M., Miura, T., Speidel, S., Cooper, R.W., Moriyama, E.N., Gojobori, T. & Hayami, M. (1989). *Nature*, **341**, 539–41.

Vanden Haesevelde, M., Peeters, M., Willems, B., Saman, E., van der Groen, G. & Van Heuverswyn, H. (1992). Abstract WeA1081, VIII Int. Conf. on AIDS/III. STD World Congress, Amsterdam.

Wolfe, K.H., Sharp, P.M. & Li, W.-H. (1989). *J. Mol. Evol.*, **29**, 208–11.

Yokoyama, S., Chung, L. & Gojobori, T. (1988). *Mol. Biol. Evol.*, **5**, 237–51.

31

Molecular evolution of papillomaviruses
MARC VAN RANST, JEFFREY B. KAPLAN, JOHN
P. SUNDBERG AND ROBERT D. BURK

Introduction

Papillomaviruses (PVs) are members of the genus *Papillomavirus* which,
along with viruses of the genus *Polyomavirus*, comprise the family
Papovaviridae (*pa*pillomavirus, *po*lyomavirus and *va*cuolating agent).
PVs are small, non-enveloped structures with icosahedral symmetry
and a circular double-stranded DNA genome. They infect humans in
addition to multiple other animal vertebrate species, resulting in a variety
of proliferative epithelial lesions and tumours. Most papillomaviruses
are species specific and have a cellular tropism for squamous epithelial
cells. In benign papillomatous lesions, the viral genomes replicate as
extrachromosomal episomes in the nuclei of basal and suprabasal
epithelial cells. Complete vegetative replication with production of
intact virions is found in the superficial and differentiated epithelial cells.
In contrast, malignant lesions do not support vegetative viral replication,
but instead often contain an integrated disrupted viral genome.

Sixty years ago, Richard Shope (1933) linked the presence of the
cottontail rabbit papillomavirus (CRPV) to cutaneous papillomatosis
in rabbits, and Francis Peyton Rous described the conversion of pap-
illoma to squamous cell carcinoma in rabbits, initiating the field of
tumour virology (Rous & Beard, 1935). Much of the current interest
in papillomavirus research can be attributed to the recent association of
human papillomaviruses (HPVs) with cervical cancer, one of the leading
causes of cancer-attributed death in women.

Most of our knowledge of PVs has been obtained during the last
decade through the use of recombinant DNA techniques. The small
double-stranded DNA genomes of PVs have been readily amenable to
cloning and sequencing. In the absence of a serological classification,
HPVs have been categorized according to their genotype. New PV

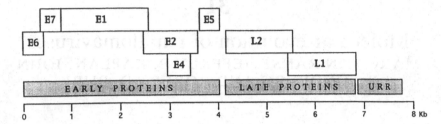

Fig. 31.1. Linearized representation of the open reading frame organization of a prototypic circular papillomavirus genome. The open reading frames of the early (E) and late (L) regions were deduced from the nucleotide sequence of HPV-58 (Kirii, Iwamoto & Matsukura, 1991).

isolates have been defined as independent types when their genomes exhibited less than 50% cross-hybridization with previously typed PVs by DNA reassociation kinetics (Coggin & zur Hausen, 1979). More recently, the definition of a new HPV type has been amended; it was agreed upon by the Papillomavirus Nomenclature Committee of the ICTV that a novel HPV isolate should have less than 90% nucleotide identity with other papillomavirus types in the E6, E7, and L1 region in order to be recognized as a new HPV type. A new classification system of the HPVs based on computer analysis of the papillomavirus genomes has been proposed (Van Ranst, Kaplan & Burk, 1992a).

The overall nucleotide sequence organization of the genomes of all known animal and human PVs is conserved. The genome is approximately 7900 bp long, and contains eight major open reading frames (Fig. 31.1). All potential protein coding sequences occur in homologous positions on one strand (sense strand) of the PV genome, and are of comparable length in all sequenced virus types. Through splicing processes, more than one protein can be derived from a given open reading frame (ORF). A typical PV genome is divided into three domains based on their putative functional properties. The 'early' (E) region contains the ORFs for proteins thought to be important in early events in infection, such as DNA replication and cell transformation. The 'late' (L) region contains two large ORFs (L1 and L2) that code for structural proteins constituting the viral capsid. The E4 ORF probably encodes a 'late' gene, which is primarily expressed in productively infected keratinocytes. A non-coding region (NCR), also known as the upstream regulatory region (URR) or long control region (LCR), is defined as the part of the genome between the stop codon of the L1 ORF and the start codon of E6. This region contains the origin of replication and numerous control signals for DNA replication and transcription.

PVs do not encode their own DNA polymerase but, instead, utilize the host cell enzymes for replication of the viral genome. Thus, they benefit from the high fidelity, proofreading capacity, and post-replication DNA repair mechanisms of the host DNA polymerase. This unquestionably contributes to the relative stability of PV genomes, in contrast to those of RNA riboviruses, retroviruses, and hepadnaviruses, which replicate their genomes through an RNA polymerase or reverse transcriptase, and whose genomes show significantly more variablity.

As recombination between different PVs has not been detected, it can be deduced that PVs have been, and are, evolving either by the slow accumulation of point mutations and/or by other non-stochastic mechanisms which result in fixation of specific genome types in a population. The slow rate of mutational events of PVs is exemplified by the finding that a Swedish BPV-1 variant showed 99.9% DNA sequence similarity to a BPV-1 genome isolated in Wisconsin (USA) 30 years earlier (Ahola *et al.*, 1983).

Relationship of papillomaviruses to other small DNA tumor viruses

Papillomaviruses, polymaviruses and adenoviruses all have relatively small double-stranded DNA genomes, and share the ability to cause malignant neoplasms. These viruses use similar strategies to take control of the host cell DNA replication and transcription processes. These functions are specified from the early genes in all three virus groups. For example, the papillomavirus E6/E7 protein, adenovirus E1A, and SV40 large T antigen are all nuclear proteins that can induce cellular DNA synthesis in growth-arrested cells, morphologically transform rodent cells, and immortalize primary cells. Recently, it has been shown that these proteins can participate in the binding of cellular regulatory proteins, including p53, p105–Rb, p107 and c-*src* (Phelps *et al.*, 1988; Villareal & Fan, 1989). The cellular division (cd) sequence motif, involved in binding of the retinoblastoma gene product p105-Rb and casein kinase II phosphorylation, is conserved in E7, E1A and large T antigen (Stabel, Argos & Philipson, 1985; Figge & Smith, 1988; Moran, 1988; Barbosa *et al.*, 1990). The pRb-domain also displays some similarity to cellular retinoblastoma binding proteins (RBP-1 and RBP-2) (Defeo-Jones *et al.*, 1991). A hypothetical alignment of this region of 23 papillomaviruses, 7 polymaviruses, 10 adenoviruses, and RBP-1 and -2 is shown in Fig. 31.2. The consensus amino acid sequence of this motif is $\{D/E\}$-X_{0-3}-$\{L/V/I/M\}$-D-L-X-C-$\{D/E/N/Q/H/Y\}$-E-$\{D/E/Q\}$-X_{2-6}-S-$\{D/E\}$-$\{D/E\}$-$\{D/E\}$ (The brackets indicate analogous alternatives at a given position.) The small size and the relatively modest conservation of this domain precludes the elucidation of its origin (i.e. common ancestry, cellular origin, or convergent evolution) with absolute confidence. However, the conservation of such small functional domains and the exploitation of similar 'master regulatory pathways' are suggestive of a common, albeit ancient, evolutionary origin for papillomaviruses, polymaviruses and adenoviruses.

Although the overall genetic organizations of the two genera comprising the family *Papovaviridae* are distinct, significant amino acid sequence similarities were observed between a part of the polyomavirus large T antigen and the carboxyterminal half of the papillomavirus E1 sequence (Clertant & Seif, 1984). Both proteins are required for viral plasmid replication, and this specific region was shown to have ATP-binding and ATPase activity (Clark *et al.*, 1983; Sun *et al.*, 1990). A 210-amino acid alignment of this region for 7 polyomaviruses and 12 papillomaviruses is shown in Fig. 31.3. In this region, 21 residues were identical and 17 residues were functionally conserved in homologous positions in all viruses. A further 69 identical or functionally conserved residues were shared by at least one polyomavirus and one papillomavirus. The striking degree of sequence conservation over more than 200 amino acids suggests a common evolutionary ancestry for papillomaviruses and polyomaviruses, rather than convergent evolution. This is not unlike the recent hypothesis for a shared ancestry of positive-strand and double-stranded RNA viruses, based on the alignment of RNA-dependent RNA polymerase sequences (Bruenn, 1991; Koonin, 1991).

Animal papillomaviruses: co-evolution with host species

PVs are found in a wide variety of animal hosts (Sundberg, 1987). More than 70 different PV types from humans have been cloned and characterized (De Villiers, 1989). However, in most other mammalian species studied to date, only one or a few different PVs have been identified. The multitude of PV types infecting humans may reflect a greater research effort rather than a special situation in our species. Six different bovine PVs have been detected in domestic cattle (*Bos taurus*) (Jarrett *et al.*, 1984). PVs have also been identified in cutaneous lesions in three avian species, i.e. chaffinch (*Fringilla coelebs*), brambling

```
                 pRB-Binding domain                    CKII domain

HPV  1    D - - - L D L Y C Y E E V P P D D I - E E E
HPV  2    E - - I V D L H C D E Q F - - - D S S E E E
HPV  5    E V L P V D L F C E E E L P N E Q E T E E E
HPV  6    D - - P V G L H C Y E Q L V - - D S S E D E
HPV  8    E V L P V D L L C E E E L P N E Q E T E E E
HPV 11    D - - P V G L H C Y E Q L - - E D S S E D E
HPV 13    D - - P V G L H C N E Q L - - - D S S E D E
HPV 16    E - - T T D L Y C Y E Q L - N - D S S E E E
HPV 18    E - I P V D L L C H E Q L - - S D - S E E E
HPV 31    E - - A T D L H C Y E Q L P - - D S S D E E
HPV 33    E - - P T D L Y C Y E Q L - - S D S S D E D
HPV 39    Q - - P V D L V C H E Q L G - - E - S E D E
HPV 42    E - T P I D L Y C Y E Q L - - - D S S D E D
HPV 47    E V L P V D L F C D E E L P N E Q Q A E E E
HPV 51    E - - - I D L Q C Y E Q F - - - D S S E E E
HPV 56    E - - - I D L Q C N E Q L - - - D S S E D E
HPV 57    E - - I V D L H C D E Q F - - - D N S E E D
HPV 58    E - - P T D L F C Y E Q L C - - D S S D E D
HPV 66    E - - - I D L Q C N E Q L - - - D - S E D E
PCPV 1    D - - P V G L H C Y E Q L - - - D S S E E D
RHPV 1    Q - - P V D L M C Y E Q L - - S D - S E D E
COPV      E - - P I D L Q C Y E Q L P - - - S S E D E
CRPV      E - - A L S L H C D E A L E N L - - S D D D

SV40 LT   E - - - - N L F C S E E M P - - - S S D D E
BK LT     E - - - - D L F C H E D M F A - - - S D E E
JC LT     E - - - - D L F C H E E M F A - - - S D D E
HA LT     E - - - - D L T C Q E E L - - - S S S E D E
MU LT     Q - - P - D L F C Y E E P L L S P S S P T D
LY LT     D - - - - D L F C S E T M - - S S S S D E D
BFDV LT   E - - - - G L R A D E T L - - - E D S D F E

E1A Ad2   E - - V I D L T C H E A G F P P - - S D D E
E1A Ad4   E - - K M D L R C Y E E C L P P - - S D D E
E1A Ad5   E - - V I D L T C H E A G F P - - S D D E
E1A Ad7   E - - - M D L R C Y E E G F P P - - S D D E
E1A Ad12  E - - - M D L L C Y E M G F P C S D S E D E
E1A Ad40  D - - - L D L K C Y E D G L P P - - S D P E
E1A Ad41  E - V N L D L K C Y E E G L P P S G S E A D
E1A Sim7  E - - - - D L F C Y E D G F P P - - S D S E
E1A Can2  D - - - - M I L L C L E E M - P T F - - D D E
E1A Mu1   D - - - M D L R C Y E Q L S P S P E S I E T

RBP-1     E - - - - T L V C H E V D L D D L - - D E K
RBP-2     E - - P - N L F C D E E I P I K - - S E E V
```

Fig. 31.2. Alignment of the pRB-binding and casein kinase II domain in the E7 protein of 23 papillomaviruses (HPV: human PV; PCPV: pygmy chimpanzee PV; RHPV: rhesus monkey PV; COPV: canine oral PV; CRPV: cottontail rabbit PV), large T(LT) antigen of 7 polyomaviruses (SV40:simian virus 40; BK: polyomavirus BK; JC: polyomavirus JC; HA: hamster; MU: murine; LY: lymphotropic polyomavirus; BFDV: budgerigar fledging disease virus), E1A protein of 10 adenoviruses (Sim: simian; Can: canine; Mu: murine), and 2 retinoblastoma binding proteins (RBP) from human lung fibroblasts. The consensus motif is shown in the shaded boxes and corresponds to the following sequence: $\{D/E\}$-X_{0-3}-$\{L/V/I/M\}$-D-L-X-C-$\{D/E/N/Q/H/Y\}$ -E-$\{D/E/Q\}$-X_{2-6}-S-$\{D/E\}$-$\{D/E\}$-$\{D/E\}$. The brackets indicate analogous alternatives at a given position (D,E,N,Q : negatively charged or acid amide; I,V,L,M : aliphatic; H,Y: aromatic residues); X_{i-j}: spacer of i to j amino acids long.

```
            **+ ++++ +*  + + +  +    ++ + ++# ++ #   ▼▼▼▼▼
HAPOLTAG    NAelFLhckqQksIcqQaAdnVlarRRlkvlEsTrqElLaeRLNKLldqlkdlspvdkh----LYlagVa
JCPOLTAG    NAqiFadSkNQksIcqQaVdtVaakqRvDsihMTreEmLveRfNfLldkmdlifgahgnavleQYmagVa
BKPOLTAG    NAAiFaeSkNQksIcqQaVdtVlakKRvDtlhMTreEmLteRfNhIldkmdlifgahgnavleQYmagVa
MUPOLTAG    NAdlFLNckaQktIcqQaAdgVlasRRlklvEcTrSQlLkeRLQQsllrlkels-----sdallYlagVa
SV40LTAG    NAAiFadSkNQktIcqQaVdtVlakKRvDslQlTreQmLtnRfNDLldrmdimfgstgsadieEWmagVa
BFDVLTAG    NASlFLhikdQkrLcqcaVdaVlaeKRfrsatMTrdErLkeRfrtVlrniqelldgetea-idDfvtata
LYPOLTAG    NAclFLesraQknIcqQaVdqVlaaKRlklvEcSriElLeeRflQLfdemddfl--hgeieilrWmagVa
BPV1E1      NAraFLaTnsQakhvkDcAtmVrhylRaEtqalSmpayIkaRckla-------------tgegsWks-Il
BPV4E1      NAAaFLkcnNQvkhvkEcAqmtryyKtaEmtEMSmgQwIkkcIgEI-------------egvgDWkq-Ic
EEPVE1      NAkaFLaStNQarLvkDcctmVkhylRaEeqslTisaffkrRcDNa-------------tgkgsWls-Im
HPV5E1      NAvawLahnNQakfvrEcAymVrfyKKgQmrDMSiSEWIytKINEV-------------egeghWsd-Iv
HPV11E1     NAraFLNSnmQakyvkDcAimcrhyKHaEmkkMSikQwIkyRgtKV-------------dsvgNWkp-Iv
HPV16E1     NAsaFLkSnsQakIvkDcAtmcrhyKRaEkkQMSmSQwIkyRcDrV-------------ddgqDWkq-Iv
HPV18E1     NAAaFLkSncQakylkDcAtmckhyRRaQkrQMnmSQwIrfRcskI-------------deggDWrp-Iv
HPV47E1     NAvawLahnNQakyvrEcAmmVryyKKgQmrDMSmSEwIytRIhEV-------------egegQWss-Iv
RHPV1E1     NAAaFLkSnaQakyvkDcAtmcrhyKRaErqQMTmSQwIkqRcEKt-------------ddgqDWrp-Iv
HPV1E1      NAraFLsSnsQekyvkDceqmVrhylRaEmaQMSmSEWIfrKLDNV-------------egsgNWke-Iv
CRPVE1      NAraFLaSnsQakyvrDccnmVrlylRaEmrQMTmSawInyRLDgM-------------nddgDWkv-Vv
FPVE1                                                                          Il


            ++    + + ++++ ++++++ +++*#  +++ *+ ##**#+++  +# +++*+ #  +*        * *
HAPOLTAG    wyqcmfpdfeMMlLdILKLftenVPKKrnVLFrGPvNSGKTSLAaaIMnLVGGvaLn-VNcpadklnFeL
JCPOLTAG    wIhcllpqmdtViydfLKcIVlnIPKKrywLFkGPiDSGKTTLAaaLLdLcGGKsLn-VNmplerlnFeL
BKPOLTAG    wLhcllpkmdsVifdfLHcIVfnVPKKrywLFkGPiDSGKTTLAagLLdLcGGKaLn-VNlpmerltFeL
MUPOLTAG    wyqcllEdfpqtlfkmLKLLtenVPKRrnILFrGPvNSGKTgLAaaLISLLGGKsLn-INcpadklaFeL
SV40LTAG    wLhcllpkmdsVvydfLKcMVynIPKKrywLFkGPiDSGKTTLAaaLLeLcGGKaLn-VNlpldrlnFeL
BFDVLTAG    ILfnmlfpDvdViVdILqtMVknpPKRryyIFkGPvNTGKTTVAaaILaLctGasLn-VNgtpdrlqFeL
LYPOLTAG    wytillDnswdVFqnILqLIttsQPKKrnVLIkGPiNSGKTTLAsaFMhFfdGKaLn-INcpadklsFeL
BPV1E1      tffnyqNiElItFinaLKLwLkgIPKKncLaFiGPpNTGKSmLcnsLIhFLGGsvLsfaWhksh----FWL
BPV4E1      kflkfqNvnfLsFMsaLKdLLhrVPKRncMVIcGPpNTGKSmfvmsFMkaLqGKvLsfVNsksh----FWL
EEPVE1      nLlkfqgiEpInFVnaLKpwLkgtPKHncIaIvGPpNSGKSllcntLMSFLGGKvLtfaNhssh----FWL
HPV5E1      kfiryqNinfIVFLtaLKefLhsVPKKncILIyGPpNSGKSfAmsLIrVLkGRvLsfVNsksq----FWL
HPV11E1     qflrhqNiEfIpFLskLKLwLhgtPKKncIaIvGPpDTGKScfcmsLIkFLGGtvIsyVNscsh----FWL
HPV16E1     MflryqgvEfMsFLtaLKrfLqgIPKKncILLyGaaNTGKSlfgmsLMkFLqGsvIcfVNsksh----FWL
HPV18E1     qflryqQiEfItFLqaLKsLLkgtPKKncLVFcGPaNTGKSyfgmsFIhFIqGavIsfVNstsh----FWL
HPV47E1     kflryqEiNfIsFLaaLKdLLhSVPKRncILFhGPpNTGKSSfgmsLIkVLrGRvLsfVNsksq----FWL
RHPV1E1     qflryqgvEfIaFLaaLKLfLkgIPKKncIVLfGPpNTGKSyfgmsLIhFLqGsiIsyVNsnsh----FWL
HPV1E1      rflrfqEvEfIsFMiafKdLLcgkPKKncLLIfGPpNTGKSmfctsLLkLLGGKvIsycNksq----FWL
CRPVE1      hflrhqrvEfIpFMvkLKafLrgtPKKncMVFyGPpNSGKSyfcmsLIrLLaGRvLsfaNsrsh----FWL
FPVE1       VfltfqhiNfkeFIsILcMwLkgrPKKscItIaGvpDSGKSmfAysLIkFLnGsvLsfaNsksh----FWL


            +  + ++#*+        +* +#*+ *#++  # ##+*#     ++ +**+#+*  *
HAPOLTAG    gvaiDkfavVfEDVkgqtgdkrhlqsglginnLD-NLRDyLDGsvkVnLEKKHvnkrsQI-FPPcIVTaN
JCPOLTAG    gvgiDqfMvVfEDVkgtgaesrdlpsghgisnLD-cLRDyLDGsvkVnLERKHqnkrtQV-FPPgIVTmN
BKPOLTAG    gvgiDqyMvVfEDVkgtgaeskdlpsghgisnLD-sLRDyLDGsvkVnLEKKHlnkrtQI-FPPgLVTmN
MUPOLTAG    gvaqDqfVvcfEDVkgqialnkqlqpgmgvanLD-NLRDyLDtsvkVnLEKKHsnkrsQL-FPPcVcTmN
SV40LTAG    gvaiDqfLvVfEDVkgtggesrdlpsgqqinnLD-NLRDyLDGsvkVnLEKKHlnkrtQI-FPPgIVTmN
BFDVLTAG    gcaiDqfMvLfEDVkgtpepdtnlpsgfgmvnLD-NLRDhLEGsvPVnLERKHqnkvsQI-FPPgIITmN
LYPOLTAG    gcaiDqfcvLLDDVkggqitlnkhlqpgqqvnnLD-NLRDhLDGTikVnLEKKHvnkrsQI-FPPVIMTmN
BPV1E1      aslaDtraaLVDDathac-----------wryfDtyLRNaLDGy-PVsIDRKHkaav-QIkaPPLLVTsN
BPV4E1      qplrgakVaVLDDatrat-----------wtyfDtyLRNgLDGT-PVsLDmKHrapl-QICFPPLVITtN
EEPVE1      apltDcrVaLIDDathac-----------wryfDtyLRNvLDGy-PVcIDRKHskav-QLkaPPLLLTsN
HPV5E1      qplsEckIaLLDDVtdpc-----------wiyMDtyLRDGh-yVsLDcKYrapt-QMkFPPLLLTsN
HPV11E1     qpltDakVaLLDDatqpc-----------wtyMDtyMRNlLDGn-PMsIDRKHralt-lIkcPPLLVTsN
HPV16E1     qplaDakIgMLDDatvpc-----------wnyIDdNLRNaLDGn-lVsMDvKHrplv-QLkcPPLLITsN
HPV18E1     epltDtkVaMLDDatttc-----------wtyfDtyMRNaLDGn-PIsIDRKHkpli-QLkcPPILLTtN
HPV47E1     qplgEckIaLLDDVtdpc-----------wvyMDqyLRNgLDGh-fVsLDcKYrapm-QtkFPPLILTsN
RHPV1E1     qplaDakVaMLDDatpqc-----------wsyIDnyLRNaLDGn-PIsVDRKHknlv-QMkcPPLLITtN
HPV1E1      qplaDakIgLLDDatkpc-----------wdyMDtyMRNaLDGn-tIcIDlKHrapq-QIkcPPLLITsN
CRPVE1      qplaDakLaVLDDatsac-----------wdfIDtyLRNaLDGn-PIsVDlKHkapi-EIkcPPLLITtN
FPVE1       qpltEckaaLIDDVtlpc-----------wdyVDtfLRNaLDGn-aIcIDcKHrapv-QtkcPPLLLTsN
```

Fig. 31.3. Alignment of a 210 amino acid region of the large T antigen of 7 polyomaviruses (POLTAG) and the carboxyterminal half of the E1 protein of 12 papillomaviruses (PV) (BPV: bovine PV; EEPV: European elk PV; FPV: chaffinch PV; other abbreviations same as in Fig. 31.2). Identical amino acids and functionally conserved amino acids found in all papillomaviruses are marked with an asterisk (*) and hash sign (#), respectively. Identical or functionally conserved residues found in at least one polyomavirus and papillomavirus are marked with a plus sign (+).

(*Fringilla montifringilla*), and African grey parrot (*Psittacus erithacus*) (Sundberg, 1987).

PVs and their lesions have been detected throughout the world in both humans and animals. This geographic distribution cannot be attributed to airborne transmission, as transmission of PVs requires close cutaneous contact with heavily infected exfoliated skin cells or muco-mucosal sexual contact. Together with the viral species-specificity, physical contact requirements make it unlikely that recent interspecies transmission can account for the global presence of PVs in many, if not all, mammalian and avian species. The well-adapted biological behaviour in a given host species, together with the worldwide distribution in numerous mammalian and avian hosts, imply that PVs are ancient viruses that may have co-evolved with their host species during vertebrate evolution.

To test the co-evolution hypothesis, two requirements have to be consecutively met. First, the parasites from evolutionarily related host species should themselves be phylogenetically closely related. The second and more important requirement is to demonstrate that PVs co-evolved and co-speciated in synchrony with their hosts. This latter point can be argued when the branching pattern of the evolutionary tree derived from the parasites is concordant with the branching pattern of the host species tree (Mitter & Brooks, 1983; Stone & Hawksworth, 1986; Hafner & Nadler, 1988; Lake *et al.*, 1988).

The first requirement seems fulfilled for PVs. For example, PVs from host species belonging to the order *Artiodactyla* (even-toed ungulates) are phylogenetically related. Fig. 31.4 is a dendrogram of the amino acid alignment of the E6 proteins of bovine (*Bos taurus*) PV type 1 (BPV-1), BPV-2, European elk (*Alces* alces) PV (EEPV), reindeer (*Rangifer tarandus*) PV (RPV), and American white-tailed or mule deer (*Odocoileus virginianus* or *hemionus*) PV (DPV), and shows their close phylogenetic relationships. The PVs isolated from the *Bovidae* and *Cervidae* form monophyletic groups. Three other bovine PVs (BPV-3, -4 and -6; subgroup B bovine PVs), could not be included in this analysis because they lack an E6 open reading frame (Jackson *et al.*, 1991). Another example of the phylogenetic relationship of PVs from related species is shown in two genera of leporids. The cutaneous cottontail rabbit (*Sylvilagus floridanus*) PV (CRPV), and the mucosal rabbit oral PV (ROPV) from domestic rabbits (*Oryctolagus cuniculus*), show more nucleotide similarity to each other than to the other cloned cutaneous or mucosal PVs (Tachezy *et al.*, personal communication).

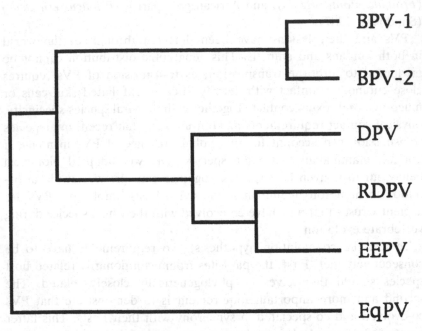

Fig. 31.4. Phylogenetic relatedness of five different papillomaviruses from even-toed ungulate host species (Artiodactyla) based on an amino acid alignment of the E6 genes using PAUP. The E6 sequence of the equine PV (EqPV; Kaplan *et al.*, unpublished observations) was used as an outgroup (BPV: bovine PV; DPV: white-tailed deer PV; RDPV: reindeer PV; EEPV: European elk PV).

The second prerequisite is more difficult to substantiate, because the phylogeny of higher mammals has not been established (Novacek, 1992). In primates, from which a comparatively large amount of sequence data are available, evolutionary relationships are well accepted and the divergence times of different primates have been approximated with a reasonable degree of confidence (Miyamoto *et al.*, 1988). Primatologists concur that the common chimpanzee (*Pan troglodytes*) and pygmy chimpanzee (*Pan paniscus*) are the two species most closely related to humans (Schmitt, Grauer & Tomink, 1990; Begun, 1992). These two chimpanzee species diverged from each other approximately 2.5 million years (Myr) ago, whereas the human branch and the ancestor of the two chimpanzee species diverged approximately 4.7 Myr ago (Horai *et al.*, 1992). The cercopithecoids (Old World primates) and the human branch were estimated to have diverged about 40 Myr ago. Non-human primate

PVs have been identified in two chimpanzee species, the common chimpanzee (PtPV) and the pygmy chimpanzee (PCPV-1), and in two more distantly related members of the family *Cercopithecidae*, the black-and-white Abyssinian colobus (*Colobus guereza*; CgPV-1 and CgPV-2), and the rhesus macaque (*Macaca mulatta*; RhPV-1) (O'Banion *et al.*, 1987; Kloster *et al.*, 1988; Van Ranst *et al.*, 1991; Sundberg, Shima & Adkinson, 1992; Favre *et al.*, unpublished observations). The RhPV-1 was isolated from a lymph node metastasis of a primary penile squamous cell carcinoma. Several of the females in the rhesus monkey colony who had mated with the RhPV-1 infected male developed preneoplastic and neoplastic lesions of the cervix, suggesting an oncogenic potential for RhPV-1 (Ostrow *et al.*, 1990). When a phylogenetic tree is generated from an alignment of human and non-human primate PVs, the latter are found among the HPVs as opposed to forming a separate outgroup (Fig. 31.5) (Van Ranst *et al.*, unpublished observations). The position of the non-human primate PVs within the HPV phylogenetic tree reflects their tropism for specific anatomical sites. For instance, RhPV-1 is found amongst the oncogenic mucosal HPVs in the phylogenetic tree. The branching points between the non-human primate PVs and their hypothetical HPV counterpart in the PV phylogenetic tree mirror the phylogeny of their respective host species. HPV-13, PCPV-1, and PtPV are extremely similar at the nucleotide level. All three viruses are associated with a similar disease, oral focal epithelial hyperplasia (Van Ranst *et al.*, 1991, 1992*b*; Sundberg *et al.*, 1992). In a 387 bp E6-alignment, there were 35 (9%) nucleotide substitutions between PCPV-1 and PtPV, 59 (15.2%) between HPV-13 and PCPV-1, and 62 (16%) between HPV-13 and PtPV. The phylogenetic relationships between these three viruses are concordant with the evolutionary relationship between their hosts, again supporting the hypothesis that PVs have co-evolved with their hosts. Given a constant evolutionary clock and the above-mentioned species divergence times, a mutation rate for the PV genome per site per year can be approximated. Using the estimated divergence time between the two chimpanzee species (PtPV and PCPV-1) (2.5 Myr), and between the two chimpanzee species and humans (4.7 Myr), we obtained mutation rates of 3.6×10^{-8} and 3.3×10^{-8} nucleotide substitutions/site in the E6 ORF/year, respectively (Van Ranst *et al.*, unpublished observations). This is about 20–30 times higher than the rate of evolution that was estimated for the human and chimpanzee β-globin gene ($1.12–1.79 \times 10^{-9}$ nucleotide substitutions/site/year) (Miyamoto *et al.*, 1988). Thus, the PVs contain

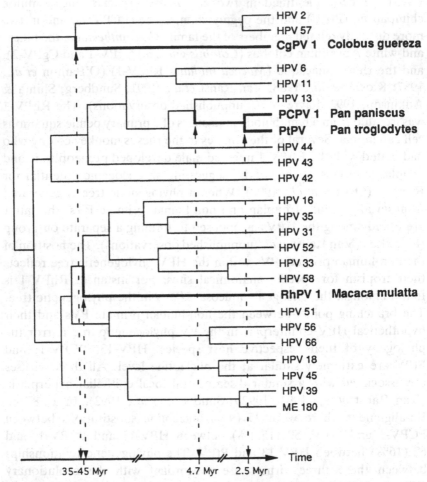

Fig. 31.5. Phylogenetic tree constructed from the alignment of 384 nucleotides in the E6 genes of the non-human primate papillomaviruses (bold) and mucosal human papillomaviruses, using the UPGMA (unweighted pair group maximum average) pairwise similarity matrix method. HPV-1, a cutaneous papillomavirus, was used as an outgroup (not shown). The time-scale is non-linear. The divergence time estimations of the primate host species (Horai *et al.*, 1992; Miyamoto *et al.*, 1988) are indicated on the horizontal axis in million years (Myr).

a higher degree of divergence than the germline genes of their respective hosts (Table 31.1). Approximately two-thirds to three-fourths of the E6 codons of PVs from different host species have a nucleotide change in the third position, compared to only 17–22% of the β-globin codons. Most of the DNA changes in the E6 genes of closely related PVs (HPV-18/45; 6/11; 5/47) occur in the third positions of the codons, similar to the pattern observed in the β-globin genes, whereas the E6 genes of distantly related PV types (HPV-8/CRPV; Cottontail rabbit PV; HPV-8/BPV-1; CRPV/BPV-1) show similar degrees of divergence at all positions. Taken together with the lack of a continuum of HPV type 'species', the mechanism responsible for the third position silent changes in closely related HPV types is unlikely to be the driving force involved in speciation of HPV types.

It is possible that different HPV-ORFs display different mutation rates based on selective pressure. Determination of the mutation rate of PVs from different ecological niches (i.e. EV-HPV-types) awaits further investigation. Although of interest, extrapolation of the mutation rate to more divergent PVs would likely underestimate the true divergence times due to the presence of an increasing number of superimposed or hidden nucleotide substitutions.

Phylogenetic analysis of human papillomaviruses

It was assumed until recently that all infectious wart-like lesions in humans were caused by a single PV which infected different anatomical sites. With the availability of more than 70 different cloned HPVs, it is now evident that specific HPV types cause anatomically distinct diseases. For example, HPVs can induce benign lesions such as common, flat or plantar warts. Genital HPV-infection, including condyloma acuminatum, is now recognized as the most common sexually transmitted disease (Rosenfeld *et al.*, 1992). Laryngeal papillomas, oral condylomata, focal epithelial hyperplasia, inverted nasal papillomas, and conjunctival papillomas have also been shown to be HPV-related. In humans, the association between PVs and cancer was first recognized in epidermodysplasia verruciformis (EV), a rare hereditary cutaneous disorder with an increased susceptibility to infection by a diversity of HPVs. Skin carcinomas eventually develop in about one third of these patients. The association between specific HPV types and invasive anogenital cancer and its precursor lesions has been extensively documented over the last decade.

Table 31.1. *Divergence between human, rabbit and bovine β-globin genes, between E6 genes from papillomaviruses that infect human, rabbit and bovine host species, and between E6 genes from closely related human papillomaviruses*

Comparison	Codons compared	% similarity		% of nucleotide substitutions by codon position			% of codons with a change in third position
		amino acid	nucleotide	1st	2nd	3rd	
β-globin[a]							
Human/rabbit	147	90	89	19	15	67	22
Human/bovine	144	85	87	27	18	55	22
Rabbit/bovine	145	88	90	30	14	56	17
Papillomavirus E6							
HPV-8/CRPV	143	27	42	33	28	39	68
HPV-8/BPV-1	133	29	31	32	32	36	74
CRPV/BPV-1	137	20	38	30	33	37	69
HPV-18/HPV-45	158	78	83	29	17	54	28
HPV-6/HPV-11	150	81	83	33	14	52	29
HPV-5/HPV-47	156	76	75	31	20	50	37

Note: [a] β-globin sequences (Van Ooyen *et al.*, 1979; Schimenti & Duncan, 1984) were manually aligned. Gaps were introduced to maximize similarity.

To investigate the evolutionary history of the HPVs, phylogenetic trees were constructed based on the alignment of the E6 sequence of 29 HPV types and 4 subtypes, using the PAUP (phylogenetic analysis using parsimony) software package (Swofford, 1991) (Fig. 31.6). This region was chosen for analysis because it allowed an unambiguous alignment of all sequences (a prerequisite for reliable phylogenetic evaluations), based on the presence of four copies of a Cys-X-X-Cys motif spaced at regular and invariant intervals. The branching pattern of the most parsimonious phylogenetic tree clustered different HPVs into groups corresponding to their known tissue tropism and oncogenic potential. The first major branch (A) consists of viruses that infect the skin. From this branch, one group (I) corresponds to HPV-1, found in skin warts, and a second group (II) corresponds to the EV-types. HPV-14, -20, -21 and -25 form one subgroup (IIa), and HPV-5, -8 and -47 form a second subgroup (IIb). HPVs in the first subgroup are found only rarely in malignant EV lesions, whereas viruses in the second subgroup have been identified frequently in EV-related skin cancers. The second major branch (B) groups HPVs which primarily infect mucous membranes. An intermediate group (III) consists of HPV-2 and HPV-57. These viruses have been found in papillomas in both the oral mucosa and the skin, causing verruca vulgaris in the latter. The PVs in the other group (IV) predominantly infect the genital and oral mucosae. One subgroup (IVa) comprises HPV-6, -11, -42, -43 and -44, and is associated with benign genital condylomata and low grade cervical intra-epithelial neoplasia (CIN). HPV-13, a virus involved in oral focal epithelial hyperplasia (Heck's disease) has also been identified in some genital lesions. The viruses in the second subgroup (IVb) are associated with low grade CIN, high grade CIN and invasive anogenital cancer. One branch contains HPV-16, -31, -33, -35, -51, -52, -56, -58 and -66. A second branch includes HPV-18, -39, -45 and ME180, which have recently been suggested to be associated with more rapidly progressive cervical neoplasias.

To test the validity of the maximum parsimony phylogeny of the HPV E6 sequences, a distance matrix approach (UPGMA) was applied on the same data set. In general, the UPGMA tree (Fig. 31.7) had the same topology as the PAUP-tree, with occasional differences in the terminal branchings. The UPGMA-tree clustered HPVs [{{5a6, 5a5}, 5a1}, 5b], [{20,21}, 14], [{6,11}, {13,44}], and [{16, {35,31}] whereas the PAUP-tree clustered HPVs [{5a1, 5a6}, {5a5, 5b}], [{20,14}, 21], [{6,11}, 13,44] and [{16,35}, 31], respectively. None

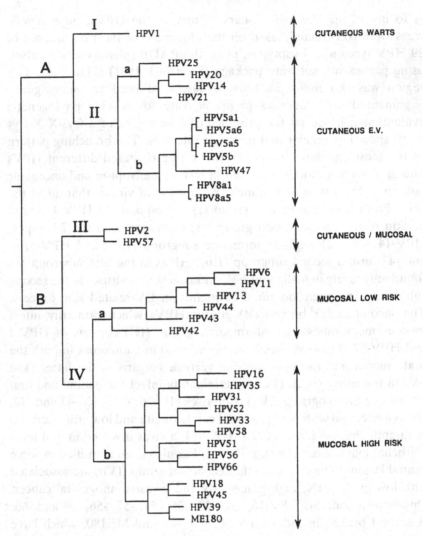

Fig. 31.6. Phylogenetic tree constructed from the alignment of 384 nucleotides in the HPV E6 genes using maximum parsimony analysis (PAUP 3.0). To root the tree, bovine papillomavirus type 1 was used as an outgroup (not shown). The labels on the right indicate the corresponding clinical category associated with each HPV-type.

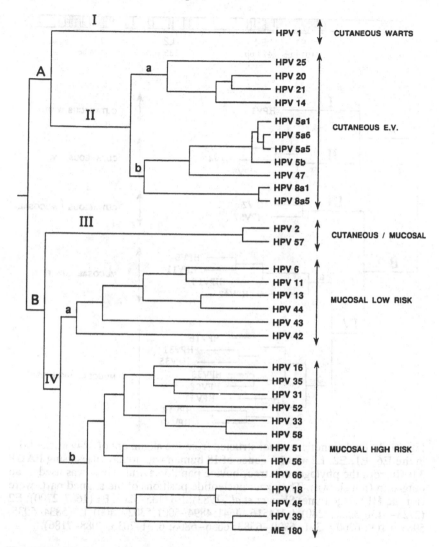

Fig. 31.7. Phylogenetic tree based on the same data as in Fig. 31.6, using the UPGMA pairwise similarity matrix method.

of these differences had any impact on the classification as described for the maximum parsimony tree.

To determine the robustness of the E6 tree, a phylogenetic tree was made based on the alignment of a compilation of 3249 bp (40% of the HPV genome) from the E6, E1, E2, L2 and L1 open reading frames of 18 HPVs for which the complete genomic sequence was

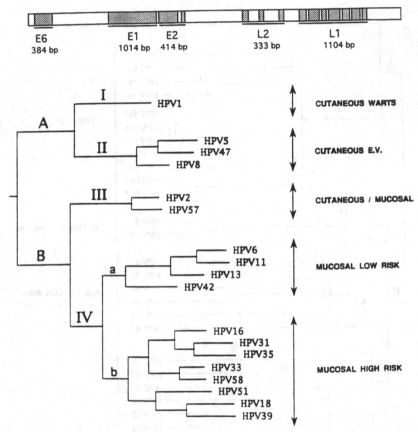

Fig. 31.8. Phylogenetic tree constructed from the alignment of 3249 nucleotides in the E6, E1, E2, L2 and L1 genes of 18 human papillomaviruses using PAUP 3.0. To root the phylogenetic tree, bovine papillomavirus type 1 was used as an outgroup (not shown). The exact nucleotide positions of the aligned parts were (for the HPV-6 genome; Shwarz *et al.*, 1983): E6 (135–520), E1 (1687–2700), E2 (2723–3136; 3221–3280), L2 (4516–4641; 4894–5001; 5302–5400), L1 (5834–5935; 5963–6169; 6200–6301; 6329–6583; 6656–6808; 6851–7036; 7088–7186).

available (Fig. 31.8). This tree had the same overall topology as the tree constructed for the E6 tree of 28 HPVs. The only difference was the position of HPV-51, which was more closely related to HPV-18 and -39 in the 3249 bp tree than on the E6 tree.

It can be inferred from the phylogenetic tree in Fig. 31.6 and Fig. 31.7 that during the molecular evolutionary history of the PVs, there was an early division into PVs that infect epithelial skin cells and PVs that infect mucosal cells. Given the relative genetic stability of the PVs, it is unlikely that this division occurred during the evolution of the human species. This is supported by the fact that a similar division into different ecological niche-specific PVs can already be found in the non-human primate PVs (Fig. 31.5). It appears that the co-evolution of the human and non-human PVs with their hosts occurs in an anatomic site-specific manner. More sequence data from a larger number of primate and other animal PVs will be required to determine when the division into cutaneous and mucosal PVs began.

Micro-evolution (variation) of HPV types

The sequence divergence between different variants of a given HPV type in the mucosal phylogenetic group is consistently less than 1%. However, in the epidermodysplasia verruciformis (EV) group, the sequence divergence between different isolates of HPV-5 and -8 has been shown to vary from 3 to 10% (Deau, Favre & Orth, 1991). The distinction between an HPV variant and a different HPV type is less clear in this group. Moreover, a larger group of distinct HPV types has been isolated from EV-patients. The molecular basis for the increased genetic variation in the EV-group is unknown. One hypothesis is that EV-HPV types either recruit a subset of host DNA polymerases with a lower fidelity or have reduced DNA-proofreading or -repair mechanisms. Therefore, a mutation in a subset of the cellular DNA processing enzymes, resulting in a lower fidelity of the viral replication process, may be hypothesized as the genetic basis for this disease. Experiments with Qβ-replicase have shown that organisms can overcome an imperfect replication fidelity, by surviving with quasi-species distributions rather than with unique genes (Eigen *et al.*, 1982). Alternatively, UV damage may contribute to a higher mutation rate for EV-PVs in sun-exposed areas of the skin, when compared to PVs with a tropism for UV-inaccessible mucosal surfaces. The EV-lesions are often infected with more than one EV-HPV type (Obalek *et al.*, 1992). Thus, co-infecting EV-HPVs may provide *trans*-acting viral proteins that would allow the persistence of other episomal EV-HPV genomes with a biologically inferior fitness. Therefore, the biological constraints that counterbalance the success of mutation events may be less rigid for EV-specific PVs.

The presence of microvariants of a given HPV type, each with characteristic sequence polymorphisms, has facilitated the investigation of the dispersion of HPV in the population. The upstream regulatory region (URR), the only lengthy non-coding part of the PV genome, differs extensively between different PV types. Ho *et al.* (1991) showed that even in a given HPV type (HPV-16), this region accumulated point mutations. The mutations were not distributed randomly over the studied region (i.e. the same nucleotides were mutated in multiple distinct isolates), suggesting that most mutations occurred as singular events and spread subsequently. This allowed Chan *et al.* (1992) to investigate the geographical distribution of HPV-16 microvariants. More than 50 different microvariants with polymorphisms in the URR were found in a series of 200 HPV-16 isolates from four different continents (Germany/Finland, Singapore, Brazil, and Tanzania). A phylogenetic tree constructed with these URR-microvariants had two major branches, a Eurasian and an African branch (Fig. 31.9). All of the German/Finnish and most of the Singaporean variants were attributed to the Eurasian branch, while most of the Tanzanian isolates were in the African branch. The Brazilian variants were divided between the Eurasian and the African clusters in approximately equal numbers. This may reflect the colonization of Brazil by European colonists, followed by the introduction of African Bantus in the eighteenth century (Fig. 31.10). The rare occurrence of Eurasian variants in Tanzania may represent the relatively recent period of European colonial immigration to Africa. The presence of some African variants in Singapore may be associated with the former Arab slave trade and the historical presence of Indonesian commercial missions in the East African coastal regions. Taken together, the data in the study by Chan *et al.* (1992) suggest that HPV-16 co-existed and co-migrated with the human species for a long period of time separately in Africa and Eurasia. The increased geographic mobility of the human race during the past centuries may have facilitated the pandemic expansion of HPV-16 microvariants, including speciation.

Conclusion

PVs are a large and heterogeneous family of species-specific viruses with a double-stranded DNA genome and that infect a wide variety of vertebrate hosts. Because they use the high-fidelity host cell DNA polymerases for viral replication, PVs are remarkably stable genetic entities.

PVs use biological strategies similar to those of other small DNA tumour viruses, such as polyomaviruses and adenoviruses, in manipulating the host cell-cycle control mechanisms to their own advantage. Although overall sequence similarity between papillomavirus,

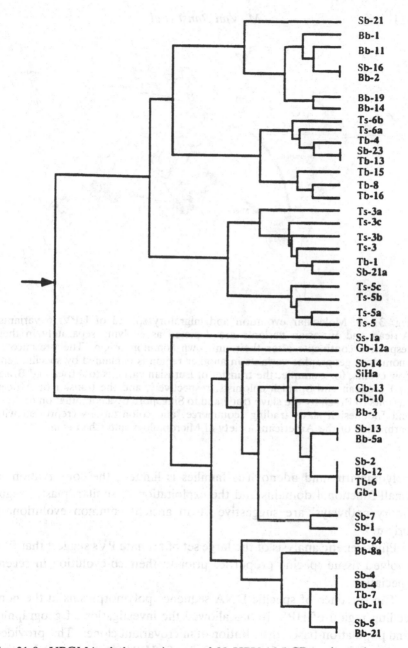

Fig. 31.9. UPGMA phylogenetic tree of 38 HPV-16 LCR variants that were identified in cervical samples from Singapore (S samples), Brazil (B samples), Tanzania (T samples), and Germany (G samples). Distinct African (top) and Eurasian (bottom) lineages divide at the root (arrow), which is placed centrally between the two groups (reprinted with permission of the American Society of Microbiology and Chan *et al.*, 1992).

Fig. 31.10. Molecular evolution and migratory spread of HPV-16 variants. African and Eurasian variants are depicted as evolving separately in their respective continents after their unknown common origin. The presence of anomalous geographic variants in another region is explained by specific gene flow events, for example, the transfer of Eurasian variants to Africa and Brazil by European and Spanish colonists, respectively, and the transfer of African variants to Brazil by the slave trade and to Singapore by a combination of Arab and Indonesian slave trading, commerce, and colonizations (reprinted with permission of the American Society of Microbiology and Chan *et al.*, 1992).

polyomavirus and adenovirus families is limited, the conservation of small functional domains and the exploitation of similar 'master regulatory pathways' are suggestive of an ancient common evolutionary origin.

Phylogenetic analysis of the large set of primate PVs suggest that PVs evolved tissue-specific properties prior to their co-evolution in recent speciation.

The presence of specific DNA sequence polymorphisms in the noncoding region of HPV-16 has allowed the investigation of geographic- and population-based distribution of microvariant clones. This provided evidence for a pandemic spread of specific HPV types via co-migration and co-evolution with humans.

Circumstantial and experimental data indicate that PVs are ancient viruses that originated early in vertebrate evolution and have co-evolved and speciated in synchrony with their host species. Resolution of the mammalian phylogeny and the availability of a wider variety of sequenced animal PV genomes are needed to produce a more precise picture of the molecular evolution and origin of the PVs. In this regard, PVs may provide an excellent model system to elucidate more basic aspects of the evolutionary process.

Acknowledgements

The authors would like to thank G. Opdenakker, R. Tachezy, A.B. Jenson, S.-Y. Chan and H.-U. Bernard for helpful discussions. M. Van Ranst is a fellow of the University of Leuven Collen Research Foundation, and the Belgian American Educational Foundation.

References

Aloha, H., Stenlund, A., Moreno-Lopez, J. & Petterson, U. (1983). *Nucl. Acids Res.*, **11**, 2639–50.

Barbosa, M.S., Edmonds, C., Fisher, C., Schiller, J.T., Lowy, D.R. & Vousden, K. H. (1990). *EMBO J.*, **9**, 153–60.

Begun, D.R. (1992). *Science*, **257**, 1929–33.

Bruenn, J.A. (1991). *Nucl. Acids Res.*, **19**, 217–26.

Chan, S.-Y., Bernard, H.-U., Ong, C.-K., Chan, S.-P., Hofmann, B. & Delius, H. (1992). *J. Virol.*, **66**, 5714–25.

Clark, R., Peden, K., Pipas, J.M., Nathans, D. & Tijan, R. (1983). *Mol. Cell. Biol.*, **3**, 220–8.

Clertant, P. & Seif, I. (1984). *Nature*, **311**, 276–9.

Coggin, J.R. & zur Hausen, H. (1979). *Cancer Res.*, **39**, 545–46.

Deau, M.-C., Favre, M. & Orth, G. (1991). *Virology*, **184**, 492–503.

Defeo-Jones, D., Huang, P.S., Jones, R.E., Haskell, K.M., Vuocolo, G.A., Hanobik, M.G., Huber H.E. & Oliff, A. (1991). *Nature*, **352**, 251–4.

De Villiers, E.-M. (1989). *J. Virol.*, **63**, 4898–903.

Eigen, M., Gardiner, W., Schuster, P. & Winkler-Oswatitisch, R. (1982). In *Evolution Now*, J. Maynard Smith (ed.). W.H. Freeman and Co., San Fransisco, pp. 10–33.

Figge, J. & Smith, T.F. (1988). *Nature*, **334**, 109.

Hafner, M.S. & Nadler, S.A. (1988). *Nature*, **332**, 258–9.

Ho, L., Chan, S.-Y., Chow, V., Chong, T., Tay, S.-K., Villa, L.L. & Bernard, H.-U. (1991). *J. Clin. Microbiol*, **29**, 1765–72.

Horai, S., Satta, Y., Hayasaka, K., Kondo, R., Inoue, T., Ishida, T., Hayashi, S. & Takahata, N. (1992). *J. Mol. Evol.*, **35**, 32–43.

Jackson, M.E., Pennie, W.D., McCaffery, R.E., Smith, K.T., Grindlay, J. & Campo, M.S. (1991). *Mol. Carcinogenesis*, **4**, 382–7.

Jarrett, W.F.H., Campo, M.S., O'Neil, B.W., Laird, H.M. & Coggins, L.W. (1984). *Virology*, **136**, 255–64.

Kirii, Y., Iwamoto, S. & Matsukura, T. (1991). *Virology*, **185**, 424–27.

Kloster, B.E., Manias, D.A., Ostrow, R.S., Shaver, M.K., McPherson, S.W., Rangan, S.R.S., Uno, H. & Faras, A.J. (1988). *Virology*, **166**, 30–40.

Koonin, E.V. (1991). *J. Gen. Virol.*, **72**, 2197–206.

Lake, J.A., de la Cruz, V.F., Ferreira, P.C.G., Morel, C. & Simpson, L. (1988). *Proc. Natl. Acad. Sci. USA*, **85**, 4779–83.

Mitter C. & Brooks, D.R. (1983). In *Coevolution*, D.J. Futuyuma and M. Slatkin (eds.). Sinauer, Sunderland, Massachusetts, pp. 65–98.

Miyamoto, M.M., Koop, B.F., Slightom, J.L., Goodman, M. & Tennant, M.R. (1988). *Proc. Natl. Acad. Sci. USA*, **85**, 7627–631.

Moran, E. (1988). *Nature*, **334**, 168–70.

Novacek, M.J. (1992). *Nature*, **356**, 121–5.

Obalek, S., Favre, M., Szymanczyk, J., Misiewicz, Jablonska, S. & Orth, G. (1992). *J. Invest. Dermatol.*, **98**, 936–41.

O'Banion, M.K., Sundberg, J.P., Shima, A.L. & Reichmann, M.E. (1987). *Intervirology*, **28**, 232–7.

Ostrow, R.S., McGlennen, R.C., Shaver, M.K., Kloster, B.E., Houser, D. & Faras, A.J. (1990). *Proc. Natl. Acad. Sci. USA*, **87**, 8170–4.

Phelps, W.C., Lee, C.L., Munger, K. & Howley, P.M. (1988). *Cell*, **53**, 539–47.

Rosenfeld, W.D., Rose, E., Vermund, S.H., Schreiber, K. & Burk, R.D. (1992). *J. Pediatr.*, **121**, 307–11.

Rous, P. & Beard, J.W. (1935). *J. Exp. Med.*, **62**, 523–48.

Schimenti, J.C. & Duncan, C.H. (1984). *Nucl. Acids Res.*, **10**, 1641–55.

Schmitt, J., Graur, D. & Tomink, J. (1990). *Primates*, **31**, 95–108.

Schwarz, E., Durst, M., Demankowski, C., Lattermann, O., Zech, R., Wolfsperger, E., Suhai, S. & Rowekamp, W.G. (1983). *EMBO J.*, **2**, 2341–8.

Shope, R.E. (1933). *J. Exp. Med.*, **58**, 607–24.

Stabel, S., Argos, P. & Philipson, L. (1985). *EMBO J.*, **4**, 2329–36.

Stone, A.R. & Hawksworth, D.L. (eds.) (1986). *Coevolution and Systematics*, Oxford University Press.

Sun, S., Thorner, L., Lentz, M., MacPherson, P. & Bothcan, M. (1990). *J. Virol.*, **64**, 5093–105.

Sundberg, J.P. (1987). In *Papillomaviruses and Human Disease*, K. Syrjänen, L. Gissmann and L.G. Koss (eds.). Springer Verlag, Berlin, pp. 40–103.

Sundberg, J.P., O'Banion, M.K. & Reichmann, M.E. (1987). *Cancer Cells*, **5**, 373–9.

Sundberg, J.P., Shima, A.L. & Adkinson, D.L. (1992) *J. Vet. Diagn. Invest.*, **4**, 70–4.

Swofford, D.L. (1991). In *PAUP: Phylogenetic Analysis Using Parsimony, Version 3.0*. Computer program and documentation. Distributed by the Illinois Natural History Survey, Champaign, Illinois, USA.

Van Ooyen A., van den Berg, J. Mantei N. & Weissmann, C. (1979). *Science*, **206**, 337–44.

Van Ranst, M., Fuse, A., Fiten, P., Beuken, E., Pfister, H., Burk, R.D & Opdenakker G. (1992b). *Virology*, **190**, 587–96.

Van Ranst, M., Fuse, A., Sobis, H., De Meurichy, W., Syrjanen, S.M., Billiau, A. & Opdenakker, G. (1991). *J. Oral Pathol. Med.*, **20**, 325–31.

Van Ranst, M., Kaplan, J.B. & Burk, R.D. (1992a). *J. Gen. Virol.*, **73**, 2653–60.

Villareal, L.P. & Fan, H. (1989). In *Common Mechanisms of Transformation by Small DNA Tumor Viruses*, L.P. Villareal (ed.). American Society for Microbiology, Washington, DC, pp. 1–20.

32

Molecular systematics of the Potyviridae, the largest plant virus family

COLIN W. WARD, GEORG F. WEILLER, DHARMA
D. SHUKLA AND ADRIAN GIBBS

Introduction

Brandes and Wetter (1959) first showed that potato virus Y together with 13 other viruses had filamentous flexuous virions about 750 nm in length, some of which were serologically related to one another. They proposed that these viruses formed a natural group, and this was subsequently named the potyvirus group by Harrison et al. (1971).

Large numbers of viruses have been added to the group, and it now contains at least 200 distinct species, or more than a fifth of all known plant viruses. It is the largest and most rapidly growing of the 50 or so families (or groups) of viruses that infect plants and is now named the Potyviridae comprising, at present, three genera, the potyviruses, rymoviruses and bymoviruses (Ward & Shukla, 1991; Barnett, 1992).

Potyviruses are found in all climatic zones, especially the tropics (Hollings & Brunt, 1981). They cause diseases in almost all crop plants and in many uncultivated species; in 1974 they were reported to infect 1112 plant species of 369 genera in 53 families (Edwardson, 1974), and that list too has grown. Their economic impact was also highlighted in a recent survey of important viruses with filamentous virions as 73% of those named were potyviruses (Milne, 1988). Thus it is important to try to understand how the potyviruses have evolved because it might give clues as to why they are such successful viruses.

Most potyviruses belong to the genus *Potyvirus*, and are transmitted in nature by aphids and through seeds of infected plants. These two properties, together with the diversity of crops they infect, assure the continuous presence of potyviruses in nature throughout the year (Hollings & Brunt, 1981). The five potyvirus species that belong to the genus *Bymovirus* are transmitted by root-infecting fungi, whereas nine species are transmitted by mites and belong to the genus *Rymovirus*

(Brunt, 1992). In addition there is another potyvirus, sweet potato mild mottle virus, that is transmitted by whitefly, and which may be placed in a fourth genus, *Ipomovirus*, when its relationships with other potyviruses have properly been established by genomic sequencing.

The virions of all potyviruses are flexuous rods 11–15 nm wide. The potyviruses that are transmitted by arthropods (*Potyvirus* and *Rymovirus*) have virions with a single modal length in the range of 680–900 nm. Each virion is constructed from about 2000 subunits of a single protein species of M_r 28 000–34 000 kD, and contains the genome which is one molecule of single-stranded, positive-sense RNA of about 10 000 nucleotides (Hari, 1981; Hollings & Brunt, 1981). By contrast, bymoviruses have a bipartite genome; two species of single-stranded, positive-sense RNA of about 3500 and 7500 nucleotides respectively. The two genome segments are contained in virions of 200–300 nm and 500–600 nm length, respectively, which are constructed from a single species of virion protein of M_r 30 000–33 000 kD (Usugi *et al.*, 1989; Kashiwazaki *et al.*, 1990; Kashiwazaki, Minobe & Hibino, 1991).

Genome organization and relationships with other virus groups

Goldbach and Wellink (1988) analysed genomic sequences of viruses with single-stranded positive-sense RNA genomes, and showed that most fell into two clusters or supergroups. The potyviruses fell into the 'picorna-like supergroup', which also includes picornaviruses of animals, and the comoviruses and nepoviruses of plants. All have genomes that:

(i) are single-stranded RNA and are messenger sense;
(ii) have a small protein (VPg) attached to the 5'-end and have a poly (A) tail at the 3'-end;
(iii) are translated as a monocistronic message to produce a single polyprotein from which individual proteins are generated by proteolytic hydrolysis.

In addition, some of their non-virion proteins, especially the polymerases, have significant (>20%) sequence identity with those of other members of the supergroup; and the genes encoding these conserved non-virion proteins are arranged in the same order in the genome (Goldbach & Wellink, 1988) as shown in Fig. 32.1.

The primary polyproteins of arthropod-borne potyviruses range in

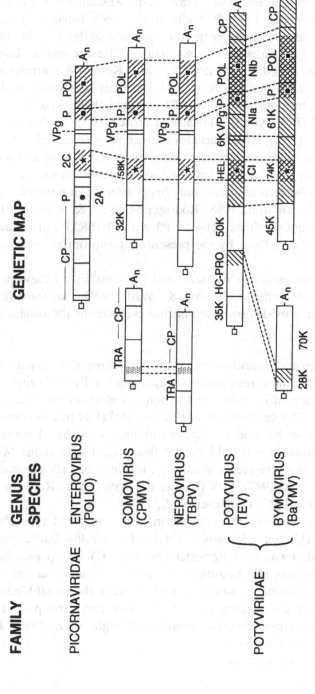

Fig. 32.1. Schematic representation of the genomes of the aphid-transmitted potyviruses, the fungus-transmitted bymoviruses and other members of the picorna-like supergroup of viruses (from Goldbach, 1992 with permission). Regions of amino acid sequence homology are shaded. All genomes have a 5'-terminal VPg and a poly(A) tail; CP, coat protein; HEL, helicase; P, protease; POL, polymerase.

size from M_r 340 000 to 370 000 (Domier *et al.*, 1986; Allison, Johnston & Dougherty, 1986; Maiss *et al.*, 1989; Lain, Reichmann & Garcia, 1989; Teycheney *et al.*, 1989; Robaglia *et al.*, 1989; Johansen *et al.*, 1991). Each is cleaved by three proteinases into at least eight proteins (Dougherty & Carrington, 1988; Garcia, Reichmann & Lain, 1989; Carrington, Freed & Oh, 1990; Verchot, Koonin & Carrington, 1991). The names of these proteins, and the order they occur in the polyprotein (N- to C-terminus), are: P1-protease, helper component (HC), protein 3 (P3), cylindrical inclusion protein (CI), small nuclear inclusion protein (NIa), which includes the genome-associated protein (VPg) at its N-terminus, large nuclear inclusion protein (NIb) and coat protein (CP). The VPg and CP are the viral proteins detected in the virions; all the other proteins have only been detected in infected cells (Dougherty & Carrington, 1988; Rodriguez-Cerezo & Shaw, 1991). Two small cleavage products between P3 and CI (6K1) and CI and NIa (6K2) are also defined by the presence of appropriate cleavage site motifs.

Current knowledge of the functions and relationships of these proteins is incomplete (Shukla, Frenkel & Ward, 1991), but the available information provides some tantalizing evidence of the origins of potyviral genes:

- P1 appears to be a multifunctional protein whose C-terminal end is a serine proteinase responsible for cleaving the P1–HC junction, but the function of its N-terminal region is still unknown (Verchot *et al.*, 1991). The cell movement proteins (MPs) of tobamoviruses and tobraviruses have some sequence identity with the P1 protein of TVMV, however it is unlikely that this is significant as the MPs show no significant sequence similarity with those of TEV (Domier, Shaw & Rhoades, 1987), PPV (Lain *et al.*, 1989), PVY (Robaglia *et al.*, 1989), or PSbMV (Johansen *et al.*, 1991).
- HC has two, possibly three, functions. It is required for aphid transmission (Govier & Kassanis, 1974) but how it fulfils that function is as yet unknown. Its C-terminal end (20 kD) is a papain-like cysteine proteinase (Carrington *et al.*, 1989). Its N-terminal end has a cysteine-rich sequence similar to that found in the metal-binding sites of nucleic acid-binding proteins and thus may form part of a cell-to-cell movement function complex (Robaglia *et al.*, 1989). Its relationships are unknown.
- P3 is of unknown function.

- CI-(VPg) NIa-NIb is an ordered set of conserved non-virion proteins, is involved in RNA replication and is characteristic of the picornavirus supergroup (Fig. 32.1; Goldbach & Wellink, 1988; Goldbach, 1992). The CI of potyviruses has been shown to be a helicase and to have sequence identity with the NS3 proteins of flaviviruses, the p80 proteins of pestiviruses and the RNA helicase proteins of prokaryotic and eukaryotic cells (Lain *et al.*, 1991). NIa is a two-domain protein. Its N-terminal half is the genome-linked, replication-primer VPg (Shahabuddin, Shaw & Rhoads, 1988; Murphy *et al.*, 1990), and its C-terminal half is a protease (Dougherty & Carrington, 1988) that hydrolyses specific sequences in the C-terminal two-thirds of the potyvirus polyprotein (Carrington, Cary & Dougherty, 1988). It is a cysteine protease of the class that is structurally related to the trypsin-like serine proteases found in a large number of animal and plant viruses (Bazan & Fletterick, 1988). The sequences of NIb of potyviruses are clearly related to those of the RNA-dependent RNA polymerases of other plant and animal viruses (Domier *et al.*, 1987). NIb is the most conserved of the potyviral gene coding regions (Robaglia *et al.*, 1989), and Koonin (1991) has constructed a possible phylogeny of them and confirmed the relationship, suggested by Goldbach and Wellink (1988), of picorna- , como- , nepo- , and potyviruses.

- CP is multi-functional. It is the major component of the virions, encapsidating the viral RNA, and is involved in vector transmission. It may also affect host specificity, infection and cross-protection (Atreya, Raccah & Pirone, 1990; Shukla *et al.*, 1991). The full sequences of the CPs of over 60 potyviruses are known. An analysis of the amino acid sequences of the CPs of various viruses with elongated particles shows that all, including that of tobacco mosaic tobamovirus (TMO), are related and probably evolved from a common ancestor (Dolja *et al.*, 1991). So the CPs of potyviruses probably have, like that of TMV, a four-stranded antiparallel α-helical bundle structure (Shukla & Ward, 1989). By contrast, viruses of the other three groups in the picornavirus supergroup have isometric virions and their CPs are 8-stranded anti-parallel β-barrel proteins (Rossman *et al.*, 1985). These two types of CP are quite unrelated and hence their genes were probably acquired from quite different sources by the ancestors of the different groups of the 'picornavirus super group'.

The preceding summary of potyviral proteins indicates that potyviruses are distinguished from other viruses by the distinctive set of genes they share. Many of these genes are related to those of otherwise unrelated viruses (Morosov, Dolja & Atabekov, 1989), and were probably acquired by introgression; the progenitor potyvirus arose when co-infection of a single host by disparate parental viruses permitted their genes to recombine (Goldbach & Wellink, 1988). However, some viral genes (e.g. the HC or P3 proteins of potyviruses) seem to be unique and their origins are not known. They could have come from unknown viral or cellular organisms or they may have arisen *de novo* (Keese & Gibbs, 1992; Chapter 6).

Evolution of the potyviruses; evidence from gene sequence relationships

Genomic sequences

Complete genome sequences are available for two isolates of barley yellow mosaic bymovirus (BaYMV: Kashiwazaki *et al.*, 1990, 1991; Davidson *et al.*, 1991; Peerenboom *et al.*, 1992), and 14 isolates of ten aphid-transmitted potyviruses, tobacco etch (TEV-NAT: Allison *et al.*, 1986), tobacco veinal mottle (TVMV: Domier *et al.*, 1986; Atreya *et al.*, 1990), potato virus Y (PVY-N: Robaglia *et al.*, 1989), pepper mottle virus (PepMoV: Vance *et al.*, 1992), pea seed-borne mosaic (PSbMV: Johansen *et al.*, 1991), papaya ringspot virus (PRSV: Yeh *et al.*, 1992), turnip mosaic virus (TuMV: Nicolas & Laliberte, 1992), Johnson grass mosaic (JGMV-JG: Gough & Shukla, 1993), two strains of soybean mosaic virus (SbMV-G2 and SbMV-G7: Jayaram, Hill & Miller, 1992) and three strains of plum pox virus (PPV-NAT, PPV-R and PPV-D : Maiss *et al.*, 1989; Lain *et al.*, 1989; Teycheney *et al.*, 1989; Palkovics, Burgayan & Balazs, 1993).

Three levels of variation can be seen when these genome sequences are compared. The four strains of PPV and the two strains of SbMV, respectively, have closely similar gene sequences, and these encode proteins that have 94–99% sequence identity whichever of them are compared (Fig. 32.2). By contrast, the proteins of different potyviral species are significantly less similar; the least conserved protein of the aphid-transmitted potyviruses is P1 (18–19% identity), and NIb is the most conserved (61–64%), and similar relative differences are seen when the identities of the nucleic acid sequences of different parts of the genome, both coding and non-coding, are compared (Shukla

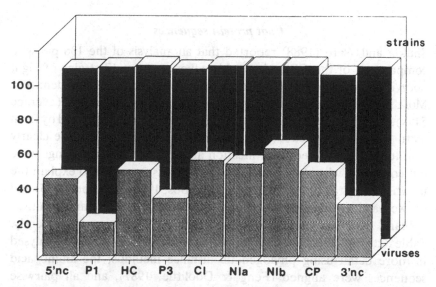

Fig. 32.2. Sequence differences in regions of the genomes of distinct species (PVY-N and PPV-NAT) and strains (PPV-NAT and PPV-R) of potyviruses. Nucleotide sequence identities are given for the 5' and 3' non-coding regions, and amino acid sequence identities for the other regions (from Ward *et al.*, 1992).

et al., 1991). However, the most distant relationships in the family are found when the genomes of species of the four genera are compared. For example, RNA-1 (7632 nucleotides) of BaYMV, corresponds to the 3' two thirds of the aphid-transmitted potyvirus genome and encodes proteins homologous to CI, NIa(VPg), NIb and CP (Kashiwazaki *et al.*, 1990). When the sequences of the most readily aligned parts of the BaYMV proteins are compared with those of TEV-NAT, TVMV, PPV-NAT and PVY-N, their identities range from 26–28% for CI, 18–19% for NIa, 31–34% for NIb and 22–27% for CP (Kashiwazaki *et al.*, 1990). There is even less discernible relatedness between the RNA-2 of BaYMV and other potyviral genes; BaYMV RNA-2 encodes a polyprotein (M_r 98 000 kD) whose N-terminal 255 amino acids are homologous to those of the cysteine proteinase domain at the C-terminal end of the HC protein of the aphid-transmitted potyviruses (Kashiwazaki *et al.*, 1991); however, the remaining 635 amino acids of the RNA-2 encoded polyprotein have no known counterpart. Thus it is clear that the bipartite genome of BaYMV is not merely a divided version of the monopartite genome of the arthropod-borne potyviruses.

Coat protein sequences

Shukla and Ward (1988) reported that an analysis of the 136 pairwise comparisons of the CP amino acid sequences from 17 strains of eight potyvirus species revealed a bimodal distribution of sequence identities. Most species differed in sequence identity by from 38 to 71% (average 54%), whereas most agreed strains of single species only differed by from 90 to 99% (average 95%); the very few exceptions to this rule clearly resulted from incorrect identification and naming. These findings were not consistent with the 'continuum' hypothesis proposed to explain the unsatisfactory taxonomy of potyviruses (Bos, 1970; Harrison, 1985) and showed a clear demarcation of the relatedness of species and strains.

Many more potyviral CP sequences have been reported recently. Table 32.1 lists details of a larger set whose sequences we have analysed in more detail using phylogenetic distance methods; the amino acid sequences were aligned (Feng & Doolittle, 1987), and all pairwise '% identity' and 'FJD distances' (Feng, Johnson & Doolittle, 1985) calculated and analysed by the neighbour-joining tree method (Saitou & Nei, 1987). These CPs comprise those analysed earlier together with CPs of a further 37 isolates of aphid-transmitted potyviruses, together with those of wheat streak mosaic rymovirus (WSMV) and BaYMV.

The CPs of potyviruses have three domains: a N-terminal region, a conserved core, which corresponds to Asp33-Arg248 in the PVY CP and a much smaller C-terminal region. The cores and the more variable N-terminal regions are probably under different selection pressures, and were analysed separately; the short C-terminal regions were not analysed. Relationships calculated from the core sequences, but not the N-terminal regions, correlate well with those indicated by other parts of the genome (Ward & Shukla, 1991, Shukla et al., 1991).

CP core sequences

The phylogenetic relationships, calculated using two distance metrics, are shown as unrooted neighbor-joining trees in Fig. 32.3; in each the basal node is drawn equidistant from the two most distant CPs. The dendrograms have closely similar topologies, and are also closely similar to those computed by Rybicki and Shukla (1992), who mostly used parsimony methods. In both of our dendrograms, the basal node occurs on the branch connecting BaYMV CP to the others in the tree, and this is also the position of the root found by Rybicki and

Table 32.1. *Summary of coat protein and gene sequence data*

	Acronym	CP Size	3'nc	Reference
Aphid transmitted				
Bean common mosaic virus	BCMV-NL4	287	256	Vetten, Lesemann & Maiss (1992)
Bean necrosis mosaic virus	BNMV-NL3	261	ND	Mills pers comm.
	BNMV-NL5	261	ND	Mills pers comm.
	BNMV-NL8	261	256	Vetten *et al.* (1992)
Bean yellow mosaic virus	BYMV-GDD	273	169	
	BYMV-CS	273	174	
	BYMV-S	273	173	Tracy *et al.* (1991)
Clover yellow vein virus	ClYVV-30	273	178	
	ClYVV-NZL	273	177	Bryan, Gardner & Forster (1992)
Johnson grass mosaic virus	JGMV-JG	303	475	
	JGMV-KS1	303	476	Jilka & Clark per comm.
	JGMV-O	303	499	Jilka & Clark per comm.
Lettuce mosaic virus	LMV-O	278	211	
Maize dwarf mosaic virus	MDMV-A	291	234	Jilka & Clark per comm.
Ornithogalum mosaic virus	OrMV	253	274	Burger, Brand & Rybicki (1990)
Papaya ringspot virus	PRSV-P	287	209	
	PRSV-W	287	211	
Passionfruit woodiness virus	PWV-K	278	250	
	PWV-TB	273	ND	
	PWV-S	273	ND	
	PWV-M	273	ND	

Table 32.1. (cont.)

	Acronym	CP Size	3'nc	Reference
Aphid transmitted (cont.)				
Peanut stripe virus	PStV	287	ND	McKern *et al.* (1991)
	PSbMV-P1	287	189	
Pea seed-borne mosaic virus	PepMoV-C	273	265	Vance *et al.* (1992)
Pepper mottle virus	PPV-D	329	218	
Plum pox virus	PPV-AT	329	208	
	PPV-NAT	314	220	
	PPV-R	329	220	
Potato virus Y	PVY-D	267	ND	
	PVY-10	267	ND	
	PVY-18	266	ND	
	PVY-43	267	ND	
	PVY-Chin	267	ND	Zhou *et al.* (1990)
	PVY-GO16	267	ND	Wefels *et al.* (1989)
	PVY-I	267	333	
	PVY-N^a	267	331	
	PVY-N^b	267	329	
	PVY-PepMo	267	335	
Sorghum mosaic virus	SrMV-SCH	329	238	Jilka & Clark per comm.
Soybean mosaic virus	SbMV-N	265	259	
	SbMV-G2	265	259	Jayaram *et al.* (1992)
	SbMV-G7	265	259	Jayaram *et al.* (1992)
Soybean virus	Unknown	264	223	
South African passiflora virus	SAPV-Natal	279	231	Brand, Burger & Rybicki (1993)

Table 32.1. (cont.)

	Acronym	CP Size	3'nc	Reference
Aphid transmitted (cont.)				
Sugarcane mosaic virus	SCMV-SC	313	235	
	SCMV-MDB	328	236	
Tamarillo mosaic virus	TaMV	268	229	Eagles, Gardner & Forster (1990)
Tobacco etch virus	TEV-NAT	263	189	
	TEV-HAT	263	189	
Tobacco vein mottling virus	TVMV	264	223	
Turnip mosaic	TuMV-Ch	288	209	Kong et al. (1990)
	TuMV-Quebec	288	667	Tremblay et al. (1990)
Watermelon mosaic virus 2	WMV2	281	251	
	WMV2-FC	281	256	
Zucchini yellow mosaic virus	ZYMV-NAT	279	207	
	ZYMV-C	279	211	
	ZYMV-F	279	215	Quemada et al. (1990)
Mite transmitted				
Wheat streak mosaic virus	WSMV	322	147	
Fungus transmitted				
Barley yellow mosaic virus	BaYMV-type	297	231	

Note: [a] References not indicated below are listed in Ward and Shukla (1991).

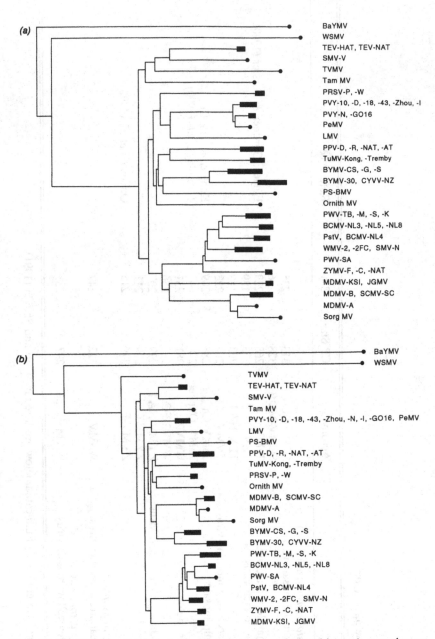

Fig. 32.3. Neighbor-joining trees showing the relationships of potyviruses calculated from the '% non-identity' (panel (*a*)) or 'FJD distance' (panel (*b*)) of the aligned core domains of their CPs.

Shukla (1992) in a cladistic analysis of the core CP sequences of nine potyviruses using the sequences of the CPs of five viruses with filamentous or rod-shaped virions as an outgroup (see Fig. 8 of Rybicki & Shukla, 1992). Thus all three taxonomies place BaYMV CP alone on the basal branch, wheat streak mosaic rymovirus alone on the second branch with a third major branch connecting to all the aphid-transmitted potyviruses. In addition, the three taxonomies have a similar basal set of aphid-transmitted potyviruses, most of which (namely TVMV, TEV and TamMV) are pathogens of solanaceous plants. The two distance matrix taxonomies are closely similar, except that whereas the '% identity' dendrogram (Fig. 32.3(a)), like the cladistic analysis of Rybicki and Shukla (1992), groups all the aphid-transmitted potyviruses of grasses together alongside the BCMVs, PWVs, WMV-2s and ZYMVs, the taxonomy calculated from FJD distances splits the MDMVs, SCMV-SC and SrMV from the JGMVs (and those listed above) and places them with the BYMVs and ClYVVs (Fig. 32.3(b)).

There is no correlation between any of the taxonomies and the type of inclusion bodies the viruses induce in infected plants (Edwardson, 1974; Makkouk & Singh, 1992); all three place PVY and PeMV (both type 4 inclusions) with LMV (type 2), and WMV-2 (type 3) with ZYMV and BCMV (both type 1). We discuss below the correlation between relationships indicated by these taxonomies and host and vector relationships.

Fig. 32.4 shows the frequency distribution of the dissimilarities of all 1485 pairwise comparisons of the CP core domains. The patterns are similar but more complex than that reported earlier (Shukla & Ward, 1988). The dissimilarities form four clusters. The largest sequence differences are around 75% 'non-identity' and correspond to intergeneric comparisons. There are then many sequence differences that cluster around 25–45% and result from inter-species comparisons. The third range of sequence diversity around 20% 'non-identity' reflects the existence of sets of related species such as: (i) BYMV and ClYVV; (ii) PWV, the BCMV serogroup A strains (NL3, NL5 and NL8); South African Passiflora virus; the BCMV serogroup B strain NL4 and PStV; WMV 2 and SbMV-N; and (iii) SCMV, MDMV and SrMV. The smallest sequence differences (1–10%) include all those comparisons between agreed strains of individual potyvirus species, and also some between isolates that were previously considered to be separate species, and that have been shown by these analyses to be, in reality, related strains. There are corresponding, but exponentially related, clusterings

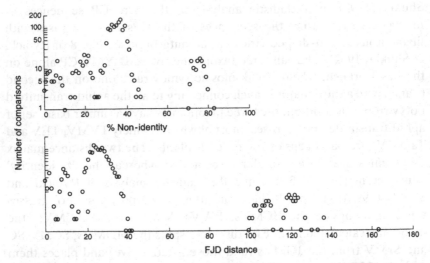

Fig. 32.4. Frequency distribution of the pairwise comparisons of sequence '% non-identity' (top panel) and 'FJD distance' (bottom panel) of the core domains of the CPs of 58 isolates of 27 potyvirus species, one bymovirus and one rymovirus.

in the frequency distribution of the 'FJD distances' of all 1485 pairwise comparisons of the CP core domains (Fig. 32.4).

CP N-terminal sequences

The N-termini of potyviral CPs vary considerably in length and sequence compared with their core domains (Shukla & Ward, 1988; Ward & Shukla, 1991). In intact virions, the N-terminal region of the CP, together with a short region of its C-terminus is exposed (Shukla *et al.*, 1988). The exposed N-terminal regions of different potyvirus species are from 22 to 97 amino acids long, while the exposed C termini are less variable and generally 18 to 20 amino acids in length (Shukla *et al.*, 1988; Ward & Shukla, 1991) except for SAPV which is 24 residues and BaYMV which is only 11 residues (see Table 32.1 for refs). These regions are immunodominant (Shukla *et al.*, 1988, 1989c) and, because of their variability, provide virus-specific epitopes (Shukla *et al.*, 1989a, b). The N- and C- termini can be removed enzymatically from the virions, and the resulting 'core virions', with proteins only 213 to 227 amino acids in length, are indistinguishable from untreated virions when examined by electron microscopy (Shukla *et al.*, 1988). This indicates that the

CP cores, together with the genomic RNA, are the major, if not sole, structural components of the virions.

It is clear that the variation of the N-terminal region of the CP reflects its special role in potyvirus biology. A recent comparison of CP sequences of two strains of sugarcane mosaic virus (SCMV-SC and SCMV-MDB, formerly MDMV-B; Frenkel *et al.*, 1991) revealed that, whereas their core domains were 92% identical, their N-terminal domains were quite different (Fig. 32.5); the SCMV-SC N-terminus was 44 residues in length whereas that of SCMV-MDB was 59 residues, and only 22% were identical. Frenkel *et al.* (1991) concluded that these N-termini had come from different sources, and that at least one of them had arisen by recent recombination, as they did not differ by simple mutations, deletions or a frameshift. The sequences of SCMV-SC and SCMV-MDB both show clear evidence of partial gene duplication (Fig. 32.5), although the duplicated sequences are different (Frenkel *et al.*, 1991). This region was subsequently examined in other SCMV strains with different host ranges (Xiao *et al.*, 1994). The results showed that the strains that infect sugarcane (SCMV-Isis and SCMV-Brisbane) are almost identical to the SCMV-SC sequence and had the same 12 residue repeat sequence motif GAQPPATGAAAQ (Fig. 32.5). These strains have very similar host ranges as shown in Table 32.2. No data are available for SCMV-A or SCMV-D, two other strains that infect

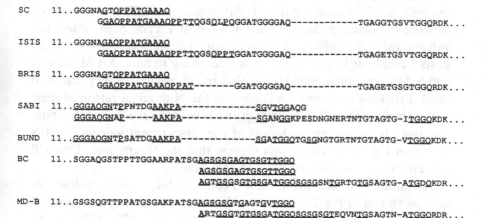

Fig. 32.5. N-terminal domains of the CPs of SCMV-SC, Isis, Brisbane, Sabi, Bundaberg, BC and MDB (formerly MDMV-B) showing the repetitive elements of the sequences. The core domain starts at residues 78 in SCMV-SC, and 93 in SCMV-MDB (based on Frenkel *et al.*, 1991; Xiao *et al.*, 1994).

Table 32.2. Host range correlations of SCMV strains[a]

Strain	Sugarcane	Maize	Wild sorghum	Sabi grass	Blue couch grass
SC	yes	yes	yes	no	no
Isis	yes	yes	yes	no	no
Brisbane	yes	yes	yes	no	no
Sabi	no	yes	yes	yes	no
Bundaberg	no	yes	yes	yes	no
BC	no	yes	yes	no	yes
MDB	no	yes	?	?	?

Note: [a] Data based on studies by Teakle and Grylls (1973), Teakle, Shukla and Ford (1989), Srisink (1989) and D.S. Teakle (unpublished observations).

sugarcane and which would be expected to resemble the SC/Isis/ Brisbane subset of strains. In contrast, SCMV-BC, which cannot infect sugarcane, resembled SCMV-MDB and contained three copies of the 17 residue repeat AGSGSGAGTGSGTTGGQ that is found twice in the SCMV-MDB sequence (Fig. 32.5). SCMV-Sabi and SCMV-Bundaberg, which are also unable to infect sugarcane, exhibited a third sequence pattern. SCMV-Sabi had an almost complete repeat of residues 11–19 and 25–35. The N-terminal region of the SCMV-Bundaberg coat protein was quite short, had no repeats but resembled the repeating sequences of SCMV-Sabi (Fig. 32.5). As shown in Table 32.2, Bundaberg and Sabi are the only SCMV strains reported to infect sabi grass. They are also both capable of infecting the dicotyledonous host French bean, *Phaseolus vulgaris* (Srisink, 1989), and are very similar in their reactions on other hosts (Teakle & Grylls, 1973). It is interesting to note that MDMV-A and SrMV, two other potyviruses that infect maize and sorghum, also exhibit striking sequence duplication in the N-terminal region of their coat proteins (Fig. 32.6), although the repeated sequences are totally unrelated to each other or to any of the other repeat sequence motifs shown in Fig. 32.5. These data suggest that the N-terminus of the CP influences, or is influenced by, host specificity.

   Partial duplications appear to be a common feature of potyvirus CPs and, as shown in Fig. 32.6, are found in the N-terminal domains of all CPs that are 287 residues or longer (Ward & Shukla, 1993; Gibbs *et al.*, 1995). All of the viruses isolated from graminaceous species have long N-terminal domains, but so too do the PPVs, PSbMV and TuMV, which form part of one cluster in all three classifications discussed above. Such

```
Bymovirus
BaYMV              AADP----LTDAQKEDARIAAADGARFEL
                           ADADRRRKVE
                           ADRVEAARVKK
                           A-ADAALKPVNLTATRTPTED---------------DGKLK---

Bymovirus
WSMV              SSQS----ASTASGSGSSQ
                           SGSGSGAA
                           GGSGSGAAQTQ---------------SNNVSVMAGLDTGG
                                              AKTGQGSGSKGTGG
                                              SFT---SNPVRTGG-----RATDVQD

Potyviruses
PSbMV-P1          AGDE----TKDDEBRRKEEE-D
                           RKKREESIDASQFGSNRDN
                           KKNKNKESDTSNKLI----------------------VKSDRDVDA

PPV-R             ADEREDEEEVDAGKPSVVTAPAATSPILQPPPVIQ
                           PAPRTTASMLNPIFT
                           PATTQ
                           PATKPVSQVSGPQLQTFGTYGNEDASPSNSNALVN-TNRDRDVDA

TuMV              AGE-----TLDAGLTEEQKAEKER
                           KERERSE
                           KEREBORQLALKKGKNAAQEE-------------GERDNEVNA

PRSV-P            SKNE----AVDAGLNEKLKEKEN
                           QKEKE
                           KEK--
                           QKEKEKDGASDGNDVSTSTKT----------------GERDRDVNV

MDMV-A            AGE-----NVDAGQKTDAQKEAEKKA
                           -AEEKKA
                           KEAEAKQ
                           KETKEKSTEKTGDGGS--------------------IGKDKDVDA

SrMV-SCH          AGGG----TVDAGATTAEATAQAQR-DAAAKA
                           QR-DADAKKKADDEAAER
                           QRODAAAKKKADDDAKAKADAIVK
                           QNQIADAKKKADDEAARKAQ-------------NQKDKDVDV

JGMV-JG           SG------NEDAGKQKSATPAANQT-ASGDGK
                           PA--QTTATADNK
                           PSSDNTSNAQGTSQTKGGGESGGTNATA---------TKKDKDVDV
```

Fig. 32.6. Repeat sequences in the N-terminal domains of some of the 'long' potyvirus CPs. The sources of sequence data are listed in Table 32.1.

duplications are readily detected by comparing a sequence with itself using an averaging dot diagram method (Gibbs & McIntyre, 1970). Fig. 32.7 shows an analysis of this sort applied to the NIb and CP genes of WSMV.

The capacity to change the exposed surface of the CP without affecting its ability to form particles may allow potyviruses to extend or change their host ranges and vector, and hence be the reason why there are so many potyvirus species. It seems that the conserved triplet DAG, or NAG, which is found in the N-terminal region (residues 5 to 13) of potyviral CPs (Harrison & Robinson, 1988; Lain *et al.*, 1988; Shukla & Ward, 1989; Shukla *et al.*, 1991) is essential for the CP/helper component interactions involved in aphid transmission (Atreya *et al.*, 1990). The CP

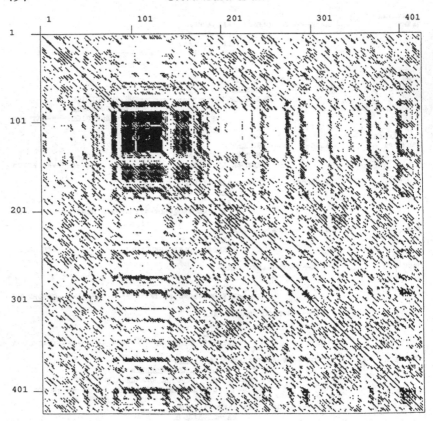

Fig. 32.7. Dot diagram of the amino acid sequence encoded by part of the WSMV genome, compared with itself. The sequence of 425 amino acids includes the C-terminus of the NIb protein and all of the coat protein. The MDM78 matrix (Chapter 36) was used to score each comparison and, after averaging in a 5 residue window, each of the largest 6.7% of the scores represented by a dot.

is also involved in cross-protection (Sherwood & Fulton, 1982) and may be involved in the interactions that activate a membrane factor on the surface of plant cells during the infection process (Salomon, 1989), although the region of the CP involved is unknown.

An attempt to classify the N-termini by their sequences using the same methods that were used for their core domains gave inconsistent dendrograms, which did not correlate with those of the CP cores. This is not only because they are more variable and much shorter than the core sequences, but, more significantly, some are internally repetitive and hence the previous methods are unsuitable. We therefore devised a version of nearest neighbour (NN) analysis (Gibbs *et al.*, 1971) to make the comparisons (Gibbs *et al.*, in preparation). However NN distances for the N-terminal CP sequences correlated only poorly with the corresponding distances of the CP core sequences. The lack of similarity may indicate that some of the N-termini and cores have independent origins, or that the N-termini have evolved faster than the cores, and now have little detectable homology. However, the poor correlation is also due to the fact that the N-termini are short, and therefore unable to contain much 'phylogenetic signal'.

Evolutionary trends among the potyviruses

The phylogenies of interacting organisms are often congruent. This usually indicates that such organisms have co-evolved (Rothschild & Clay, 1952). Any incongruities in their phylogenies indicate that one or more of the lineages acquired a novel partner at some stage. Some virus groups (e.g the poxviruses, tobamoviruses and tymoviruses) show clear evidence of having retained or acquired their natural hosts in such a way. For example, the tobamoviruses that naturally occur in different species of the Solanaceae form a single major branch of the group, whereas those from legumes and those from cucurbits form the other branches. However there are some anomalies, for example the solanaceous branch also includes a virus of orchids, and another from a cactus (Gibbs, 1986; Chapter 23).

The consistent features of the taxonomy of potyviral CPs correlates similarly with the taxonomy of their hosts (Chapter 17). For example bymoviruses and rymoviruses form the basal groups of all classifications of the Potyviridae. Fourteen species of these two groups are known and all are only found naturally in monocotyledonous angiosperms, mostly grasses. The major groupings of the aphid-transmitted potyviruses also show clear host clusterings. For example, the basal species in the taxonomies are mostly from Solanaceae (TVMV, TEV and TaMV). Two of the major clusters are mostly from 'rosid/caryophyllid' hosts (Chapter 17); one group includes PPV, TVMV and BYMV, and the other PWV, BNMV, BCMV, PStV, SbMV, WMV-2, SAPV and ZYMV. A third major cluster is from 'asterids' and includes PVY and LMV. The aphid-transmitted potyviruses that infect grasses also cluster together, although their affinities with the other major clusters are uncertain.

There is also phylogenetic patterning in the vector specificities of the potyviruses. The bymoviruses are the basal group of the Potyviridae and are transmitted by plasmodiophoraceous fungi. The next branch, the rymoviruses, are transmitted by the evolutionarily 'more advanced' vectors, eriophyid mites, whereas the third major branch comprises all the potyviruses transmitted by aphids, the most 'advanced' of the three vector types. Plasmodiophoraceous fungi infect angiosperms, gymnosperms, algae and fungi (Karling, 1968) and, although little is known of their origins, they are probably as ancient as their hosts. By contrast, eriophyid mites are clearly more recent and are a specialized phytophagous group of four-legged acarines. Most infest angiosperms, a few gymnosperms and one a fern. The oldest acarine fossil is from Devonian sandstone (400 Myr old), and specimens resembling most modern genera have been found in Tertiary amber (less than 65 Myr old); however, the oldest eriophyid fossil is only 37 Myr old, but clearly close to modern leaf vagrant eriophyids indicating that the group is older (Jeppson, Keifer & Baker, 1975). Of the three potyvirus vector groups, most is known about aphids. Eastop (1973) analysed the taxonomy, host ranges and specificities of living aphid taxa and concluded that, during evolution, aphids have progressively adapted to different groups of plants. The original hosts of aphids were probably conifers, some adapted to angiosperms, then moved secondarily to monocotyledons, mosses, horsetails and ferns, and some even returned to conifers; the earliest aphids were host specific and fed on trees, whereas more recent ones are heteroecious, polyphagous and many have herbaceous hosts. The extensive fossil record of aphids confirms this picture (Heie, 1987). The oldest fossil is a wing with typical aphid venation from Triassic deposits (around 200 Myr old), when there were probably no angiosperms and land plants were mostly ferns and gymnosperms. Some intact fossil aphids of modern appearance have been found in Jurassic deposits, but many more, mostly in amber, from the Cretaceous (around 80 Myr ago). The modern aphid fauna started to develop in the early to mid Tertiary (40 Myr ago) when angiosperms came to dominate the flora. Modern groups then appeared, notably the Aphididae, which includes most of the vectors of potyviruses (Eastop, 1973; Shaposhnikov, 1987).

Thus, the simplest picture of the evolution of Potyviridae, which is consistent with what is known of the taxonomy of their CPs, and the taxonomies and fossils of their natural hosts and vectors, is that the first potyviruses were bymoviruses infecting grasses or their immediate ancestors, and were transmitted by plasmodiophoraceous fungi; the grasses are a modern group of monocotyledons, which themselves probably evolved from primitive angiosperms within the last 100 Myr (Duvall *et al.*, 1993). A separate lineage of potyviruses with eriophyid mite vectors was then established, which was also confined to monocotyledons, mostly grasses; information about grasses and eriophyids suggests that this may have occurred around 50 +/ − 25 Myr ago. Finally, the aphid-transmitted potyviruses arose, probably after herbaceous angiosperms and modern polyphagous aphids diversified explosively and came to dominate the biota of their respective niches in the Miocene less than 25 Myrs ago (Heie, 1987). The first of the aphid-transmitted potyviruses was probably a virus of solanaceous plants, but potyviruses that infected other angiosperms soon arose. This radiation probably occurred over a relatively short period and, as a result, the branching order of the major groups of potyviruses is uncertain.

Although the paleovirological ideas outlined above are speculation, they are internally consistent, despite being based on data of several independent types. Minimally they establish a likely limit to the age of the Potyviridae, but we hope that they also provide a framework for devising tests of their accuracy. However, they do not produce any clues about the biological links that resulted in the potyviruses and picornaviruses having related 'replicase casettes'.

References

Allison, R.F., Johnston, R.E. & Dougherty, W.G. (1986). *Virology*, **154**, 9–20.
Atreya C.D., Raccah, B. & Pirone, T.P. (1990). *Virology*, **178**, 161–5.
Barnett, O.W. (ed.) (1992). In *Potyvirus Taxonomy, Arch. Virol.*, **5** [Suppl], 435–44.
Bazan, J.F. & Fletterick, R.J. (1988). *Proc. Natl. Acad. Sci. USA*, **85**, 7872–82.
Bos, L. (1970). *Neth. J. Plant Path.*, **76**, 8–46.
Brand, R.J., Burger, J.T. & Rybicki, E.P. (1993). *Arch. Virol.*, **128**, 29–41.
Brandes, J. & Wetter, C. (1959). *Virology*, **8**, 99–115.
Brunt, A.A. (1992). In *Potyvirus Taxonomy*, O.W. Barnett (ed.) *Arch. Virol.*, **5**, [Suppl], 3–16.
Bryan, G.T., Gardner, R.C. & Forster, R.L.S. (1992). *Arch. Virol.*, **124**, 133–46.
Burger, J.T., Brand, R.J. & Rybicki, E.P. (1990). *J. Gen. Virol.*, **71**, 2527–34.
Carrington, J.C., Cary, S.M. & Dougherty, W.G. (1988). *J. Virol.*, **62**, 2313–20.
Carrington, J.C., Cary, S.M., Parks, T.D. & Dougherty, W.G. (1989). *EMBO J.*, **8**, 365–70.
Carrington, J.C., Freed, D.D. & Oh, C.-S. (1990). *EMBO J.*, **9**, 1347–53.
Davidson, A.D., Prols, M., Schell, J. & Steinbiss, H.-H. (1991) *J. Gen. Virol.*, **72**, 989–93.

Dolja, V.V., Boyko, V.P., Agranovsky, A.A. & Koonin, E.V. (1991). *Virology*, **184**, 79–86.

Domier, L.L., Franklin, K.M., Shahabuddin, M., Hellman, G.M., Overmeyer, J.H., Hiremath, S.T., Siaw, M.E.E., Lomonossoff, G.P., Shaw, J.G. & Rhoads, E. (1986). *Nucl. Acids Res.*, **14**, 5417–30.

Domier, L.L., Shaw, J.G. & Rhoades, E. (1987). *Virology*, **158**, 20–7.

Dougherty, W.G. & Carrington, J.C. (1988). *Ann. Rev. Phytopath.*, **26**, 123–43.

Duvall, M.R., Learn, G.H., Eguiarte, L.E. & Clegg, M.T. (1993). *Proc. Natl. Acad. Sci. USA*, **90**, 4641–4.

Eagles, R.M., Gardner, R.C. & Forster, R.L.S. (1990). *Nucl. Acids Res.*, **18**, 7166.

Eastop, V.F. (1973). *Symp. Roy. Ent. Soc. Lond.*, **6**, 157–74.

Edwardson, J.R. (1974). *Florida Agric. Exp. St. Monogr. Ser.*, **4**, 1–398.

Feng, D.-F. & Doolittle, R.F. (1987). *J. Mol. Evol.*, **25**, 351–60.

Feng, D.-F., Johnson, M.S. & Doolittle, R.F. (1985). *J. Mol. Evol.*, **21**, 112–25.

Frenkel, M.J., Jilka, J., McKern, N.M., Strike, P.M., Clark Jr, J.M. Shukla, D.D. & Ward, C.W. (1991). *J. Gen. Virol.*, **72**, 237–42.

Garcia, J.A., Reichmann, J.L. & Lain, S. (1989). *Virology*, **170**, 362–9.

Gibbs, A. (1986). In *The Plant Viruses, Vol. 2; the Rod-Shaped Plant Viruses*. M.H.V. van Regenmortel and H. Fraenkel-Conrat (eds.) Plenum Press, New York, pp. 167–180.

Gibbs, A.J., Dale, M.B., Kinns, H.R. & MacKenzie, H.G. (1971). *Syst. Zool.*, **20**, 417–25.

Gibbs, A.J. & McIntyre, G.A. (1970). *Eur. J. Biochem.*, **16**, 1–11.

Gibbs, A.J., Weiller, G., Armstrong, J.A., Shukla, D.D. & Ward, C.W. (1995). In preparation.

Goldbach, R. (1992). In *Potyvirus Taxonomy*, O.W. Barnett (ed.) *Arch. Virol.*, **5** [Suppl.], 299–304.

Goldbach, R. & Wellink, J. (1988). *Intervirology*, **29**, 260–7.

Gough, K.H. & Shukla, D.D. (1993). *Intervirology.*, **136**, 181–92.

Govier, D.A. & Kassanis, B. (1974). *Virology*, **61**, 420–6.

Hari, V. (1981). *Virology*, **112**, 391–99.

Harrison, B.D. (1985). *Intervirology*, **25**, 71–8.

Harrison, B.D., Finch, J.T., Gibbs, A.J., Hollings, M., Shepherd, R.J., Valenta, V. & Wetter, C. (1971). *Virology*, **45**, 356–63.

Harrison, B.D. & Robinson, D.J. (1988). *Phil. Trans. R. Soc. Lond. B*, **321**, 447–62.

Heie, O.E. (1987). In *Aphids: Their Biology, Natural Enemies and Control, Volume 2A*. A.K. Minks and P. Harrewijn (eds). Elsevier, pp. 367–391.

Hollings, M. & Brunt, A.A. (1981). *CMI/AAB Descript. Plant Viruses*, No. 245.

Jayaram, Ch., Hill, J.H. & Miller, W.A. (1992). *J. Gen. Virol.*, **73**, 2067–77.

Jeppson, L.R., Keifer, H.H. & Baker E.W. (1975). *Mites Injurious to Economic Plants*, California Press, Berkeley and Los Angeles.

Johansen, E., Rasmussen, O.F., Heide, M. & Borkhardt, B. (1991). *J. Gen. Virol.*, **72**, 2625–32.

Karling, J.S. (1968). *The Plasmodiophorales*, Hafner, New York.

Kashiwazaki, S., Minobe, Y. & Hibino, H. (1991). *J. Gen. Virol.*, **72**, 989–93.

Kashiwazaki, S., Minobe, Y., Minobe, T. & Hibino, H. (1990). *J. Gen. Virol.*, **71**, 2781–90.

Keese, P. & Gibbs, A. (1992). *Proc. Natl. Acad. Sci. USA*, **89**, 9489–93.

Kong, L-J., Fang, R-X., Chen, Z-H. & Mang, K-Q. (1990). *Nucl. Acids Res.*, **18**, 5555.

Koonin, E.V. (1991). *J. Gen. Virol.*, **72**, 2197–206.

Lain, S., Martin, M.T., Reichmann, J.I. & Garcia, J.A. (1991). *J. Virol.*, **18**, 7003–6.

Lain, S., Reichmann, J.I., & Garcia, J.A. (1989). *Virus Res.* **13**, 157–72.

Lain, S., Reichman, J.L., Mendez, E. & Garcia, J.A. (1988). *Virus Res.*, **10**, 325–42.

Maiss, E., Timpe, U., Brisske, A., Jelkmann, W., Casper, R., Himmler, G., Mattanovich, D. & Katinger, H.W.D. (1989). *J. Gen. Virol.*, **70**, 513–24.

Makkouk, K.M. & Singh, M. (1992). In *Potyvirus Taxonomy*, O.W. Barnett (ed.) *Arch. Virol.* 5 [Suppl], 177–82.

McKern, N.M., Edskes, H.K., Ward, C.W., Strike, P.M., Barnett, O.W. & Shukla, D.D. (1991). *Arch. Virol.*, **119**, 25–35.

Milne, R.G. (1988). In *The Plant Viruses. Vol.4*, R.G. Milne (ed), Plenum Press, New York pp. 3–50.

Morosov, S.Y., Dolja, V.V. & Atabekov, J.G. (1989). *J. Mol. Evol.*, **29**, 52–62.

Murphy, J.F., Rhoads, R.E., Hunt, A.G. & Shaw, J.G. (1990). *Virology*, **178**, 285–8.

Nicolas, O. & Laliberte, J-F. (1992). *J. Gen. Virol.*, **73**, 2785–93.

Palkovics, I., Burgayan, J. & Balazs, E. (1993). Comparative sequence analysis of four complete primary structures of plumpox virus strains. *Virus Genes*, **7**, 339–47.

Peerenboom, E., Prols, M., Schell, J., Steinbi, B. & Davidson, A.D. (1992). *J. Gen. Virol.*, **73**, 1303–8.

Quemada, H., Sieu, L.C., Siemieniak, D.R., Gonsalves, D. & Slighton, J.L. (1990). *J. Gen. Virol.*, **71**, 1451–60.

Robaglia, C., Durand-Tardif, M., Tronchet, M., Boudazin, G., Astier-Manifacier, S. & Casse-Delbart, F. (1989). *J. Gen. Virol.*, **70**, 935–47.

Rodriguez-Cerezo, E. & Shaw, J.G. (1991). *Virology*, **185**, 572–9.

Rossman, M.G., Arnold, E., Erickson, J.W., Frankenberger, E.A., Griffith, J.P., Hecht, H.J., Johnson, J.E., Kamer, G., Luo, M., Mosser, A.G., Rueckert, R.R. Sherry, B. & Vriend, G. (1985). *Nature*, **317**, 145–53.

Rothschild, M. & Clay, T. (1952). *Fleas, Flukes and Cuckoos: A Study of Bird Parasites*. Collins, London.

Rybicki, E.P. & Shukla, D.D. (1992). In *Potyvirus Taxonomy*, O.W. Barnett (ed.) *Arch. Virol.* 5 [Suppl], 139–70.

Saitou, N. & Nei, M. (1987). *Mol. Biol. Evol.*, **4**, 406–25.

Salomon, R. (1989). *Res. Virol.*, **140**, 453–60.

Shahabuddin, M., Shaw, J.G. & Rhoads, R.E. (1988). *Virology*, **163**, 635–7.

Shaposhnikov, G.Ch. (1987). In *Aphids: Their Biology, Natural Enemies and Control, Volume 2A*. A.K. Minks and P. Harrewijn (eds). Elsevier, pp. 409–414.

Sherwood, J.L. & Fulton, R.W. (1982). *Virology*, **119**, 150–8.

Shukla, D.D. & Ward, C.W. (1988). *J. Gen. Virol.*, **69**, 2703–10.

Shukla, D.D. & Ward, C.W. (1989). *Adv. Virus Res.*, **36**, 273–314.

Shukla, D.D., Strike, P.M., Tracy, S.L., Gough, K.H. & Ward, C.W. (1988). *J. Gen. Virol.*, **69**, 1497–508.

Shukla, D.D., Jilka, K., Tosic, M. & Ford, R.E. (1989a). *J. Gen. Virol.*, **70**, 13–23.

500   C.W. Ward et al.

Shukla, D.D., Tosic, M., Jilka, J., Ford, R.E., Toler, R.W. & Langham,
    M.A.C. (1989b). Phytopathology, 79, 223–9.
Shukla, D.D., Tribbick, G., Mason, T.J., Hewish, D.R., Geysen, H.M. &
    Ward, C.W. (1989c). Proc. Natl. Acad. Sci. USA, 86, 8192–6.
Shukla, D.D., Frenkel, M.J. & Ward, C.W. (1991). Can. J. Plant Path.,
    13, 78–91.
Srisink, (1989). MSc Dissertation, University of Queensland, St Lucia,
    Australia, pp. 111.
Teakle & Grylls, (1973). Aust. J. Agric. Res., 24, 465–77.
Teakle, D.S., Shukla, D.D. & Ford, R.E. (1989). Sugarcane mosaic virus.
    AAB Descriptions of Plant Viruses No. 342.
Teycheney, P.Y., Tavert, G., Delbos, R., Ravelondandro, M. & Dunez, J.
    (1989). Nucl. Acids Res., 23, 10115–16.
Tracy, S.L., Frenkel, M.J., Gough, K.H., Hanna, P.J. & Shukla, D.D.
    (1991). Arch. Virol., 122, 249–61.
Tremblay, M-F., Nicolas, O., Sinha, R.C., Lazure, C. & Laliberte, J-F.
    (1990). J. Gen. Virol., 71, 2769–72.
Usugi, T., Kashwazaki, S., Omura, T. & Tsuchizaki, T. (1989). Ann.
    Phytopath. Soc. Japan, 55, 26–31.
Vance, V.B., Moore, D. Turpen, T.H., Bracker, A. & Hollowell, V.C.
    (1992). Virology, 191, 19–30.
Verchot, J., Koonin, E.V. & Carrington, J.C. (1991). Virology, 185, 527–35.
Vetten, H.J., Lesemann, D.E. & Maiss, E. (1992). In Potyvirus Taxonomy,
    O.W. Barnett (ed.) Arch. Virol., 5 [Suppl], 415–431.
Ward, C.W., McKern, N.M., Frenkel, M.J. & Shukla, D.D. (1992). In
    Potyvirus Taxonomy, O.W. Barnett (ed.) Arch. Virol. 5 [Suppl], 283–97.
Ward, C.W. & Shukla, D.D. (1991). Intervirology, 32, 269–96.
Ward, C.W. & Shukla, D.D. (1993). Structure and variation of potyviruses.
    In Virology in the Tropics, Rishi, N., Ahuja, K.L. and Singh, B.P.
    (eds), Malhotra Publishing House, New delhi, pp. 43–61.
Wefels, E., Sommer, H., Salamini, F. & Rohde, W. (1989). Arch. Virol.,
    107, 123–234.
Xiao, X.W., Frenkel, M.J., Teakle, D.S., Ward, C.W. & Shukla, D.D.
    (1993). Arch. Virol., 136, 3811–7.
Yeh, S-D., Jan, F-J., Chian, C-H., Doong, T-J., Chen, M-C., Chung, P-H.
    & Bau, H-J. (1992). J. Gen. Virol., 73, 2531–41.
Zhou, X-R., Fang, R-X., Wang, C-Q. & Mang, K-Q. (1990). Nucl. Acids
    Res., 18, 5554.

33

Evolution of alphaviruses
SCOTT C. WEAVER

Alphavirus structure, replication and transmission cycles

Structure and replication

Alphavirus is a genus of arthropod-borne viruses (arboviruses) in the family *Togaviridae* (Calisher *et al.*, 1980; Calisher & Karabatsos, 1988). These viruses have unsegmented, single-stranded RNA genomes of 11–12 kb which are of positive or messenger sense. Alphavirus genomic RNA encodes four non-structural proteins, designated nsP1–4, as well as three structural proteins: the capsid, and two envelope glycoproteins, E1 and E2 (Fig. 33.1). A sub-genomic 26S RNA species, which encodes only the structural proteins, is also produced in infected cells.

Alphaviruses replicate in the cytoplasm of infected cells. Entry into vertebrate cells occurs via receptor-mediated endocytosis (Kielian & Helenius, 1986). The high-affinity laminin receptor serves as a mammalian and mosquito cell (*in vitro*) receptor for the alphavirus, Sindbis (SIN) (Wang *et al.*, 1992), whereas other protein receptors for SIN have been identified in mouse neural (Ubol & Griffin, 1991) and chicken cells (Wang *et al.*, 1991). Wang *et al.* (1992) believe that the wide host range of Sindbis virus is the result of its utilization of a highly conserved receptor, found in a variety of vertebrates and in mosquito cells (laminin receptor), as well as its use of multiple receptors. The mechanism of alphavirus entry into mosquito cells has not been identified, although the very specific infectivity patterns of some alphaviruses for their mosquito vectors suggest use of a less conserved protein receptor during initial infection of the midgut.

Genomic alphavirus RNA is translated by cellular components to produce the non-structural proteins, and also serves as a template for minus-strand RNA synthesis involving the non-structural proteins.

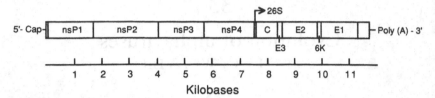

Fig. 33.1. Cartoon depicting the organization of the alphavirus genome, showing the location of genes encoding the non-structural proteins (nsP1–4) and structural [capsid (C) E3, E2, 6K and E1] proteins. The 26S sub-genomic RNA is identical to the 3′ one-third of the genomic RNA and encodes only the structural proteins.

Minus-strand RNA is used as a template for production of progeny genomic plus-strand RNA. The sub-genomic 26S RNA is also trans-lated as a structural polyprotein; the capsid protein is cleaved in the cytoplasm and the remaining polyprotein is processed and cleaved in the endoplasmic reticulum and Golgi complex to yield the E1 and E2 glycoproteins, which are inserted into the plasma membrane as a heterodimer. The E3 protein has been found in virions of only one alphavirus examined, Semliki Forest (SF) virus. Following encapsidation of genomic plus-strand RNA in the cytoplasm, enveloped virions 60–65 nm in diameter mature when nucleocapsids bud through the plasma membrane (Strauss & Strauss, 1986; Schlesinger & Schlesinger, 1990).

Most alphaviruses replicate in, and cause extensive cytopathic effects (CPE) in, nearly all cultured vertebrate cells. In contrast, infection of cultured mosquito cells leads to persistent infection and is usually not accompanied by CPE. Alphavirus maturation in mosquito cell cultures occurs within cytoplasmic membrane-bound 'virus factories' which are presumably extruded from infected cells to release progeny virus (Brown & Condreay, 1986). However, 'virus factories' have not been observed in cells in infected mosquitoes, where virus maturation occurs via plasma membrane budding (Houk *et al.*, 1985; Weaver, Scott & Lorenz, 1990).

Transmission cycles

A typical transmission cycle for mosquito-borne alphaviruses is depicted in Fig. 33.2. Transmission cycles of alphaviruses include replication in both vertebrate and invertebrate (vector) hosts. The vast majority of alphaviruses are transmitted among avian or small mammalian hosts

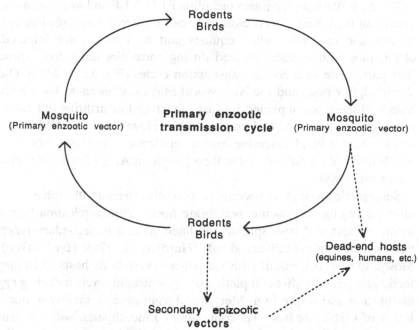

Fig. 33.2. Cartoon depicting a generalized alphavirus transmission cycle. Solid arrows indicate portions of the enzootic transmission cycle, while dashed arrows show epizootic transmission cycles which occur for some alphaviruses.

by mosquito vectors; the only exceptions to primary transmission by mosquito vectors are Fort Morgan, Bijou Bridge and Buggy Creek viruses, which are transmitted by the cliff swallow bug, *Oeciacus vicarius* (Calisher *et al.*, 1980; Calisher & Karabatsos, 1988). Most alphaviruses utilize mosquito vectors in the genus *Culex*. In the New World, mosquitoes in the subgenus *Culex (Melanoconion)* are the principal enzootic vectors of most tropical alphaviruses in the eastern equine encephalomyelitis (EEE) and Venezuelan equine encephalitis (VEE) complexes. The only alphavirus transmitted by *Anopheles* mosquitoes is o'nyong-nyong (ONN) virus. Usually, one or a few mosquito species serve as primary enzootic vectors of alphaviruses in a given geographic region. The high degree of specificity in virus-vector relationships, characterized by wide variation in blood meal titre infection thresholds for different mosquito species (or even geographic strains), contrasts with the broad vertebrate host specificity exhibited by many alphaviruses (Hardy, 1988).

The New World alphaviruses including EEE, VEE and western equine encephalitis (WEE) virus occasionally cause extensive encephalitic epizootics and epidemics when equines and humans become infected; often, mosquito vectors involved during epizootics differ from those that participate in enzootic transmission cycles (Fig. 33.2). Many Old World alphaviruses and the New World Mayaro virus cause less severe human disease accompanied by rash, fever and/or arthritis but rarely encephalitis (Peters & Dalrymple, 1990). However, some Old World viruses have caused extensive human epidemics, such as ONN virus which infected an estimated 2 million people in Africa from 1959–1962 (Johnson, 1988).

Susceptible mosquitoes become persistently infected with alphaviruses after engorging on viraemic vertebrate hosts. Virus replication begins in the midgut and later spreads via haemolymph to secondary target tissues including the salivary glands (Hardy et al., 1983; Hardy, 1988). Mosquito vectors transmit virus via saliva to vertebrate hosts when they feed again, usually after completion of a gonotrophic cycle including egg maturation and oviposition. Mechanical transmission, involving inoculation of vertebrate hosts via contaminated mouthparts, without virus replication in the mosquito vector, has been documented occasionally. Upon infection, some susceptible vertebrate hosts develop viraemias sufficient in magnitude for subsequent infection of engorging mosquito vectors, to complete the transmission cycle.

Mechanisms for alphavirus survival during seasons not conducive to active transmission (e.g. dry seasons in the tropics and winter in temperate regions) are poorly understood. Adult mosquitoes that survive such seasons in quiescent states or diapause may be capable of reintroducing virus into the transmission cycle upon blood feeding (Hardy, 1988). However, alphavirus titres in mosquito saliva and rates of transmission to susceptible vertebrates decline with time during persistent infections (Chamberlain & Sudia, 1961; Weaver et al., 1990). Vertical (transovarial) transmission from infected mosquitoes to offspring has been demonstrated for two alphaviruses, Ross River (Kay, 1982) and WEE (Fulhorst et al., 1994). Because mosquito vectors of some alphaviruses (e.g. Culiseta melanura which transmits EEE virus in North America) overwinter as larvae, persistent infections in vertebrate or insect hosts may be involved in virus overwintering; recent information suggests that EEE virus overwinters in some temperate transmission foci (Weaver et al., 1994) observations.

Evolutionary relationships among alphaviruses

Relationships derived using serological methods

The first evidence suggesting evolutionary relationships among alphaviruses came from serological studies. Reactivity between antisera
prepared against a given arbovirus and heterologous viral antigens
formed the basis for the original designation of type A arboviruses;
these viruses were later reclassified to form the genus *Alphavirus*
(Calisher *et al.*, 1980). Antigenic similarities among alphaviruses remain
valuable for rapid serological identification and characterization of newly
discovered virus strains. The first formal classification of alphaviruses
included 26 members (Calisher *et al.*, 1980); currently, 36 alphaviruses
are placed in seven complexes, based on antigenic relationships (Table
33.1). Several complexes such as EEE, Middelberg (MID), Ndumu
(NDU) and Barmah Forest (BF) contain only one species, whereas
WEE includes six species (Calisher *et al.*, 1988).

Relationships derived from molecular data

The first genetic evidence of evolutionary relationships among alphaviruses was obtained by Wengler, Wengler and Filipe (1977) using
an RNA–RNA hybridization technique. They showed that SF, SIN,
chikungunya (CHIK) and ONN viruses share detectable RNA sequence
homology, while T_1RNA fingerprinting studies showed common oligonucleotides among different strains of each virus, but not among these
four viruses.

The first use of molecular sequence data to determine evolutionary
relationships among alphaviruses was reported by Bell *et al.* (1984), who
examined N-terminal amino acid sequences of E1 and E2 glycoproteins
obtained by the Edman degradation technique. They concluded that all
alphaviruses are closely related and probably descended from a common
ancestor. The first evolutionary tree for representative alphaviruses was
generated from these sequences; however, more recent work has revised
the relationships depicted in that tree.

Recently, the development of techniques for rapid determination of
nucleotide sequences of viral RNA and cDNA clones has revolutionized
the study of alphavirus evolution. Complete genomic RNA sequences
have been determined for the alphaviruses ONN (Levinson, Strauss
& Strauss, 1990), Ross River (RR) (Faragher *et al.*, 1988), Semliki

Table 33.1. *Classification of alphaviruses*[a]

Complex	Species	Subtype	Variety	Distribution
BF	BF			Australia
EEE	EEE		North American	North America, Caribbean
			South American	South, Central American
MID	MID			Africa
NDU	NDU			Africa
SF	SF			Africa, former USSR
CHIK	CHIK	(I) CHIK	several	Africa, Asia
		(II) ONN	Igbo Ora	Africa
	GET	(I) GET		Asia, Australia
		(II) SAG		Asia
		(III) BEB		Malaysia
		(IV) RR		Australia, Oceania
	MAY	(I) MAY		South America
		(II) UNA		South America
VEE	VEE	(I) VEE	AB	South, Central, North America
			C	South, Central America
			D	South, Central America
			E	Central America
			F	Brazil
		(II) EVE		Florida, USA
		(III) MUC	A (MUC)	Brazil
			B (TON)	French Guiana
			B (BB)	Western North USA
			C (71D-1252)	Peru
		(IV) PIX		Brazil
		(V) CAB		French Guiana
		(VI) (AG80-663)		Argentina

Table 33.1. (*cont.*)

Complex	Species	Subtype	Variety	Distribution
WEE	WEE		several	North, South America
			Y62–33	Russia
	HJ			Eastern North America
	FM			Western North America
	BC			Oklahoma, USA
	SIN	(I) SIN		Africa, Asia, Europe, Australia
		(II) BAB		Africa
		(III) OCK		Europe
		(IV) WHA		New Zealand
		(V) KZL		Azerbaijan
	AURA			South America

Note: [a] Abbreviations are as follows: Aura – AURA; Babanki – BAB; Barmah Forest – BF; Bebaru – BEB; Bijou Bridge – BB; Cabassou – CAB; chikungunya – CHIK; eastern equine encephalitis – EEE; Everglades – EVE; Fort Morgan – FM; Getah – GET; Highlands J – HJ; Kyzylagach – KZL; Mayaro – MAY; Middelburg – MID; Mucambo – MUC; Ndumu – NDU; Ockelbo – OCK; o'nyong nyong – ONN; Pixuna – PIX; Ross River – RR; Sagiyama – SAG; Semliki Forest – SF; Sindbis – SIN; Tonate – TON; Una – UNA; Venezuelan equine encephalitis – VEE; western equine encephalitis – WEE; Whataroa – WHA. Information is from Calisher *et al.* (1980) and Calisher & Karabatsos (1988).

Forest (SF) (Garoff *et al.*, 1980*a,b*; Takkinen, 1986), SIN (Strauss, Rice & Strauss, 1984), VEE (subtype IAB: Kinney *et al.*, 1989) and EEE virus (Chang & Trent, 1987; Volchkov, Volchkova & Netesov, 1991); partial sequences have also been determined for WEE (Hahn *et al.*, 1988) and MID (Strauss, Rice & Strauss, 1983) viruses. Complete amino acid sequences for the structural and non-structural proteins, deduced from these nucleotide sequences, have been used to construct alphavirus phylogenetic trees. Unrooted trees of the capsid, E2 and E1 structural proteins first placed the New World EEE, WEE and VEE viruses in one cluster along with the Old World SIN virus; the remaining Old World viruses (ONN, RR, SF) fell into a distinct group (Levinson *et al.*, 1990). The different placement of WEE virus in capsid vs. E1 and E2 trees supported the conclusion that WEE is a recombinant virus descended from EEE- and SIN-like ancestors (Hahn *et al.*, 1988). The recombination event, presumably a copy choice switching of templates by the viral polymerase, occurred within the E3 region of the genome.

Amino acid sequence homology identified between the nsP4 protein of SIN virus, and polymerases of several other positive strand RNA viruses (Haseloff *et al.*, 1984; Ahlquist *et al.*, 1985; Koonin, 1991), has now made it possible to construct a rooted phylogenetic tree of alphaviruses. This tree (Fig. 33.3) confirms the monophyletic nature of alphaviruses; it places the New World alphaviruses in one cluster and those from the Old World in another. Similar phylogenetic trees generated with nsP1 and nsP2 amino acid sequences also include Old and New World alphavirus groups (Weaver *et al.*, 1993*b*). This topology implies that alphaviruses originated in either the Old or New World, followed by an early intercontinental introduction. A second introduction into the New World of an ancestor of Mayaro and Uno viruses probably occurred later (see below).

The rooting of the alphavirus tree (Fig. 33.3) differs from the implied rooting of previous capsid and E2 trees (trees generated from the structural protein genes were unrooted because homologous outgroup sequences have not been identified) (Levinson *et al.*, 1990). Because unrooted trees generated with most phylogenetic analysis programs operate under the assumption of a molecular clock (constant rate of amino acid substitution), they arrange the internal branches such that all taxa have similar total branch lengths (midpoint rooting). The fact that the rooted nsP4 tree (Fig. 33.3) is arranged differently from the unrooted structural protein trees of Levinson *et al.* (1990) implies that proteins of alphaviruses may not evolve at a uniform rate.

Distances (branch lengths) in the rooted alphavirus nsP4 tree (Fig.

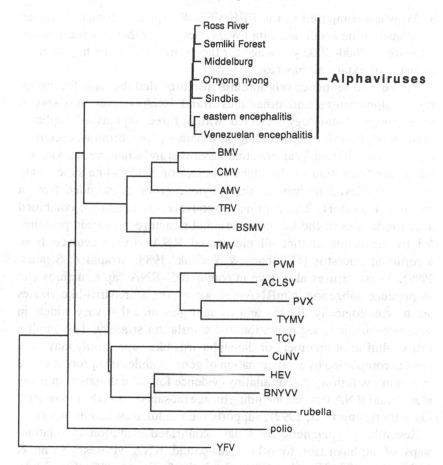

Fig. 33.3. Phylogenetic tree of viruses in the alphavirus-like superfamily obtained from alphavirus nsP4 amino acid sequences and homologous polymerase sequences of other plus-strand RNA viruses. The homologous sequences of poliovirus and yellow fever viruses were included as outgroups. Abbreviations for alphaviruses are found in Table 33.1; AMV: alfalfa mosaic virus; ACLSV: apple chlorotic leaf spot virus; BMV: brome mosaic virus; BNYVV: beet necrotic yellow vein virus; BSMV: barley stripe mosaic virus; CMV: cucumber mosaic virus; CuNV: cucumber necrosis virus; HEV: hepatitis E virus; PVM – potato virus M; PVX – potato virus X; TCV – turnip crinkle virus; TMV – tobacco mosaic virus; TRV – tobacco rattle virus; TYMV – turnip yellow mosaic virus; YFV – yellow fever virus. Branch lengths reflect the number of amino acid substitutions separating viruses from hypothetical ancestors. This tree was obtained using the TREE program of Feng and Doolittle (1990).

33.3), when compared to the EEE-VEE divergence estimate obtained from nucleotide sequence data (Weaver *et al.*, 1992*a*) provide a rough estimate of 2000–3000 years ago for the occurrence of the hypothetical ancestor of extant alphaviruses.

Amino acid sequence relationships also provided the basis for grouping of alphaviruses and other plus-strand RNA viruses into several supergroups. Homology detected within three regions of replicase proteins of plant viruses, including the tobamo-, ilar-, bromo-, cucumo-, furo-, tobra-, hordei-, carmo-, tombusviruses and alfalfa mosaic viruses, led to the formation of the alphavirus-like or Sindbis-like supergroup which is believed to form a monophyletic group (descended from a common ancestor). These primary sequence relationships, combined with similarities in the three-dimensional structures of capsid proteins, led to the proposal that all plus-strand RNA viruses evolved from a common ancestor (Goldbach & Wellink, 1988; Strauss & Strauss, 1988). These viruses also have in common 5' RNA cap structures and all produce subgenomic mRNAs. However, the alphavirus-like viruses are morphologically diverse and contain genomes that vary widely in size, organization, segmentation and translation strategy. This implies that evolution of members of the alphavirus-like superfamily may have been accompanied by recombination of gene modules in a process called 'modular evolution'. Accumulating evidence for recombination in many plus-strand RNA viruses, including brome mosaic virus in the alphavirus-like supergroup (Lai, 1992), supports the modular evolution theory.

Recently, phylogenetic trees have confirmed evolutionary relationships of alphaviruses to other plus-strand RNA viruses (Xiong & Eickbush, 1990; Koonin, 1991; Doolittle & Feng, 1992; Fig. 33.3). The closest relatives of alphaviruses are probably alfalfa mosaic, barley stripe mosaic, brome mosaic, cucumber mosaic, tobacco mosaic and tobacco rattle viruses (Fig. 33.3). This suggests that alphaviruses may have evolved from a plant virus ancestor. Another possibility is that the alphavirus-like superfamily evolved from an insect virus ancestor. Evolution from an insect virus is supported by the following information: (1) the host ranges of both plant and animal viruses in the Sindbis-like supergroup overlap in insects (vectors); (2) insects are hosts for a wide variety of RNA viruses; (3) segmented plus-strand RNA genomes are found primarily among plant and insect (not animal) viruses (Goldbach & Wellink, 1988). Furthermore, insects are by far the most diverse and abundant eukaryotes, surpassing all other terrestrial animals in numbers comprising over half of all eukaryotic species (Daly, Doyen & Ehrlich,

1978); they therefore may represent the optimal host taxon in which virus diversification could occur.

More distantly related to alphaviruses are animal picornaviruses and coliphages. Polymerase sequences of members of the plus-strand RNA virus monophyletic group that includes alphaviruses also show homology with those of some minus-strand RNA viruses, retroviruses and transposable elements. Doolittle and Feng (1992) as well as Xiong and Eickbush (1990) have suggested that RNA viruses are as old or older than retroelements, which probably evolved about 1.5–2.0 billion years ago when endosymbiotic prokaryotes invaded primitive eukaryotes (Doolittle, Anderson & Feng, 1989).

Evolutionary relationships within alphavirus complexes

VEE complex

Many of the earliest studies of genetic relationships among alphaviruses utilized the T_1 RNA oligonucleotide fingerprinting technique. This method is useful for comparing genomes of closely related viruses with 90% or greater nucleotide sequence identity (Kew, Nottay & Obijeski, 1984). Thus, fingerprinting is useful for determining relationships of alphaviruses within a given species, subtype or variety.

Fingerprinting of RNA from alphaviruses in the VEE complex first demonstrated a close genetic relationship between epizootic varieties IAB and IC, and enzootic variety ID; a more distant relationship was observed with enzootic variety IE viruses. Subtypes II, III and IV are distantly related to subtype I viruses and to each other (Trent *et al.*, 1979). Additional studies identified a close relationship among varieties IIIA, IIIB and IIIC, as well as a distant relationship between subtype V and other VEE viruses (Kinney, Trent & France, 1983). Later, Rico-Hesse *et al.* (1988) showed a closer genetic relationship between Colombian variety ID VEE viruses and some variety of IC epizootic viruses, than between Panamanian and Colombian ID isolates.

Recently, nucleotide sequence data from representative isolates of VEE viruses have been used to construct a phylogenetic tree of the VEE complex (Weaver *et al.*, 1992a; Fig. 33.4). Some VEE antigenic subtypes, such as III, form monophyletic groups, while subtype I viruses are interspersed with subtype II Everglades virus and subtype III in the tree. Phylogenetic analysis indicated that epizootic variety IAB and IC viruses evolved two or more times from enzootic variety ID-like viruses during this century. Everglades (variety II) virus also shares a ID-like ancestor which probably colonized Florida during the last century from South or Central America, while Bijou Bridge virus (variety IIIB) probably resulted from recent colonization of western North America by a Tonate-like South American virus. Multiple emergences of epizootic viruses during this century implies that these equine-virulent VEE viruses may arise again in the future from variety ID, and possibly subtype II Everglades virus (Weaver *et al.*, 1992*a*).

EEE complex

Initial serological studies of EEE virus strains from North, Central and South America, as well as the Caribbean, revealed the presence of two

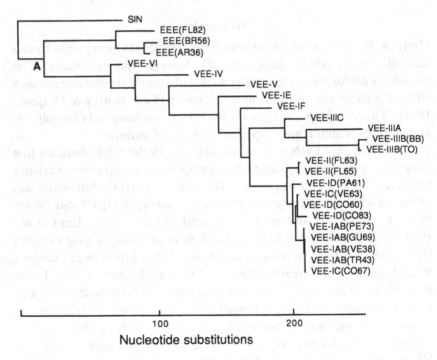

Fig. 33.4. Phylogenetic tree depicting relationships among members of the VEE complex derived from nucleotide sequences (Weaver *et al.*, 1992*a*). Homologous sequences of SIN and EEE virus are used as an outgroup to root the tree. Scale below shows nucleotide substitutions separating viruses from ancestral nodes. Virus strains are listed by abbreviated country or state (USA) followed by year of isolation. Node A indicates hypothetical ancestor of EEE and VEE complex viruses.

antigenic groups (Casals, 1964) which now comprise the North and South American antigenic varieties (Calisher *et al.*, 1980); most strains isolated from North America and the Caribbean belong to the North American variety while those from South and Central America are members of the South American variety (Table 33.1). Walder, Rosato and Eddy (1981) later compared EEE strains isolated throughout the Americas and identified seven distinct patterns of structural protein migration by SDS–PAGE. However, this grouping, based on protein migration, does not correlate with antigenic relationships (Casals, 1964) or any known biological or epidemiological characteristics (Walder *et al.*, 1981).

The lack of serological distinction among North American variety EEE virus strains (Casals, 1964) is also reflected in genetic conservation among these viruses. Roehrig *et al.* (1990) first demonstrated genetic conservation by comparing oligonucleotide fingerprints of North American strains isolated from 1933–1982. Later, nucleotide sequences from additional isolates suggested a single North American EEE lineage or monophyletic group, which also includes isolates from the Caribbean. More recent work has shown that two closely related groups or sublineages have been circulating in North America since the early 1970s. Isolation of two EEE virus genotypes, representing these two sublineages, in a sentinel bird demonstrated that these sublineages are sometimes sympatric and competing in nature (Weaver *et al.*, 1994).

Two additional EEE lineages or groups occur in South and Central America; one includes isolates from the Amazon basin in Brazil and Peru, while the other has been isolated from Argentina to Panama (Fig. 33.5). These two South/Central American lineages diverged roughly 450 years ago, while the North American lineage diverged from a hypothetical ancestor of the South/Central American lineages *c*. 1000 years ago. Weaver *et al.* (1992*a*) estimated that the EEE and VEE complexes diverged *c*. 1400 years ago from a common ancestral alphavirus in South or Central America.

Each of the three main EEE monophyletic groups or lineages also includes closely related sublineages. Geographically independent evolution of viruses within a given lineage can occur for several years; however, overall relationships (based on time of isolation) within the main lineages indicate that viruses are periodically exchanged among transmission foci. This could reflect 1) periodic extinction of EEE viruses in some transmission foci and geographic regions, followed by reintroduction of a consensus genotype and/or 2) displacement of regionally divergent genotypes by a more fit consensus genotype,

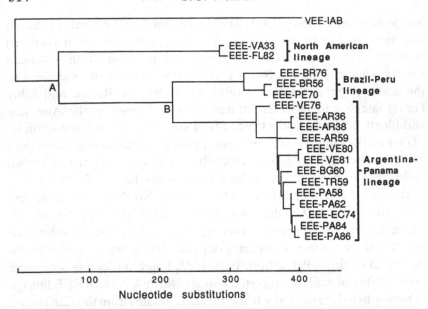

Fig. 33.5. Phylogenetic tree depicting relationships among members of the EEE complex derived from nucleotide sequences (Weaver *et al.*, in preparation). Homologous sequences of the Trinidad donkey (variety IA) strain of VEE virus are used as an outgroup to root the tree. Nodes A and B represent hypothetical ancestral EEE viruses. Virus strains are listed by abbreviated country or state (USA) followed by year of isolation. Scale below indicates numbers of nucleotide substitutions represented in branch lengths.

following competition. Alphavirus lineages may frequently undergo extinction due to instability inherent in their complex, two host transmission cycles (Weaver *et al.*, 1992c; Scott *et al.*, 1994).

Rates of EEE virus evolution have been estimated for the Argentina–Panama and North American lineages (Weaver *et al.*, 1994). In limited portions of the nsP4, E2 and 3' untranslated regions of the genome, the Argentina–Panama lineage underwent a relatively uniform evolutionary rate of 4.3×10^{-4} substitutions per nucleotide per year from 1938–1986. The complete 26S region of the North American viruses underwent an average evolutionary rate of 1.6×10^{-4} substitutions per nucleotide per year from 1945–1991. However, the North American evolutionary rate appears to have increased roughly 20-fold (from 2×10^{-5} to 4×10^{-4}) concurrent with divergence of two sublineages during the early 1970s. This non-uniform rate of EEE virus evolution contrasts with the constant, higher rate of evolution reported for the NS genes of influenza viruses (Buonagurio *et al.*, 1986). The neutral theory of molecular evolution predicts that genomes acquire primarily neutral or synonymous nucleotide substitutions in a clock-like, constant manner over time (Kimura, 1983). However, non-uniform rates of EEE virus evolution indicate that some alphaviruses may not follow the molecular clock pattern of evolution when studied over a relatively short time frame.

The reason(s) for the increased evolutionary rate of North American EEE virus during the 1970s have not been determined. One possibility is that virus dispersal and/or population sizes have been affected by changes in mosquito and/or vertebrate host populations. Populations of most passerine birds that breed in forests of eastern North America and migrate to the neotropics declined from 1978–1987 after a period of stability or increasing abundance (Robbins *et al.*, 1989). Since many of these birds serve as hosts for EEE virus (Scott & Weaver, 1989), reductions in populations of migrant birds may have affected dispersal of viruses among transmission foci and/or amounts of virus circulating in North America.

Another possible explanation for the change in the EEE virus evolutionary rate is that divergence into two sublineages affected rates of evolution in a manner similar to that proposed by the punctuated equilibrium theory of evolution. This theory predicts that organisms (eukaryotic) will undergo rapid evolutionary change during peripatric speciation events (Eldredge & Gould, 1972); founder effects operating in small, peripheral, isolated populations will cause rapid genetic drift, leading to reproductive isolation from the parent population.

Haploid alphaviruses are primarily asexual, and recombination has only been detected in the evolution of one alphavirus, WEE (Hahn *et al.*, 1988); therefore reproductive isolation mechanisms believed to accompany speciation of sexual eukaryotes cannot function. However, spatial or ecological partitioning may be required to prevent competitive exclusion of incipient or parental alphavirus lineages. Potential partitioning mechanisms for alphavirus lineages must therefore be examined before this hypothesis can be evaluated. Frequent geographic isolation in isolated transmission foci, accompanied by reductions in the virus population size, could lead to genetic drift, accelerating the rate of EEE evolution. Changes in migrant avian host populations (see above) could have altered patterns of virus dispersal during the 1970s and 1980s, resulting in greater isolation of genetically distinct viruses, and a reduction in the effective virus population size. Incipient EEE virus genotypes (sublineages) originating within geographically isolated transmission foci may be periodically reintroduced to occupy the overall geographic range of the virus, resulting in an apparent sympatric distribution. Sympatric divergence could also occur if some form of ecological partitioning prevents competitive exclusion. More detailed information on patterns of EEE virus dispersal and modes of competition is needed to evaluate these hypotheses. If worldwide degradation of bird habitats continues to alter avian populations, the evolution of EEE and other alphaviruses utilizing these hosts could be unpredictably affected.

Another possible mechanism for an increase in the EEE virus evolutionary rate, coincident with divergence of distinct sublineages, is a reduction in the constraining effect of selective pressure associated with diminished population sizes. A weakly selected trait is affected by selection (i.e. changes in frequency) only if the population is relatively large (Wright, 1937). Therefore, reductions in population size might allow for the accumulation of mutations previously suppressed by weak selective pressure. Most of the nucleotide substitutions that have accumulated in the two North American EEE virus sublineages are 'neutral' or synonymous (Weaver et al., 1994). However, natural selection for specific primary RNA structure to accommodate codon usage preferences, RNA secondary structure, or packaging requirements may occur. More information on the role of primary RNA sequence in the fitness of alphavirus genotypes is needed to evaluate this hypothesis.

WEE complex

Three species within the WEE complex have been studied using genetic methods. Trent and Grant (1980) compared isolates of WEE virus and Highlands J (HJ) virus using RNA oligonucleotide fingerprinting. These two species were relatively distantly related, while conspecific isolates were genetically very similar. WEE virus strains isolated from 1941–1975 in North and South America differed by no more than 9 T_1 resistant oligonucleotides, implying a single WEE virus lineage and/or a slow rate of evolution.

Rentier-Delrue and Young (1980) first studied isolates of SIN virus using a heterologous viral RNA-RNA hybridization technique. They identified two main SIN groups: a European–African group and a Indian–Far Eastern–Australian group. Later, Olson and Trent (1985) identified four distinct (Paleoarctic, Ethiopian, Oriental and Australian) genotypes using T_1 RNA oligonucleotide fingerprinting. Recently, Shirako et al. (1991) used nucleotide sequence data to show that Ockelbo

(OCK) virus is closely related to the prototype AR339 strain of SIN virus (5.7% nucleotide sequence divergence). They identified two groups of SIN viruses: an Indian-Australian group and one including isolates from South Africa and Europe. Shirako *et al.* (1991) suggested that OCK virus, which is more closely related to South African strains of SIN virus than to the prototype strain AR339 from Egypt, may have been introduced from South Africa into northern Europe via migratory birds.

Sequence data from other WEE complex species including Aura, Highlands J, Fort Morgan and Buggy Creek would be useful in determining if all New World WEE complex viruses are recombinant descendents, and in estimating when and where recombination between EEE- and SIN-like viruses occurred. Levinson *et al.* (1990) speculated that the New World Aura virus may have been the SIN-like ancestor which recombined with EEE virus. If so, the nsP1, nsP2 and nsP4 amino acid sequences for Aura could be used to determine whether the recombination event occurred prior to divergence of the Old and New World alphavirus groups (Fig. 33.3).

Other alphaviruses

Few molecular data are available to determine relationships among members of the remaining alphavirus complexes. More information on alphaviruses in the SF antigenic complex would be particularly useful in delineating events that led to the occurrence of these viruses in both the Old and New World. The close antigenic relationship between ONN and CHIK viruses (Calisher *et al.*, 1988) suggests that ONN may have evolved recently from a CHIK-like ancestor in Africa. The lack of confirmed transmission since 1978 (Johnson, 1988) suggests that ONN virus may be extinct. Sequence data for CHIK virus could also be used to estimate the time of divergence of these viruses. It is possible that ONN represents a CHIK variant which caused an epidemic when it acquired the ability to be transmitted by peridomestic *Anopheles* mosquitoes (ONN is the only alphavirus to be transmitted by mosquitoes of this genus) but could not be sustained in an endemic/enzootic transmission cycle. Sequence differences between CHIK and ONN viruses might also be useful in determining which alphavirus genes define mosquito host range, a critical factor in understanding the evolution of these pathogens.

Mayaro and Una viruses of the SF complex, which occur in South America, are antigenically very similar, suggesting that they share a recent New World ancestor. As pointed out by Levinson *et al.* (1990), the association of these viruses with birds (like their Old World SF complex relatives), and their antigenic and genetic similarity to other viruses in the SF complex imply that a transcontinental introduction accompanied diversification of the SF complex. Sequence data for Getah (GET), Sagiyama (SAG) and Bebaru (BEB) viruses would also be useful in determining if the SF complex is indeed monophyletic and where its ancestor occurred. Antigenic relationships of these three viruses to RR (Table 33.1), and the genetic relationship of RR and SF to ONN (Fig. 33.3), suggest that GET, SAG, BEB and RR may share a common ancestor which emerged from the SF lineage in Africa and was introduced into Asia and the Pacific.

Mechanisms of alphavirus diversification

Recombination

Several mechanisms have been described for diversification of arboviruses. One is recombination, which led to the generation of WEE virus (or its ancestor) from SIN- and EEE-like ancestors (Hahn *et al.*, 1988). Sequence data on all members of the WEE complex are needed to determine which alphaviruses are descendents of the recombinant ancestor. If many WEE complex viruses are recombinants, this would suggest that adaptive radiation of an ancestral recombinant virus may have contributed to diversification of the WEE complex; recombination may have created a new, highly successful genotype which diversified as it occupied new transmission cycles and habitats.

Phylogenetic analyses of alphaviruses in the EEE and VEE complexes have uncovered no evidence of recombination within the 26S region of their genomes (Weaver, Bellew & Rico-Hesse, 1992a). However, recombination between closely related alphavirus genomes would be difficult to detect and might therefore go unnoticed (Scott *et al.*, 1994). Dual EEE virus infections have been identified in two of three bird isolates from North American transmission foci, while dual infections were not identified in seven mosquito isolates (Weaver *et al.*, 1993a). This high rate of dual avian infections suggests that dual alphavirus infections of vertebrate hosts may occur frequently, probably resulting from multiple mosquito bites during a short time period. This implies the possibility of recombination if individual cells become infected with both genotypes. Dual infections of vectors may be uncommon because mosquitoes usually take a single blood meal during a gonotrophic cycle, and become resistant to superinfection with homologous alphaviruses within 24 hr after infection (Davey, Mahon & Gibbs, 1979).

Adaptive radiation

Adaptive radiation of alphaviruses could occur when viruses diversify by occupying new spatial habitats and/or transmission cycles (niches), followed by mutational change leading to genetic distinction. Generation of new viruses (speciation in the broad sense) could be allopatric when viruses are dispersed to new geographical locations. The non-overlapping distribution of many subtypes and varieties of VEE virus (Walton & Grayson, 1988) as well as the three main EEE lineages (Fig. 33.5) is consistent with allopatric speciation. Alternatively, sympatric speciation could occur when an alphavirus undergoes host switching within its current range; allopatric speciation may also involve host switching.

Mosquito and vertebrate host switching has been implicated in the evolution of alphaviruses in the EEE and VEE complexes. Colonization of North America by an ancestral South or Central American EEE virus, within the past 1000 years, probably involved vector switching from a *Culex* (*Melanoconion*) mosquito to *Culiseta melanura*, and vertebrate host switching from small mammals to birds (S.C.W., unpublished observations). The recent colonization of Florida by an ancestral VEE variety II Everglades virus also involved mosquito host switching, and introduction of Bijou Bridge virus into North America from a Tonate-like ancestor involved switching from a mosquito to cimicid bug host (Weaver *et al.*, 1992*a*). Fort Morgan and Buggy Creek viruses in the WEE complex, which also are transmitted by these bugs, probably also switched from mosquito hosts during their evolution.

Effects of alphavirus dispersal, population partitioning and genetic drift

Ecological concepts related to the regulation of alphavirus evolution are reviewed in greater detail elsewhere (Weaver *et al.*, 1992*c*; Scott *et al.*, 1994). Recent information on evolutionary relationships and genetic diversity among viruses in the EEE (Weaver *et al.*, 1994) and VEE complexes (Weaver *et al.*, 1992*a*) suggests that founder effects, genetic drift and virus population partitioning may play important roles in alphavirus diversification (see also above). Most South and Central American EEE and VEE viruses are transmitted primarily among small mammals, while North American EEE viruses utilize avian hosts. The limited mobility of mammalian vs. avian hosts probably results in a greater degree of isolation among tropical EEE and VEE virus populations (transmission foci) relative to EEE virus in North America. Differences in virus dispersal may have contributed to the greater diversity among VEE and tropical EEE viruses compared to EEE virus in North America, as well as the slower rate of evolution of North American EEE virus. Differences in virus diversity may also reflect a greater diversity in mosquito host taxa; the *Culex (Melanoconion)* subgenus, which includes all known enzootic vectors of EEE and VEE viruses in tropical America, is very diverse compared to the genus *Culiseta* which includes the North American EEE virus vector (Weaver *et al.*, 1992*c*).

Long-term co-evolution

Another proposed mechanism for arbovirus diversification is long-term co-evolution (co-diversification) in concert with mosquito vectors. Eldridge (1990) has suggested that mosquito-borne bunyaviruses in the California serogroup co-evolved with vectors in the genus *Aedes*. Long-term co-evolution is an attractive hypothesis for some alphaviruses such as those in the VEE complex, which are transmitted by several species of mosquitoes in the subgenus *Culex (Melanoconion)*. The more specific relationships between alphaviruses and mosquito vectors, compared to those between alphaviruses and vertebrate hosts, imply a greater likelihood of diversification with mosquito than vertebrate hosts. However, the estimated time frames for alphavirus diversification (see above) are inconsistent with those of eukaryotic mosquitoes with DNA genomes, and alphavirus-host relationships indicate host switching during diversification of the EEE and VEE complexes (Weaver *et al.*, 1992*a*; see above). As pointed out by Strauss and Strauss (1988), considering the high rates of RNA virus evolution, owing to their high mutation frequencies (Steinhauer & Holland, 1987; Holland, De La Torre & Steinhauer, 1992), it would be surprising if any RNA arboviruses have diversified in concert with their eukaryotic hosts.

Co-evolution of alphaviruses and their hosts

Long-term co-evolution (co-diversification) of alphaviruses with their hosts has not been supported by phylogenetic analyses or estimates of alphavirus evolutionary rates (see above). However, the term co-evolution has been defined in other ways. Janzen (1980) described co-evolution as 'an evolutionary change in a trait of the individuals in one population in response to a trait of the individuals of a second population, followed by an evolutionary response by the second population to the change in the first'. For a mosquito-borne alphavirus, this reciprocal evolutionary relationship requires that the virus impose selective pressure on its mosquito and/or vertebrate hosts, and that the hosts, likewise, affect evolution of the virus.

Although traditional views regarded arbovirus infections of mosquito vectors as benign, and the virus–vector relationship as commensalistic (for review see Scott & Weaver, 1989), recent studies have shown that EEE (Weaver *et al.*, 1988) and WEE (Weaver *et al.*, 1992*b*) viruses cause pathological changes in the alimentary tract of their natural enzootic mosquito vectors. Pathological changes in the midgut may lead to rapid dissemination of alphaviruses within their mosquito vectors and thereby facilitate transmission; mosquito virulence may therefore be favoured to some degree during alphavirus evolution (Weaver *et al.*, 1988; Scott & Weaver, 1989). However, in addition to causing cytopathology in the midgut, EEE virus has been shown to cause reductions in survival and fecundity of *Culiseta melanura* mosquitoes (T.W. Scott, personal communication). This implies that EEE and other alphaviruses impose some selective pressure towards insusceptibility of mosquito vectors in nature. A form of group selection like that described by Wilson (1980) may also limit alphavirus virulence in nature; virus populations (virions within a mosquito) with high average mosquito virulence may contribute less to the overall viral gene pool than less virulent populations if high virulence is accompanied by reduced transmission by adversely affected mosquito hosts. Thus, virulent populations may be selected against even if virulence is coupled with high virus reproductive (replication) rates.

This type of group selection may also limit vertebrate host pathogenicity of EEE and other alphaviruses. Several arboviruses are also known to cause mortality in vertebrate hosts (Scott, 1988). Avian mortality from natural infection with EEE virus has been reported (Stamm, 1958; Williams *et al.*, 1971; McLean *et al.*, 1985), as well as mortality among nestling house sparrows infected with Fort Morgan virus (Scott, Bowen & Monath, 1984). These results suggest that alphaviruses can exert selective pressure on susceptibility of vertebrate hosts in nature.

Possible effects of mosquito and vertebrate hosts on evolution of alphaviruses have received little attention. However, the highly specific susceptibility patterns of some vectors to the viruses they transmit suggest that some alphaviruses may have evolved (adapted) to efficiently infect their mosquito hosts. For example, *Culex (Melanoconion) taeniopus*, the Central American mosquito vector of variety IE VEE virus, is highly susceptible to infection by IE viruses but is refractory to infection with closely related, allopatric VEE virus varieties (Scherer *et al.*, 1987). Experimental transmission studies are needed to determine if alphaviruses undergo genetic changes associated with increased infectivity, when introduced into transmission cycles involving a different vector species. To demonstrate co-evolution, this kind of genetic change, caused by natural selection, must be distinguished from chance 'pre-adaptation' of viruses to mosquitoes or other insects not originally involved in natural transmission cycles.

Rates of alphavirus evolution

RNA viruses replicate their genomes with low fidelity due to lack of proofreading associated with their RNA-dependent RNA polymerases (Holland *et al.*, 1992). These high mutation frequencies (*c.* one million times those of eukaryotic DNA genomes) are manifested in genetically heterogeneous quasi-species RNA virus populations (Domingo & Holland, 1988) which consist of a 'mutant distribution centered around one or several master sequences' (Eigen & Biebricher, 1988). Genetic diversity in RNA virus genomes is maintained at the quasi-species level due to the counteraction of mutational pressure vs. selective forces on the virus population. Under appropriate conditions (e.g. selection for resistance to host immunity, or founder effects) RNA viruses are therefore capable of undergoing rapid evolution (Steinhauer & Holland, 1987).

Evolutionary rates of the order of 10^{-2}–10^{-3} substitutions per nucleotide per year have been reported for several animal RNA viruses (for review see Steinhauer & Holland, 1987; Weaver *et al.*, 1992c). In contrast, natural evolutionary rates for several alphaviruses have been estimated at 2–7×10^{-4} from RNA fingerprint and sequence data. Some arboviruses in the family *Flaviviridae* evolve at similar rates (Weaver *et al.*, 1992c), and the mosquito-borne bunyavirus, La Crosse virus, exhibits a high degree of genetic stability during experimental transmission cycles (Baldridge, Beaty & Hewlett, 1989).

The reasons for slow rates of evolution of arboviruses in general and alphaviruses in particular, relative to many other RNA viruses, are unknown. A simple explanation could be higher fidelity replication of alphaviruses relative to other RNA viruses. Most temperature sensitive mutants of SIN virus, believed to involve single base substitutions, revert to wild-type at frequencies of 10^{-3}–10^{-5}. However, a few revert at frequencies of 10^{-7}–10^{-8} (Strauss *et al.*, 1977; Hahn *et al.*, 1989; Durbin & Stollar, 1986), suggesting the possibility of high fidelity replication of some portions of the alphavirus genome. However, natural mutation frequencies reported for North American isolates of EEE virus, which has the slowest rate of evolution of the alphaviruses examined in detail, are similar to those of other RNA viruses (Weaver *et al.*, 1993a). This suggests that EEE virus populations exist in nature as quasi-species, implying that mutation rates are not the limiting factor in EEE virus evolution.

Several lines of evidence suggest that selection for efficient replication in two evolutionarily divergent hosts (vertebrate and invertebrate) may constrain evolution of mosquito-borne viruses (for review see Weaver, Scott & Rico-Hesse, 1991; Weaver *et al.*, 1992c). Selection may act to conserve alphaviral proteins; the predominance of synonymous nucleotide substitutions during evolution of SIN, EEE, and VEE viruses (Shirako *et al.*, 1991; Weaver *et al.*, 1991, 1992a) suggests that alphaviral proteins are under strong selective pressure in nature. Proteins under multiple selective constraints imposed by two hosts may include envelope glycoproteins which interact with different host cell receptors, and require specific sorting in polarized mosquito cells for efficient dissemination within the vector. In addition to requiring host cell translational machinery for replication, additional unidentified components differing in vertebrate vs. invertebrate cells also appear to participate in alphavirus replication (Brown & Condreay, 1986). Interaction with such components may constrain alphavirus protein and RNA structure. The slower rate of evolution (amino acid substitution) in non-structural alphavirus proteins relative to that in structural proteins may reflect optimization of their enzymatic activities for the relatively static intracellular environment (Levinson *et al.*, 1990).

Primary RNA sequence may also be under selection to preserve codon usage for efficient replication in both vertebrate and invertebrate hosts (Strauss & Strauss, 1986). Because of methylation and resultant instability of cytosine residues in the dinucleotide CpG, vertebrate (Bird, 1987) and plant (Murray, Lotzer & Eberle, 1989) genomes are CpG deficient and often underutilize codons containing CpG. In contrast, invertebrate genomes exhibit little or no CpG deficiency. Alphaviruses are slightly but significantly deficient in the dinucleotide CpG, as are most other viruses in the alphavirus-like superfamily (Weaver *et al.*, 1993*b*). However, viruses which utilize only vertebrate hosts generally show greater levels of CpG deficiency. For example, poliovirus utilizes CpG in codons at a frequency only 45% of that expected, considering base composition and the proportion of synonymous codons containing CpG; this value compares with a mean value of 84% for six alphaviruses examined and 75% for eleven alphavirus-like plant viruses (Weaver *et al.*, 1993*b*). The lesser CpG deficiency in alphavirus-like viruses, relative to vertebrate viruses (e.g. poliovirus), may reflect compromise codon usage to permit efficient replication in insect cells which do not have CpG-deficient genomes. The non-random codon usage seen for other, non-CpG-containing amino acids in the genomes of alphaviruses (Strauss & Strauss, 1986) may reflect additional selective pressures not based on CpG content.

Interestingly, yellow fever virus exhibits even greater CpG deficiency than alphaviruses or poliovirus, with observed usage only 38% of expected (Rice *et al.*, 1985). This CpG deficiency may reflect the evolutionary history of yellow fever virus. Flaviviruses share a common ancestor with other supergroup II RNA viruses; the most closely related group includes only two viruses, both non-arthropod-borne vertebrate viruses (hepatitis C and bovine viral diarrhea virus; Koonin, 1991; see Fig. 33.3). This suggests that flaviviruses may have evolved from a vertebrate virus, and their extreme CpG deficiency may reflect ancestral adaptation for optimal replication in vertebrate hosts. However, a more likely explanation is that flaviviruses have adopted a replication strategy which favours viral protein expression in vertebrate cells at the expense of replication efficiency in mosquitoes. The slow replication and dissemination of flaviviruses within mosquito vectors, relative to that of alphaviruses (Scott & Weaver, 1989) supports this hypothesis.

Another factor that may contribute to slow rates of evolution of alphaviruses and other arboviruses is decreased rates of virus replication in poikilothermic mosquito vectors. Alphavirus host viraemias tend to be short lived, lasting 2 to 5 days (Scott, 1988); in contrast, mosquito vectors usually are infected 7 days or more before transmission occurs. Production of progeny EEE virions in BHK cells requires a minimum of 2–3 h at 37 °C, and 9–10 h in C6/36 mosquito cells at the ecologically relevant temperature of 26 °C (Weaver *et al.*, 1993*b*). Assuming a typical transmission cycle including two days of replication in a bird and seven in a mosquito, and assuming that the time required to produce progeny virions is inversely related to the number of genome replication cycles completed per unit time at a given temperature, slower replication in the mosquito may diminish the number of replication cycles by *c.* 60% relative to an equivalent, single homeothermic vertebrate host cycle (9 days of replication at 37 °C). The number of replication cycles occurring in the mosquito may be even fewer because replication of alphaviruses in mosquitoes is diminished by unknown factors several days after infection (Hardy *et al.*, 1983; Hardy, 1988). The effect on rates of arbovirus evolution of slow replication in poikilothermic mosquito hosts could be distinguished from the effects of host alternation (see below) by comparing rates of alphavirus evolution with those of single host RNA insect viruses.

Selection for neutralizing antibody resistance, which can lead to antigenic selection and rapid evolution of some RNA viruses such as influenza A (Gorman, Bean & Webster, 1992) probably has little influence on evolution of alphavirus envelope proteins (Weaver *et al.*, 1992*c*). Avian and small mammalian hosts of alphaviruses tend to be short lived, which limits the proportion of immune hosts in the population. Mosquito vectors, which become persistently infected with alphaviruses, are not known to place any immune pressure on the viruses that they transmit. Lack of immune pressure may allow alphaviruses to maintain protein structures optimally fit for replication in and transmission among hosts.

The influence of the two host cycle on slow rates of arbovirus evolution has been conceptualized using an adaptive landscape model (Weaver *et al.*, 1992*c*). The adaptive landscape describes the fitness of different viral genotypes; horizontal axes define viral genotypes (actually, the landscape should be N-dimensional, defining all possible genome sequences) while the vertical axis defines the fitness of each genotype (Fig. 33.6). Viruses evolve by moving across the landscape between fitness peaks; movement from low to higher fitness levels occurs via natural selection. An organism will therefore tend to remain on the currently occupied fitness peak (driven upward by natural selection) even if higher peaks are present in different regions of the landscape. Eigen and Biebricher (1988) have argued that, for RNA viruses, the target of selection is actually the quasi-species distribution of genotypes and corresponding phenotypes rather than individual virus genotypes and phenotypes.

Different selective constraints are imposed by the vertebrate and invertebrate hosts of arboviruses. Therefore, alphaviruses must evolve within the intersection of two adaptive landscapes which may be somewhat independent (Fig. 33.6A,B). The two-host landscape (Fig. 33.6D), which includes only overlapping mosquito and vertebrate fitness peaks (these peaks must overlap in order to allow host alternation without subjecting the virus to movement through a low fitness valley during each transmission cycle) contains wider fitness valleys than either single host landscape. This makes movement from one peak to another (evolution) less likely if selective constraints are constant (Weaver *et al.*, 1992*c*). To move across this wide valley of low fitness, the alphavirus must acquire a relatively large amount of genetic change. Therefore,

Fig. 33.6. A–D. Limited hypothetical two-dimensional region of adaptive landscapes for evolution of mosquito-borne viruses. Horizontal axes define different virus genotypes (actually, the field is N-dimensional and represents all possible genotypes of the virus). The vertical axis indicates the relative fitness of any genotype defined by the horizontal axis. A. Landscape for mosquito infections. B. Landscape for vertebrate infections. C. Mosquito and vertebrate landscapes superimposed to represent landscape for evolution of mosquito-borne (alternating host) viruses. D. Alternating host landscape containing only mosquito and vertebrate landscape peaks which overlap or coincide.

arbovirus evolution is theoretically constrained relative to that of a single host virus.

As discussed previously (Weaver *et al.*, 1992c), disruptive factors including high mutation rates, landscape instability (ecological changes in the transmission cycle or host changes), population partitioning, and founder effects can overcome the stabilizing effect of selection, while efficient virus dispersal (gene flow) can enhance the stabilizing effect of selection by increasing effective population sizes. Therefore, better understanding of the population genetics and ecology of alphavirus transmission cycles will undoubtedly lead to greater understanding of the evolution of mosquito-borne viruses.

Acknowledgements

This chapter benefited from discussions with Charles Calisher, David Clarke, Russell Doolittle, John Holland and James Strauss. This research was supported by National Institutes of Health grants AI26787 and AI14627. The literature review for this chapter was completed in June, 1992.

References

Ahlquist, P., Strauss, E.G., Rice, C.M., Strauss, J.H., Haseloff, J. & Zimmern, D. (1985). *J. Virol.*, **53**, 536–42.

Baldridge, G.D., Beaty, B.J. & Hewlett, M.J. (1989). *Arch. Virol.*, **108**, 89–99.

Bell, J.R., Kinney, R.M., Trent, D.W., Strauss, E.G. & Strauss, J.H. (1984). *Proc. Natl. Acad. Sci. USA*, **81**, 4702–6.

Bird, A.P. (1987). *Trends in Genetics*, **3**, 342–47.

Brown, D.T. & Condreay, L.D. (1986). In *The Togaviruses and Flaviviruses*, S. Schlesinger and M. Schlesinger (eds.). Plenum Press, New York, pp. 171–207.

Buonagurio, D.A., Nakada, S., Parvin, J.D., Krystal, M., Palese, P. & W.M. Fitch. (1986). *Science*, **232**, 980–2.

Calisher, C.H. & Karabatsos, N. (1988). In *The Arboviruses: Epidemiology and Ecology, Vol. I*. T.P. Monath (ed.). CRC Press, Boca Raton, Florida, pp. 19–55.

Calisher, C.H., Karabatsos, N, Lazuick, J.S., Monath, T.P. & Wolff, K.L. (1988). *Am. J. Trop. Med. Hyg.*, **38**, 447–52.

Calisher, C.H., Shope, R.E., Brandt, W., Casals, J., Karabatsos, N., Murphy, F.A., Tesh, R.B. & Wiebe, M.E. (1980). *Intervirology*, **14**, 229–32.

Casals, J. (1964). *J. Exp. Med.*, **119**, 547–65.

Chamberlain, R.W. & Sudia, W.D. (1961). *Annu. Rev. Entomol.*, **6**, 371–90.

Chang, G-J.J. & Trent, D.W. (1987). *J. Gen. Virol.*, **68**, 2129–42.

Daly, H.V., Doyen, J.T. & Ehrlich, P.R. (1978). *Introduction to Insect Biology and Diversity*. McGraw Hill, New York.

Davey, M.W., Mahon, R.J. & Gibbs, A.J. (1979). *J. Gen Virol.*, **42**, 641–3.

Domingo, E. & Holland, J.J. (1988). In *RNA Genetics, Vol. III*. E. Domingo, J.J. Holland and P. Ahlquist (eds.) CRC Press, Boca Raton, pp. 3–36.

Doolittle, R.F., Anderson, K.L. & Feng, D.F. (1989). In *The Heirarchy of Life*. F. Fermholm, K. Bremer and H. Jornvall (eds.) Exerpta medica, Amsterdam, pp. 73-85.

Doolittle, R.F. & Feng, D.-F. (1992). *Curr. Topics Microbiol. Immunol.*, **176**, 195-211.

Durbin, R.K. & Stollar, V. (1986). *Virology*, **154**, 135-43.

Eigen, M. & Biebricher, C.K. (1988). In *RNA Genetics, Vol. III*. E. Domingo, J.J. Holland and P. Ahlquist (eds.). CRC Press, Boca Raton, Florida, pp. 211-245.

Eldridge, B. (1990). *J. Med. Entomol.*, **27**, 738-49.

Eldredge, N. & Gould, S.J. (1972). In *Models in Paleobiology*, T.J.M. Schopf (Ed.) Freeman, Cooper and Co., New York, pp. 82-115.

Faragher, S.G., Meek, A.D.J., Rice, C.M. & Dalgarno, L. (1988). *Virology*, **163**, 509-26.

Feng, D.-F. & Doolittle, R.F. (1990). *Methods in Enzymology*, **183**, 375-87.

Fulhorst, C.F., Hardy, J.L., Eldridge, B.F., Presser, S.B. & Reeves, W.C. (1994). *Science*, **263**, 676-8.

Garoff, H., Frischauf, A.M., Simons, K., Lehrach, H. & Delius, H. (1980*a*). *Proc. Natl. Acad. Sci. USA*, **77**, 6376-80.

Garoff, H., Frischauf, A.M., Simons, K., Lehrach, H. & Delius, H. (1980*b*.) *Nature*, **288**, 236-41.

Goldbach, R. & Wellink, J. (1988). *Intervirology*, **29**, 260-7.

Gorman, O.T., Bean, W.J. & Webster, R.G. (1992). *Curr. Topics Microbiol. Immunol.*, **176**, 75-97.

Hahn, C.S., Lustig, S., Strauss, E.G. & Strauss, J.H. (1988). *Proc. Natl. Acad. Sci. USA*, **85**, 5997-6001.

Hahn, C.S., Rice, C.M., Strauss, E.G., Lenches, E.M. & Strauss, J.H. (1989). *J. Virol.*, **63**, 3459-65.

Hardy, J.L. (1988). In *The Arboviruses: Epidemiology and Ecology, Vol. I*. T.P. Monath (ed.). CRC Press, Boca Raton, Florida, pp. 87-126.

Hardy, J.L., Houk, E.J., Kramer, L.D. & Reeves, W.C. (1983). *Ann. Rev. Entomol.*, **28**, 229-62.

Haseloff, J., Goelet, P., Zimmern, D., Ahlquist, P., Dasgupta, R. & Kaesberg, P. (1984). *Proc. Natl. Acad. Sci. USA*, **81**, 4358-62.

Holland, J.J., De La Torre J.C. & Steinhauer, D.A. (1992). *Curr. Topics Microbiol. Immunol.*, **176**, 1-20.

Houk, E.J., Kramer, L.D., Hardy, J.L. & Chiles, R.E. (1985). *Virus Research* 2, 123-38.

Janzen, D.H. (1980). *Evolution*, **34**, 611-12.

Johnson, B.K. (1988). In *The Arboviruses: Epidemiology and Ecology, Vol. III*. T.P. Monath (ed.). CRC Press, Boca Raton, Florida, 217-223.

Kay, B.H. (192). *Aust. J. Exp. Biol. Med. Sci.*, **60**, 339-44.

Kielian, M. and Helenius, A. (1986). In *The Togaviruses and Flaviviruses*, S. Schlesinger and M. Schlesinger (eds.). Plenum Press, New York. 91-119.

Kew, O.M., Nottay, B.K. & Obijeski, J.F. (1984). *Methods in Enzymology*, **8**, 41-84.

Kimura, M. (1983). *The Neutral Theory of Evolution*. Cambridge University Press, London.

Kinney, R.M., Johnson, B.J.B., Welch, J.B., Tsuchiya, K.R. & Trent, D.W. (1989). *Virology*, **170**, 19-31.

Kinney, R.M., Trent, D.W. & France, J.K. (1983). *J. Gen. Virol.*, **64**, 135-47.

Koonin, E.V. (1991). *J. Gen. Virol.*, **72**, 2197–06.
Lai, M.M.C. (1992). *Curr. Topics Microbiol. Immunol.*, **176**, 21–32.
Levinson, R.S., Strauss, J.H. & Strauss E.G. (1990). *Virology*, **175**, 110–23.
Mclean, R.G., Frier, G., Parham, G.L., Francy, D.B., Monath, T.P.,
    Campos, E.G., Therrien, A., Kerschner J. & Calisher, C.H. (1985).
    Am. J. Trop. Med. Hyg., **34**, 1190–202.
Murray, E.E., Lotzer, J. & Eberle, M. (1989). *Nucl. Acids Res.*, **17**, 477–98.
Olson, K. & Trent, D.W. (1985). *J. Gen. Virol.*, **66**, 797–810.
Peters, C.J. & Dalrymple, J.M. (1990). In *Virology*, 2nd edn, B.N. Fields,
    D. M. Knipe, (eds.) Raven Press, New York, pp. 713–761.
Rentier-Delrue, F. & Young, N.A. (1980). *Virology*, **106**, 59–70.
Rice, C.M., Lenches, E.M., Eddy, S.R., Shin, S.J., Sheets, R.L. & Strauss,
    J.H. (1985). Nucleotide sequence of yellow fever virus: implications for
    flavivirus gene expression and evolution. *Science*, **229**, 726–33.
Rico-Hesse, R., Roehrig, J.T., Trent, D.W. & Dickerman, R.W. (1988). *Am.
    J. Trop. Med. Hyg.*, **38**, 195–204.
Robbins, C.S., Sauer, J.R., Greenberg, R.S., & Droege, S. (1989). *Proc.
    Natl. Acad. Sci. USA*, **86**, 7658–662.
Roehrig, J.T., Hunt, A.R., Chang, G-J., Sheik, B., Bolin, R.A., Tsai, T.F.
    & Trent, D.W. (1990) *Am. J. Trop. Med. Hyg.*, **42**, 394–8.
Scherer, W.F., Weaver, S.C., Taylor, C.A., Cupp, E.W., Dickerman, R.W.
    & Rubino, H.H. (1987). *Am. J. Trop. Med. Hyg.*, **36**, 194–7.
Schlesinger, S. & Schlesinger, M.J. (1990). In *Virology*, 2nd edn, B.N. Fields,
    D.M. and Knipe, (Eds.) Raven Press, New York, pp. 697–711.
Scott, T.W. (1988). In *The Arboviruses: Epidemiology and Ecology, Vol. I.*
    T.P. Monath (ed.). CRC Press, Boca Raton, Florida, pp. 257–280.
Scott, T.W., Bowen, G.S. & Monath, T.P. (1984). *Am. J. Trop. Med. Hyg.*,
    33, 981–91.
Scott, T.W. & Weaver, S.C. (1989). *Adv. Virus Res.*, **37**, 277–328.
Scott, T.W., Weaver, S.C. & Mallampalli, V. (1994). In *The Evolutionary
    Biology*, S.S. Morse (ed.). Raven Press, New York, pp. 293–324.
Shirako, Y., Niklasson, B., Dalrymple, J.M., Strauss, E.G. & Strauss, J.H.
    (1991). *Virology*, **182**, 753–64.
Stamm, D.D. (1958). *Am. J. Publ. Health.*, **48**, 328–35.
Steinhauer, D.A. & Holland, J.J. (1987). *Ann. Rev. Microbiol.*, **41**, 409–33.
Strauss, E.G., Birdwell, C.R., Lenches, E.M., Staples, S.E. & Strauss, J.H.
    (1977). *Virology*, **82**, 122–49.
Strauss, E.G., Rice, C.M. & Strauss, J.H. (1983). *Proc. Natl. Acad. Sci.
    USA*, **80**, 5271–5.
Strauss, E.G., Rice, C.M. & Strauss, J.H. (1984). *Virology*, **133**, 92–110.
Strauss, E.G. & Strauss, J.H. (1986). In *The Togaviruses and Flaviviruses*,
    S. Schlesinger and M. Schlesinger (eds.). Plenum Press, New York,
    pp. 35–90.
Strauss, J.H. & Strauss, E.G. (1988). *Ann. Rev. Microbiol. Immunol.*,
    42, 657–683.
Takkinen, K. (1986). *Nucl. Acids Res.*, **14**, 5667–82.
Trent, D.W., Clewey, J.P., France, J.K. & Bishop, D.H.L. (1979). *J. Gen.
    Virol.*, **43**, 365–81.
Trent, D.W. & Grant, J.A. (1980). *J. Gen. Virol.*, **47**, 261–82.
Ubol, S. & Griffin, D.E. (1991). *J. Virol.*, **65**, 6913–21.
Volchkov, V.E., Volchkova, V.A. & Netesov, S.V. (1991). *Mol. Gen.
    Mikrobiol. Virusol.*, **5**, 8–15.

Walder, R., Rosato, R.R. & Eddy, G.A. (1981). *Arch. Virol.*, **68**, 229–37.
Walton, T.E. & Grayson, M.A. (1988). In *The Arboviruses: Epidemiology and Ecology, Vol. I*. T.P. Monath (ed.). CRC Press, Boca Raton, Florida, pp. 59–85.
Wang, K.-S., Kuhn, Schmalijohn, A.L., Kuhn, R.J. & Strauss, J.H. (1991). *Virilogy*, **181**, 694–702.
Wang, K.-S., Kuhn, R.J., Strauss, E.G., Ou, S. & Strauss, J.H. (1992). *J. Virology*, **66**, 4992–5001.
Weaver, S.C., Bellew, L.A. & Rico-Hesse, R. (1992*a*). *Virology*, **191**, 282–90.
Weaver, S.C., Bellew, L.A., Gousset, L., Repik, P.M., Scott, T.W. & Holland, J.J. (1993*a*). *Virology*, **195**, 700–9.
Weaver, S.C., Hagenbaugh, A., Bellew, L.A., Netesov, S.V., Volchkov, V.E., Chang, G.-J.J., Clarke, D.K., Gousset, L., Scott, T.W., Trent, D.W. & Holland, J.J. (1993*b*). *Virology*, **197**, 375–90.
Weaver, S.C., Hagenbaugh, A., Bellew, L.A., Gousset, L., Mallampalli, V., Holland, J.J. & Scott, T.W. (1994). *J. Virol.*, **68**, 158–69.
Weaver, S.C., Lorenz, L.H. & Scott, T.W. (1992*b*). *Am. J. Trop. Med. Hyg.*, **47**, 691–701.
Weaver, S.C., Rico-Hesse, R. & Scott, T.W. (1992*c*). *Curr. Topics Microbiol. Immunol.*, **176**, 99–117.
Weaver, S.C., Scott, T.W. & Lorenz, L.H. (1990). *J. Med. Entomol.*, **27**, 878–91.
Weaver, S.W., Scott, T.W., Lorenz, L.H., Lerdthusnee, K. & Romoser, W.S. (1988). *J. Virol.*, **62**, 2083–90.
Weaver, S.C., Scott, T.W. & Rico-Hesse, R. (1991). *Virology*, **182**, 774–84.
Wengler, G., Wengler, G. & Filipe, A.R. (1977). *Virology*, **78**, 124–34.
Williams, J.E., Young, O.P., Watts, D.M. & Reed, T.J. (1971). *J. Wildl. Dis.*, **7**, 188–94.
Wilson, D.S. (1980). *The Natural Selection of Populations and Communities*. Benjamin Cummings, Menlo Park, California.
Wright, S. (1937). *Proc. Natl. Acad. Sci. USA*, **23**, 307–20.
Xiong, Y. & Eickbush, T.H. (1990). *EMBO J.*, **9**, 3353–62.

34

Evolution of influenza viruses: rapid evolution and stasis

R.G. WEBSTER, W.J. BEAN AND O.T. GORMAN

Introduction

Influenza is the paradigm of a viral disease that relies on continued evolution of the virus to cause annual epidemics and occasional pandemics of disease in humans. The gene pool of influenza A viruses in aquatic birds provides all the genetic diversity required for the emergence of pandemic influenza virus for humans, lower animals and birds. In humans, pigs and horses, influenza A viruses show both antigenic drift and genetic shift (Webster *et al.*, 1992). In contrast, there is emerging evidence that avian influenza viruses are in evolutionary stasis. This review reports that rapid evolution in influenza A viruses in humans and other mammals has continued since the beginning of recorded medical history and depends on periodic introductions of gene segments or entire influenza viruses from the avian influenza virus gene pool. In aquatic wild birds, influenza virus appears to be fully adapted to its host and causes no disease signs. Thus, understanding the ecology of influenza viruses in the reservoir of aquatic birds is essential if we wish to find ways to intervene and reduce or prevent the occasional catastrophic pandemics such as the one that decimated the human population of the world in 1918 after the appearance of 'Spanish' influenza.

Host-specific evolution of influenza viruses

It is likely that the NP protein gene determines host range (Scholtissek *et al.*, 1985; Tian *et al.*, 1985; Snyder *et al.*, 1987); consequently NP gene evolution and host taxonomy has been studied (Gorman *et al.*, 1990*a*, 1991; Gammelin *et al.*, 1990). Using RNA hybridization techniques, Bean (1984) showed that NP genes of influenza A viruses fall into five host-specific groups. Subsequently, Gorman *et al.* (1990*a*, 1991) showed in phylogenetic analyses of NP gene sequence that they have evolved into five major host-specific lineages that correspond to five NP gene RNA hybridization groups (Bean, 1984) (Fig. 34.1). These lineages are: (a) EQPR56 (Equine/Prague/56) (equine 1 viruses); (b) recent equine viruses, i.e. those related to Equine/Miami/63 (equine 2 viruses); (c) human viruses together with classical swine viruses, i.e, those related to Swine/Iowa/15/30; (d) H13 gull viruses; (e) all other avian viruses. Geographic patterns of evolution are evident in avian virus NP genes; North American, Australian, and Old World isolates form separate sublineages.

Positive phylogenies for the six internal protein influenza virus genes that include H1N1 and H3N2 human and swine viruses have also been determined (PB1: Kawaoka, Krauss & Webster, 1989; PA: Okazaki, Kawaoka & Webster, 1989; PB2: Gorman *et al.*, 1990*b*; NP: Gammelin *et al.*, 1990; Gorman *et al.*, 1990*a*, 1991; M: Ito *et al.*, 1991; NS: Kawaoka, personal communication), and also for the surface protein subtype genes H1, H3, and H4 HAs (Kawaoka *et al.*, 1990; Bean *et al.*, 1992; Donis *et al.*, 1989; Kawaoka, unpublished data). Generalized phylogenetic trees (cladograms) showing the major branching topologies for the phylogenies of these genes are shown in Fig. 34.1. Of the eight gene phylogenies shown in Fig. 34.1, none show identical topologies. This probably indicates that there have been re-assortment events in the past, and that differential extinctions have affected patterns of evolution for each gene, although other explanations are possible.

For those gene phylogenies that include EQPR56 (equine 1), this virus always shows an outgroup relationship to other virus lineages. This pattern indicates that the EQPR56 (equine 1) is the most divergent and possibly the oldest of the nonavian influenza A virus lineages. As such, its relationship to other gene lineages permits us to estimate the relative order of ages of divergence (vicariance) among the gene lineages. Cladograms for the NP and PA genes are very similar and would be identical if the EQPR56 PA gene showed the same relative position (Fig. 34.1), but its sequence is not known. The close match in the phylogenies of NP and PA genes indicates that they share a common evolutionary history in each host; they may be linked and apparently have not reassorted independently.

The M gene cladogram shows significant differences from cladograms of the NP and PA genes. First, the equine 2 M genes appear to have been recently derived from North American avian viruses in contrast to the much older origin of equine 2 NP and PA genes. Second, the

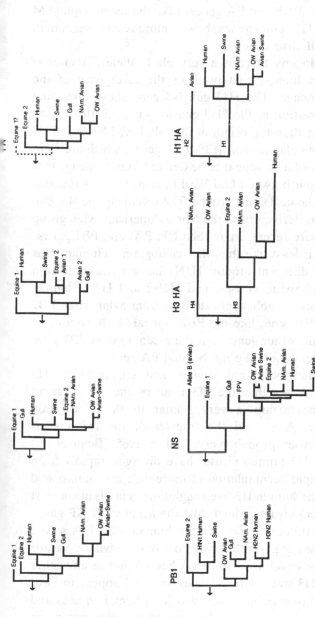

Fig. 34.1. Generalized phylogenies (cladograms) of influenza virus genes. Phylogenies were determined with PAUP software version 2.4 (David Swofford, Illinois Natural History Survey), which uses a maximum parsimony algorithm to find the shortest length trees. Horizontal distance is proportional to the minimum number of nucleotide changes needed to join the gene sequences (no scale is given, but all cladograms are on the same scale). Vertical lines merely space the branches and labels. The arrow at the left of each tree represents the node connecting the influenza B virus homologue. Equine 1, Eq/Prague/56 (H7N7) virus isolate; Equine 2, recent H3N8 equine viruses; Human (unless noted otherwise), human H1N1, H2N2, and H3N2 viruses; Swine, classical swine viruses (i.e. those related to Sw/Iowa/15/30); Gull, H13 gull viruses; FPV, fowl plague viruses; NAm, Avian, North American avian viruses; OW Avian, Old World or Eurasian avian viruses; Avian Swine, avian-like H1N1 swine viruses. There are two distinct avian lineages in the PB2 tree, Avian 1 and Avian 2 which contain Eurasian and North American avian viruses. Nucleotide phylogenies represented are taken from Kawaoka et al. (1989; PB1); Gorman et al. (1990b; PB2), Okazaki et al. (1989; PA), Gorman et al. (1990a, 1991; NP); Kawaoka (unpublished data, H1HA), Bean et al. (1992; H3HA), Ito et al. (1991; M) Kawaoka et al. (unpublished observations; NS).

H13 gull virus M gene is closer to EQPR56 and more distant from other virus M genes suggesting that the gull M gene may have an older origin compared to the H13 gull NP and PA genes. Like the recent equine M genes, the PB2 gene of H13 gull viruses shows a more recent origin than the NP, PA, and M gull virus genes.

In the NS gene phylogeny there is a very old b allele (Treanor *et al.*, 1989) in the avian lineage that predates the divergence of the EQPR56 (equine 1) lineage. The H13 gull NS gene shows the same relative evolutionary position as the H13 gull M gene, i.e. the origins of these genes are older than the origin of the NP, PA, PB1, and PB2 gull H13 genes. The fowl plague virus (FPV) NS genes, which are from the earliest avian virus isolates, appear to be derived from a lineage that diverged prior to the split between Old World avian, North American avian, and human and classical swine groups. The recent equine NS, like the M and PB2 genes, is derived from the North American avian group and seems to have a more recent origin than NP, PA, and PB1 genes.

The PB1 cladogram is least like the other cladograms. Human virus PB1 genes show three different origins: H1N1 human viruses form a sister group to classical swine viruses, and H2N2 and H3N2 human viruses each form separate sublineages derived from avian virus PB1 genes. The H13 gull PB1 gene, like the PB2, appears to have a more recent origin than for the other genes, but the recent equine PB1 gene appears to have an older origin like the NP and PA genes.

A comparison of nucleotide and amino acid sequences of H3 haemagglutinin genes shows that the H3 gene of the 1968 human pandemic (Hong Kong) strain is very similar to those currently circulating in ducks in Asia, and the transfer of the avian virus H3 gene to human viruses probably occurred in 1965 (Bean *et al.*, 1992). Since then, the H3 human viruses have diverged rapidly from this progenitor. This rapid accumulation of nucleotide and amino acid changes, especially in the human H3 haemagglutinin gene, is in contrast to those of the avian H3 viruses which, like the avian virus NP genes appear to be in evolutionary stasis. Unlike human H3 viruses, the equine 2 (H3N8) viruses apparently diverged from an avian ancestor much earlier and no close relatives have yet been found in any other species. Each of four H3 swine virus isolates analysed appear to have been independently introduced into pigs, two from human viruses and two from avian viruses. This supports the concept that pigs may serve as intermediates in the transmission of avian influenza viruses or their genes to the human virus gene pool.

Evolutionary stasis in avian influenza viruses

In contrast to the rapid, progressive changes of both the nucleotide and amino acid sequences of the mammalian virus gene lineages, avian virus genes show far less variation and, in most cases, there is no clear relationship between dates of isolation and position in the phylogenetic tree. Divergent cumulative evolution observed in the nucleotide sequences of avian virus genes has been correlated with geographic separation of host populations, e.g. North American avian virus strains versus Old World virus strains (Donis *et al.*, 1989; Gorman *et al.*, 1990*a*; Figs. 34.1 and 34.2). The absence of a corresponding lineage divergence among the avian virus proteins (e.g, Fig. 34.2) indicates that the genetic differences are accumulating in redundant non-coding changes and also indicates that there is a remarkable degree of phenotype stability among geographically separated virus gene pools.

A comparison of the patterns of coding and non-coding changes in mammalian and avian virus gene lineages shows fundamental differences in evolution (Gorman *et al.*, 1990*a*, *b*, 1991; Bean *et al.*, 1992). For example, terminal and internal branches in the mammalian virus NP gene lineages show similar numbers of coding and noncoding changes (compare nucleotide and amino acid trees in Fig. 34.2). This pattern suggests that mammalian viruses undergo positive directional selection; immunological selection by the mammalian hosts continuously select for new phenotypes which results in the continuous elimination of virus predecessors. In contrast, in the avian virus NP gene lineage (Fig. 34.2), most coding changes occur in the terminal branches whereas the internal branches show none or few coding changes (as reflected by their absence from the amino acid tree, Fig. 34.2*b*). This pattern suggests that avian viruses are subjected to stabilizing selection that maintains the ancestral phenotype. This conservation of phenotype (evolutionary stasis) probably indicates that the virus has reached a long-established adaptive optimum for virus proteins in the avian host. A possible mechanism for evolutionary stasis in avian viruses has been proposed (Bean *et al.*, 1992). If long-term survival favours those virions that do not change, then virus populations in environments that undergo relatively few replication cycles would be more likely to yield progeny that do not have deleterious mutations. Those replicating in other environments or mutants in the original population might have a temporary selective advantage in a particular host or environment, but the accumulation of mutations in these subpopulations would be deleterious in other hosts.

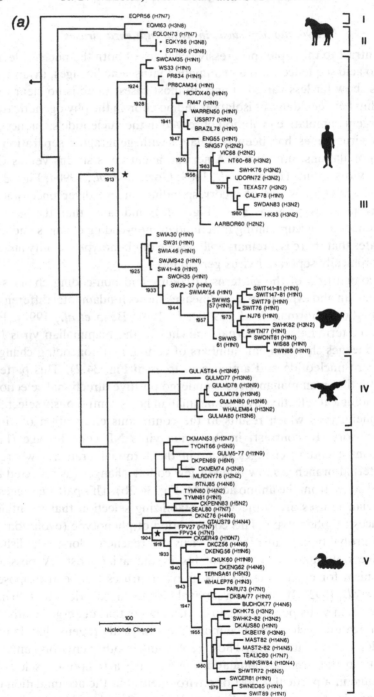

Fig. 34.2. (*a*) Phylogenetic tree of 89 influenza A virus NP gene nucleotide sequences rooted to the NP of B/Lee/40. Sequences were analysed with PAUP software (David Swofford, Illinois Natural History Survey) which uses a maximum parsimony algorithm to find the shortest length trees. The arrow indicates the direction of the B/Lee/40 NP from the root. Horizontal distance is proportional to the minimum number of nucleotide changes needed to join nodes and NP sequences. Vertical lines are for spacing branches and labels. Roman numerals indicate Bean's (1984) RNA NP hybridization groups: I, Equine/Prague/56; II, recent equine; III, human and classical swine; IV, H13 gull; V, avian. Animal symbols indicate host specificities of the lineages. Dates for hypothetical ancestor nodes were derived by dividing branch distance by evolutionary rate estimates. Stars indicate the hypothetical ancestors for the human-swine NP lineage (upper) and avian NP lineage (lower).

Fig. 34.2 (*b*) (*next page*) Phylogenetic tree of influenza A virus NP protein amino acid sequences. The amino acid phylogeny conforms to the topology of the nucleotide tree (*a*). (Sequences represented in these trees are listed in Gorman *et al.*, 1991.)

Thus the original population, or perhaps a small part of it, would have a selective advantage as hosts or environmental conditions change.

Rapid evolution in human influenza A viruses

The modes of evolution of influenza viruses in avian and human hosts appear to be very different. If, as argued below, all influenza A viruses have an avian origin, the rapid cumulative evolution in human virus genes and proteins reflects a lack of current equilibrium between virus and host, probably because of significant host-specific differences in the epidemiology, immune responses and the sites of replication in the avian and human hosts. The rapid evolution of virus surface proteins of human (as well as those of swine and equine viruses) is undoubtedly driven by immunoselection as the virus circulates in a partially immune population (Both *et al.*, 1983). Human influenza A viruses may be viewed as fugitive species that are able to survive and reinfect a population that has become immune; the virus must evolve rapidly enough to evade the immune protection of the host species. As a result of the rapid antigenic evolution of human influenza A viruses, vaccines must be changed regularly in order to provide protection.

Phylogenetic analyses have shown that internal viral proteins undergo a dramatic increase in evolutionary rate after avian virus genes were introduced into gene pools of mammalian viruses (e.g. the NP gene, Gorman *et al.*, 1990*a*, 1991). One plausible explanation is that evolution

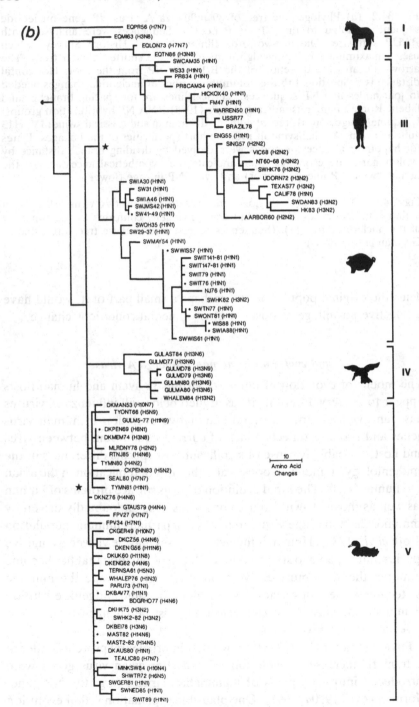

of internal proteins is carried 'piggy back' by the rapid antigenic evolution of the surface proteins, i.e. following periodic epidemic episodes and subsequent collapses of the virus population, random mutations in the genes of internal proteins are fixed by the cloning involved in the next epidemic episode (Buonagurio *et al.*, 1986). Another possibility is that these mutations may be associated with the adaptation of the virus to the new host or are driven by T cell immune selection. Of course, all of these mechanisms could operate simultaneously. However, if the first predominates, the random mutations might be expected to lead to a decrease in the relative fitness of the virus, whereas the latter mechanisms should lead to an increase in the relative fitness of the virus as mutations accumulate. Which of these possibilities has operated in past epidemics has not yet been distinguished.

Evidence for an avian origin of influenza viruses

The common features of influenza A and B viruses leave little doubt that they share a common ancestor. Although the record of the past is incomplete, enough evidence is available to provide some insight into the origin of these viruses. It would be interesting to know: (a) in what host did the first influenza virus evolve?; and (b) in what host(s) are the nearest common ancestors of each of the influenza virus gene segments? We believe there is strong evidence that all of the current gene segments circulating in both mammals and birds originated from avian influenza viruses.

In any long-continued host–parasite relationship there will be selection in the host to eliminate or lessen the deleterious effects of the parasite. In the parasite, there will be reciprocal selection for mutations that reduce the deleterious effects of the parasite. At equilibrium, the virus would be expected to replicate efficiently, cause minimal disease to the host, infect a high proportion of the host population, be perpetuated in host populations, and show increased genetic diversity with time. In relationships where the generation time of the parasite is much shorter than that of the host we can expect the parasite to evolve rapidly towards an adaptive equilibrium or optimum. Failure of this reciprocal host–parasite adaptation process is likely to lead to the extinction of one or the other.

A number of features of avian influenza viruses suggest that waterfowl may be the original hosts. Influenza viruses in wild waterfowl populations are ubiquitous, infection is nearly always asymptomatic, and large amounts of virus are shed by infected birds (Webster *et al.*, 1978). In addition, there is considerable genetic diversity in avian viruses; 14 haemagglutinin and 9 neuraminidase subtypes persist and circulate in the avian host reservoir. Each haemagglutinin and neuraminidase subtype appears to be antigenically and genetically homogeneous when compared with the antigenic and genetic differences between the subtypes. The selective pressure that would have caused diversification and the reasons for the continued co-existence of this array of avian virus subtypes remain unknown. The conservation of the proteins of avian viruses suggests that an adaptive optimum has been achieved. The apparent evolutionary stasis of these proteins suggests further that, within the normal avian host population, any modification of the protein sequence has proved to be detrimental. Therefore, avian influenza viruses and their waterfowl hosts appear to be a classic example of an optimally adapted system. The very small amount of evolution of avian virus proteins suggests that many centuries have been required to generate the current genetic diversity and distinct separation of avian virus haemagglutinin and neuraminidase subtypes.

There is only one known report of an avian virus causing disease in a wild bird population (Tern/South Africa/61; Becker, 1966). In domestic fowl and mammals, outbreaks and epidemics of influenza viruses are relatively frequent but unpredictable and are usually accompanied by disease symptoms and mortality. Only a few of the numerous avian virus subtypes have been observed in non-avian hosts. Evolution of virus proteins in non-avian hosts typically shows a rapid accumulation of mutations so that over time they become less like the avian viruses and this indicates an avian origin for these viruses. In summary, influenza viruses are probably long-established pathogens of wild birds and more transient in other hosts.

Summary of influenza virus evolution

A summary of the major vicariant events (those that lead to lineage divergence) in the evolution of influenza viruses is presented in Fig. 34.3. The relative order of the vicariances was deduced from the collective examination of influenza virus gene phylogenies (e.g. as summarized in Fig. 34.1), antigenic and genetic data, and historical records. Divergence of the major lineages of HA and NA subtypes and the split between NS 'a' and 'b' alleles represent possibly the oldest vicariant events in the evolution of influenza A viruses. The diversity of HA and NA subtypes among the older virus isolates (1902–1956) and the distinct pattern of evolutionary divergence among all known HA subtypes (Kawaoka *et al.*, 1990) indicate that their divergence predates the twentieth century. Moreover, the large differences among some avian virus gene lineages and the slow or negligible evolution of avian virus genes (Gorman *et al.*, 1990*a*, *b*, 1991; Bean *et al.*, 1992) suggests that some HA and NA groups and NS a and b alleles diverged at least centuries ago. The divergence of human influenza B viruses from avian influenza viruses must also be quite old, probably occurring after the divergence of the major avian HA subtypes but before the appearance of more recent HA subtypes (e.g. H13, H4, H14) (Fig. 34.3).

The oldest vicariant event among mammalian influenza A viruses is the divergence of EQPR56 (equine 1) viruses. Gorman *et al.* (1990*a*) provide 1800 as an estimate of the latest date of divergence of the EQPR56 virus. The divergence of other host-specific virus gene lineages occurred after this (during the nineteenth century). For example, the origin of the HA, M, NS, NP, and PA genes of present-day H13 gull viruses probably occurred in two separate re-assortment events during the nineteenth century. Other major nineteenth century events include the origin of North American avian virus strains, recent equine strains and FPV viruses (Fig. 34.3).

Present-day human viruses originated in the early twentieth century, just before the 1918 pandemic, and classical swine viruses were probably derived from human viruses during the 1918 pandemic (Gorman *et al.*, 1991). Re-assortment in human, recent equine, and possibly H13 gull viruses have occurred since the 1920s. The most recent events have been the appearance of a new avian-derived H1N1 virus in European swine populations in 1979 and the H3N8 virus in horses in Northeastern China in 1989. The new swine influenza virus has continued to evolve and circulate in European swine populations for more than 10 years. The most recent host-specific virus strain to appear is the new avian-like H3N8 equine influenza virus in China (Guo *et al.*, 1991).

Acknowledgements

Work in our laboratory is supported by National Institute of Allergy and Infectious Diseases grants AI-29680, AI-08831 and AI-20591 from the National Institutes of Health, Cancer Center Support (CORE) Grant CA-21765 and American Lebanese Syrian Associated Charities.

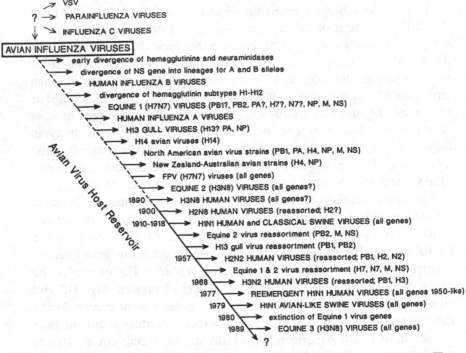

Fig. 34.3. Historical summary of the evolution of influenza viruses. The vicariant events (those that lead to lineage divergence) depicted in this Figure are in rough chronological order and have been deduced from collective analysis of influenza virus gene phylogenies, antigenic and genetic data, and historical records. The dashed line indicates an indeterminant time frame. The genes involved in evolutionary events are included in parentheses. The Equine 3 viruses at the bottom of the Figure refer to avian-like H3N8 viruses that have appeared in Northeastern China (Guo *et al.*, 1991). At the top, VSV stands for vesicular stomatitis virus.

References

Bean, W.J. (1984). *Virology*, **133**, 438–42.

Bean, W., Schell, M., Gorman, O., Kawaoka, Y., Katz, J. & Webster, R. (1992). *J. Virol.*, **66**, 1129–38.

Becker, W.B. (1966). *J. Hyg.*, **64**, 309–20.

Both, G.W., Sleigh, M.J., Cox, N. & Kendal, A.P. (1983). *J. Virol.*, **48**, 52–60.

Buonagurio, D.A., Nakada, S., Parvin, J.D., Krystal, M., Palese, P. & Fitch, W.M. (1986). *Science*, **232**, 980–2.

Donis, R.O., Bean, W.J., Kawaoka, Y. & Webster, R.G. (1989). *Virology*, **169**, 408–17.

Gammelin, M., Altmuller, A., Reinhardt, U., Mandler, J., Harley, V.R., Hudson, P.J., Fitch, W.M. & Scholtissek, C. (1990). *Mol. Biol. Evol.*, **7**, 194–200.

Gorman, O.T., Donis, R.O., Kawaoka, Y. & Webster, R.G. (1990*b*). *J. Virol.*, **64**, 4893–902.

Gorman, O.T., Bean, W.J., Kawaoka, Y. & Webster, R.G. (1990*a*). *J. Virol.*, **64**, 1487–97.

Gorman, O.T., Bean, W.J., Kawaoka, Y., Donatelli, I., Guo Y. & Webster, R.G. (1991). *J. Virol.*, **65**, 3704–14.

Guo, Y., Wang, M., Kawaoka, Y., Gorman, O., Ito, T. & Webster, R.G. (1991). *Virology*, **188**, 245–55.

Ito, T., Gorman, O., Kawaoka, Y., Bean, W. & Webster, R.G. (1991). *J. Virol.*, **65**, 5491–8.

Kawaoka, Y., Krauss, S. & Webster, R.G. (1989). *J. Virol.*, **63**, 4603–8.

Kawaoka, Y., Yamnikova, S., Chambers, T.M., Lvov, D.K. & Webster, R.G. (1990). *Virology*, **179**, 759–67.

Okazaki, K., Kawaoka, Y. & Webster, R.G. (1989). *Virology*, **172**, 601–8.

Scholtissek, C., Burger, H., Kistner, O. & Shortridge, K.F. (1985). *Virology*, **147**, 287–94.

Snyder, M.H., Buckler-White, A.J., London, W.T., Tierney, E.L. & Murphy, B.R. (1987). *J. Virol.*, **61**, 2857–63.

Tian, S.F., Buckler-White, A.J., London, W.T., Reck, L.J., Channock, R.M. & Murphy, B.R. (1985). *J. Virol.*, **53**, 771–5.

Treanor, J.J., Snyder, M.H., London, W.T. & Murphy, B.R. (1989). *Virology*, **171**, 1–9.

Webster, R.G., Bean, W.J., Gorman, O.T., Chambers, T.M. & Kawaoka, Y. (1992). *Microbiol. Rev.*, **56**, 152–79.

Webster, R.G., Yakhno, M.A., Hinshaw, V.S., Bean, W.J. & Murti, K.G. (1978). *Virology*, **84**, 268–78.

Part VII
Techniques for viral systematics

Part VII
Techniques for virus resistance

35

The RNase A mismatch method for the genetic characterization of viruses

C. LÓPEZ-GALÍNDEZ, J.M. ROJAS
AND J. DOPAZO

Introduction and background

Point mutation has gained a lot of attention in molecular biology because of the implication of this genomic alteration in medicine and pathology, such as genetic disorders and cancers. The RNase A mismatch method was developed for the detection of point mutations related to the activation of the K-ras oncogene in colon tumours using RNA:RNA hybrids (Winter *et al.*, 1985) and to the diagnosis of genetic disorders with RNA:DNA hybrids (Myers, Larin & Maniatis, 1985).

RNA viruses are characterized by great genetic variability. This implies the occurrence of frequent mutation in the genome of different isolates. Some of these mutations are involved in phenotypic properties, such as virulence, tropism, resistance to antiviral drugs and other characteristics (Domingo *et al.*, 1985). They are also the basis for evolutionary studies and strain comparison. Mutations in RNA viruses have been detected by T1 oligonucleotide fingerprinting and lately by sequencing through cDNA.

Our group adapted the system of RNase A mismatch for studies on genetic variability of RNA viruses using influenza orthomyxovirus as a model (López-Galíndez *et al.*, 1988) and Owen and Palukaitis (1988) to plant viruses. The system is based on the comparison by hybridization of a riboprobe from a reference strain with RNAs from different strains. Each one will give a complex pattern of bands resistant to the RNase A digestion which is specific for each one as a fingerprint. Comparing the different patterns we are able to draw a qualitative estimate of genetic relatedness and evolution of field strains (López-Galíndez *et al.*, 1988, 1991).

Although the technique does not provide a comparable amount of information to nucleic acid sequencing, it is more accurate and has technically some advantages over other methods because it gives more information than the classical restriction fragment length polymorphism technique (RFLP), used for DNA viruses, that analyses around 1% of the genome, and T1 oligonucleotide fingerprinting employed for RNA viruses.

Application of RNase A mismatch method in virology

Molecular epidemiology studies

Genetic analysis of influenza

To study the possibilities of the method in virology, we analysed the influenza virus haemagglutinin gene, with a 1200 nucleotide long antisense riboprobe, in different reference strains and field isolates from an outbreak of the H3N2 subtype. The complexity of the band patterns obtained correlated with the genetic relationships and the isolation date of each virus to the reference strain used for the synthesis of the probe, allowing a qualitative estimation of their genetic distance. The isolates obtained from the same outbreak were very similar in their banding patterns (López-Galíndez et al., 1988).

Genetic analysis of human immunodeficiency virus (HIV)

In HIV we have used the RNase A mismatch method to study different aspects of genetic variability: molecular epidemiology studies, comparison of sequential isolates of infected individuals, analysis of distinct viral clones and appearance of phenotypic variants. We have used multiple gene analysis in each virus looking for conserved regions of the genome, the reverse transcriptase (rt) gene, gag gene, the env gene coding for GP41 and vif-vpu region covering altogether around 17–20% of the genome. The samples were from two geographical locations Madrid (Spain) and San Diego (California, USA) and the Lai strain was the template for the riboprobe synthesis. Looking at the env gene, we have detected the presence of two genotypes, one related to SF-2/RF reference strains and another closer to Lai, cocirculating in San Diego. This was concluded from the presence of common bands (from 2–5) in some viruses. On the other hand, the Spanish viruses are a group of genetically homogeneous viruses and related to Lai. We also detected the presence of the AZT resistant genotype among the isolates.

When analysing biological clones from a viral population, it permits an estimation of the composition of variants within viral populations, by comparing the patterns observed and the band intensity of the global population in relation to those obtained from cloned viruses.

Genetic analysis of other RNA viruses

Using a multigenic approach, Cristina *et al.* (1990) were able to detect the presence of two subgroups of respiratory syncytial virus (RSV) circulating in one area and Owen and Palukaitis (1988) divided different strains of cucumber mosaic virus (CMV) into two groups. García-Arenal (personal communication) has used the technique to study satellite RNAs of isolates of CMV and also to analyse the genetic variability of different regions of tobacco mild green mosaic virus (TMGMV).

Genetic analysis of herpes simplex virus (HSV)

This method is not only limited to RNA viruses; it can also be used to DNA viruses because RNase A also digests RNA:DNA hybrids. The first application of this methodology to DNA virus has been for the analysis of genetic variability of herpes simplex virus type 1 and 2 (HSV-1 and HSV-2).

Twenty-five isolates of each subtype, obtained from epidemiologically unrelated patients of the Madrid area (Spain), were studied using riboprobes of thymidine kinase (TK) and gᴮ genes of both types. The results showed the possibility of clustering viruses from the same geographical area in different groups or subtypes (Rojas *et al.*, 1995). Also, the genetic variability detected was similar in the two analysed genes. This approach was used to study HSV latency, by the analysis of sequential HSV isolates of the same patient. As all HSV-1 isolates obtained from brain biopsy (from a patient with herpetic encephalitis) showed similar RNase A mismatch pattern with TK riboprobe, this could indicate a tropism relation.

Detection of phenotypic variants

When setting up the method with influenza virus, we analysed different resistant viruses against one neutralizing monoclonal antibody. Using different and overlapping probes, a possible mutation related to the resistant phenotype was detected, mapped and then sequenced (López-Galíndez *et al.*, 1988).

In HIV, when we analysed field isolates of San Diego and Madrid in the RT gene a set of specials bands appeared in the gels that were identified with the mutations related to the AZT resistant phenotype (Larder, Darby & Richman, 1989; López-Galíndez et al., 1991). The RNase A offers a simple approach to this problem because it always detects the double mutation in codon 215 of the RT gene, the most important one out of the four related to this phenotype (Larder et al., 1987).

In HSV, different point mutations in the TK gene of HSV are associated with phenotypes resistant to nucleoside analogue drugs. Although in the majority of cases we did not find any mutation in relation to the wild type, some of these mutations were detected by the RNase A cleavage method (Rojas, 1991). In TK mutants, the most frequent mutations are transitions G-A (giving a G-U mismatch) and these mismatches are very poorly recognized by RNase A.

Technical considerations

One of the disadvantages of the technique is that only a fraction of all possible mismatches are cleaved by RNase A and even those which are recognized vary in the extent of hydrolysis. This depends not only on the bases forming the mismatch but also on the surrounding nucleotides (Perucho, 1989). The percentage of the recognized mismatch has been estimated around 60% of the total. This aspect is nevertheless more important in the detection of phenotypic variants than in molecular epidemiology studies. It is worthy to note that RNase A digestion is done in partial conditions, in order to obtain fewer fragments, but nevertheless it is a reproducible reaction.

All the previous studies have been done with total RNA from infected cultures hybridized to a P^{32} labelled antisense riboprobe. In HSV, purified viral DNA was used with the riboprobe in an RNA:DNA hybrid.

The cultivation of a virus is a very time-consuming process and moreover it may allow mutation and selection of the viral population (Meyerhans et al., 1989). Therefore, we have checked the use of the amplification by the polymerase chain reaction (PCR) of the RT HIV gene coupled to the RNase A mismatch analysis in the detection of the AZT resistance phenotype. The results showed that the AZT resistant phenotype could be easily detected by mismatch analysis from total RNA of lymphocytes after first strand synthesis by RT and two nested PCR as in the cultivated virus. The PCR amplification procedure can also be used to study other viruses starting from direct material of the infected host.

Conclusions and future perspectives

In summary, the RNase A mismatch method is a technique easy to perform which allows the analysis of many samples at the same time. Using distinct riboprobes that cover different regions of the viral genome it permits the study of large and/or specific parts of the nucleic acid. It is a good technique for a primary genetic characterization of viruses and in molecular epidemiology studies of RNA and DNA viruses. It is very helpful in the description of different subgroups within viruses circulating such as in RSV, HIV and to plant viruses like CMV. It gives easily a valuable information on different genes that could be used in order to decide which strain or genes are interesting to be studied further by nucleotide sequencing. To draw relationships between viruses it is convenient to select more conserved regions of the genome for the riboprobes. It is also a good complement to information coming from other sources such as antigenic analysis.

For detecting phenotypic variants it can give a primary map of mutations that may be related to the characteristic being studied. For detecting AZT resistant isolates in which the mutations are well characterized, it gives a rapid and simple detection of the phenotype coupled to a PCR amplification of lymphocyte material. This approach, using PCR amplification of nucleic acids from primary tissues or from isolation material, could be generalized and avoid the need for virus cultivation.

In spite of the sensitivity of this technique, it suffers an important drawback: although it clearly distinguishes between two different sequences, it does not give a quantitative estimate of the extent of this difference. Indeed, in the few attempts made to quantify its application (Cristina *et al.*, 1991) it has been specifically stated that the classification obtained for the sequences analysed was based on pattern resemblance, which might not agree with the actual genetic kinship. Computer simulation experiments (Dopazo *et al.*, 1993) demonstrated that, when comparing patterns of RNase A digestion of hybrids, a good correlation can be found between the number of non-shared bands and the number of mismatches that the sequences have. Where this relationship holds, it is possible to obtain reliable estimations of genetic distances from the comparison of the digestion patterns, and a wide field of application for the RNase A mismatch cleavage method in epidemiological and evolutionary studies is possible.

References

Cristina J., López, J.A., Albó, C., Garcia-Barreno, B., Garcia, J., Melero, J.A. & Portela, A (1990). *Virology*, **174**, 126–34.

Cristina, J., Moya, A., Arbiza, J., Russi., Hortal, M., Albo, C., Garcia-Barreno, B., Garcia, O., Melero, J.A. & Portela, A. (1991). *Virology*, **184**, 210–18.

Domingo, E., Martinez-Salas, E., Sobrino, F., de la Torre, J.C., Portela, A., Ortin, J., López-Galíndez, C., Pérez-Breña, P., Villanueva, N., Nájera, R., Vandepol, S., Steinhauer, D., dePolo, N. & Holland, J.J. (1985). *Gene*, **40**, 1–8.

Dopazo, J., Sobrino, F. & López-Galindez, C. (1993). *J. Virol. Methods*, **45**, 73–82.

Larder, B.A., Darby, G. & Richman, D.D. (1989). *Science*, **243**, 1731–34.

Larder, B.A., Purifoy, D.J.M., Powell, K.L. & Darby, G. (1987). *Nature*, **327**, 716–17.

López-Galíndez, C., López, J.A., Melero, J.A., de la Fuente, L., Martinez, C., Ortin, J. & Perucho, M (1988). *Proc. Natl. Acad. Sci. USA*, **85**, 3522–6.

López-Galindez, C., Rojas, J.M., Najera, R., Richman, D.D. & Perucho M. (1991). *Proc. Natl. Acad. Sci. USA*, **88**, 4280–4.

Meyerhans, A., Cheynier, R., Albert, J., Seth, M., Kwok, S., Sninsky, J., Morfeldt-Mason, L., Asjo, B. & Wain-Hobson, S. (1989). *Cell*, **58**, 901–10.

Myers, R.M., Larin, Z. & Maniatis, T. (1985). *Science*, **230**, 1242–6.

Owen, J. & Palukaitis, P. (1988). *Virology*, **166**, 495–502.

Perucho M. (1989). *Strategies*, **2**, 37–41.

Rojas, J.M. (1991). PhD Thesis, Univ. Autonoma, Madrid.

Rojas, J.M., Dopazo, J., Santana, M., López-Galíndez, C. & Tabarés, E. (1995). *Virus Research*, in press.

Winter, E., Yamamoto, F., Almoguera, C. & Perucho, M. (1985). *Proc. Natl. Acad. Sci. USA*, **82**, 7575–9.

36

Molecular phylogenetic analysis

GEORG F. WEILLER, MARCELLA A. McCLURE AND ADRIAN J. GIBBS

Introduction

When the key biological molecules, proteins and nucleic acids, were first studied, they were characterized by their sizes, compositions and, for proteins, their antigenicity. These simple characters were often used to infer the relationships of the organisms from which the molecules were obtained, and the results seemed to be sensible. However, it was not until the sequences of the amino acids in the proteins were determined, and more recently, the nucleotides in the nucleic acids, that the molecular basis of the relationships were shown to reside in the sequences and their three-dimensional structures.

In this chapter we outline some of the methods used for inferring the phylogenetic or other relationships of sequences. We distinguish the various components of analysis; the ways that nucleotide and amino acid sequences are recorded for analysis, the means by which that data can be transformed to improve the chances of discovering the true relationships, and the ways by which the relationships are inferred and displayed. We have omitted to discuss the more traditional ways for indirectly comparing sequences using, for example, their composition, RFLP similarity or nearest neighbour frequency. However, it is worth emphasizing that, if such methods indicate relationships that correlate with those inferred from sequence data, then a much larger body of older, simpler data may become available and permit broader comparisons and extrapolations.

Sequence analysis

Aims of sequence analysis

The most appropriate way to analyse a sequence depends on the goal of the analysis. Most often the reason for determining whether a new sequence is significantly similar to other sequences is either to predict its function by analogy with those of known function, or to understand its evolutionary relationships. However, sequences probably contain several types of information, and the main problem in their analysis is to extract the information of choice; 'one person's signal is another person's noise'.

When first determined, a new gene sequence is usually checked for base composition, internal repetition, known motifs and control regions and, if single-stranded, its potential for folding and forming secondary structures. If the sequence contains open reading frames (ORFs), the implied amino acid sequence may be analysed for its codon usage, hydropathy profile and possible secondary structure as this may, for instance, help to identify transmembrane regions, etc. Almost always the nucleotide and implied amino acid sequences are used to search databases for related sequences. If the intention is to try to understand its evolutionary relationships, it is aligned with related sequences, their relationships are calculated and represented in a simplified form, usually as a tree or network. The likelihood that the relationships are significant may be assessed and, finally, attempts may be made to correlate those relationships with other characteristics of the organisms from which the sequences were obtained.

Sequence variation of viral genes

When analysing new sequences, it helps to be aware of the commonest types of molecular evolutionary change already found. For example, the genomes of closely related viruses, such as isolates of a single viral species, usually only differ by point mutations. These are usually found in the third (redundant) codon positions of open reading frames (ORFs); thus redundancy of the genetic code ensures that nucleotide sequences usually evolve more quickly than the proteins they encode. The sequences may also have a few inserted or deleted nucleotides (indels). The genomes of different species of each viral genus usually differ by a greater number of changes of the same type as those found in different isolates of the same species. Some genes are conserved more than others, especially those parts encoding, for example, catalytic sites or the core of proteins. Other genes may have little or no similarity.

Viruses placed in separate genera usually have discernible sequence relatedness only in the genes that encode viral enzymes or virion protein genes (Chapters 4 and 5). These similarities, when found, are usually very distant, involve only short segments (motifs) interspersed with large regions with no similarity and of variable length, indicating that many mutations and indels have occurred since they had evolved from their common ancestor. Some of the viral proteins from separate genera may have no significant sequence similarity but clearly similar secondary and tertiary structures; primary structure is lost more quickly than secondary and tertiary structure during evolutionary change. For example, the virion proteins of all viruses with RNA genomes and isometric particles about 30 nm in diameter, except those of the leviviruses, show such distant similarities (Rossmann & Johnson, 1989).

Viruses from different genera usually have different combinations, and order of occurrence, of shared genes in their genomes. Taxonomies calculated separately for such shared genes are often not congruent. These facts indicate that viral genera probably acquired the shared genes by recombination from two or more different ancestral viruses (Chapters 4, 9, 25 and 28). This feature clearly distinguishes the evolution of viruses from that of cellular organisms, where genetic recombination between distantly related organisms seems to be rare.

In addition to the genes shared with other viral genera, there are, in most viral genomes, one or more genes that seem to be shared only with species of the same genus. Some of these may be related to genes that are as yet unknown, others may be related to known genes but have changed so much during evolution that no similarity remains, and some may have arisen *de novo* (Chapter 6). Thus differences between isolates of a single viral population are assessed most sensitively by analysis of their nucleotide sequences. More distant relationships, between viral species and genera, are best analysed by comparing amino acid sequences, and may only be revealed by parts of some genes and their encoded proteins.

Searching for homologous sequences

There are several monographs on the theory, methods and practice of analysing molecular sequence information (e.g. Von Heijne, 1987; Waterman, 1988; Doolittle, 1990); that by Gribskov and Devereux (1991) is particularly helpful. There are also valuable ideas to be found in more general texts on molecular evolution, such as those by Nei (1987), Li and Graur (1991) and Hillis and Moritz (1990). The multitude of sequence alignment and database search programs and scoring schemes that are available have been reviewed by Meyers (1991) and States and Boguski

(1990). Some of these methods have been compared by Chan, Wong and Chiu (1992) and McClure, Vasi and Fitch (1993).

Comparing pairs of sequences

The simplest method for comprehensively comparing two sequences and finding regions of sequence similarity is the dot diagram (Gibbs & McIntyre, 1971), often also called the dot plot. This is worth describing briefly as, when understood, it is easier to understand other methods devised for comparing and analysing sequences. In the simplest dot diagram, the two sequences are placed at right angles to form the adjacent edges of a rectangular matrix, and a dot is put in the matrix wherever a row and column with the same sequence element intersect. Sequence similarities then appear as diagonal runs of dots in the diagram (Fig. 36.1). The diagonal runs can be seen by eye, and can be tested for significance in various ways. Closely similar sequences usually give diagrams with a run of dots along a single diagonal, mutations put gaps into this run, and indels cause the run to 'jump' from one diagonal to another. Sequence repetitions give parallel diagonal runs of dots. Hence it is always useful to compare a sequence with itself by this method as the first stage of an analysis.

There are many ways in which this simple method can be made more sensitive (States & Boguski, 1990). Segments of the sequences (windows) rather than single elements may be compared, and quantitative estimates of the similarities of element types (see 'Similarity Matrices' below) used.

When a dot diagram has runs of similarities in more than one diagonal, it may be difficult to decide which is the optimum pathway. This can be done objectively by the elegant 'dynamic programming algorithm' that was devised by Needleman and Wunsch (NW) (1970). This uses a matrix of similarities to weight independently each pairwise comparison in the primary dot diagram. Then it calculates a second matrix, which is a progressive diagonal summation function of the primary matrix, with weights added for gaps. Finally, a search for diagonal runs with the largest scores in the second matrix reveals the optimum pathway. There are many variants of the NW algorithm that seek to improve its speed and resolution. For example, gaps can be weighted in different ways (Sellers, 1974), the search may be restricted to parts of the primary matrix close to the main diagonal (Sankoff & Kruskal, 1983), or only the most similar parts of the sequences aligned (Smith & Waterman, 1981) so that significant similarities do not become 'diluted' by unrelated sequences.

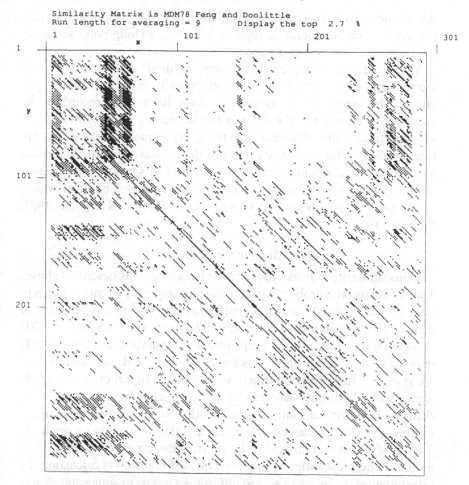

Fig. 36.1. Dot diagram comparing the virion proteins of the A and B strains of maize dwarf mosaic potyvirus (*x* and *y* axes, respectively). Similarities were scored using Dayhoff's MDM78 matrix averaged over nine residues; the top 2.7% of the resulting diagram entries are represented by dots. Note that the N-terminal regions of these two viruses differ in sequence and in length, and contain sequence repetitions.

Sequences can also be compared in fragments. Simplest of all is an analysis of doublets, nearest neighbours or Markov chains (Gibbs *et al.*, 1971; Blaisdell, 1986, 1989); these give sensible taxonomies of phages even from non-homologous genes (B.E. Blaisdell, personal communication). Each sequence is converted to a doublet matrix and the patterns of doublet frequencies in the matrices are compared. Obviously, this method can be increased in specificity by comparing longer runs (i.e. triplets, etc, or, in general, 'k-tuples'); however, all long-range sequence information is lost in this method.

A related, but more discriminatory, method is to compare both k-tuples and their position, a method that uses 'hash' or 'neighbourhood' tables. One sequence is converted into a hash table recording the positions of each type of k-tuple in the sequence. The position of the same k-tuples is then determined in the query sequence, and, if they are found to be in related relative positions, this indicates that the sequences are related. The k-tuple method is increased in sensitivity by using a similarity matrix to quantify the k-tuple similarities. The most widely used k-tuple programs are FASTA and FASTP (Wilbur & Lipman, 1982; Pearson, 1990).

Similarity matrices

Various similarity matrices are used in the calculations described above. This is because the probability that different sequence elements (nucleotides or amino acids) occur in homologous positions in related genes or their encoded proteins varies greatly; transitions (G<>A or C<>T/U changes) usually occur more frequently than transversions (G<>C/T/U or A<>C/T/U) in the early stages of divergence, and some amino acids are replaced frequently by others with similar properties, for example serine with threonine, whereas others, like cysteine, seem rarely to change. Thus for calculations, such as the Needleman–Wunsch alignment algorithm, one can use these probabilities tabulated as a similarity matrix, of which that devised by Dayhoff, Schwartz and Orcutt (1978) is probably the most widely used. The Dayhoff matrix was calculated, by extrapolation, from the observed differences in sets of somewhat closely related sequences. However it has been found to be biased in some ways, and new similarity matrices, like the ones calculated by Gonnet, Cohen and Benner (1992) from a very large set of sequences, may prove superior in some cases. Incidentally, during the analyses preparing the 'Gonnet' matrices, it was found that the probability of occurrence of a gap (or insertion) of length k decreased as an exponential function, $k^{-1.7}$, not a linear one, as is used in most algorithms.

Other matrices based on physical or chemical similarities or, for amino acids, the number of mutations required to change from one to another, have also been used (Feng, Johnson & Doolittle, 1985; Taylor, 1986; Rao, 1987; Risler et al., 1988). Fortunately, the results obtained using different matrices usually correlate well.

Databases

At present sequence information is stored in databases at the National Centre for Biotechnology Information (NCBI), the National Library of Medicine (NLM), the European Molecular Biology Organisation (EMBO) and the DNA Database of Japan (DDJ). Most of these databases are 'in the public domain' (i.e. available for use without charge) and all commercially developed databases are derived from these sources. The host associations for these databases change from time to time, and the latest information about them is best acquired through the international computer networks, using Gopher, Mosaic or World Wide Web facilities.

Currently the main primary nucleic acid databases are GenBank, EMBL and HIVNUC, and the analogous protein ones Genpept, Swissprot and HIVPRO. Only those sequences that are clearly designated as genes are translated and present in the protein databases; some gene sequences that are poorly annotated when submitted to the databases may not have been translated and included in the protein databases. The original PIR database is also available, but has not been updated for several years. All of the databases have redundant data, list sequence fragments and have inaccurate titles and descriptions. There are non-redundant protein databases, such as OWL; however, note that such secondary databases may have had to decide whether a difference between two sequences is real, and should be recorded, or is just an error; even a single difference could be crucial in a statistical analysis.

Individual databases are searched in various ways, and some organizations provide a search service via the international computer network, for example BLAST at NCBI, which uses a FASTA like program to search the GenBank databases (Altschul, Carroll & Lipman, 1989), and BLITZ, which searches the Swissprot database by the Smith and Waterman method (1981). One should be aware, however, that no search method is perfect, and related sequences may be missed.

Multiple sequence alignment

The two most common reasons for aligning several sequences are either to reconstruct the evolutionary history of a protein family, or to find the structural similarities of members of a protein family. There is considerable overlap in these approaches but they are different and require different programs.

Within a multiple sequence alignment conserved patterns or motifs that define the entire family may be found (Day & McMorris, 1993). These can then be used to search the entire database for additional members of the family. Among families of conserved sequences (greater than 50% identity) it is very difficult to deduce which residues of a set of aligned sequences are important for their function or structure. However, among distantly related proteins (less than 30% identical residues) conserved residues often indicate the regions of a protein that are important.

Multiple alignment methods

Two types of methods are used for aligning sequences: global or local. *Global methods* attempt to find an 'optimal' alignment throughout the length of sequences (Barton & Sternberg, 1987a,b; Feng & Doolittle, 1987; Taylor, 1987; Lipman, Altschul & Kececioglu, 1989; Subbiah & Harrison, 1989; Higgins & Fuchs, 1992). A sub-class of global methods attempts first to identify an ordered series of motifs and then proceeds to align the remaining regions (Martinez, 1988; Vingron & Argos, 1991). *Local methods*, such as the Multiple Alignment Construction Workbench (MACAW; Schuler, Altschul & Lipman, 1991) and PIMA (Smith & Smith, 1990; Waterman & Jones, 1990), which use the Smith–Waterman algorithm, only attempt to identify motifs and do not attempt to align regions between motifs. Global methods are best for both evolutionary and structural studies of unequivocally related sequences, whereas local methods are best for seeking distant relationships and for motif identification.

Global methods rarely attempt to compare simultaneously all sequences because of the computational expense. Usually the two most closely related sequences (determined by pairwise comparisons) are aligned and the others, in order of similarity, are added progressively. This method of progressive multiple alignment was introduced in 1987 (Feng & Doolittle, 1987; Taylor, 1987). The alignment continues in an iterative fashion, adding gaps where required to achieve alignment, but only to all members of each growing cluster. The most widely used variant of this approach is CLUSTAL V (Higgins & Fuchs, 1992). The shortcoming of this approach is the inherent bias when one subset of sequences is over represented, but this can be corrected (Altschul *et al.*, 1989). Other methods directly produce a progressive multiple alignment or an optimal alignment defined in some way (Subbiah & Harrison, 1989; Lipman *et al.*, 1989; Vingron & Argos, 1991). Barton and Sternberg's method (1987*b*) allows the user to analyse sequences following either strategy.

A comparative study of global and local methods was made by McClure *et al.* (1993). Four data sets, each containing sequences from the globin, kinase, retroid aspartic acid protease and ribonuclease H sequence families, were aligned and used to test the ability of twelve global and local methods to accurately identify their characteristic motifs. It was found that CLUSTAL V and DFalign (also called PILE-UP in the GCG package) perform better than the others; the global methods were better than the local methods; PIMA was slightly better than MACAW. CLUSTAL V is flexible and user friendly, its instructions and interactive interface are clear, it can align both nucleic acids and proteins and there are DOS, VMS and UNIX versions.

Other new approaches to multiple sequence alignment are being tested; one such is the use of Hidden Markov Models (HMM) with adaptive algorithms for parameter training, which has also been used for speech recognition (Baldi *et al.*, 1994; Krogh *et al.*, 1994). The method creates a stochastic model using the sequences being studied as a 'training set'. At each position of a sequence there is the probability that a residue, a deletion or an insertion will occur. The model starts with a uniform probability of each type; one chance out of three for a nucleotide, deletion or insertion and one out of 20 for a particular amino acid for the residue or insertion states. As the model 'learns' from each sequence, these probabilities are adjusted. Thus the differences in a family of sequences are incorporated into the model, so there is no need for a separate dissimilarity matrix. The final model is used as a template for aligning individual sequences to produce a multiple alignment, and for identifying related sequences.

Excellent though these methods may be, it is clear that none of the currently available computer methods is foolproof, and manual refinement of the alignment may still be required. A skilled scientist can best identify and correct small regions of similarities not detected by multiple alignment programs. Multiple alignment editors such as the Genetic Data Environment (GDE), that runs in X-Windows on Unix systems, greatly help a human operator with such final corrections.

Motifs

Each protein family is characterized by a set of conserved or semi-conserved amino acids that are always found in the same order in the sequences of that family. The spacing between the motifs may vary, perhaps because these are regions of the protein that have fewer functional constraints, or perhaps because these regions have different functions in different members of the family. Some members of a protein family may not have all the motifs present in others; furthermore motif sequences may vary, so searching for them and aligning them may be difficult. Various motif families are well studied, and are stored in specialized databases, such as PROSITE. The motifs commonly found in viral sequences include the -GDD- polymerases, the -GCGK/T- helicases, and -PEST- (degradable) sequences. Many of the motif families probably represent the progeny of single ancestral sequences, but this is not proven for all, especially as the sequences are so short, and it is likely that some represent convergences.

Profile analysis

The search for additional members of a gene family, especially its distant ones, is done most sensitively by profile analysis (Gribskov, McLachlan & Eisenberg, 1987). First the sequences known to be related are aligned and the residues at each position in them are represented in a single matrix, the profile, which is then used instead of a single sequence to search for related sequences in a database using a dynamic alignment algorithm.

Statistical analyses

It is always desirable to assess whether an alignment is correct and to estimate the statistical confidence in that alignment. Some of the tools available in general statistical packages, such as NTSYS (Version 1.80; Rohlf, 1993), are useful, but some programs, such as MEGA (Kumar, Tamura & Nei, 1993), are designed specifically for molecular phylogeny.

Methods of statistical analysis of pairwise alignments are difficult to apply directly to multiple alignments, but Monte Carlo methods are useful. The sequences in each pairwise alignment are randomly shuffled and the mean score and standard deviation (SD) calculated from these randomized sets. Different shuffling regimes test different hypotheses (Faith & Cranston, 1991); shuffling within each sequence tests the amino acid or nucleotide order, whereas shuffling the nucleotides but retaining their position within each codon is an alternative, and shuffling amino acids within each sequence position can test the tree structure. If potential relationships are judged to be significant, >3–5

SD above the mean (Schwartz & Dayhoff, 1978), then the sequences are probably homologous and multiple alignment is statistically justified. There are limits to these methods, however, for determining distant protein relationships, and for evaluating the probability of the chance distribution of an ordered series of conserved motifs along the entire length of the multiple alignment of a given set of sequences, because they depend on the reality of the randomization pattern applied.

When two protein sequences share more than 30% identity, the methods of analysis outlined above are straightforward and unambiguous. However, there is less confidence in relationships with less than 30% identity, and these must be corroborated by other experimental data. Many viral protein relationships fall into this latter category. There are numerous instances, for example when aligning the retroviral and *E. coli* ribonuclease H segments to other more distantly related retroid ribonuclease H sequences (McClure, 1992), where there is no statistical evidence that the sequences are related, though there is biological/functional evidence that they are. In practice, when sufficient biological information exists, and it is obvious to the human eye that they contain an ordered series of motifs, then the lack of statistical significance in the relationship suggests that the method used for testing it was inappropriate.

Reconstructing phylogeny

Reconstructing phylogeny from gene sequences is, unfortunately, not as straightforward as one might hope, and it is rarely possible anyway to verify that one has arrived at the right conclusion. There are no uniquely correct methods for inferring phylogenies, and many methods are in use. These are sometimes grouped according to whether they adopt a phenetic or cladistic approach.

Phenetic methods try to reconstruct phylogeny using phenotypic data without attempting a complete understanding of the evolutionary pathway. In contrast, *cladistic methods* concentrate on the evolutionary pathway; they attempt to predict the ancestors by assuming certain criteria (e.g. parsimonious evolution and the recognition of 'shared derived' characters). Trees obtained by either method have sometimes been called *phenograms* or *cladograms*, respectively (see below). At first, the cladistic approach might appear superior but in practice the criteria used to predict the ancestors do not always apply during evolution and thus may lead to wrong conclusions. We doubt that the terms cladistic and phenetic are helpful to classify the various methods, some of which use a combination of both (Hein, 1990), and other classifications (Penny, Hendy & Steel, 1992) are equally useful.

Tree terms

Phylogenetic relationships (pedigrees) often resemble the structure of a tree. Hence the various diagrams used for depicting these relationships are commonly called *phylogenetic trees*, and the terms referring to the various parts of these diagrams (*root, stem, branch, node, leaf*) are also reminiscent of trees.

Phylogenetic trees can also be described using the terms of *graph theory* (see Harary, 1969; Gould, 1988), which describes the connectivity of objects. Thus it can describe any system of connected objects, such as road maps, data structures used in computing, the index of this book, the maze at Hampton Court Palace and, last but not least, phylogenetic trees. Some mathematically oriented phylogenists have suggested the exclusive use of its terminology, so we will describe it briefly.

On your last flight you probably looked at the 'In Flight' magazine, and found a map showing the cities between which the airline flies and the routes taken. This map is a typical example of a *graph*. A graph consists of *vertices* (cities) and *edges* (routes). An edge always connects two vertices, but as the airline does not fly to all cities, some vertices have no edge. If the map not only shows the routes but also the mileage between the cities then it is a *weighted graph*, as opposed to a map only showing routes, which is *unweighted*. A vertex that has only one edge is called *external* or *pendant*, as opposed to *internal* edges. The sequence of edges that connect two non-adjacent vertices is called a *path*, of which there may be more than one between any two vertices. In an unweighted graph the *path length* is simply the number of edges that make up the path. Airlines usually fly their routes in either direction; if the edges of a graph do not specify any direction the graph is called *undirected* otherwise the graph is called *directed* or a digraph. A vertex that starts a directed path is called an *origin*. A *cyclic* path, as opposed to an *acyclic* one, visits several cities before returning to the first, and if all points are on a path, the graph is said to be *connected*. Thus a rooted phylogenetic tree is an *acyclic connected digraph*.

Unfortunately the terminology used in graph theory also varies. Frequent synonyms for vertex are: *node, point* and *dot*. External vertices are also called *pendant dots, terminal nodes, outer points* or *leaves* and distinguished from *internal* or *inner* vertices. Origins are also termed *roots*. Synonyms for edge are *line, link, arc, segment, interval, internode* and *branch*. We prefer to use the terms 'vertex', and 'edge' when referring to a graph, but 'node' and 'branch' when referring to a tree.

External nodes, the extant taxa, are often called *operational taxonomic units* (OTUs), a generic term that can represent many types of comparable taxa, for example, a family of organisms, individuals of a single species, a set of related genes or even gene regions. Similarly internal nodes may be called *hypothetical taxonomic units* (HTU) to emphasize that they are the hypothetical progenitors of OTUs.

Styles of dendrograms

Various styles of dendrograms are used to depict phylogenetic trees. Fig. 36.2 demonstrates some by showing the same rooted tree in different styles; we find the styles in 1(*a*)–(*b*), 2(*a*)–(*b*) and 3(*a*) best for particular tasks.

Each graph is sometimes called a *network* as it only positions the individual taxa relative to each other without indicating the direction of the evolutionary process. In order to indicate the direction of a graph it must have a root, which leads to the common ancestor of all the OTUs in it. The graph can be rooted if one or more of the OTUs forms an *outgroup*, because they are known, or believed to be the most distantly related of the OTUs. The remainder then forms the *ingroup*, and the root is joined to the edge that joins the ingroup and the outgroup. If it is uncertain which OTU is an outgroup it is still possible to assign a root, if we believe that the rate of evolution in the different lineages is similar, because the root will then lie either at the midpoint of the path joining the two most dissimilar OTUs, or the mean point of the paths that join the most dissimilar OTUs connected through a single edge.

Methods for tree construction

Many methods for constructing phylogenetic trees from molecular data have been developed, and most methods originally designed for morphological data can be applied to sequences as well. We will briefly describe some of the more important methods and outline their strengths and weaknesses as different methods are best suited for a particular study.

We have grouped them firstly according to whether the method uses

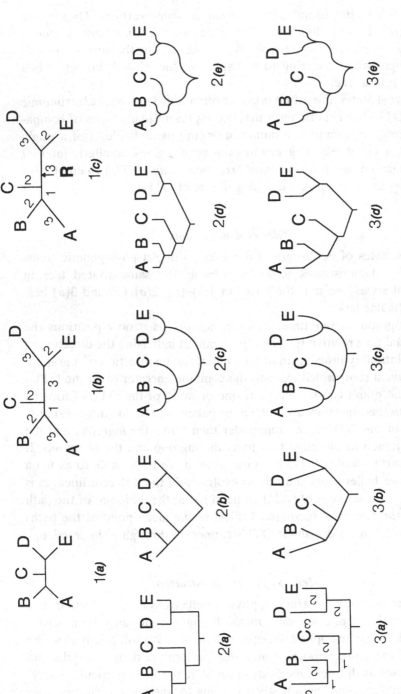

Fig. 36.2. Styles of dendrograms. Series 1(a–c) represents unrooted trees (networks) as undirected graphs: 1(a) is unweighted and merely indicates relationships, 1(b) is weighted and gives the relative edge lengths, 1(c) also shows the possible root 'R', which is at the midpoint of the path between the most dissimilar OTUs, A and D. Series 2 (unweighted) and 3 (weighted) are rooted dendrograms representing 1(c), (a)s are phenograms, (b)s are cladograms, (c)s are curvograms, (d)s are eurograms and (e)s are swoonograms!

discrete character states or a *distance matrix* of pairwise dissimilarities, and secondly whether the method clusters OTUs stepwise or considers all theoretically possible trees.

Character state methods can use any set of discrete characters such as morphological characters, physiological properties, restriction maps or sequence data. When comparing sequences, each sequence position in the aligned sequences is a 'character', and the nucleotides/amino acids at that position are the 'states'. All characters are analysed separately and usually independently from each other.

By contrast *distance matrix* methods start by calculating some measure of the dissimilarity of each pair of OTUs to produce a pairwise distance matrix, and then estimate the phylogenetic relationships of the OTUs from that matrix. These methods seem particularly well suited for analysing sequence data. There are many metrics for estimating evolutionary distances from sequence differences (Gojobori, Moriyama & Kimura, 1990; Nei, 1987). Although it is possible to calculate distances directly from pairwise aligned sequences, more consistent results are obtained when all sequences are aligned; gaps can be treated as an additional state or omitted.

All *exhaustive search* methods examine all or a large number of the theoretically possible tree topologies and use certain criteria to choose the best one. Their main advantage is that they produce a large number of different trees together with a relative estimate of the likelihood that they represent the phylogeny. This allows the investigator to compare the support for the best tree with the support for the second best, and thus estimate the confidence in the tree obtained. Unfortunately the number of possible trees and thus computing time grows very quickly as the number of taxa increases; the number of bifurcated rooted trees for n OTUs is given by $(2n-3)!/(2^{n-2}(n-2))!$ (Cavalli-Sforza & Edwards, 1967). This means that, for a data set of more than 10 OTUs, only a subset of possible trees can be examined, so various strategies are used to search the 'tree space' but there is no algorithm that guarantees that the best possible tree was examined.

By contrast, the *stepwise clustering* methods avoid this problem by examining local subtrees first. Typically the two closest related OTUs are combined to form a cluster. The cluster is then treated like a single OTU representing the ancestor of the OTUs it replaces. Thus the complexity of the data set is reduced by one OTU. This process is repeated, clustering the next closest related OTUs, until all OTUs are combined. The various stepwise clustering algorithms differ in their methods of determining the relationship of OTUs and in combining OTUs to clusters. They are usually very fast and can accommodate large numbers of OTUs. As they produce only one tree, the confidence estimators of the exhaustive search methods are not available, although various other statistical methods have been developed for estimating the confidence in the correctness of a tree obtained (see Li & Gouy, 1990).

The majority of distance matrix methods use a stepwise clustering to compute trees to represent their relationships, while many character state methods adopt the exhaustive search approach.

Maximum parsimony (MP) and weighted parsimony (WP) (Method: character state, exhaustive search)

The MP method was first proposed by Camin and Sokal (1965) as a method for reconstructing phylogenetic trees from morphological data. Later modifications adjusted the MP method for amino acid (Eck & Dayhoff, 1966) and nucleotide sequences (Fitch 1971, 1977). For a comprehensive review of the many facets of MP methods see Felsenstein (1982). The latest modification utilizes dynamically modified character weighting (Patrick & Fitch, 1990) and is called 'weighted parsimony' (WP). The most popular parsimony program is PAUP (phylogenetic analysis under parsimony; Swofford, 1991).

MP aims to find the tree topology for a set of aligned sequences that can be explained with the smallest number of character changes (mutations). The MP algorithm starts by considering a tree with a particular topology. It now infers the minimum number of character changes required to explain all nodes of the tree at every sequence position. Another topology is then evaluated. When all reasonable topologies have been evaluated, the tree that requires the minimum number of changes is chosen as the best tree.

Not all positions of a set of aligned sequences are used by the MP process to discriminate different tree topologies. Only sites that have at least two different characters, each represented at least twice, can be used, and these are called *informative sites* or, by some, *parsimony sites*. This usually restricts drastically the number of positions used in the analysis. The standard MP method weights all changes equally although, for sequence data, this is not always desirable. First, it is sometimes desirable to weight different sequence positions differentially in order, for example, to emphasize the few changes at conserved positions rather than the many at variable positions. Secondly, it is often advisable to weight different types of changes differently, for example, transversions are usually more significant than transitions, and certain amino acid substitutions more than others. A general method for differentially weighting nucleotide substitutions before parsimony has been proposed, and PAUP version 3.0 allows the user to specify different weights to different types of substitutions as well as to different positions.

Recently Patrick and Fitch (1990) introduced the dynamically weighted parsimony method (WP). Their method uses an initial tree (seed tree) to assign different weights to different types of substitution and/or sequence positions. It then uses this weighting scheme to generate a new tree. This process is repeated until the same best tree is obtained in two consecutive runs.

Maximum likelihood (ML)
(Method: character state, exhaustive search)

ML was originally proposed by Cavalli-Sforza and Edwards (1967) for gene frequency data. Later Felsenstein (1981) developed a ML algorithm for nucleotide sequences. ML is similar to the MP method in that it examines every reasonable tree topology and evaluates the support for each by examining every sequence position.

In principle, the ML algorithm calculates the probability of expecting each possible nucleotide (amino acid) in the ancestral (internal) nodes and infers the likelihood of the tree structure from these probabilities. This allows for various statistical corrections, such as the Jukes and Cantor correction (see below), that are not available to MP. The most likely tree is then defined by the product of the likelihoods of all positions. In addition, ML maximizes the likelihood value by varying branch lengths. The likelihood of all reasonable tree topologies are searched in this way, and the most likely tree is chosen as the best tree. The actual process is complex, especially as different tree topologies require different mathematical treatments, and so it is computationally very demanding. Furthermore statistical analysis of ML trees is difficult, but is being developed (Hasegawa & Kishino, 1994).

UPGMA / WPGMA method
(Method: distance matrix, stepwise clustering)

U(W)PGMA is an acronym for unweighted (weighted) pair group method with arithmetic means. These are probably the oldest and simplest methods used for reconstructing phylogenetic trees from distance data, first described by Sokal and Michener (1958). Clustering is done by searching for the smallest value in the pairwise distance matrix.

The newly formed cluster replaces the OTUs it represents in the distance matrix. The distances between the newly formed cluster and each of the remaining OTUs are then calculated. UPGMA and WPGMA are distinguished only by the formula used to calculate these distances; i.e. for the tree topology (((ab)c)d), the distance $D_{(abc)d}$ between cluster (abc) and OTU (d) is $1/3(D_{ad} + D_{bd} + D_{cd})$ in WPGMA, and $1/2(D_{(ab)d} + D_{cd})$ in UPGMA. This process is repeated until all OTUs are clustered.

Fitch and Margoliash method (FM)
(Method: distance matrix, exhaustive search)

The FM algorithm (Fitch & Margoliash, 1967) initially uses the same clustering method as UPGMA and thus the initial topology is always the same as in UPGMA, but the branch lengths are calculated differently. All trees with closely related topologies are then explored and compared in terms of the so called 'percent standard deviation'. This is essentially a measure that assesses how well the distances in the matrix correlate with the branch length obtained in the tree (patristic distances). The tree with the smallest 'standard deviation' is chosen.

Distance Wagner method (DW)
(Method: distance matrix, clustering-search hybrid)

As the network reconstructed from character state data is often called a *Wagner network*, Farris (1972) called his method for reconstructing unrooted trees from distance data *Distance Wagner*. Like the FM method, the DW method uses a distance matrix to analyse many possible trees. Like many clustering algorithms, the DW method first combines the two most closely related OTUs. As the network grows it tries to fit all remaining OTUs in turn into any of the possible edges, choosing the one that can be connected with the shortest branch. This method minimises the total length of each subtree as it is built. As it depends very much on correct estimation of every branch length, it is very susceptible to sampling errors. Improvements to the algorithm have been suggested (Tateno, Nei & Tajima, 1982; Faith, 1985) and these yield the modified Farris method (MF).

Neighbour-joining method (NJ)
(Method: distance matrix, clustering)

The NJ method (Saitou & Nei, 1987) constructs the tree by sequentially finding pairs of *neighbours*, which are the pairs of OTUs connected by a single interior node. The clustering method used by this algorithm is quite different from those described above as it does not attempt to cluster the most closely related OTUs but minimizes the length of all internal branches and thus the length of the entire tree. So it can be regarded as parsimony applied to distance data.

The NJ algorithm starts by assuming a bush-like tree that has no internal branches. In the first step it introduces the first internal branch and calculates the length of the resulting tree. The algorithm sequentially connects every possible OTU-pair and finally joins the OTU-pair that yields the shortest tree. The length of a branch joining a pair of neighbours, X and Y, to their adjacent node is the average distance between all OTUs and X and all to Y, minus the average distances of all remaining OTU-pairs. This process is then repeated, always joining two OTUs (neighbours) by introducing the shortest possible internal branch.

Spectral analysis
(Method: character state, exhaustive search)

Spectral analysis of character state data is a newly devised and very promising way for inferring phylogenies (Hendy & Penny, 1993; Hendy & Charleston, 1993; Penny *et al.*, 1993). The method examines all possible ways to divide the data set into two *bipartitions* or *splits*. Each such split corresponds to an edge in a tree since each edge separates the data into two subgroups. For this, a tree constitutes a set of compatible splits or bipartitions. For example, in a set of four aligned sequences (t1–4), each position can only have one of eight bipartition patterns; {t1,t2,t3,t4}; {t1} {t2,t3,t4}; {t2} {t1,t3,t4}; {t3} {t1,t2,t4}; {t4} {t1,t2,t3}; {t1,t2} {t3,t4}; {t1,t3} {t2,t4}; {t1,t4} {t2,t3}. The support for each pattern (split) by each position of a sequence alignment is evaluated and the relative supports of all splits is called the sequence spectrum. The sequence spectrum is transformed into the conjugate spectrum using the *Hadamard* (discrete Fourier) transformation, and in the process corrected for unobserved changes in the data. Selection criteria are then used to find the tree with the strongest support using an inverse Hadamard conjugation. The fit between the observed sequence spectrum and that of the inferred tree is usefully presented as a ranked spectrum, or 'lentoplot', to indicate the support for the inferred tree, and for the splits that support alternative trees.

Thus spectral analysis is a way of analysing the strength of the different kinds of signals indicated by different positions in a set of aligned sequences; the phylogenetic signal indicating a shared ancestry is often the strongest signal, but others indicating convergence, recombination and the independent appearance of particular features may also be revealed.

Choosing suitable methods for difficult data sets

There have been some reports of comparisons of different sets of algorithms using different sets of data. However it is very difficult to decide which method or methods are best, perhaps because different data sets seem to favour different algorithms. The reason is that different tree-making algorithms are based on different assumptions, and if these assumptions are met by the data, the algorithm will perform well.

Some studies have used real data, but the results are inconclusive as the real tree is usually not known. Other authors have therefore used artificial data by applying specific rules to create a set of (progenitor) sequences from a single (ancestral) sequence, and then used different tree-making algorithms to test their ability to reconstruct the model tree. The problem with this approach is that its results mainly depend on whether the assumptions on which the tree-making algorithm is based reflect the rules applied to simulate evolution and create the data set. In this way, every algorithm can be shown to perform best. Not only can the tree-constructing technique be inappropriate, but also the data. Here, we give a short list of typical traps that may be encountered when collecting data for tree reconstruction.

Base composition bias

When some sequences of a data set have pronounced differences in their nucleotide composition, most methods will tend to incorrectly group sequences with similar base composition. Correcting for the biased use, or survival, of different nucleotides in genes is currently receiving much attention as it causes great problems in assessing the relationships of some ancient lineages, such as protists and organelles (Lockhart *et al.*, 1993), and also viruses. One promising method of correcting this bias is the *logdet* method (Steel, Lockhart & Penny, 1993; Lockhart *et al.*, 1994), in which the difference between each pair of aligned nucleotide sequences is represented as a transition matrix, and the logarithms of the determinants of the matrices are used as measures of the distances between the sequences for tree construction. Another promising method, that works well with some data, is to correct each type of nucleotide change separately assuming that composition alone determines the probability of change (Weiller & Gibbs, in preparation).

It is worth noting that biases of all types differ greatly in intensity in the different codon positions of ORFs, and hence should be corrected separately, although this is rarely done.

Unequal frequency of different types of mutations

Different types of mutation often occur with different frequencies; for instance more than 90% of the substitutions found in human mitochondrial genes are transitions (Brown *et al.*, 1982). Thus different types of substitutions can provide different phylogenetic information. This problem is best dealt with using either WP or distance matrix methods, provided the appropriate corrections are applied. For example, in human mitochondrial sequences, each transversion difference could be given ten times the score of a transition. For strategies to differentially weight different mutational changes see Nei (1987) or Li and Graur (1991).

Parallel, backward and multiple mutations

In distantly related sequences, where many sites have mutated more than once, the MP strategy of inferring ancestral sequences fails. The estimates of distances and branch length must be transformed to compensate for multiple mutations (Gojobori *et al.*, 1990). Jukes and Cantor (1969) correction as well as the two parameter correction Kimura (1980) are also available in some versions of the ML algorithm (e.g. Phylip) and for spectral analysis using the Hadamard transformation (Hendy & Charleston, 1993; Hendy & Penny, 1993).

Unequal mutation frequency in different sites

It seems that not all sites are equally free to mutate and survive. Some sequences like rRNA genes or immunoglobulin proteins have both conserved and (hyper-) variable regions. Thus it is possible for conserved regions to have no significant differences, or for very variable regions to have changed so much that no similarity can be detected and, as a result, neither region would provide phylogenetic information. Problems of this kind can best be detected by consulting a heterozygosity profile of the aligned sequences. It is feasible to circumvent this sort of error specifying a phylogenetic weight for every sequence position based on the variability at each position. This correction can be applied to distance data (Weiller & Gibbs, in preparation) and there is a similar correction mechanism in some implementations of the MP algorithm (e.g. PAUP 3.0). WP also provides for positional weighting.

Unequal rates of mutations in different parts of the tree

The UPGMA, WPGMA and related methods assume that neighbouring OTUs are equidistant from the ancestral node and thus that the rate of mutational changes has been equal in all parts of the tree. When this is not the case, these methods produce wrong results. Parsimony methods are less affected by this problem, but ML and NJ seem to be best suited to provide the correct answer despite unequal rates of mutational change.

Interdependency of sites

Most methods assume the independence of sites. However this is not always so. For instance, hairpin structures require nucleotide changes in one part of the sequence to be matched by complementary changes elsewhere. Furthermore deletions or repetitions of short sequence motifs often reflect single mutational events.

When distance matrices are used, the distance values can sometimes be corrected accordingly, whereas character-state methods normally cannot accommodate this problem and the sequences must be edited to remove misleading changes.

Non-additive distance values

Branch lengths (distances) are said to be strictly *additive* if the length of the path between any two nodes equals the sum of the individual edge lengths. Methods that try to minimize the tree length (e.g. FM, NJ) depend to some extent on distances being additive. If distances are not additive, because the data represent relationships that are not 'tree-like', no method will perform well. This is observed, for instance, in distantly related sequences when the appropriate correction for multiple hits has not been applied. How additive distance data are is also reflected in their correlation with the patristic distances calculated from the final tree. This is called *co-phenetic correlation* and has been used as an estimator of the tree quality. We suggest that this correlation should be used to estimate the quality of data, whereby the maximum divergence value should have a higher resolution power than the correlation coefficient. Computer programs like DiPloMo or NTSYS (see below) can be used to compare the two distance matrices. A big divergence indicates a systematic error in the treatment of the data and is sometimes confined to single OTUs. These errors are most obvious with long edges and outgroups, so that when the OTU is removed from the tree the topology of the remaining subtree may change. This can be tested by omitting one OTU at a time, a technique called jackknifing. This test for taxon stability can be also applied to character-state methods.

Insufficient information

When the sequences do not provide enough phylogenetic information (e.g. sequences are too short or lacking in variation), no algorithm will produce sensible answers. One way to evaluate whether there is enough phylogenetic signal in the data is to apply tests like jackknifing or bootstrapping (Efron, 1982; Felsenstein, 1985; Li & Gouy, 1990; 1991). Both methods indicate whether even smaller sample sizes would result in the same tree. As both methods need to repeat the tree generation process many hundreds of times, they are preferably used with the faster clustering methods.

Comparing phylogenies

After a tree has been calculated, boot-strapping, jack-knifing or Monte Carlo techniques are often used to test the statistical significance of branchings within it; however, more useful biological information is obtained by comparing the new tree with other trees obtained from the same group of organisms, to see how well they correlate, and in what way. For example, comparing trees of hosts and their parasites will indicate which parasites co-evolved with their hosts and which have swapped hosts; comparing different parameters of evolutionary change within a single gene, say nucleotide and amino acid differences, may indicate if, and when, there have been changes in selection; comparing the distances used to compute a tree with the patristic distances of that tree will indicate how well the tree represents the real relationships of the OTUs.

Tree comparison

Comparing different tree topologies is often difficult and the use of computer programs like MacClade (Maddison & Maddison, 1993) or COMPONENT (Page, 1993) is very useful. The latter includes a wide range of tree mapping and comparison measures including the partition metric (Penny & Hendy, 1985), quartet measures (Estabrook,

McMorris & Meacham, 1985) and the nearest neighbour interchange metric (Robinson, 1971; Waterman & Smith, 1978).

Distance comparison (DiPloMo)

An alternative new method is DiPloMo (distance plot monitor), an interactive computer program for comparing pairs of distance measures for a set of OTUs (Weiller & Gibbs, 1993). DiPloMo plots pairs of distance measures with each other, displaying them as a scatter plot. It then helps the user identify which points within the plot correspond with particular comparisons or groups of comparisons by plotting them in various colours and symbols. This principle is extremely simple and versatile and, depending on the distance matrices compared, can provide insights into evolutionary trends and constraints, degrees of mutational freedom, effects of various data corrections, additivity and tree fitting of distance data, and also correlations with other features of the OTUs, such as their time or place of origin.

Fig. 36.3 shows a DiPloMo plot for the envelope gene of various lentiviruses, and illustrates the power of the technique. The plot compares two distance measures: the proportions of different amino acids and the proportions of different nucleotides in the third codon position. The plot shows two specific trends in the proportion of mutations that have resulted in amino acid changes, a clear indication of differences in selection pressure against (or for) amino acid changes. The leftmost and steeper diagonal cluster contains comparisons of virus isolates that have a larger proportion of nucleotide changes resulting in amino acid changes in the envelope protein, than in the other cluster. The clusters are not uniquely correlated with the major lineages revealed by tree analysis; clusters contain isolates from subsets of the lineages.

The clustering pattern in the DiPloMo plot correlates with other features of the isolates:

1. All comparisons between human HIV1 sequences are located in the left most cluster, indicating the greater acceptance of amino acid change in their envelope proteins;
2. Most comparisons between SIV viruses are located in the rightmost cluster, possibly because their 'environment' has selected more actively against amino acid change.
3. The HIV2 viruses occupy an intermediate position. Thus it seems that viruses that changed their host range recently are located further to the left.

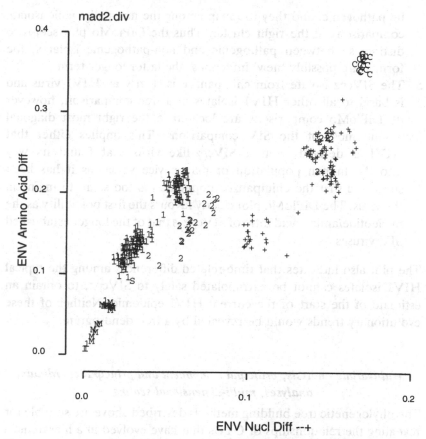

Fig. 36.3. DiPloMo plot of distance matrices of pairwise amino acid and nucleotide (third codon position) sequence differences of the envelope protein (and gene) of various lentiviruses. Comparisons between HIV1 isolates are marked '1', between HIV2's '2', between SIVmacs 'M', between SIVcpz and other HIV1's 'C' and between other combinations of isolates '+'.

4. Most of the viruses in the left-most cluster come from animals with immunodeficiency diseases, whereas those in the right-most cluster are from apparently healthy animals. This difference is clearest with the SIV isolates; the SIV_{mac} isolates that map on the very left are particular pathogenic, kill macaques within a few months of infection, and have probably only recently been infected with this virus, probably for the first time, whereas other SIV isolates seem not to harm their hosts, and probably have had a longer term relationship with their hosts. In addition SIV cpz and HIV2d205 are not known to

be pathogenic, and they too map among the non-pathogenic isolate comparisons of the right cluster. Thus the DiPloMo plot seems to distinguish between pathogenic and non-pathogenic isolates; the former are possibly 'new' infections, the latter longer term.

5. The SIVcpz isolate from chimpanzee is clearly an HIV1 virus and is basal to all other HIV1 isolates in a tree comparison; however its DiPloMo comparisons are located in the right most diagonal among most of the SIV comparisons. This implies either that HIV1 is derived from a SIVcpz-like virus that found its way into the human population or maybe vice versa, as it has been suggested that the chimpanzee population is too small to maintain the virus. The DiPloMo plot clearly favours the first possibility as the nucleotide/amino acid ratio of SIVcpz is that of the longer established SIV viruses.

The plot also indicates that time-related differences among the typical HIV1 isolates cannot be extrapolated safely to SIVcpz, to obtain an estimate of the start of the current HIV1 epidemic. Neither of these evolutionary trends would be revealed by a tree dendrogram.

Multivariate analysis; principal component and principal coordinate analyses, multi-dimensional scaling

The phylogenetic tree building methods described above are suitable for revealing the relationships of OTUs that have evolved in a hierarchical tree-like manner, but not all OTUs have evolved in this way. For example, most viral genera have clearly originated by the recombination of genes from more than one source and, in populations, some of the observed features shared by individuals have probably arisen independently and hence are not hierarchically related to one another. A tree diagram is inappropriate for representing such data, and some form of ordination based on *multivariate analysis* will be better.

Multivariate methods aim to detect general patterns and to indicate potentially interesting relationships in data. The most valuable methods are those that display their results graphically, rather than just examining statistical properties derived from the data; Reyment (1991) has reviewed the application of such methods; Everitt and Dunn (1992) discuss cluster analysis and multi-dimensional scaling, and the monograph by Sneath and Sokal (1973) gives a general introduction to numerical taxonomy.

One of the oldest and most widely used multivariate analytical and graphical techniques is *principal component analysis*, which was devised by Pearson (1901) and independently by Hotelling (1933), and which, with the other similar ordination techniques, such as *principal coordinate analysis* (PCA) and *multi-dimensional scalings* (MDS), are available in a user-friendly format in NTSYS (Rohlf, 1993). These techniques should be more widely used than they are in phylogenetic analysis, to check whether the groupings found by tree-building methods represent the strongest signal, and also whether there are any other consistent signals.

The basis of these methods is that a matrix of phylogenetic distances between n OTUs can be used to position the OTUs in n-1 dimensional space. This is easy to comprehend when there are only 3 OTUs (2 dimensions) as in Fig. 36.4(a), or even 4 OTUs (3 dimensions), as in Fig. 36.4(b), where the third dimension is indicated by shading, but is more difficult for larger numbers of OTUs. Ordination techniques overcome this problem by using statistical regression techniques to calculate the two or more variables (dimensions, components or co-ordinates) that best represent the relationships in a large multi-dimensional data set; most information in the first variable, and decreasing amounts of information in subsequent variables. The process is illustrated in Fig 36.4(c), using the 4 OTUs of Fig 36.4(b), and with the resulting variables used to plot the OTUs in two dimensions in Fig 36.3(d).

Ordination methods, such as PCA or MDS, group taxa with shared properties. The different coordinates usually correlate with different properties of the OTUs, and, if the ordination was calculated from quantitative or multi-state characters, such as sequences, rather than directly from distances, such as serological titres, it is easy to calculate the correlation between each dimension and the original data. The major trends in the data set are expressed by the first few coordinates, while the finer details between closely related taxa are often hidden in minor coordinates; over two-thirds of the information in a well-structured data set, such as the phylogenetic relationships of a group of OTUs with clear lineages, will be represented by the first three variables.

Ordination methods provide a different view of the relationships of a set of OTUs from that provided by tree-building analyses; the former reveal general trends in shared characters, whereas the latter are dominated by pairwise similarities. This is illustrated in a study by Kuiken *et al.* (1993) of sequence variation in the V3 loop region of the envelope gene of a large population of HIV1 viruses collected from homosexuals, intravenous drug users and haemophiliacs. Phylogenetic

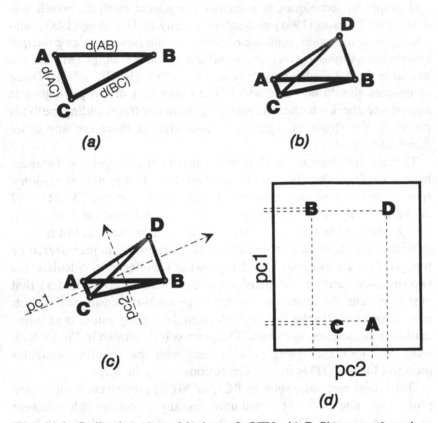

Fig. 36.4. Ordination plots. (a) shows 3 OTUs (A,B,C) arranged as three vertices in a 2D graph, where the distances d(AB), d(AC) and d(BC) represent measures of relatedness; (b) shows that 3Ds are required to display 4 OTUs; (c) shows the first two principal coordinates (PCs) calculated from the distances between the OTUs in (b) by statistical regression and with the vectors orthogonal to one another; the first PC is the vector that distinguishes OTUs A and C from OTUs B and D; it is closest to all four vertices in the direction in which the average distances between vertices are maximal, thus it represents most of the distances in the data; (d) is a scatter plot of the 4 OTUs using their positions in the two PCs. Note that successive PCs, especially in biological data, usually represent decreasing amounts of information, but that, had the OTUs been equidistant or randomly distributed, the PCs would have had approximately equal variances.

trees constructed using various nucleotide and amino acid distance measures failed to produce clear and sensible clusterings of isolates from the different risk groups; however a PCA analysis (Higgins, 1992) of the same data set clearly distinguished those groups (Fig 36.5).

PLOT OF FIRST (horizontal) AND SECOND (vertical) AXES

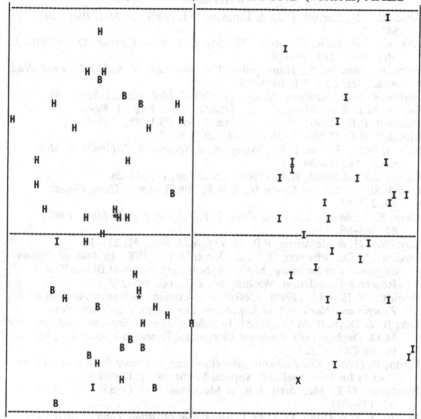

Fig. 36.5. Principal coordinate analysis of the sequence of the V3 loop region of the virion envelope protein of HIV1 isolates from homosexuals (H), intravenous drug users (I) and haemophiliacs (B) (from Kuiken *et al.*, 1994).

Conclusions

Gene sequence data is being acquired at an ever increasing rate, especially for viruses. There are several ways of constructing trees from such data. However, there is no evidence that any one method is superior to others, so we consider it advisable to use more than one method with each set of data, and compare the results using tools such as DiPloMo.

582   G. Weiller, M.A. McClure and A. Gibbs

References

Altschul, S.F., Carroll, R.J., & Lipman, D.J. (1989). *J. Mol. Biol.*, **207**, 647–53.

Altschul, S.F., Gish, W., Miller, W., Myers, E.W. & Lipman, D.J. (1990). *J. Mol. Biol.*, **215**, 403–10.

Baldi, P., Chauvin, Y., Hunkapiller, T. & McClure, M.A. (1994). *Proc. Natl. Acad. Sci. USA*, **91**, 1059–63.

Barton, G.J. & Sternberg, M.J.E. (1987a). *J. Mol. Biol.*, **198**, 327–37.

Barton, G.J. & Sternberg, M.J.E. (1987b). *Prot. Eng.*, **1**, 89–94.

Blaisdell, B.E. (1986). *Proc. Natl. Acad. Sci. USA*, **83**, 5155–9.

Blaisdell, B.E. (1989). *J. Mol. Evol.*, **29**, 526–37.

Brown, W.M., Prager, E.M., Wang, A. & Wilson, A.C. (1982). *J. Mol. Evol.*, **18**, 225–39.

Camin, J.H. & Sokal, R.R. (1965). *Evolution*, **19**, 311–26.

Cavalli-Sforza, L.L. & Edwards, A.W.F. (1967). *Am J. Hum. Genet.*, **19**, 233–57.

Chan, S.C., Wong, A.K.C. & Chiu, D.K.Y. (1992). *Bull. Math. Biol.*, **54**, 563–98.

Day, W.H.E. & McMorris, F.R. (1993). *NZJ. Bot.*, **31**, 211–18.

Dayhoff, M.O., Schwartz, R.M. & Orcutt, B.C. (1978). In *Atlas of Protein Sequence and Structure*, M.O. Dayhoff (ed). National Biomedical Research Foundation, Washington, DC., pp. 345–352.

Doolittle, R.F. (ed.). (1990). *Molecular Evolution: Computer Analysis of Protein and Nucleic Acid Sequences*. San Diego: Academic Press.

Eck, R. & Dayhoff, M.O. (1966). In *Atlas of Protein Sequence and Structure*, M.O. Dayhoff (ed). National Biomedical Research Foundation, Washington DC, p. 327.

Efron, B. (1982). *The Jackknife, the Bootstrap and other Resampling Plans*. Society for Industrial and Applied Mathematics, Philadelphia.

Estabrook, G.F., McMorris, F.R. & Meacham, C.C. (1985). *Syst. Zool.* **34**, 193–200.

Everitt, B.S. & Dunn, G. (1992). *Applied Multivariate Data Analysis*. Oxford Univ. Press: New York, p. 304.

Faith, D.P. (1985). *Syst. Zool.*, **34**, 312–25.

Faith, D.P. & Cranston, P.S. (1991). *Cladistics*, **7**, 1–28.

Farris, J.S. (1972). *Am. Natur.*, **106**, 645–8.

Felsenstein, J. (1981). *J. Mol. Evol.*, **17**, 368–76.

Felsenstein, J. (1982). *Quart. Rev. Biol.*, **57**, 379–404.

Felsenstein, J. (1985). *Evolution*, **39**, 783–91.

Feng, D.F. & Doolittle, R.F. (1987). *J. Mol. Evol.*, **25**, 351–60.

Feng, D.F., Johnson, M.S. & Doolittle, R.F. (1985). *J. Mol. Evol.*, **21**, 112–25.

Fitch, W.M. (1971). *Syst. Zool.*, **20**, 406–16.

Fitch, W.M. (1977). *Am. Natur.*, **111**, 147–64.

Fitch, W.M. & Margoliash, E. (1967). *Science*, **155**, 279–84.

Gibbs, A.J., Dale, M.B., Kinns, H.R. & MacKenzie, H.G. (1971). *Syst. Zool.*, **20**, 417–25.

Gibbs, A.J. & McIntyre, G.A. (1971). *Eur. J. Biochem.*, **16**, 1–11.

Gojobori, T., Moriyama, E.N. & Kimura, M. (1990). *Methods in Enzymology*, **183**, 531–50.

Gonnet, G.H., Cohen, M.A. & Benner, S.A. (1992). *Science*, **256**, 1443–5.

Gould, R. (1988). *Graph Theory*. Benjamin/Cummings, Menlo Park, California

Gribskov, M. & Devereux, J. (ed.). (1991). *Sequence Analysis Primer*. New York, M Stockton Press.

Gribskov, M., McLachlan, A.D. & Eisenberg, D. (1987). *Proc. Natl. Acad. Sci. USA*, **84**, 4355–8.

Harary, F. (1969). *Graph Theory*. Addison-Wesley, Reading, Mass.

Hasegawa, M. & Kishino, H. (1994). *Mol. Biol. Evol.*, in press.

Hein, J.H. (1990). *Methods in Enzymology*, **138**, 626–45.

Hendy, M.D. & Charleston, M.A. (1993). *NZJ. Bot.*, **31**, 231–7.

Hendy, M.D. & Penny, D. (1993). *J. Classif.*, **10**, 5–24.

Higgins, D.G. (1992). *CABIOS*, **8**, 12–22.

Higgins, D.G. & Fuchs, R. (1992). *CABIOS*, **8**, 189–91.

Hillis, D.M. & Moritz, C. (eds) (1990). *Molecular Systematics*. Sinauer, Mass.

Hotelling, H. (1933). *J. Educ. Psychol.*, **24**, 417–41.

Jukes, T.H. & Cantor, C.R. (1969). In *Mammalian Protein Metabolism*, H.N. Munro (ed.), Academic press, New York, pp. 21–132.

Kimura, M. (1980). *J. Mol. Evol.*, **16**, 111–20.

Krogh, A., Brown, M., Mian, I.S., Olander, K. & Haussler, D. (1994). *J. Mol. Biol.*, **235**, 1501–31.

Kuiken, C.L., Nieselt-Struwe, K., Weiller, G.F. & Goudsmit, J. (1993). *Meth. Mol. Genet.*, 4, in press.

Kumar, S., Tamura, K. & Nei, M. (1993). *Mega: Molecular Evolutionary Genetic Analysis – Version 1.0*. Pennsylvania State University.

Li, W.H. & Gouy, M. (1990). *Meth. Enzymol.*, **183**, 645–59.

Li, W.H. & Gouy, M. (1991). In *Phylogenetic Analysis of DNA Sequences*. Oxford University Press, M.M. Miyamoto, and J. Cracraft, (eds) pp. 249–77.

Li, W.H. & Graur, D. (1991). *Fundamentals of Molecular Evolution*. Sinauer, Mass.

Lipman, D.J., Altschul, S.F. & Kececioglu, J.D. (1989). *Proc. Natl. Acad. Sci. USA*, **86**, 4412–15.

Lockhart, P.J., Penny, D., Hendy, M.D. & Larkum, A.D.W. (1993). *Photosynth. Res.*, **37**, 61–8.

Lockhart, P.J., Steel, M.A., Hendy, M.D. & Penny, D. (1994). *Mol. Biol. Evol.*, in press.

McClure, M.A. (1992). *Mathematical and Computer Modelling*, **16**, 121–36.

McClure, M.A., Vasi, T.K. & Fitch, W.M. (1993). *Comparative Analysis of Multiple Protein-Sequence Alignment Methods*, in press.

Maddison W.P. & Maddison D.R. (eds.) (1993). MacClade, Version 3. Sinauer Associates, Inc. Sunderland, Massachusetts, USA

Martinez, H.M. (1988). *Nucl. Acids Res.*, **16**, 1683–691.

Meyers, E.W. (1991). *An Overview of Sequence Comparison Algorithms in Molecular Biology*, Technical Report No. TR 91–92. University of Arizona.

Needleman, S.B. & Wunsch, C.D. (1970). *J. Mol. Biol.*, **48**, 443–53.

Nei, M. (1987). *Molecular Evolutionary Genetics*. Columbia University Press, New York.

Page, R.D.M. (1993). COMPONENT. Version 2.0. The Natural History Museum, London.

Patrick, L.W. & Fitch W.M. (1990). *Meth. Enzymol.*, **183**, 615–26.

Pearson, K. (1901). *Phil. Mag.*, **2**, 559–72.
Pearson, W.R. (1990). In *Molecular Evolution: Computer Analysis of Protein and Nucleic Acid Sequences* vol. 183, R.F. Doolittle (eds.), Academic Press, San Diego, pp. 63–98.
Penny, D. & Hendy, M.D. (1985). *Syst. Zool.*, **34**, 75–82.
Penny, D., Hendy, M.D. & Steel, M.A. (1992). *Trends in Ecol. Evol.*, **7**, 73–9.
Penny, D., Watson, E.E., Hickson, R.E. & Lockhart, P.J. (1993). *NZJ. Bot.*, **31**, 275–88.
Rao, J.K.M. (1987). *Int. J. Pept. Protein Res.*, **29**, 276–81.
Reyment, R.A. (1991). *Multidimensional Paeleobiology*. Pergamon Press, New York p. 377.
Risler, J.L., Delorme, M.O., Delacroix, H. & Henaut, A. (1988). *J. Mol. Biol.*, **204**, 1019–29.
Robinson, D.F. (1971). *J. Comb. Theor.*, **11**, 105–19.
Rohlf, F.J. (1993). *NTSYS-pc. Numerical Taxonomy and Multivariate Analysis System* (Version 1.80) Exeter Software, Setauket, New York.
Rossmann, M.G. & Johnson, J.E. (1989). *Ann. Rev. Biochem.*, **58**, 533–73.
Saitou, N. & Nei, M. (1987). *Mol. Biol. Evol.*, **4**, 406–25.
Sankoff, D. & Kruskal, J.B. (1983). *Time Warps String Edits and Macromolecules: The Theory and Practice of Sequence Comparison*. Addison-Wesley, Reading Mass.
Schuler, G.D., Altschul, S.F. & Lipman, D.J. (1991). *Proteins, Struct. Funct. Genet.*, **9**, 180–90.
Schwartz, R.M. & Dayhoff, M.O. (1978). In *Atlas of Protein Sequence and Structure*, M.O. Dayhoff (Ed.) National Biomedical Research Foundation, Washington, DC. 353–358.
Sellers, P. (1974). *SIAM J. Appl. Math.*, **26**, 787–93.
Smith, R.F. & Smith, T.F. (1990). *Proc. Natl. Acad Sci. USA*, **87**, 118–22.
Smith, T.F. & Waterman, M.S. (1981). *J. Mol. Biol.*, **147**, 195–7.
Sneath, P.H.A. & Sokal, R.R. (1973). *Numerical Taxonomy*. Freeman, San Francisco, p. 573.
Sokal, R.R. & Michener, C.D. (1958). *Sci. Bull.*, **28**, 1409–38.
States, D.J. & Boguski, M.S. (1990). *Similarity and Homology in Sequence Analysis Primer* M. Gribskov and J. Devereux (eds.) W.H. Freeman, New York, Chapter 3, pp. 89–157.
Steel, M.A., Lockhart, P.J. & Penny, D. (1993). *Nature*, **364**, 440–2.
Subbiah, S. & Harrison, S.C. (1989). *J. Mol. Biol.*, **209**, 539–48.
Swofford, D.L. (1991). PAUP: Phylogenetic Analysis Using Parsimony, Computer program distributed by the Illionois Natural History Survey, Champain, Illionois.
Tateno, Y., Nei, M. & Tajima, F. (1982). *J. Mol. Evol.*, **18**, 387–404.
Taylor, W.R. (1986). *J. Mol. Biol.*, **188**, 233–58.
Taylor, W.R. (1987). *CABIOS*, **3**, 81–7.
Vingron, M. & Argos, P. (1991). J. Mol. Biol. 218:33–43.
Von Heijne, G. (1987). *Sequence Analysis in Molecular Biology. Treasure Trove or Trivial Pursuit*. Academic Press. New York, p. 188.
Waterman, M.S. (ed.). (1988). *Mathematical Methods for DNA Sequences*. CRC Press, Boca Raton Fla.
Waterman, M.S. & Jones, R. (1990). In *Molecular Evolution: Computer Analysis of Protein and Nucleic Acid Sequences* R.F. Doolittle (ed.), **183**, 221–37, Academic Press, San Diego

Waterman, M.S. & Smith, T.F. (1978). *J. Theor. Biol.*, **73**, 789–800.
Weiller, G.F. & Gibbs A.J. (1993). DIPLOMO: Distance Plot Monitor, Computer program distributed by the Australian National University, Canberra.
Wilbur, W.J. & Lipman, D.J. (1982). *Proc. Natl. Acad. Sci. USA*, **80**, 726–30.

Index